Lecture Notes in Computer Science   10046

Commenced Publication in 1973
Founding and Former Series Editors:
Gerhard Goos, Juris Hartmanis, and Jan van Leeuwen

More information about this series at http://www.springer.com/series/7409

Emma Spiro · Yong-Yeol Ahn (Eds.)

Social Informatics

8th International Conference, SocInfo 2016
Bellevue, WA, USA, November 11–14, 2016
Proceedings, Part I

 Springer

Editors
Emma Spiro
University of Washington
Seattle, WA
USA

Yong-Yeol Ahn
Indiana University
Bloomington, IN
USA

ISSN 0302-9743   ISSN 1611-3349   (electronic)
Lecture Notes in Computer Science
ISBN 978-3-319-47879-1   ISBN 978-3-319-47880-7   (eBook)
DOI 10.1007/978-3-319-47880-7

Library of Congress Control Number: 2016954470

LNCS Sublibrary: SL3 – Information Systems and Applications, incl. Internet/Web, and HCI

Printed on acid-free paper

This Springer imprint is published by Springer Nature
The registered company is Springer International Publishing AG
The registered company address is: Gewerbestrasse 11, 6330 Cham, Switzerland

Preface

This volume contains the papers presented at SocInfo 2016, the 8th International Conference on Social Informatics, held during November, 2016, in Bellevue, WA, USA. After the conferences in Warsaw, Poland, in 2009, Laxenburg, Austria, in 2010, Singapore in 2011, Lausanne, Switzerland, in 2012, Kyoto, Japan, in 2013, Barcelona, Spain, in 2014, and Beijing, China in 2015, the International Conference on Social Informatics came to United States for the first time.

SocInfo is an interdisciplinary venue for researchers from diverse fields including computer science, informatics, and the social sciences to share ideas and opinions, and present results from their research at the intersection of social sciences and information sciences. The ultimate goal of social informatics is to facilitate and promote multidisciplinary research that transcends the boundaries between social sciences, computer science, and information sciences, so that researchers can better leverage the power of informatics, computing, and social theories to advance our understanding of society and social phenomena. We envision SocInfo as a venue that attracts open-minded researchers who can cross the disciplinary boundaries and talk to other researchers regardless of their background and training. In doing so, we have invited and selected highly interdisciplinary keynote speakers and papers, which integrate social concepts and theories with large-scale datasets, algorithms, or other concepts and methods of computing.

We were delighted to present a strong technical program, which was a result of the hard work of the authors, reviewers, and conference organizers. We received 120 submissions, an increase from the last SocInfo. From these, 36 papers were accepted as full papers (30.0 %), and 39 were accepted as poster papers (32.5 %). This year, we decoupled the presentation format and the paper format; papers that are accepted as posters are published as is (with the same page limit as the full papers), without enforcing the shorter page limit. We also allowed the authors of accepted papers to opt for a "presentation only" mode with no inclusion in the proceedings: The authors of eight papers chose that option. Finally, a lightning talk option was offered to all authors of papers accepted as posters, to give interested authors the opportunity to present results with a brief oral control initial.

We were also pleased to have Joshua Blumenstock (University of California, Berkeley), Meeyoung Cha (KAIST and Facebook), Tina Eliassi-Rad (Northeastern University), Adam Russell (DARPA), Matthew Salganik (Princeton University), and Hanna Wallach (Microsoft Research and University of Massachusetts Amherst) give exciting keynote talks.

This year we hosted eight satellite workshops, namely, on Data Visualization (SocInfo VIZ: Actionable I From Visualization to Research Narratives); Virality and Memetics; Cultural Analytics; Activity-Based Networks; Social Media for Older Adults (SMOA); Urban Homelessness and Wise Cities; Web, Social Media, and Cellphone Data for Demographic Research; Computational Approaches to Social Modeling (ChASM), and Online Experimentation with Large and Diverse Samples.

We would like to thank all authors and participants for making the conference and the workshops a success. We express our gratitude to the Program Committee members and reviewers for their hard and dedicated work that ensured the highest-quality papers were accepted for presentation. We are extremely grateful to the program co-chairs, Y.Y. Ahn and Emma Spiro, for their tireless efforts in putting together a high-quality program and for directing the activity of the Program Committee. We owe special thanks to Nathan Hodas, our local co-chair, who had a vital role in all the stages of the organization. We thank our publicity chairs, Munmun De Choudhury and Brian Keegan, our Web chair Farshad Kooti, and workshop chairs, Tim Weninger and Emilio Zagheni. Also, last but not least we are grateful to Adam Wierzbicki for his continuous support.

Lastly, this conference would not be possible without the generous help of our sponsors and supporters: Leidos, University of Washington eScience Institute, Facebook, Microsoft Research, and MDPI.

September 2016

Emilio Ferrara
Kristina Lerman
Katherine Stovel

Organization

Organizing Committee

General Co-chairs

Emilio Ferrara — University of Southern California, Los Angeles, USA
Kristina Lerman — University of Southern California, Los Angeles, USA
Katherine Stovel — University of Washington, Seattle, USA

Steering Committee Chair

Adam Wierzbicki — Polish-Japanese Institute of Information Technology, Poland

Program Co-chairs

Yong-Yeol Ahn — Indiana University, Bloomington, USA
Emma Spiro — University of Washington, Seattle, USA

Workshop Co-chairs

Tim Weninger — University of Notre Dame, South Bend, USA
Emilio Zagheni — University of Washington, USA

Publicity Co-chairs

Munmun De Choudhury — Georgia Institute of Technology, Atlanta, USA
Brian Keegan — University of Colorado, Boulder, USA

Local Chair

Nathan Hodas — PNNL, Richland, WA, USA

Web Chair

Farshad Kooti — University of Southern California, Los Angeles, USA

Program Committee

Palakorn Achananuparp — Singapore Management University, Singapore
Luca Maria Aiello — Yahoo Labs, UK
Leman Akoglu — Stony Brook University, USA
Harith Alani — KMi, The Open University, UK
Fred Amblard — IRIT, University of Toulouse 1 Capitole, France

Kyomin Jung	Seoul National University, South Korea
Andreas Kaltenbrunner	Barcelona Media, Spain
Kazuhiro Kazama	Wakayama University, Japan
Brian Keegan	University of Colorado, USA
Andreas Koch	University of Salzburg, Germany
Farshad Kooti	USC Information Sciences Institute, USA
Tobias Kuhn	VU University Amsterdam, The Netherlands
Haewoon Kwak	Qatar Computing Research Institute, Qatar
Sang Hoon Lee	Korea Institute for Advanced Study, South Korea
David Liben-Nowell	Carleton College, USA
Yu-Ru Lin	University of Pittsburgh, USA
Huan Liu	Arizona State University, USA
Yabing Liu	Northeastern University, USA
Vera Liao	University of Illinois at Urbana-Champaign, USA
Jared Lorince	Indiana University, USA
Tyler McCormick	University of Washington, USA
Matteo Magnani	Uppsala University, Sweden
Winter Mason	Facebook, USA
Pasquale De Meo	VU University, Amsterdam, The Netherlands
Rosa Meo	University of Turin, Italy
Eric Meyer	University of Oxford, UK
Stasa Milojevic	Indiana University, USA
Mikolaj Morzy	Poznan University of Technology, Poland
Tsuyoshi Murata	Tokyo Institute of Technology, Japan
Mirco Musolesi	University College London, UK
Shinsuke Nakajima	Kyoto Sangyo University, Japan
Keiichi Nakata	University of Reading, UK
Alice Oh	KAIST, South Korea
Diego Fregolente Mendes de Oliveira	Northwestern University, USA
Nuria Oliver	Telefonica Research, Spain
Anne-Marie Oostveen	University of Oxford, UK
Symeon Papadopoulos	Information Technologies Institute, Greece
Luca Pappalardo	University of Pisa, Italy
Jaimie Park	KAIST, South Korea
Orion Penner	École polytechnique fédérale de Lausanne, Switzerland
Ruggero G. Pensa	University of Turin, Italy
Nicola Perra	Northeastern University, USA
Alexander Petersen	IMT Lucca Institute for Advanced Studies, Italy
Georgios Petkos	Information Technologies Institute/CERTH, Greece
Giovanni Petri	ISI Foundation, Italy
Gregor Petrič	University of Ljubljana, Slovenia
Hemant Purohit	George Mason University, USA
Alessandro Provetti	University of Messina, Italy
Jose J. Ramasco	IFISC, Spain
Georgios Rizos	CERTH-ITI, Greece

John Robinson	University of Washington, USA
Fabio Rojas	Indiana University, USA
Giancarlo Ruffo	University of Turin, Italy
Mostafa Salehi	University of Tehran, Iran
Claudio Schifanella	RAI research centre, Italy
Rossano Schifanella	University of Turin, Italy
Xiaoling Shu	UC Davis, USA
Philipp Singer	GESIS - Leibniz Institute for the Social Sciences, Germany
Frank Schweitzer	ETH Zurich, Switzerland
Tom Snijders	University of Oxford, UK
Rok Sosic	Stanford University, USA
Steffen Staab	University of Koblenz-Landau, Germany
Bogdan State	Stanford University, USA
Markus Strohmaier	University of Koblenz-Landau, Germany
Cassidy Sugimoto	Indiana University Bloomington, USA
Bart Thomee	Yahoo Labs, USA
George Valkanas	University of Athens, Greece
Daniel Villatoro	IIIA-CSIC, Spain
Claudia Wagner	GESIS-Leibniz Institute for the Social Sciences, Germany
Dashun Wang	Northwestern University, USA
Ingmar Weber	Qatar Computing Research Institute, Qatar
Katrin Weller	GESIS Leibniz Institute for the Social Sciences, Germany
Brooke Foucault Welles	Northeastern University, USA
Robert West	Stanford University, USA
Christo Wilson	Northeastern University, USA
Emilio Zagheni	University of Washington, USA
Li Zeng	University of Washington, USA
Arkaitz Zubiaga	University of Warwick, UK

Sponsors

Leidos (www.leidos.com/)
University of Washington eScience Institute (escience.washington.edu/)
Facebook (www.facebook.com)
MDPI (www.mdpi.com)
Microsoft Research (www.microsoft.com)

Microsoft

Contents – Part I

Markets, Crowds, and Consumers

Privacy, Health and Wellbeing

Contents – Part II

Networks, Communities, and Groups

How Well Do Doodle Polls Do?

Danya Alrawi[1], Barbara M. Anthony[2], and Christine Chung[1(✉)]

[1] Department of Computer Science, Connecticut College, New London, CT, USA
{dalrawi,cchung}@conncoll.edu
[2] Math and Computer Science Department,
Southwestern University, Georgetown, TX, USA
anthonyb@southwestern.edu

Abstract. Web-based Doodle polls, where respondents indicate their availability for a collection of times provided by the poll initiator, are an increasingly common way of selecting a time for an event or meeting. Yet group dynamics can markedly influence an individual's response, and thus the overall solution quality. Via theoretical worst-case analysis, we analyze certain common behaviors of Doodle poll respondents, including when participants are either more generous with or more protective of their time, showing that deviating from one's "true availability" can have a substantial impact on the overall quality of the selected time. We show perhaps counter-intuitively that being more generous with your time can lead to inferior time slots being selected, and being more protective of your time can lead to superior time slots being selected. We also bound the improvement and degradation of outcome quality under both types of behaviors.

1 Introduction

Online scheduling tools such as Doodle (www.doodle.com) are a popular way of scheduling events or meetings, with Doodle reporting in 2011 that "online scheduling is used by 67 % of the Swiss and 21 % of the rest of the world".[1] More recent data indicate that in 2014 Doodle had over 20 million monthly users worldwide, with more than 17 million polls created in 2013.[2]

In a Doodle poll, the goal of the poll initiator is to determine the most suitable time for an event or meeting. The initiator selects a set of possible meeting times and sends the Doodle poll invitation to the potential participants. Each participant then checks the boxes for the times they are available to meet; with the default Doodle options, full information about the responses is available to both the initiator and all participants.

Figure 1 shows an example of an open yes-no Doodle poll where three participants have each indicated availability for one or two of the six time slots proposed by the poll initiator; a fourth participant can now enter her name and check boxes for her availability. She can easily see previous responses and that

---

[1] https://en.blog.doodle.com/2011/07/13/.
[2] https://en.blog.doodle.com/2014/01/29/.

© Springer International Publishing AG 2016
E. Spiro and Y.-Y. Ahn (Eds.): SocInfo 2016, Part I, LNCS 10046, pp. 3–23, 2016.
DOI: 10.1007/978-3-319-47880-7_1

the most popular slot thus far is 1:00 PM on Saturday April 30, 2016, indicated both via the frequency counter at the bottom of the poll, and the boldface number showing the currently most popular time. As seen here, the Doodle algorithm simply recommends the time slot(s) with the most checked boxes, or "yes" responses.

This social choice mechanism employed by Doodle is equivalent to *approval voting*, where each voter in an election chooses to approve or disapprove each of the candidates. In a Doodle poll, the "voters" are the participants and the "candidates" are the time slots.

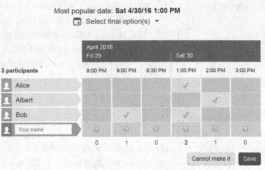

Fig. 1. An example open Doodle poll after three participants have indicated their availability.

While approval voting is the mechanism adopted by a number of professional societies, including the AMS (American Mathematical Society) and the MAA (Mathematical Association of America), such a mechanism clearly has limitations. For one, a voter has no way to express her preference for one candidate she approves over another candidate that she also approves. To be fair, Arrow's classic impossibility theorem has long established that when choosing among three or more candidates, all voting mechanisms have flaws [1]. But approval voting in particular has been a point of controversy, called by Saari and Van Newenhizen [15] a "cure worse than the disease", because, as summarized by [7], "the same voter profile can produce many different results, depending on where each voter decides to draw the line between approved and non-approved candidates." On the other hand, this "feature" of approval voting can be viewed as an advantage, as, according to Brams et al. [3] as interpreted by [7], "it gives each voter 'sovereignty' over the way she expresses her preferences." It is precisely the variation in the location of this "line" drawn by each voter that we model and give a preliminary theoretical analysis for in this work.

We assume each voter has a privately-held, normalized, utility value for each candidate time slot. Intuitively, the utility can be thought of as a quantification of how much the voter expects to benefit from attending the meeting at that time (even if derived simply by satisfying some professional obligation) minus any inconvenience/cost of attending the meeting at that time. To measure the "goodness" of a time slot, we consider the social welfare, or total utility of all voters, for that slot. The fundamental question we ask is, "How well does a Doodle poll work for selecting a time?" We proceed using a standard theoretical worst-case analysis approach common in the algorithms research community.

First we ask, how bad can the time slot chosen by the Doodle mechanism be in comparison to the time slot with maximum social welfare? We show that if an event organizer wishes to maximize social welfare in selecting a time slot, they naturally should not choose one with few "yes" votes. But we also show that, perhaps counter-intuitively social welfare of the chosen slot can be low if people are being very generous with their time and are therefore voting yes too readily. We further show that when voters are protective of their time, voting "no" on slots for which they are available (i.e., slots for which they have utility above the typical voter's "yes"-threshold), the social welfare can worsen in some cases, while improving in others. We define the notion of a positive (resp. negative) *welfare impact factor* and bound the positive (resp. negative) welfare impact factor of voting protectively. We also show that when voters are cooperative, voting "yes" on slots for which they are not easily available (i.e., slots for which they have a utility below the typical "yes"-threshold), the welfare can also both improve in some cases, while worsening quite dramatically in others. We also provide bounds on the the positive (resp. negative) welfare impact factor of voting cooperatively.

1.1 Related Work

Doodle polls are just one of the group scheduling tools available, and previous research has studied these more generally, considering the conditions under which they are used and useful, and the implications thereof [9,12,16].

There has been extensive research done in approval voting dating back to the 1970s. For surveys on approval voting from the voting theory literature see the book by Brams and Fishburn [2] and the article by Weber [17]. We note that while many researchers have accepted for decades that strategic and manipulative voting behavior is "inevitable" and have continued to seek to quantify the negative effects of it [4], even with respect to approval voting in particular [5,10], others have long argued that the notion of self-interested voting in any large-scale election is implausible, since the act of voting itself is "irrational" [8,14]. In contrast to these large-scale political elections, Doodle polls are usually conducted on a small scale; a sample of over 340,000 polls from the US in a three-month period in 2011 had a median of about 5 respondents and 12 time slots [19], so it is fair to assume that strategic voting indeed takes place.

A recent work of Zou et al. analyzes real Doodle poll data and demonstrates that indeed, Doodle poll participants seem to vote strategically. They hypothesize and give positive evidence for a theory of "social voting" where voters are more likely to say yes to popular time slots, perhaps in an effort to be cooperative [19]. Our model does not attempt to address the "social voting" behavior of the voters. Instead we present a simple model that focuses on the aspect of Doodle polls where some participants generally lean toward being more generous with their time and others lean toward being more protective of their time.

Reinecke et al. [13] also analyze anonymized Doodle poll data, this time from countries around the world, and showed that voting behavior is indeed informed by cultural norms and societal expectations, which supports our model's notion of an externally-imposed default "yes"-threshold value.

More recently, Obraztsova et al. [11] model the Doodle poll as a game, where players have utilities for each slot, similar to our model. Their paper focuses instead on identifying and showing that trembling hand perfect Nash equilibria (under the assumption that voters derive a utility bonus when they act cooperatively) behave consistently with the "social voting" theory [19] of Zou et al. And in an earlier work, Xu [18] proposes the use of auctions as an alternative to Doodle polls for selecting a good time slot, citing the benefit of allowing participants to specify a valuation for each slot in an auction setting, as well as the tendency of participants to give false or incomplete information in Doodle polls.

One way to quantify the effects of strategic voting is to use welfare as a metric. Lehtinen [6] studies the welfare of approval voting outcomes using a simulation-based approach, concluding that the percentage of simulated voting games where the welfare-maximizing candidate is chosen is rather high, whether voters are sincere or utility-maximizing. While our work also uses social welfare to measure the effects of voting behaviors, our methodology is purely via theoretical worst-case analysis. And rather than assuming the traditional utility-maximizing voters, we consider what happens when sincere voters either vote cooperatively (a la the social voting model of Zou et al. [19]) or are more protective of their time.

2   Theoretical Framework

2.1   Formalizing a Yes-No Doodle Poll

We first formalize the activities encompassed in a yes-no Doodle poll, generally following the notation of [19]. A poll initiator creates a poll with a set of time slots, namely $A = \{a_1, a_2, \ldots, a_m\}$, for consideration. The poll is then made available to the n participants or *voters* in a given poll, denoted by $V = \{v_1, v_2, \ldots, v_n\}$. Each voter's *response* or *vote* is a binary vector r_i for voter i, over the m time slots in A, with $r_i(a) = 1$ if voter v_i approves slot a, and $r_i(a) = 0$ otherwise. When it is clear from the context, we use *vote* to either refer to the full vector, or to the binary value the voter assigns to a specific time slot. A vote of 1 is considered a *yes* vote, and a vote of 0 is a *no* vote.

We thus define a Doodle poll instance to be a 4-tuple $I = (A, V, U, R)$, where A is the set of time slots, V is the set of voters, U is the matrix of utility values each voter has for each time slot, and R is the response matrix of votes that each voter enters for each time slot. In this work, we assume for simplicity that yes and no are the only options for voters. While Doodle does have an "if-need-be" option than can be added, the empirical data of [13,19] provided by Doodle on polls from July-September 2011 contained very few three-option polls [19]. Their dataset likewise showed that the vast majority of polls were *open*, where voters can see the responses of participants who have already responded, as opposed to *closed*, where only the poll initiator can see the responses.

Let $s(j) = \sum_{i=1}^{n} r_i(j)$, or the total count of yes votes for slot j, be the reported *score* for a slot a_j. Note that in Doodle, all voters are given equal consideration; there is no weighting of the votes. The default Doodle algorithm is simply to determine the one or more slots which maximize the total reported score, that is $\max_{j \in A} s(j)$. Thus, Doodle may report multiple maximum-score time slots, and the poll initiator then ostensibly chooses among those slots. (While the poll initiator is of course free to choose a slot with a lower score, in an open poll, which is by far the most commonly-used kind [19], the participants can all see which slots have the most votes, so it is reasonable to assume that the poll initiator will generally choose among the slots recommended by Doodle.) Doodle provides no tie-breaking mechanism, but human poll initiators may certainly have biases (e.g. preferring slots selected by board members or senior personnel; time slots that are personally convenient; the earliest time slot; etc.), and so we assume that when there is a tie, any of the tied slots may be selected.

2.2   Valuations and Voter Types

We now consider the assumptions we make about how a voter determines his or her vote. We assume that for each time slot a_j a voter v_i has a utility u_{ij} with $0 \leq u_{ij} \leq 1$ indicating her valuation of attending the meeting or event during that time slot. This utility value may be thought of as somehow representative of or derived from how much monetary value a voter would place on attending the event at a given time.

We assume that there is a *yes-threshold* $0 \leq t \leq 1$ that represents the utility beyond which a voter "typically" votes yes, so each participant or voter v_i is expected to say yes (i.e. $r_i(a_j) = 1$) to a time slot a_j when her utility for that slot $u_{ij} \geq t$. We note that we are assuming this typical yes-threshold is an externally-imposed or socially-determined global value for all voters. Incorporating the possibility of individual default yes-thresholds t_i for each voter i is a direction for future work.

Notice that unlike with Doodle polls, in approval voting, regardless of where a person chooses to "draw their line" as long as they are voting *sincerely* (never voting "no" on one candidate while simultaneously voting "yes" on a less pre-ferred candidate), they are considered to be voting honestly. Whereas in Doodle polls, there is some notion and expectation that the participants will not only be sincere, but also be "forthcoming" about their "true" availability.

Indeed, other studies have often assumed that the most straightforward, "honest" behavior of a voter is simply to vote "yes" on those time slots for which she is available, and "no" on those she is unavailable. However, we note that availability is not so black and white, and in theory, people can make them-selves available for *any* time slot, at varying degrees of cost. Our model accounts for the fact that a person's degree of "availability" is in fact on a continuum. For example, if one wished to consider negative utilities (for time slots where costs outweigh the benefits of attending), then a natural yes-threshold would be at utility 0. In this interpretation of the model, the voting behavior of the players can be seen as "honest" (when they vote yes if and only if their utility is posi-tive), or "dishonest" (when they either vote no for positive-utility time slots, or vote yes on negative-utility slots). Re-scaling utilities to the interval $[0, 1]$ allows the previous threshold value of 0 to also be accordingly mapped to a value t in the interval $[0, 1]$.

On the other hand, if the community culture or larger social/societal expectations imply a different "default" threshold t for voting yes on a time slot, where here we think of t as the utility threshold beyond which a participant is "ordinarily expected" to agree on a time slot, then our model still applies. Normalizing the utility values so that they lie between 0 and 1 and using some non-specified default threshold t makes our model general enough to capture multiple interpretations of the utility values and voting.

We assume all participants are *sincere* in their completion of polls, i.e., if v_i says yes to a time slot a_j which has utility u_{ij} then they also say yes to all slots a_k with utility $u_{ik} > u_{ij}$. Note that the social voting hypothesis arising from empirical data analyzed by Zou et al. [19] supports the expectation of sincere participants. (See their Proposition 2.)

Yet in reality some people are either more protective of their time, voting no on a slot even when their utility for it is above the yes-threshold t, or more cooperative, voting yes on a slot even when their utility is below the yes-threshold t. While our analysis does not assume that a poll is open or closed, there are certainly plausible reasons why either variant could lead voters to be protective or cooperative. Note that while such terms may sometimes have associated positive or negative connotations, we merely use them to categorize participants, and no judgment of the voters' behavior is intended.

We define an *ordinary* voter to be one who votes according to the yes-threshold t, as expected, voting yes to exactly the time slots a_j for which her utility $u_{ij} \geq t$, and no to all others. It might be helpful to think of ordinary voters as those who are responding "honestly" in some sense, akin to how other works have discussed a participant's "availability" in a black and white way [13,19]. But the term ordinary more impartially allows our model to apply to the idea that one's availability is on a continuous spectrum and t is the threshold beyond which social convention dictates one should respond yes.

We define a *cooperative* voter to be one who agrees to slots that are below the yes-threshold t, ostensibly trying to make more slots viable options at one's own expense. Since we assume voters are sincere, this is in practice the same as the voter "lowering" the value of the yes-threshold t for her votes. So she effectively uses a different threshold $t' < t$ such that she says yes to a time slot a_j if and only if her utility $u_{ij} \geq t'$.

We define a *restrictive* voter as one who votes no on slots that are above the yes-threshold, perhaps trying to be more protective of her time. Due to our assumption of sincerity, this is equivalent to the voter "raising" the value of the yes-threshold t for her votes. So she uses an alternative threshold $t' > t$ such that she says yes to a time slot a_j when her utility $u_{ij} \geq t'$.

2.3 Analysis Model

We now present the metric we use for the overall quality of each time slot as well as the framework we use for our analysis of the effects of the above-defined voting tendencies.

The *social welfare* of a given slot a_j is $u(a_j) = \sum_{i=1}^{n} u_{ij}$, the total utility assigned to that slot by all voters. Note that the social welfare is a measure of the theoretical goodness of a time slot; it does not account for the actual attendance of the participants, who may ultimately not attend a time slot for which they had voted yes, or may in fact attend at a time slot for which they had voted no.

We use $OPT(I)$ to denote a slot which maximizes the social welfare in a given Doodle poll instance I, and $u(OPT(I))$ to denote the utility (welfare) of an optimal slot. Hence

$$OPT(I) = \arg\max_{a_j \in A} \sum_{i=1}^{n} u_{ij}, \qquad u(OPT(I)) = \max_{a_j \in A} \sum_{i=1}^{n} u_{ij}.$$

Let $DDL(I)$ likewise denote a time slot returned by the default Doodle algorithm, and let $u(DDL(I))$ denote the utility (welfare) of $DDL(I)$.

In the spirit of worst-case analysis, the conventional approach of the theoretical algorithms (and algorithmic game theory) communities, we aim to determine a quantity that captures how far from optimal the Doodle poll mechanism may be. We therefore define the *welfare approximation ratio* of an algorithm DDL for choosing a time slot to be the maximum over all possible Doodle poll instances of the ratio $u(OPT(I))/u(DDL(I))$. I.e., if $\mathcal{I}$ is the set of all possible Doodle poll instances, the welfare approximation ratio of the default Doodle algorithm DDL is

$$\max_{I \in \mathcal{I}} \frac{u(OPT(I))}{u(DDL(I))}.$$

We also consider in this work the effect of cooperative and restrictive voting on welfare. To quantify this effect, we again employ worst-case analysis. Let a partial Doodle poll instance I' be just the first three elements (A, V, U) from the 4-tuple of a complete Doodle poll instance. Let $\mathcal{I}'$ be the set of all partial instances, and let $R_O(I')$ denote the response matrix that results from a given partial instance $I' \in \mathcal{I}'$ when all voters are ordinary. Let $\mathcal{R}_C(I')$ and $\mathcal{R}_S(I')$ be the set of all possible response matrices when a positive number of voters are cooperative and restrictive, respectively. (We drop the I' when the instance is clear from context.) Then we define the *positive welfare impact factor* (resp. *negative*) of cooperative voting to be

$$\max_{I' \in \mathcal{I}'} \max_{R_C \in \mathcal{R}_C} \frac{u(DDL(I', R_C))}{u(DDL(I', R_O))}, \qquad \max_{I' \in \mathcal{I}'} \max_{R_C \in \mathcal{R}_C} \frac{u(DDL(I', R_O))}{u(DDL(I', R_C))}.$$

Intuitively, these quantities represent the extreme limits of how many times better (resp. worse) social welfare can become when voters are cooperative (as opposed to ordinary).

We define welfare impact factors for restrictive voting analogously. To succinctly specify a partial Doodle poll instance (A, V, U), we use a table such as Table 1a to indicate the utility values of different categories of participants for each of the possible time slots in a Doodle poll.

Table 1. A template for displaying participants' utilities, voter types (ordinary, cooperative, or restrictive), and the number of voters in each group is given in (a), (b) is an example instance using this table format yielding a welfare approximation ratio of $\frac{1}{t} + \frac{n-x}{x}$. (See Lemma 2, below, for more details on (b).)

Participants	Time Slot 1	...	Time Slot m
# voter type	utility	...	utility
⋮ ⋮	⋮	⋮	⋮
# voter type	utility	...	utility

(a)

Participants	1	2
x ordinary	t	1
$n - x$ ordinary	0	$t - \epsilon$

(b)

3 Ordinary Voting

We start by evaluating how well the selected slot optimizes social welfare when all participants are ordinary voters. Throughout the examples and analysis, let $\epsilon > 0$ be a fixed constant, which may be arbitrarily small. We begin with an upper bound on the welfare approximation ratio of Doodle, and then we give an instance that demonstrates this upper bound is tight.

Lemma 1. *The welfare approximation ratio of the default Doodle algorithm with only ordinary voters is strictly less than $\frac{1}{t} + \frac{n-s^*}{s^*}$, where s^* is the score of the winning time slot.*

Proof. Consider any Doodle poll instance, I. We define $s^* = s(DDL(I))$, i.e., the score of the winning time slot, with $1 \leq s^* \leq n$. (We exclude an s^*-value of 0 because that would mean that all voters voted no for every time slot in the poll.) A reported score of s for the slot picked by the algorithm, meaning exactly s^* yes votes, ensures that $u(DDL(I)) \geq s^* t$, since ordinary voters vote yes precisely when their valuation is greater than or equal to t.

Since the time slot OPT was not picked, it must have an equal or smaller reported score than that of DDL. Note that if OPT and DDL had equal reported scores, the poll initiator has no additional information from the poll about voters' utilities for tie-breaking, and thus could have picked either OPT or DDL. Thus, the OPT time slot has at most s^* voters who voted yes (their valuation is at most 1) and the rest, at most $n - s^*$, have valuation strictly less than t. Thus the optimal social welfare is $u(OPT) < s^* + (n - s^*) \cdot t$. Hence, the ratio of the social welfare of OPT compared to the social welfare of the solution selected by Doodle is $u(OPT)/u(DDL) < \frac{s+(n-s^*)t}{s^* t} = \frac{1}{t} + \frac{n-s^*}{s^*}$.

The approximation ratio is largest when t is small, or when $n - s^*$ is large: these observations illustrate some of the inherent limitations of Doodle polls. The first limitation, a poor ratio when the yes-threshold t is small, shows that if people are "too willing" to say yes to a time (perhaps trying to be more agreeable), the chosen slot may be far from the best. Explicitly, as the yes-threshold decreases, the approximation to optimal welfare worsens. The second limitation, that the ratio is largest when $n - s^*$ is large, is perhaps less concerning in practice. When $n - s^*$ is large, that means that s^*, the reported score, is small, and many poll initiators would expect worse results in terms of overall social welfare when the 'most popular' slot has a small number of people voting yes for it.

The delicate dependence on t we have established also points to ways voters may exploit the system, intentionally or unintentionally. Suppose that there are only two slots, and most people value slot a_1 at or just above t, but strongly prefer slot a_2, with a valuation near 1. Most people thus vote yes to both slots. A single person who values both slots at above t but would rather have slot a_1 can now sincerely vote yes for a_1 and no to a_2 to get their preferred slot, harming the social welfare. Alternatively, a different individual could vote yes for their preferred slot a_2, even if their utility for both is below t, causing the slot with the overall better social welfare to be selected.

The formulation of the above proof gives rise to the following instance, showing that the upper bound of Lemma 1 is tight.

Lemma 2. *The welfare approximation ratio of the default Doodle algorithm with only ordinary voters is at least $\frac{1}{t} + \frac{(n-s^*)(t-\epsilon)}{s^*t}$, where $\epsilon > 0$ may be arbitrarily close to 0.*

Proof. Consider the instance represented in Table 1b. The utility of the first slot is $u(a_1) = xt$, while $u(a_2) = x + (n-x)(t-\epsilon)$. The reported scores with ordinary voters are $s(a_1) = x$ and $s(a_2) = x$, and thus $s^* = x$. Thus, with the tie, either spot may be chosen, and if a_1 is chosen, the indicated ratio is achieved.

While ties such as the instance in Table 1b may in fact be a reality in Doodle polls, if the tie-breaking aspect seems disconcerting, consider that one additional ordinary voter with valuation t for slot a_1 and 0 for a_2 can be added, so that the reported scores are now no longer tied, but the achieved ratios are comparable for sufficiently large n. Likewise, the instance need not have only two time slots; there can be many more slots, all with reported score less than s^*, and lower total social welfare. Since Lemmas 1 and 2 give matching bounds, we have the following theorem.

Theorem 1. *The welfare approximation ratio of the default Doodle algorithm with only ordinary voters is arbitrarily close to $\frac{1}{t} + n - 1$.*

4 Restrictive Voting

In this section, we make the assumption that some subset of the participants of size ℓ are restrictive voters, while the rest, $n - \ell$, are ordinary voters. Though the ℓ must all vote restrictively, they need not have identical valuations. We show that restrictive voting can not only harm, but also improve the social welfare.

4.1   Restrictive Voting Can Improve Social Welfare

We begin by giving an upper bound on the positive welfare impact factor of restrictive voting. Then we demonstrate that this upper bound is tight by providing a lower bound instance showing that restrictive voters can indeed have that degree of positive welfare impact.

Consider an arbitrary instance I' with ℓ restrictive voters. If everyone voted according to the yes-threshold as ordinary voters, slot $a = DDL(I', R_O)$ would be selected. But since the ℓ restrictive voters vote restrictively, slot b is selected. We assume that slots a and b are not the same, since otherwise there is no change in welfare, but make no other assumptions about these slots; they are just two of possibly many. We let $s^*(a)$ indicate the reported score of slot a when all participants are ordinary voters, that is, vote according to the yes-threshold t. Let $s'(a)$ indicate the reported score of slot a when the restrictive voters use an adjusted yes-threshold greater than t.

Fact 1. $u(a) \geq s^*(a)t$.

Proof. When everyone votes according to the yes-threshold, then all yes votes correspond to voters with valuations of at least t.

Fact 2. $u(b) < s^*(b) + (n - s^*(b)) \cdot t$.

Proof. When everyone votes according to the yes-threshold, yes votes correspond to valuations of at most 1, and no votes correspond to valuations strictly less than t.

By Facts 1 and 2, the welfare approximation ratio

$$\frac{u(b)}{u(a)} < \frac{s^*(b) + (n - s^*(b))t}{s^*(a)t}.$$

Since slot a is selected when everyone votes according to the yes-threshold, $s^*(a) \geq s^*(b)$. Suppose that $s^*(a) - s^*(b) = k$ for some fixed constant k. Then

$$\frac{u(b)}{u(a)} < \frac{s^*(a) - k + (n - s^*(a) + k)t}{s^*(a)t} = \frac{1}{t} + \frac{n - s^*(a)}{s^*(a)} - \frac{(1-t)k}{s^*(a)t}.$$

Observe that the second term is largest when $s^*(a)$ is smallest [a perhaps dissatisfying solution to an initiator]. Since k appears only in the final term (and $t \leq 1$), the ratio is largest when k is smallest. If $k = 0$, the ratio is thus $\frac{1}{t} + \frac{n - s^*(a)}{s^*(a)}$. The above discussion gives the following lemma.

Lemma 3. *The positive welfare impact factor of restrictive voters on any instance I' is strictly less than $\frac{1}{t} + \frac{n - s^*}{s^*}$, where s^* is the winning slot score when all voters are ordinary.*

Table 2. Participant types and valuations where restrictive voting improves the social welfare by a factor of $\approx \frac{1}{t} + (n - \ell)/\ell$, $\frac{1}{t} + 1$, and $\approx \frac{1}{t}$, respectively, with the last requiring only a single restrictive voter. See Lemmas 4, 14, and 15 for more details.

Participants	1	2
ℓ restrictive	t	1
$n - \ell$ ordinary	0	$t - \epsilon$

(a)

Participants	1	2
1 restrictive	t	1
$n/2 - 1$ ordinary	t	1
$n/2$ ordinary	0	$t - \epsilon$

(b)

Participants	1	2
1 restrictive	1	t
$n - 1$ ordinary	1	t

(c)

Note that the case where $k = 0$ may be dissatisfying because it involves a tie and tie-breaking procedure. Thus, we also note that when $k = 1$, the ratio becomes

$$\frac{1}{t} + \frac{n - s^*(a)}{s^*(a)} - \frac{1 - t}{s^*(a)t} = \frac{1}{t} + \frac{n + 1 - s^*(a)}{s^*(a)} - \frac{1}{s^*(a)t}.$$

For a matching lower bound, consider the instance illustrated in Table 2a, with valuations identical to that of Table 1b, but now the first group of voters vote restrictively.

Lemma 4. *The positive welfare impact factor of restrictive voting is at least $\frac{1}{t} + \frac{n-\ell}{\ell}$, suppressing epsilons.*

Proof. Consider the instance represented in Table 2a. The utilities of the time slots are $u(a_1) = t\ell$ and $u(a_2) = \ell + (n - \ell)(t - \epsilon)$. When all participants vote according to the yes-threshold t, the reported scores are $s^*(a_1) = \ell$ and $s^*(a_2) = \ell$, with the tie meaning either slot can be chosen. When the ℓ restrictive voters vote restrictively, the reported scores are $s'(a_1) = 0$ and $s'(a_2) = \ell$, ensuring that slot a_2 is chosen. Thus, restrictive voting yields a factor $\frac{1}{t} + \frac{n-\ell}{\ell}$ improvement of the social welfare, ignoring epsilons.

Taken together, Lemmas 3 and 4 give the following theorem.

Theorem 2. *The positive welfare impact factor of restrictive voters is $\frac{1}{t} + n - 1$.*

Noting that the example in Table 2a requires the poll initiator to select a slot with low reported score when there are few restrictive voters (and indeed that the ratio of $1/t + n - 1$ is only achieved when the winning slot has a score of 1 and most people vote no to both slots), we provide instances in which a single restrictive voter can still have a positive impact on the social welfare, in situations that are more satisfying to a poll initiator.

In Table 2b, the structure is similar to Table 2a with $\ell = 1$, and gives a $\frac{1}{t} + 1$ improvement of the social welfare but now an additional set of ordinary voters ensures that reported scores are at least half the number of participants. While the instance in Table 2b allows a poll initiator to select a time slot for which half of the participants are available, it suffers from the limitation that half of

Table 3. Participant types and valuations where restrictive voting harms the social welfare by a factor of $\approx \frac{1}{t}$, $\approx \frac{1}{t} + \frac{\ell t'}{(n-\ell)t}$, and ℓ, respectively. See Lemmas 6 and 7 for details.

Participants	1	2
1 restrictive	$1-\epsilon$	1
$n-1$ ordinary	1	t

(a)

Participants	1	2
ℓ restrictive	$t < t' < 1$	0
$n - \ell$ ordinary	1	t

(b)

Participants		1	2	$\cdots$	$n-1$	n	$n+1$
ℓ {	1 restrictive	$t+\epsilon$	0	$\cdots$	0	0	t
	1 restrictive	0	$t+\epsilon$	$\cdots$	0	0	t
	: restrictive	0	0	$\ddots$	0	0	t
$n-\ell$ {	1 ordinary	0	0	$\ldots$	$t+\epsilon$	0	0
	1 ordinary	0	0	$\ldots$	0	$t+\epsilon$	0

(c)

the participants vote no on both slots, yet social expectations may make that an unlikely response for most participants. Thus, we provide a different instance in Table 2c with a single restrictive voter, reported scores indicating all participants are available, and a $1/t$ improvement of the social welfare. Full proofs of these observations are deferred to the Appendix, in Lemmas 14 and 15.

4.2   Restrictive Voting Can Harm Social Welfare

We first provide an instance (Table 3a) showing that a single restrictive voter can harm the welfare by a factor of $\approx \frac{1}{t}$. We assume $t < 1 - \epsilon$. The utilities of the time slots are $u(a_1) = n - \epsilon$ and $u(a_2) = (n-1)t + 1$. When all participants vote according to the yes-threshold t, the reported scores are $s^*(a_1) = n$ and $s^*(a_2) = n$, with the tie meaning either slot can be chosen. When the one restrictive voter votes restrictively, the reported scores are $s'(a_1) = n - 1$ and $s'(a_2) = n$, ensuring that slot a_2 is chosen. Thus, restrictive voting decreases the social welfare from $n - \epsilon$ to $(n-1)t + 1$, which for n large is $\approx 1/t$.

Note that the instance in Table 3a is both plausible from a restrictive voter's perspective (choosing to say no to a less preferred slot), and satisfying to a poll initiator (the reported score indicates availability of all participants). We now show that with additional restrictive voters, social welfare can be harmed further, but first provide an upper bound on the negative welfare impact factor of restrictive voting. The proof mirrors that of Lemma 3, and is deferred to the appendix. We then provide a matching lower bound instance in Lemma 6, as portrayed in Table 3b, where $s'(a) = n - \ell$.

Lemma 5. *The negative welfare impact factor of restrictive voters on any instance I' is strictly less than $\frac{1}{t} + \frac{n-s'}{s'} \frac{t'}{t}$, where s' is the reported score with restrictive voters on the slot that wins when all voters are ordinary.*

Lemma 6. *The negative welfare impact factor of restrictive voting is at least* $\frac{1}{t} + \frac{\ell}{(n-\ell)} \frac{t'}{t}$.

Proof. Consider the instance represented in Table 3b. The utilities of the time slots are $u(a_1) = \ell t' + n - \ell$ and $u(a_2) = (n - \ell)t$. When all participants vote according to the yes-threshold t, the reported scores are $s^*(a_1) = n$ and $s^*(a_2) = n - \ell$, so slot a_1 is chosen. When the ℓ restrictive voters vote restrictively, the reported scores are $s'(a_1) = (n - \ell)$ and $s'(a_2) = n - \ell$, so a_2 may be chosen. Thus, restrictive voting decreases the social welfare from $\ell t' + n - \ell$ to $(n - \ell)t$, giving the stated ratio.

We now provide an instance of a different nature exhibiting a negative welfare impact factors linear in the number of restrictive voters, and, when all participants are restrictive voters, that is, $n = \ell$, the corollary is immediate.

Lemma 7. *The negative welfare impact factor of restrictive voting is at least ℓ, where ℓ is the number of restrictive voters.*

Proof. Consider the instance represented in Table 3c. The utilities of the first n time slots are all equal, with $u(a_1) = \cdots = u(a_n) = t + \epsilon$, while $u(a_{n+1}) = \ell t$. When all participants vote according to the yes-threshold t, the reported scores are $s^*(a_1) = \cdots = s^*(a_n) = 1$ and $s^*(a_{n+1}) = \ell$. When the ℓ restrictive voters vote restrictively, that is, no to slot $n+1$ (but still yes to the slot with valuation $t + \epsilon$), the reported scores are $s'(a_1) = \cdots = s'(a_n) = 1$ and $s'(a_{n+1}) = 0$. Thus, restrictive voting changes the selected time slot from slot $n + 1$ to any of the others, decreasing the social welfare by a factor of ℓ, suppressing epsilons.

Corollary 1. *The negative welfare impact factor of restrictive voting is at least n.*

The example in Table 3c has some nice features, but also some limitations. It is certainly possible that a restrictive voter who dislikes most of the time slots, and has similar valuations for two of the slots, may in fact say yes to only one of those slots. However, many poll initiators, when faced with the reported scores when all participants vote restrictively (namely that all slots have reported scores of 0 or 1) are likely to declare none of the options viable rather than selecting a time to which only one participant voted yes. This concern motivates related examples whose details are deferred to Appendix A where the reported score is now a constant fraction of the number of participants (Table 6) and where the effect of restrictive voting depends on the number of time slots (Table 7).

5   Cooperative Voting

In this section, we make the assumption that some subset of the participants of size c are cooperative voters, while the rest, $n - c$, are ordinary voters. Note that the c must all vote cooperatively, but need not have identical valuations. We show that cooperative voting can greatly improve the social welfare, but also can substantially harm it.

Due to space considerations, all proofs in this section are deferred to the appendix.

Table 4. Participant types and valuations where cooperative voting improves the social welfare by a factor of c and $\frac{1}{t} + \frac{c}{n-c}$, respectively. See the proofs of Lemmas 8 and 9 for details.

Participants	1	2	$\cdots$	$n-1$	n	$n+1$
c { 1 cooperative	t	0	$\cdots$	0	0	$t-\epsilon$
1 cooperative	0	t	$\cdots$	0	0	$t-\epsilon$
$\vdots$ cooperative	0	0	$\ddots$	0	0	$t-\epsilon$
$n-c$ { 1 ordinary	0	0	$\cdots$	t	0	0
1 ordinary	0	0	$\cdots$	0	t	0

(a)

Participants	1	2
c cooperative	$t-\epsilon$	0
$n-c$ ordinary	1	t

(b)

5.1 Cooperative Voting Can Improve Social Welfare

We present some instances of how cooperative voting can help social welfare, many of which arise from slightly altering the valuations (and changing the participant types) of instances illustrating how restrictive voting can harm social welfare. More precisely, by switching the restrictive voters from Table 3c to cooperative, and decreasing valuations of $t+\epsilon$ to t, and those of t to $t-\epsilon$, we get the instance in Table 4a, which gives Lemma 8. In addition, analogously to Corollary 1, when $n = c$, cooperative voting can help social welfare by a factor of n.

Lemma 8. *The positive welfare impact factor of cooperative voting is at least c, the number of cooperative voters.*

We give an instance in Table 4b that exhibits a positive impact factor as detailed in Lemma 9. Notice that when $c = n$ in the instance of Table 4b the welfare improvement factor becomes unbounded. We then give a matching upper bound in in Lemma 10 on the welfare improvement factor of cooperative voting. The proof is analogous to the proof of Lemma 3.

Lemma 9. *The positive welfare impact factor of cooperative voting is at least $\frac{1}{t} + \frac{c}{n-c}$ (suppressing epsilon terms).*

Lemma 10. *The positive welfare impact factor of cooperative voting for any instance I' is strictly less than $\frac{1}{t} + \frac{n-s^*}{s^*}$, where $s^* = s(DDL(I', R_O))$ is the score of the winning slot when all voters are ordinary.*

Let s^* denote the winning slot when respondents are ordinary and vote according to the yes-threshold. Substituting $s^* = n - c$ in Lemma 10 and taking that together with Lemma 9 gives the following theorem.

Theorem 3. *The positive welfare impact factor of cooperative voting is $\frac{1}{t} + \frac{n-s^*}{s^*}$.*

5.2 Cooperative Voting Can Harm Social Welfare

Though cooperative voting may be quite beneficial, it can likewise be quite harmful. As illustrated in the instance in Table 5a, even a single cooperative voter can harm welfare by a factor of $\approx 1/t$. The instance in Table 5b gives Lemma 11, showing the effects of cooperative voting can be even more harmful to social welfare.

Lemma 11. *The negative welfare impact factor of cooperative voting is at least $1/t'$, where $t' < t$ is the adjusted yes-threshold of the cooperative voters. (This ratio is unbounded when $t' = \epsilon$.)*

The same negative welfare impact factor can also be achieved with n cooperative voters, all of whom value slot 1 at 1 and slot 2 at $t' < t$. Note that the situation in Table 5b does not necessarily seem problematic to an initiator selecting a result. The selected slot has reported score of half of the participants, which may in fact be appropriate in some settings. If the tie-breaking aspect is concerning, having one more cooperative participant (with the same valuations) yields essentially the same results. The default yes-threshold t does not play a role in this instance. And while the results are most striking when $t' = \epsilon$ is small, a person who has valuation 0 for one slot and any amount for another slot, no matter how small, may in fact be inclined (socially) to be a cooperative voter, thus saying yes to the slot for which they have some marginal value.

Table 5. Participant types and valuations where cooperative voting harms the social welfare by a factor of $\approx 1/t$ (with a single cooperative voter), $1/t'$, and $\frac{1}{t'} + \frac{n-c}{c}\frac{t}{t'}$, respectively. See the proofs of Lemmas 11 and 12 for details.

Participants	1	2
1 cooperative	0	$t - \epsilon$
$n - 1$ ordinary	1	t

(a)

Participants	1	2
$n/2$ cooperative	0	$t' < t$
$n/2$ ordinary	1	0

(b)

Participants	1	2
c cooperative	$t' < t$	1
$n - c$ ordinary	0	$t - \epsilon$

(c)

As alarming as Lemma 11 may be, we show in Lemma 12 based on the instance in Table 5c that cooperative voting can have an even more harmful impact. The moral here for Doodle poll participants is perhaps as follows: if you think you are being helpful by voting yes generously in a Doodle poll, don't be so sure: you might actually be making things worse overall.

Lemma 12. *The negative welfare impact factor of cooperative voting is at least $\frac{1}{t'} + \frac{n-c}{c}\frac{t}{t'}$.*

We then upper bound the negative welfare impact factor of cooperative voting in Lemma 13. Observe that the Lemma 12 instance precisely matches the upper bound since $s^* = c$.

Lemma 13. *The negative welfare impact factor of cooperative voters on any instance I' is strictly less than $\frac{1}{t'} + \frac{n-s^*}{s^*}\frac{t}{t'}$, where s^* is the score of the winning slot when all voters are ordinary, i.e., $s^* = s(DDL(I', R_O))$.*

6   Conclusion

People often assume that a Doodle poll is a mechanism for finding the best time slot for a meeting, yet we show in this work that the optimal social welfare is not always achieved. Under ordinary voting, a Doodle-recommended slot may have social welfare $1/t$ times worse than the optimal. This means that we might want voters to (perhaps counter-intuitively) have a higher yes-threshold t. We also show the Doodle-recommended slot may be as bad as $(n - s^*)/s^*$ times worse than the optimal one, where s^* is the score of the winning slot. So, naturally, a winning slot with a large number of yes votes is preferred. We then show that cooperative voters may in fact harm the overall social welfare, while restrictive voters can improve the overall social welfare. In fact, both cooperative and restrictive voting are capable of harming or improving the overall social welfare. We prove worst-case bounds on both the positive and negative welfare impact of both cooperative and restrictive voting in Doodle polls. We find that even with cooperative and restrictive voting, a lower default yes-threshold, while perhaps conventionally thought of as desirable so that the response matrix is more easily filled with yes votes, can in fact be detrimental to the quality of the winning slot.

The impacts on social welfare naturally suggest future work in this area, including the impacts of having both cooperative and restrictive voters in a single poll. Another direction of investigation would be to use an objective function that considers not just total utility of the winning slot but also its number of yes-votes (which presumably predicts the level of attendance at the event). It would also be interesting to incorporate the social voting hypothesis of [19] into our model. An analysis that includes Doodle's "if-need-be" option, though infrequently used, may demonstrate benefits to poll initiators and participants alike, as it allows participants to have more power to express their preferences over the slots, which may result is improved overall social welfare of selected times. Respondents also then have an added ability to appear more cooperative. It would also be interesting to investigate alternate mechanisms that may lead to improved social welfare of the chosen time slot. Additionally, we could ask what tactics the standard game-theoretic utility-maximizing participant could employ in the Doodle game model we have proposed here, and perhaps study the quality of the Nash equilibria outcomes of such a game. Finally, we would like to acquire and experiment with real Doodle data to see how often these welfare impact effects play out. Since we would not have users' private utility values in this case we would have to simulate the utilities and run what-if scenarios to determine how likely and how often we see such effects.

A Appendix

A.1 Restrictive Voting

Lemma 14. *Even if the score of the winning slot must be at least $n/2$ and there is only one restrictive voter, the positive welfare impact factor of restrictive voting is still at least $\frac{1}{t} + 1$ (suppressing epsilons).*

Proof. Consider the instance represented in Table 2b. The utilities of the time slots are $u(a_1) = nt/2$ and $u(a_2) = n/2 + n(t - \epsilon)/2$. When all participants vote according to the yes-threshold t, $s^*(a_1) = n/2 = s^*(a_2)$, with the tie meaning that slot 1 could be chosen. When the one restrictive voter votes restrictively, still saying yes to slot 2 but now saying no to slot 1, the reported scores become $s'(a_1) = n/2 - 1$ and $s'(a_2) = n/2$, so that slot 2 is now chosen. Thus, suppressing epsilons, the social welfare improves by a factor of $1/t + 1$.

Lemma 15. *Even if the score of the winning slot is n and there is only one restrictive voter, the positive welfare impact factor of restrictive voting is still at least $\frac{1}{t}$.*

Proof. Consider the instance represented in Table 2c. The utilities of the time slots are $u(a_1) = n$ and $u(a_2) = nt$. When all participants vote according to the yes-threshold t, $s^*(a_1) = n = s^*(a_2)$, with the tie meaning that slot a_2 could be chosen. When the one restrictive voter votes restrictively, still saying yes to slot 1 but now saying no to slot 2, the reported scores become $s'(a_1) = n$ and $s'(a_2) = n - 1$, so that slot 1 is now chosen. Thus, the social welfare improves by a factor of $1/t$.

Proof (of Lemma 5). Mirroring the proof of Lemma 3, define $a = DDL(I', R_O)$ to be the slot selected when everyone votes according to the yes-threshold, and b is the slot selected when the ℓ restrictive voters vote restrictively. Since we are analyzing the negative welfare impact factor, we must upper bound $u(a)/u(b)$. Observe that by the definitions of a and b and that since restrictive voting can only lower reported scores, we have that $s'(a) \leq s'(b) \leq s^*(b) \leq s^*(a)$. With that observation, and noting that, similarly to Fact 1, $u(b) \geq s^*(b)t$, we then have that $u(b) \geq s'(a)t$. Since $t < t'$ in restrictive voting, and a restrictive yes vote indicates a valuation of at most 1, while a restrictive no vote indicates a valuation less than t', Fact 2 now becomes $u(a) < s'(a) + (n - s'(a))t'$. A comparable remaining argument to that of Lemma 3 thus gives the resulting upper bound.

Consider the instance represented in Table 6. Let $k > 2$ be a fixed constant, and without loss of generality, assume $k \mid n$ and $2 \mid n$, for ease of analysis. The instance has $n/2$ restrictive voters with valuations as before, but also $n/2$ ordinary voters, numbered $i = 1$ to $n/2$, who all value slot $n/2 + 1$ at $t - \epsilon$, the slots i to $i + n/k - 1$ (wrapping around for slots exceeding $n/2$) at t, and the rest at 0. The utilities of the time slots are $u(a_1) = \ldots = u(a_{n/2}) = ((n/k) + 1)t + \epsilon$ and $u(a_{n/2+1}) = nt - n\epsilon/2$. When all participants vote according to the yes-threshold t, the reported scores are $s^*(a_1) = \cdots = s^*(a_{n/2}) = n/k + 1$ and

Table 6. Participant types and valuations where restrictive voting harms the social welfare, with a reported score that is a constant fraction k of the participants.

Participants		1	2	3	$\cdots$	$n/2+1$
	1 restrictive	$t+\epsilon$	0	0	0	t
	1 restrictive	0	$t+\epsilon$	0	0	t
$n/2$	$\vdots$ restrictive	0	0	$\ddots$	0	t
	1 restrictive	0	0	0	$t+\epsilon$	t
	$n/2$ ordinary	for ordinary voter $i=1\ldots n/2$,				$t-\epsilon$
		t for slots i to $i+n/k-1$,				
		wrapping around; 0 otherwise				

$s^*(a_{n/2+1}) = n/2$. When the restrictive voters vote restrictively, saying yes to their one slot with valuation $t+\epsilon$ but no to the slot with valuation t, the reported scores are $s'(a_1) = \cdots = s'(a_{n/2}) = n/k + 1$ and $s^*(a_{n/2+1}) = 0$. Thus, since $k > 2$, restrictive voting changes the selected time slot from slot $n/2 + 1$ to any of the other slots, decreasing the social welfare from nt to $(n/k+1)t$, suppressing epsilons. Note that while this example does have a more plausible reported score, it does require the number of time slots to be about half of the number of participants.

Table 7. Participant types and valuations where restrictive voting harms the social welfare by a factor of $\approx m$.

Participants	1	2	3	$\cdots$	$m = \lceil\sqrt{n}\rceil + 1$
$\lfloor\sqrt{n}\rfloor$ restrictive	$t+\epsilon$	0	0	0	t
$\lfloor\sqrt{n}\rfloor$ restrictive	0	$t+\epsilon$	0	0	t
$\lfloor\sqrt{n}\rfloor$ restrictive	0	0	$t+\epsilon$	0	t
$\vdots$ restrictive	0	0	0	$\ddots$	t

Lemma 16. *The negative welfare impact factor of restrictive voting is at least $\approx m$.*

Proof. Consider the instance represented in Table 7. The last slot has utility $u(a_m) = nt$, while the other slots have utilities $\lfloor\sqrt{n}\rfloor(t + \epsilon)$, except possibly for slot $m - 1$ which may have smaller utility, due to the square root and truncation with the floor operation. When all participants vote according to the yes-threshold t, most of the slots likewise have reported score $\lfloor\sqrt{n}\rfloor$, again with slot $m - 1$ possibly lower, and slot m having $s^*(a_m) = n$. When all n restrictive voters vote restrictively, that is, no to slot m, the reported scores of the first $m - 1$ slots are unchanged, with most at $\lfloor\sqrt{n}\rfloor$, but slot m now has $s'(a_m) = 0$.

Thus, restrictive voting changes the selected time slot from slot m to one of the earlier ones (except perhaps for $m - 1$), decreasing the social welfare from nt to $\lfloor \sqrt{n} \rfloor (t + \epsilon)$, giving the desired result.

Similarly, the instance in Table 7 with restrictive voting can be transformed to an instance showing that cooperative voting can improve social welfare by a factor of $\approx m$ by making all voters cooperative, changing valuations of t to $t - \epsilon$, and valuations of $t + \epsilon$ to t.

A.2 Cooperative Voting

We again define $s^*(a) = s(DDL(I', R_O))$ to be the score of the winning slot in an instance I' when all voters are ordinary. We now let $s'(a)$ indicate the reported score of slot a when the cooperative voters use an adjusted yes-threshold $t' < t$.

Proof (of Lemma 8). Consider the instance represented in Table 4a. The utilities of the time slots are $u(a_1) = \cdots = u(a_n) = t$ and $u(a_{n+1}) = c(t - \epsilon)$. When all participants vote according to the yes-threshold t, the reported scores are $s^*(a_1) = \cdots = s^*(a_n) = 1$ and $s^*(a_{n+1}) = 0$. When the c cooperative voters vote cooperatively, the reported scores become $s'(a_1) = \cdots = s'(a_n) = 1$ and $s'(a_{n+1}) = c$. Thus, cooperative voting changes the selected time slot from any of the first n to time slot $n + 1$, increasing the social welfare by a factor of c, suppressing epsilons.

Proof (of Lemma 9). Consider the instance represented in Table 4b. The utilities of the time slots are $u(a_1) = c(t - \epsilon) + n - c$, and $u(a_2) = (n - c)t$. When all participants vote according to the yes-threshold t, the first group of c participants report no for both slots, while the second group report yes for both slots. Thus, $s^*(a_1) = n - c = s^*(a_2)$, with the tie meaning either slot can be chosen. When the c cooperative voters vote cooperatively, they vote yes for slot 1 but still no on slot 2. The ordinary voters are unchanged in their votes. Hence, $s'(a_1) = n$ and $s'(a_2) = n - c$, ensuring that slot a_1 is chosen. The improvement in social welfare when slot a_1 is chosen due to cooperative voters rather than when slot a_2 can be chosen when all voters vote ordinarily is thus a factor of $\frac{1}{t} + \frac{c}{n-c}$ (suppressing epsilon terms).

Proof (of Lemma 10). The proof is analogous to the proof of Lemma 3, except rather than ℓ restrictive voters, we have c cooperative voters. Note that Facts 1 and 2 which lower bound the utility of slot a, the slot that is chosen when everyone is an ordinary voter, and upper bound the utility of slot b, the slot that is chosen when c of the n voters vote cooperatively, still stand as they are established purely on the reported scores of the two time slots when all voters are ordinary. We therefore still have the established upper bound on the welfare approximation ratio of

$$\frac{u(b)}{u(a)} < \frac{s^*(b) + (n - s^*(b))t}{s^*(a)t} < \frac{1}{t} + \frac{n - s^*(a)}{s^*(a)}.$$

Proof (of Lemma 11). Consider the instance represented in Table 5b. If we set $t' = \epsilon$, the utilities of the time slots are $u(a_1) = n/2$ and $u(a_2) = n\epsilon/2$. When all participants vote according to the yes-threshold t, the reported scores are $s^*(a_1) = n/2$ and $s^*(a_2) = 0$. When the first group (half of the participants) vote cooperatively, the reported scores are $s'(a_1) = n/2$ and $s'(a_2) = n/2$. Thus, with cooperative voting, slot a_2 may be chosen instead of a_1. Hence, the utility goes from $n/2$ to $n\epsilon/2$.

Proof (of Lemma 12). Consider the instance represented in Table 5c. The utilities of the slots are $u(a_1) = ct'$ and $u(a_2) = c + (n - c)(t - \epsilon)$. When all participants vote according to the yes-threshold t, the reported scores are $s^*(a_1) = 0$ and $s^*(a_2) = c$. When the c cooperative voters vote cooperatively, the reported scores become $s'(a_1) = c$ and $s'(a_2) = c$. Thus, with cooperative voting, slot 1 may be chosen instead of slot 2, resulting in the indicated change in social welfare.

Proof (of Lemma 13). Mirroring the proof of Lemma 3, define $a = DDL(I', R_O)$ to be the slot selected when everyone votes according to the yes-threshold, and b is the slot selected when the c cooperative voters vote cooperatively. Since we are analyzing the negative welfare impact factor, we must upper bound $u(a)/u(b)$. Observe that since cooperative voters have a lowered threshold of t', the claim paralleling Fact 1 is $u(b) \geq s'(b)t'$. We also know that $s'(b) \geq s^*(a)$ since with cooperative voting there can only be more yes votes than under ordinary voting, so the winning score of b must be at least that of a. Taking these two inequalities together gives us $u(b) \geq s^*(a)t'$. Fact 2 now becomes $u(a) < s^*(a) + (n - s^*(a))t$. A comparable argument to that in the restrictive voting section thus gives the resulting upper bound.

References

1. Arrow, K.J.: A difficulty in the concept of social welfare. J. Polit. Econ. **58**, 328–346 (1950)
2. Brams, S.J., Fishburn, P.C.: Approval Voting. Birkhauser, Boston (1983)
3. Brams, S.J., Fishburn, P.C., Merrill, S.: The responsiveness of approval voting: comments on Saari and van Newenhizen. Public Choice **59**(2), 121–131 (1988)
4. Brânzei, S., Caragiannis, I., Morgenstern, J., Procaccia, A.D.: How bad is selfish voting? In: desJardins, M., Littman, M.L. (eds.) Proceedings of the Twenty-Seventh AAAI Conference on Artificial Intelligence, 14–18 July Bellevue, Washington, USA. AAAI Press (2013)
5. Laslier, J.-F.: The leader rule a model of strategic approval voting in a large electorate. J. Theoret. Polit. **21**(1), 113–136 (2009)
6. Lehtinen, A.: The welfare consequences of strategic behaviour under approval and plurality voting. Eur. J. Polit. Econ. **24**(3), 688–704 (2008)
7. Mackenzie, D.: Making sense out of consensus. SIAM News, Philadelphia (2000)
8. Meehl, P.E.: The selfish voter paradox and the thrown-away vote argument. Am. Polit. Sci. Rev. **71**(01), 11–30 (1977)
9. Mosier, J.N., Tammaro, S.G.: When are group scheduling tools useful? Comput. Support. Coop. Work **6**(1), 53–70 (1997)

10. Myerson, R.B., Weber, R.J.: A theory of voting equilibria. Am. Polit. Sci. Rev. 87(1), 102–114 (1993)
11. Obraztsova, S., Elkind, E., Polukarov, M., Rabinovich, Z.: Doodle poll games. AGT@IJCAI (2015)
12. Palen, L.: Social, individual and technological issues for groupware calendar systems. In: Proceedings of the SIGCHI Conference on Human Factors in Computing Systems, CHI 1999, pp. 17–24. ACM, New York (1999)
13. Reinecke, K., Nguyen, M.K., Bernstein, A., Näf, M., Gajos, K.Z.: Doodle around the world: online scheduling behavior reflects cultural differences in time perception and group decision-making. In: Proceedings of the Conference on Computer Supported Cooperative Work, CSCW 2013, pp. 45–54. ACM, New York (2013)
14. Riker, W.H., Ordeshook, P.C.: A theory of the calculus of voting. Am. Polit. Sci. Rev. 62(01), 25–42 (1968)
15. Saari, D.G., Van Newenhizen, J.: Is approval voting an 'unmitigated evil'? A response to brams, fishburn, and merrill. Public Choice 59(2), 133–147 (1988)
16. Thayer, A., Bietz, M.J., Derthick, K., Lee, C.P.: I love you, let's share calendars: calendar sharing as relationship work. In: Proceedings of the ACM Conference on Computer Supported Cooperative Work, CSCW 2012, pp. 749–758. ACM, New York (2012)
17. Weber, R.J.: Approval voting. J. Econ. Perspect. 9(1), 39–49 (1995)
18. Xu, C.: Making doodle obsolete: applying auction mechanisms to meeting scheduling. Bachelor's thesis, Harvard University (2010)
19. Zou, J., Meir, R., Parkes, D.: Strategic voting behavior in doodle polls. In: Proceedings of the 18th ACM Conference on Computer Supported Cooperative Work, CSCW 2015, pp. 464–472. ACM, New York (2015)

Bring on Board New Enthusiasts! A Case Study of Impact of Wikipedia Art + Feminism Edit-A-Thon Events on Newcomers

Rosta Farzan[1(✉)], Saiph Savage[2], and Claudia Flores Saviaga[2]

[1] University of Pittsburgh, Pittsburgh, USA
rfarzan@pitt.edu
[2] West Virginia University, Morgantown, USA
saiph.savage@mail.wvu.edu, saviaga@gmail.com

Abstract. Success of online production communities such as Wikipedia highly relies on a continuous stream of newcomers to replace the inevitable high turnover and to bring on board new sources of ideas and labor. However, these communities have been struggling with attracting newcomers, especially from a diverse population of users. In this work, we conducted a case study on how organizing an offline co-located event over a short period of time contributes to involving newcomers in the online community. We present results of our multiple-source quantitative analysis of Wikipedia Art+Feminism edit-a-thon as a case of such events. The results of our analysis shows that such offline events are successful in attracting a large number of newcomers; however, retention of the newcomers stays as a challenge.

1 Introduction

Online production communities such as Wikipedia have been enjoying omnipresent success stories; however, the success stories are accompanied by significant challenges. An important challenge identified within a range of online production communities is ensuring a stream of newcomers to replace an inevitable high turnover they face and to attract sources of new ideas and new labor [20]. This problem is even more intensified when such communities try to recruit and retain newcomers from a more diverse population of users. For example, Wikipedia not only has faced a plateaued growth of new editors over the recent years [31], it has been particularly struggling with the challenge of attracting female editors [19].

In response to the challenge of attracting new members and developing commitment, a number of online production communities such as open source software communities and Wikipedia have tried to organize offline co-located gatherings to foster recruitment and integration of newcomers. The importance and occurrence of offline interactions in conjunction with online communications have been acknowledged by a number of studies [2,13,17,23,32]. A few studies have attempted to assess and quantify the impact of offline gatherings on online participation. While there is strong evidence in support of supplementing online

© Springer International Publishing AG 2016
E. Spiro and Y.-Y. Ahn (Eds.): SocInfo 2016, Part I, LNCS 10046, pp. 24–40, 2016.
DOI: 10.1007/978-3-319-47880-7_2

interactions with offline gatherings, there is also evidence that highlights the challenges arising as a result of offline connections. In some cases, offline connections can lead to weakening of the online interactions as a result of creating stronger clicks among those who can meet offline, or by shifting the interactions offline thereby reducing online interactions [28]. It has been argued that offline gatherings can promote stronger bonds that lead to stronger bonding social capital, but this is accomplished at the expense of decreasing weak ties and bridging social capital [30]. More recently, multiple case studies of open source software hackathons presents results on how the structure of such events influences the outcomes; especially in terms of advancing the production goals of the community as well as building social ties [33,34].

In this work, we are specifically interested to investigate how offline co-located gatherings affect participation of newcomers in online production communities. Hackathons-like events often have been viewed as onboarding programs. It has been argued that the intense training and social bonding opportunities provided in such offline events can particularly benefit newcomers and socialize them to the community by teaching them the performing and social rules [33]. However, no prior work has particularly investigated the impact of these offline gatherings on newcomers' socialization into the online community and the mechanism with which they can influence newcomers. As more resources and times are dedicated to these collective efforts, it becomes critical to understand the impact such events have on online production communities.

2   Socialization of Newcomers in Online Communities

Prior research has investigated socialization of newcomers in a variety of online communities, including open-source software (e.g. [9]), social media (e.g. [22]), and peer-production (e.g. [10]). They have particularly compared and contrasted socialization approaches in traditional organizations with strategies being employed online. While the results are not conclusive, often they have documented that many online communities lack specific strategies for socializing newcomers into the community [6]. An important factor identified by a number of studies as positively influencing newcomers' commitment to the online community, is interaction between newcomers and existing members [1,4,22]. Newcomers who receive feedback and communication from the existing members of the community, even if the feedback is criticism, are more likely to continue participating in the community.

Several investigations have focused on creating computational systems and methods to increase volunteer workforces [3], especially for political causes [27]. Other systems bootstrap off social media to access large pools of people to facilitate the recruitment process [12,14]. Brady et al. [3] showed that it was feasible to recruit volunteers from people's Facebook friends to help the blind. Savage et al. [27] showed the potential of using online bots to recruit people from Twitter to do micro-volunteering for a cause. Nevertheless, these platforms are at present incomplete and have mostly focused on recruitment of newcomers and not longer term commitment of newcomers [29]. While these approach can bring an influx of volunteers to a collaborative effort, they rarely maintain the volunteers long-term [15]. This is especially because the recruitment is often not followed by any longer-term engagement mechanism.

Other approaches have focused precisely on creating work flows that encourage long-term engagement of volunteers. Such platforms have sometimes sandboxes where newly recruited volunteers can have personalized and detailed feedback on their work from experts [24]. The sandboxes let newly engaged volunteers to become integrated into the cause under a friendly welcoming environment. This can help in their retention. However, the effectiveness of such approaches on socialization and retention of newcomers has not been researched. Other approaches have engaged new crowds of volunteers with simple lightweight feedback processes [7]. These approaches showcase how new volunteers can be retained through lightweight guided contribution. In this work we take a look at newcomer socialization processes that are offline and take consequently more time from both longer term community members and the newcomers. We analyze and contrast such processes with these other methods to engage newcomers.

3   Wikipedia Art+Feminism Edit-A-Thon

Wikipedia has often been named as one of the most successful examples of online production communities and product of collective wisdom. Despite its enormous success, it has also been facing a great deal of challenges over time. In particular, as highlighted by researchers at Wikimedia foundation, attracting newcomers is one of the key challenges faced by the Wikipedia community [11]. While some argue that committed members of the community exhibit different behavior and signs of commitment from early on [25], it has been shown that active strategies employed by the community and by the newcomers, such as friendly interactions with experienced members [24], active socialization approaches within Wikipedia [6], constructive feedback and avoiding undermining of their goodwill efforts [16,35] can increase the likelihood of future commitment.

The challenge of attracting newcomers becomes even more demanding when trying to target a more diverse population and those who have been underrepresented in the existing community. At the same time, it has been shown that diversity can play an important role in success of production communities such as Wikipedia [5]. Since 2011, several studies have highlighted a phenomenon of gender imbalance in Wikipedia that indicates only around 15 % of Wikipedia contributors identify as female and a very small percentages of Wikipedia contributions are made by female Wikipedians [8,21]. It has further been documented that this gender inequality has resulted in quantitative and qualitative inequality in representation of topics more attractive to female readers as well as inequality in representation of biographies of notable women [21]. This inequality happens despite the fact that women are generally more likely to participate in volunteer and community based activities and they are more likely to participate in social sites such as Facebook [21].

In response to this challenge, since 2014, a group of Wikipedia and feminism enthusiasts have been organizing Wikipedia Art+Feminism edit-a-thon events.[1] Edit-a-thon events are collocated all-day events bringing together novices and experienced Wikipedia editors. The goal of these events is to increase the coverage of female representation in Wikipedia and to encourage female editorship. The events are advertised on the Web and through various social media platforms.[2,3] It is particularly highlighted in the advertisement of the event that no prior editing experience is required and as one of the first activity of the day, a tutorial on editing Wikipedia is presented to the participants. By 2015, 75 Art+Feminism edit-a-thon events had been organized which attracted 1,500 participants and resulted in creating or improving 900 Wikipedia articles. In the current work, we focus on investigating the impact of the most recent set of edit-a-thon events, organized in 2016, on attraction and retention of newcomers into Wikipedia. We hope that our research can provide insights to organizers of such events to better understand how these event contribute to their goals. We also hope that by studying these events we can better understand the process of integrating more minorities into the production process.

4   Research Questions

We argue that an event such as edit-a-thon can influence newcomers' socialization process through two different mechanisms: (1) the focused gathering of an edit-a-thon event can provide newcomers with intense training opportunities to learn how to get work done in Wikipedia that can lead into more effective and consequently higher level of participation; (2) the collocated gatherings can build connections between newcomers and experienced Wikipedians helping them to build strong identification with Wikipedia that can lead to higher level of commitment and participation. Therefore, we have formulated the following research questions in better understanding of the impact of edit-a-thons on newcomers:

RQ1: how does attending an edit-a-thon event influence subsequent participation of newcomers on Wikipedia?

RQ2: does attending an edit-a-thon lead to bond and connections among participants?

RQ3: how do production and social interaction factors influence the retention of edit-a-thon newcomers in editing Wikipedia articles?

---

[1] https://en.wikipedia.org/wiki/Wikipedia:Meetup/ArtAndFeminism.
[2] http://art.plusfeminism.org/.
[3] https://www.facebook.com/events/876331705807795/.

5   Research Methods and Data Collection

To address our research questions, we conducted quantitative analysis on archival log data available on Wikipedia and Twitter. Using crawling approaches, Wikipedia API,[4] and Twitter API,[5] we collected data related to 59 edit-a-thon events happening in the US in the period of Jan 2016 until March 2016.

To study our first research question, we collected information on newcomers' logged behavior on Wikipedia during and after the edit-a-thon events to assess at what level they participated on the event day and subsequently after the event. To better understand their level of commitment, we attempted to contrast subsequent Wikipedia editing participation of newcomers who attended an edit-a-thon event with comparable newcomers who joined Wikipedia independent of edit-a-thon events. We identified specific editing tasks that newcomers performed on the day of the event, including creating user pages, editing in Sandbox pages, and editing article pages. Based on our experience with Wikipedia and Wikipedians community, we classify each of these editing activities as representing different familiarity and identification with Wikipedia. We collected information on newcomers' activities with respect to each of these categories.

- Creating a user page: It serves as the first step to belonging to the Wikipedia community and gets the users started with editing a Wikipedia page which includes personal information. It provides a practice experience without too much concern regarding the content of the page.
- Editing in Sandbox: Wikipedia provides Sandbox pages as a practice environment for users to practice with syntax of the Wiki Markup language to edit Wikipedia pages as well as organizing the content of the page before editing the main article page.
- Editing article pages: It indicates a stronger level of readiness for editing in Wikipedia and a stronger involvement in Wikipedians community

To study our second research question, we identified interactions happening on Wikipedia talkpages among edit-a-thons newcomers. Following the approach in [18], we excluded talkpage posts made by automatic Wikipedia bots. We constructed a communication network based on the talkpage interactions. In addition to interactions happening on Wikipedia, participants are encouraged to post about the event and communicate on Twitter using #ArtAndFeminism and #NowEditingAF hashtags. We utilized the Twitter interactions as representation of the social interactions as well. To study our third research question, we conducted a regression analysis to predict the relationship between various factors of the edit-a-thon events and subsequent participation of newcomers in Wikipedia.

---

[4] https://www.mediawiki.org/wiki/API:Main_page.
[5] https://dev.twitter.com/rest/public.

5.1   Wikipedia Dataset

In addition to the Wikipedia data related to the face-to-face edit-a-thons, we constructed two additional dataset as analogous group of newcomers to compare against edit-a-thons newcomers participants. Below, we provide information about each data collection.

– **Face-to-face Art+Feminism edit-a-thons**: This dataset included data from 59 edit-a-thons event happening from Jan 1, 2016 until March 5, 2016. Each edit-a-thon has a dedicated Wikipedia page associated to the event which includes the list of the participants.[6] We collected the list of participants from the Wikipedia pages. Using the Wikipedia API, for each participants of the edit-a-thons, we collected the day they had registered on Wikipedia, and all of their Wikipedia edits until April 2, 2016 (last day of our data collection).
– **Virtual Art+Feminism edit-a-thons**: In addition to face-to-face events, Wikipedians interested in improving representation of Feminism related articles and female editors, have been organizing virtual edit-a-thons.[7] Anyone from anywhere in the world can participate in the virtual events. The virtual edit-a-thon events were organized over period of two weeks or a month. We collected data from four virtual edit-a-thons happening around the same time as our face-to-face edit-a-thon collection from Dec 2015 until March 2016. Those interested in participation were encouraged to sign up online on the hosting page which included information about the facilitators and the list of articles to work on.
– **Randomly selected newcomers**: Our last collection of newcomers was a set of randomly selected newcomers on Wikipedia. For each newcomers attending a face-to-face edit-a-thon event we randomly selected a group of 10 newcomers who joined Wikipedia on the same day, then for each newcomer in this dataset, we collected all their editing activities on Wikipedia.

5.2   Twitter Dataset

Finally, we collected the interactions happening on Twitter related to the ArtandFeminism Edit-A-Thons. For this purpose we collected all tweets from 2016 that contained the hashtags of #ArtAndFeminism, #NowEditingAF, as well as all of the hashtags that were reported on the Wikipedia page of each event. These hashtags were the official ones that people were advised to use during the event. This dataset included a total of 3,341 tweets from 1,171 different users related to 59 edit-a-thons events happening from Jan 1, 2016 until March 5, 2016.

We also gathered information about other twitter users mentioned in the tweet, the text of the tweet, the date in which the tweet was posted, and any additional hashtag associated with those tweets. We looked particularly on who

---

[6] For example: https://en.wikipedia.org/wiki/Wikipedia:Meetup/Florida/ArtAnd Feminism_2016.

[7] https://en.wikipedia.org/wiki/Wikipedia:WikiProject_Women_in_Red/Meetup/8.

people tagged or mentioned in tweets because research has shown that people tag each other to denote friendships [26]. We were particularly interested in studying metrics that could show that people were creating strong connections. We wanted to understand whether this might relate to how much the continued editing after the event. We also included the general Twitter interactions that users of the targeted hashtags were using before, during, and after the event, other than their Art+Feminisim related tweets. We collected all the tweets they had posted for a period of 15 days before and after the event. Since we were not able to match Wikipedia username with Twitter username, the data is collected based on hashtags that were used in the tweets other than by users. We collected these tweets to be able to assess the level of connections between the participants on Twitter before and after the events by identifying user mentions in the tweets.

6 Results

6.1 RQ1: Impact of Edit-A-Thons on Subsequnet Participation of Newcomers

Among the edit-a-thon participants, we defined users as newcomers if they had not edited Wikipedia before the edit-a-thon event. The dataset included total of 1,018 participation from 985 unique participants with 586 (60 %) of them identified as newcomers. The number of participants per each event ranged from 3 to 131 with average of 17.25 (Std. Dev = 19.47). The proportion of newcomers in each event ranged from 0 to 100 % with average of 57 % (std. Dev. 23.5 %). All together, a total of 793 articles were edited during these edit-a-thons events with 475 (60 %) articles edited by newcomers. There were total of 2,928 edits made to these 793 articles with 1,579 (54 %) of them made by newcomers. 119 out of 793 articles were edited at least once after the events by one of the participants. Overall, an important observation of the data is the large percentage of newcomers attending each of the events. In fact, as mentioned earlier, we also collected data on virtual edit-a-thons for Art+Feminism. The dataset includes a total of 182 participation from 118 unique users. All except two of these users had been already registered on Wikipedia prior to the virtual edit-a-thon events and only 4 who had less than one edit prior to the events that we could consider as newcomers. These results suggest that the face-to-face events are much more likely to attract newcomers than the online events.

Our data shows that among the group of newcomers joining Wikipedia on the same time period as participants of the edit-a-thons, only 1 % of the newcomers edit Wikipedia a week after registering on Wikipedia. To confirm our results, we repeated this analysis with three different randomly selected group of newcomers. In each case, we randomly selected a group of 5,233 newcomers and among them between 48 (0.9 %) to 53 (1 %) newcomers had edited any Wikipedia pages at least one week after registering on Wikipedia. On the other hand, among the 586 newcomers attending our target Edit-A-Thons event, 50 (9 %) of them continued editing Wikipedia pages a week after the edit-a-thon event. Our results

Table 1. Newcomers activities on Edit-A-Thon event day

	% of users	Average	Std. Dev	Median
User page	33 %			
Article edits	48 %	2.45	4.36	1
Sandbox edits	21 %	.86	3.56	0
Other edits	11 %	2.54	3.49	2

suggest that, compared to randomly selected newcomers on Wikipedia, a significant larger percentage of edit-a-thons participants continue editing Wikipedia pages. However, we would also like to acknowledge that randomly selected newcomers provides a baseline benchmark for comparison but it does not provide a fair comparison in terms of motivational factors and identification with the topic of Wikipedia articles. It is very likely that edit-a-thons participants have a stronger identification with topics represented during the edit-a-thon events that can encourage their subsequent participation as well.

Table 1 shows descriptive statistics on newcomers' activities on the day of the edit-a-thon event. The results shows that editing article pages was the most common activity and that a large number of newcomers did not get involve in editing any other Wikipedia pages, including creating their user pages.

Summary of Results. In summary, in response to our first research question on impact of edit-a-thons on newcomers' participation, our results shows that overall face-to-face edit-a-thons are very successful in attracting and recruiting a large number of newcomers who are more engaged than a random group of newcomers on Wikipedia; however, still a very small percentage of them stay engaged with Wikipedia after the event. Given somewhat limited activity level of newcomers on the event day, one potential solution to achieve more sustained engagement can involve encouraging newcomers to get involved on various Wikipedia editing activities during the event, especially activities such creating user pages that is more an entry level activity while connecting newcomers to the community.

6.2   RQ2: Impact of Edit-A-Thons on Newcomers' Community Connections

To further our understanding of the impact of Edit-A-Thons, beyond production mechanisms, we were interested to investigate the social interactions of the participants during and after the events. As mentioned earlier, we employed talkpage interactions and Twitter interactions to construct the interaction networks of newcomers.

Figure 1 presents newcomers' outgoing talkpage communication network; i.e. newcomers and all those individuals with whom the newcomers communicated on their talkpages and Fig. 2 presents newcomers' incoming talkpage communication network; i.e. communication they received from others on their talkpages. The network in Fig. 1 is generated by extracting all the posts made to talkpages by any of the newcomers. The target could be another newcomer or an existing member. The network in Fig. 2 is generated by extracting all the posts made to newcomers' talkpages and the source can be either a newcomer or an existing member. In each network, the light green nodes indicate the newcomers and the dark green nodes represent the existing members of Wikipedia. The thickness of the edges represent the number of talkpage post by that user. As presented in Fig. 1, very few newcomers post on others' talkpages. Furthermore, as presented in Fig. 2, communication between existing members and newcomers is also very limited and majority of newcomers have received very few messages on their talkpages and from a very few number of existing members. Since talkpages are the major place for communication and coordination among Wikipedia editors, this results suggest very little followup and engagement strategies employed by the existing Wikipedia members to keep these newcomers population engaged. The incoming network includes 665 nodes with 2.1 average number of neighbors and network density of .003. The outgoing network includes 64 nodes with 1 as average number of neighbors and network density of .02.

Fig. 1. Newcomers' outgoing communication network - newcomers represented in light green, existing members of Wikipedia represented in dark green, an edge indicate exchange of messages on talkpages from an existing member to a newcomer. (Color figure online)

Next, we used the Twitter data to further unravel how an edit-a-thon connected with people online. For each Twitter user, we constructed their connection network by building a link between them and another person, when either one of them mentioned the other user explicitly in their Tweets of #ArtAndFeminism, #NowEditingAF or tweets using any of the official edit-a-thon hashtags. For instance, if user @Bob had a tweet with: "#ArtAndFeminism we can change

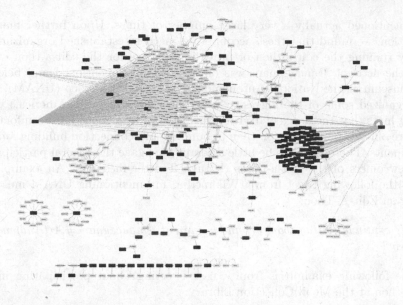

Fig. 2. Newcomers' incoming communication networks - newcomers represented in light green, existing members of Wikipedia represented in dark green, an edge indicate exchange of messages on talkpages from a newcomer to an existing member. (Color figure online)

Wikipedia! Go @Alice keep editing!" We would create link between @Bob and @Alice. Figure 3b presents a visualization of people's different Twitter connections for one of the edit-a-thons with the most online interactions, the March 5th, 2016 edit-a-thon.

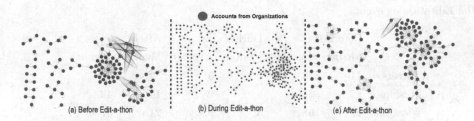

Fig. 3. Twitter network for March 5, 2016 Edit-A-Thon - red notes represent account organization. An edge represents that at least one of the users mentioned the other in at least one tweet related to the Edit-A-Thons. (Color figure online)

Figure 3b shows that only small groups of people interacted with each other online. We observe that a fair number of isolate nodes appeared. These are individuals who were tweeting a lot during the event without connecting to others. We also observed that very few accounts were mentioned, and these were mentioned actually a very large number of times. Upon further manual inspection, we found that these accounts belonged to established organizations usually running the edit-a-thon or providing a space for the edit-a-thon event (e.g., the account @muac_unam was one of the most mentioned and belongs to a museum at the National Autonomous University of Mexico (UNAM), and has organized some of the biggest edit-a-thon events in Latin America.) This finding hints that the usage of Twitter during the edit-a-thons has been focused on more official communication than in bonding and connection building among participants. There seems to be little attempt to engage the general participants and newcomers of the edit-a-thons through Twitter messages. An example of this is the following tweet from @Wikimedia_ mx mentioning UNAM museum during an Edit-A-Thon:

> *"We continue editing about art and women at @muac_unam #ArtAndfeminism"*

The following example is from a user tweeting about participating in an edit-a-thon at the Menil Collection Library:

> *"Art + Feminism wiki Edit-A-Thon. #NowEditingAF in the @menilcollection library"*

Table 2 presents the descriptive statistics about the general Twitter activity of users tweeting the related hashtags. We observe that users were fairly active on Twitter, connecting also with other individuals; however, their presence on Twitter seems to have not been utilized in relation to edit-a-thon events and connecting with other users who utilized the hashtags.

Table 2. Summary of general Twitter activity of participants before, during, and after the Edit-a-thons events

	Before		During		After	
	Median	Average	Median	Average	Median	Average
Tweets	11	26	2	3	9	26
Hashtags	12	36	4	9	14	42
Mentions	22	68	3	9	21	67

Figures 3a and c present the Twitter communication network before and after the edit-a-thon events. The network is generated by considering all Twitter interactions before and after the event. The after-network highlights that only the main organizations (highlighted in red) involved in the edit-a-thons are the ones that people most reach out to; all other Twitter users seem to be lost from the communication network. This further confirms our observations that although the events are able to attract a large number of newcomers, the interaction among

Table 3. Associated hashtags used before, during, and after by Twitter users.

	Hashtags
Before	#editathon, #editathon, #5womenartists, #Wikipedia, #8demarzo, #gendergap, #feminist, #artandfeminism, #WomenInRed
During	#editathon, #editathon, #5womenartists, #Wikipedia, #8demarzo, #gendergap, #feminist, #artandfeminism, #WomenInRed
After	#microaggressions, #feministsplaining, #homosinherstory, #artlibrariessowhite, #archivistproblems

people is very low, and communication stays within these more influential and well established accounts that represent organizations.

We also analyzed the use of hashtags before, during, and after the edit-a-thon. Perhaps, although the users were not reaching out to each other, they might keep a certain bond and connection to the group by tweeting using edit-a-thon related hashtags. Table 3 presents some of the most popular hashtags used before, during, or after the edit-a-thons. One of the most popular hashtags was #ArtAndFeminism. However, we observed that the hashtag was used primarily before and during the edit-a-thon and not after the event which confirms our prior observation of lack of follow through after the event.

On the other hand, we observed that some of the most popular hashtags used by involved users in the tweets, were not associated with the official hashtags (before, during, and after the event), but still appeared to be related to Feminism (e.g., *#MyFeminismIs, #FeministFriday, InternetFeminista, #SadFeministCat*). However, those hashtags were not adopted by the edit-a-thon organizers that could have been utilized to further engage and motivate participation of highly-motivated individuals in Wikipedia. As shown in Table 2 our results suggests that Twitter might be a useful platform to be employed for engaging edit-a-thon participants after the event, espeically on topics of their interest.

Summary of Results. In summary, in response to our second research question on impact of edit-a-thon on forming connections, we observed very little evidence on that on either Wikipedia talkpages or Twitter. We observed very little social connections created among the participants and very little followup after the events to further engage the newcomers. At the same time, we observed that many of Twitter users who were using the official hashtags for the edit-a-thon were very active during the edit-a-thon and exhibited particular interest on the topic of feminism. These results together hints missed opportunity that can be utilized by Wikipedia community and organizers of edit-a-thons to increase the likelihood of newcomers' engagement after those events.

6.3   RQ3: Predicting Subsequent Participation

While overall we observed low retention rate among edit-a-thon participants, we were interested to assess whether any of the production and social mechanisms were related to higher likelihood of retention and subsequent participation. To do so, we conducted a repeated measure logistic regression analysis to predict whether the newcomers would edit any Wikipedia pages at least a week after the event. The model nested individual users within the edit-a-thon event they attended. We included the production measures of creating a user page, editing article pages, or sandbox pages, the number of participants in the event, as well as the proportion of newcomers attending the event as independent variables in the model. In terms of interaction mechanisms, we utilized the talkpage communication network and for each newcomers, we calculated common network measures of closeness centrality, betweenness centrality, degree, and clustering coefficient. Degree represents the number of immediate connections a node has in the network and indicates how well-connected a node is; Closeness centrality represents a more global level of connectedness in the network through considering the distance of a node to all others in the network. Betweenness centrality on the other hand focuses on favored bridging positions of nodes and how many paths of connections in the network rely on this particular node. Clustering coefficient represent how close a node's neighbors are to being completely connected graph. However, in our dataset the measure of degree was significantly correlated with all other network measures (Table 4; therefore, to avoid multicollinearity, we only included degree in the regression model. The degree includes any connection between the two users based on the exchange of messages on their Wikipedia pages independent of the direction of the message. Since the size of the network and degree can be related, we ensured that degree was not correlated to the number of participants (coef $= .02$, Sig. $= .71$).

Table 4. Correlation between network measures

	Betweenness		Closeness		Clustering	
	coef	Sig	coef	Sig	coef	Sig.
Degree	.84	<.001	.23	<.001	.29	<.001

The result of the regression analysis for the significant factors is presented in Table 5. The results show that number of participants, editing articles, and receiving talkpage messages are correlated with higher likelihood of continuing to edit Wikipedia. Any one additional person attending the event leads to 1% increase in the odds of having a newcomer continue editing a Wikipedia page. An additional talkpage message leads to 68% increase in the odds of having a newcomer continue editing a Wikipedia page and one additional edit in article pages on the event day increases the odds of having a newcomer continue editing a Wikipedia page by 7%. The effect of editing articles is only marginally

Table 5. Analysis result

	Odds ratio	Std. Error	Sig.
Number of participants	1.01	.004	.021
Degree	1.68	.16	.001
Article edits	1.07	.03	.079

significant. Overall the results suggest that on-event support in terms of editing during the edit-a-thon events and social interactions can lead to higher level of subsequent commitment. Other factors, including proportion of newcomers, editing sandbox pages, and creating user pages were not significant factors in the model.

7   Conclusion and Discussion

In this work, we presented a case study of the impact that a short-term collocated event, had on onboarding newcomers into the community. We studied this in the context of Art+Feminism Wikipedia edit-a-thons focusing on increasing representation of female editors. To understand the production and social mechanism of these edit-a-thons, we triangulated different sources of log data on Wikiepdia and Twitter. Our results show that these events are very successful in attracting new members. A significant percentage of participants in each of those events are individual with no prior Wikipedia experiences, at a much higher rate in comparison to their parallel virtual events; however, they are not very successful in retaining these motivated individuals. In fact, retention has been identified as a major challenge by Wikimedia adminstrators involved in edit-a-thon events.[8] We speculate that higher level of hands-on activities on the event day, and followup communications and engagement mechanisms can play significant role in increasing retention.

Our analysis of newcomers talkpage communication network and participants' Twitter communication speak to these speculation. We found very little communication happening on Wikipedia and Twitter. On the other hand, similar to previous research we observed the importance of interaction between existing members and newcomers on encouraging their future participation. Receiving messages on talkpages was associated with a significant increase in odds of future contribution of a newcomer.

Additionally, we observed that when the general participants tweeted they use a wider range of hashtags which were somewhat disjoint from the hashtags utilized by administration members. This can signify that possible strong motivation of participants which might have not been capitalized by those in charge of these events.

---

[8] https://meta.wikimedia.org/wiki/Grants:Evaluation/Evaluation_reports/2013/
Edit-a-thons#Recruitment_and_retention_of_new_editors.

At the same time, our results also shows that activities on the day of the event can be important in encouraging future participation. While we did not observe any support that creating user pages or practicing editing in Sandbox pages impacts future participation; those actions were not common among the participants and there were very small occurrence of those cases that could affect the result of our analysis. Encouraging to take on such actions might be a good starting point to encourage further participation and foster a sense of belonging to the community.

Our analysis are based on archival log data and in the future we plan to conduct interviews with organizers and participants to gain a deeper insight into results highlighted in our current work and better understanding of the goals of each event and their satisfaction with the extent they have achieved their goals. It is possible that in some of these events, on-event activities were of higher importance to the organizers than future participation of a large number of participants. Our initial contact with a few of organizers has been received with high enthusiasm, especially with regards to the issue of engagement of newcomers that they acknowledges as a challenge. Informed by the results of our current work, in the future, we plan to work closely with organizers of these events in experimenting various followup strategies to increase newcomers' retention into the online production community. We should also acknowledge that our results are in a more of a correlational nature and without a true random experiment, we cannot make a causal conclusion about the relationship between attending edit-a-thon events and future editing of Wikipedia.

Nevertheless, our current work highlights the value of these offline gatherings on attracting a new stream of newcomers while providing insight on the challenges they face and potential approaches in addressing such challenges. While our work focused on the context of Wikipedia Art+Feminism edit-a-thon events and limited number of events, similar methodology can potentially be applied in studies of other similar offline gatherings such as other edit-a-thon events or open-source software hackathon events. We hope to extend our work in those areas in the future to be able to generalize our findings to a broader context.

References

1. Ahuja, M.K., Galvin, J.E.: Socialization in virtual groups. J. Manag. **29**(2), 161–185 (2003)
2. Angelopoulos, S., Merali, Y.: Bridging the divide between virtual and embodied spaces: exploring the effect of offline interactions on the sociability of participants of topic-specific online communities. In: International Conference on System Sciences, pp. 1994–2002. IEEE (2015)
3. Brady, E., Morris, M.R., Bigham, J.P.: Gauging receptiveness to social microvolunteering. In: Proceedings of the 33rd Annual ACM Conference on Human Factors in Computing Systems, pp. 1055–1064. ACM (2015)
4. Burke, M., Marlow, C., Lento, T.: Feed me: motivating newcomer contribution in social network sites. In: Proceedings of the SIGCHI Conference on Human Factors in Computing Systems, pp. 945–954. ACM (2009)

5. Chen, J., Ren, Y., Riedl, J.: The effects of diversity on group productivity and member withdrawal in online volunteer groups. In: Proceedings of the SIGCHI Conference on Human Factors in Computing Systems, pp. 821–830. ACM (2010)
6. Choi, B., Alexander, K., Kraut, R.E., Levine, J.M.: Socialization tactics in wikipedia and their effects. In: Proceedings of the 2010 ACM conference on Computer Supported Cooperative Work, pp. 107–116. ACM (2010)
7. Ciampaglia, G.L., Taraborelli, D.: Moodbar: Increasing new user retention in wikipedia through lightweight socialization. In: Proceedings of the 18th ACM Conference on Computer Supported Cooperative Work & Social Computing, pp. 734–742. ACM (2015)
8. Cohen, N.: Define gender gap? look up Wikipedia's contributor list. New York Times 30(362), 1050–1056 (2011)
9. Ducheneaut, N.: Socialization in an open source software community: a sociotechnical analysis. Comput. Support. Coop. Work (CSCW) 14(4), 323–368 (2005)
10. Farzan, R., Kraut, R.E.: Wikipedia classroom experiment: bidirectional benefits of students' engagement in online production communities. In: Proceedings of the SIGCHI Conference on Human Factors in Computing Systems, pp. 783–792. ACM (2013)
11. Faulkner, R., Walling, S., Pinchuk, M.: Etiquette in Wikipedia: weening new editors into productive ones. In: Proceedings of the Eighth Annual International Symposium on Wikis and Open Collaboration, p. 5. ACM (2012)
12. Flores-Saviaga, C., Savage, S., Taraborelli, D.: Leadwise: using online bots to recruite and guide expert volunteers. In: Proceedings of the 19th ACM Conference on Computer Supported Cooperative Work and Social Computing Companion, pp. 257–260. ACM (2016)
13. Ganglbauer, E., Fitzpatrick, G., Subasi, Ö., Güldenpfennig, F.: Think globally, act locally: a case study of a free food sharing community and social networking. In: Proceedings of the 17th ACM Conference on Computer Supported Cooperative Work & Social Computing, pp. 911–921. ACM (2014)
14. Grevet, C., Gilbert, E.: Piggyback prototyping: using existing, large-scale social computing systems to prototype new ones. In: Proceedings of the 33rd Annual ACM Conference on Human Factors in Computing Systems, pp. 4047–4056. ACM (2015)
15. Halfaker, A., Geiger, R., Morgan, J., Riedl, J.: The rise and decline of an open collaboration system: how wikipedia's reaction to popularity is causing its decline. American Behavioral Scientist, 28 December 2012
16. Halfaker, A., Kittur, A., Riedl, J.: Don't bite the newbies: how reverts affect the quantity and quality of Wikipedia work. In: Proceedings of the 7th International Symposium on wikis and open collaboration, pp. 163–172. ACM (2011)
17. Jin, L., Robey, D., Boudreau, M.C.: The nature of hybrid community: an exploratory study of open source software user groups. J. Community Inform. 11(1) (2015)
18. Johnson, I., Lin, Y., Jia-Jun Li, T., Hall, A., Halfaker, A., Schöning, J., Hecht, B.: Not at home on the range: peer production and the urban/rural divide (2016)
19. Lam, K., S., Uduwage, A., Dong, Z., Sen, S., Musicant, D., Terveen, L., Riedl, J.: Wp:clubhouse?: an exploration of Wikipedia's gender imbalance. In: WikiSym 2011, ACM (2011)
20. Kraut, R., Burke, M., Riedl, J., Resnick, P.: Dealing with newcomers. In: Evidence-based Social Design Mining the Social Sciences to Build Online Communities, pp. 1–42 (2010)

21. Lam, S.T.K., Uduwage, A., Dong, Z., Sen, S., Musicant, D.R., Terveen, L., Riedl, J.: WP: clubhouse?: an exploration of Wikipedia's gender imbalance. In: Proceedings of the 7th International Symposium on Wikis and Open Collaboration, pp. 1–10. ACM (2011)

22. Lampe, C., Johnston, E.: Follow the (slash) dot: effects of feedback on new members in an online community. In: Proceedings of the 2005 International ACM SIG-GROUP Conference on Supporting Group Work, pp. 11–20. ACM (2005)

23. Lin, H.F.: The role of online and offline features in sustaining virtual communities: an empirical study. Internet Res. **17**(2), 119–138 (2007)

24. Morgan, J.T., Bouterse, S., Walls, H., Stierch, S.: Tea and sympathy: crafting positive new user experiences on Wikipedia. In: Proceedings of the 2013 conference on Computer Supported Cooperative Work, pp. 839–848. ACM (2013)

25. Panciera, K., Halfaker, A., Terveen, L.: Wikipedians are born, not made: a study of power editors on Wikipedia. In: Proceedings of the ACM 2009 International Conference on Supporting Group Work, pp. 51–60. ACM (2009)

26. Savage, S., Monroy-Hernandez, A., Bhattacharjee, K., Höllerer, T.: Tag me maybe: perceptions of public targeted sharing on facebook. In: Proceedings of the 26th ACM Conference on Hypertext & Social Media, pp. 299–303. ACM (2015)

27. Savage, S., Monroy-Hernandez, A., Hollerer, T.: Botivist: calling volunteers to action using online bots (2015). arXiv preprint arXiv:1509.06026

28. Sessions, L.F.: How offline gatherings affect online communities: when virtual community members 'meetup'. Inf. Commun. Soc. **13**(3), 375–395 (2010)

29. Shaw, A., Zhang, H., Monroy-Hernández, A., Munson, S., Hill, B.M., Gerber, E., Kinnaird, P., Minder, P.: Computer supported collective action. Interactions **21**(2), 74–77 (2014)

30. Shen, C., Cage, C.: Exodus to the real world? Assessing the impact of offline meetups on community participation and social capital. New Media Soc. **17**(3), 394–414 (2015)

31. Suh, B., Convertino, G., Chi, E.H., Pirolli, P.: The singularity is not near: slowing growth of Wikipedia. In: Proceedings of the 5th International Symposium on Wikis and Open Collaboration, p. 8. ACM (2009)

32. Tewksbury, D.: Online-offline knowledge sharing in the occupy movement: Howtooccupy. org and discursive communities of practice. Am. Commun. J. **15**(1), 11–23 (2013)

33. Trainer, E.H., Chaihirunkarn, C., Kalyanasundaram, A., Herbsleb, J.D.: Community code engagements: summer of code & hackathons for community building in scientific software. In: Proceedings of the 18th International Conference on Supporting Group Work, pp. 111–121. ACM (2014)

34. Trainer, E.H., Kalyanasundaram, A., Chaihirunkarn, C., Herbsleb, J.D.: How to hackathon: socio-technical tradeoffs in brief, intensive collocation (2015)

35. Zhu, H., Zhang, A., He, J., Kraut, R.E., Kittur, A.: Effects of peer feedback on contribution: a field experiment in Wikipedia. In: Proceedings of the SIGCHI Conference on Human Factors in Computing Systems, pp. 2253–2262. ACM (2013)

The Social Dynamics of Language Change in Online Networks

Rahul Goel[1], Sandeep Soni[1], Naman Goyal[1], John Paparrizos[2],
Hanna Wallach[3], Fernando Diaz[3], and Jacob Eisenstein[1(✉)]

[1] Georgia Institute of Technology, Atlanta, GA, USA
jacobe@gatech.edu
[2] Columbia University, New York, NY, USA
[3] Microsoft Research, New York, NY, USA

Abstract. Language change is a complex social phenomenon, revealing pathways of communication and sociocultural influence. But, while language change has long been a topic of study in sociolinguistics, traditional linguistic research methods rely on circumstantial evidence, estimating the direction of change from differences between older and younger speakers. In this paper, we use a data set of several million Twitter users to track language changes in progress. First, we show that language change can be viewed as a form of social influence: we observe complex contagion for phonetic spellings and "netspeak" abbreviations (e.g., *lol*), but not for older dialect markers from spoken language. Next, we test whether specific types of social network connections are more influential than others, using a parametric Hawkes process model. We find that tie strength plays an important role: densely embedded social ties are significantly better conduits of linguistic influence. Geographic locality appears to play a more limited role: we find relatively little evidence to support the hypothesis that individuals are more influenced by geographically local social ties, even in their usage of geographical dialect markers.

1   Introduction

Change is a universal property of language. For example, English has changed so much that Renaissance-era texts like *The Canterbury Tales* must now be read in translation. Even contemporary American English continues to change and diversify at a rapid pace—to such an extent that some geographical dialect differences pose serious challenges for comprehensibility [36]. Understanding language change is therefore crucial to understanding language itself, and has implications for the design of more robust natural language processing systems [17].

Language change is a fundamentally social phenomenon [34]. For a new linguistic form to succeed, at least two things must happen: first, speakers (and writers) must come into contact with the new form; second, they must decide to use it. The first condition implies that language change is related to the structure of social networks. If a significant number of speakers are isolated from a potential change, then they are unlikely to adopt it [40]. But mere exposure is

E. Spiro and Y.-Y. Ahn (Eds.): SocInfo 2016, Part I, LNCS 10046, pp. 41–57, 2016.
DOI: 10.1007/978-3-319-47880-7_3

not sufficient—we are all exposed to language varieties that are different from our own, yet we nonetheless do not adopt them in our own speech and writing. For example, in the United States, many African American speakers maintain a distinct dialect, despite being immersed in a linguistic environment that differs in many important respects [23,45]. Researchers have made a similar argument for socioeconomic language differences in Britain [49]. In at least some cases, these differences reflect questions of identity: because language is a key constituent in the social construction of group identity, individuals must make strategic choices when deciding whether to adopt new linguistic forms [11,29,33]. By analyzing patterns of language change, we can learn more about the latent structure of social organization: to whom people talk, and how they see themselves.

But, while the basic outline of the interaction between language change and social structure is understood, the fine details are still missing: What types of social network connections are most important for language change? To what extent do considerations of identity affect linguistic differences, particularly in an online context? Traditional sociolinguistic approaches lack the data and the methods for asking such detailed questions about language variation and change.

In this paper, we show that large-scale social media data can shed new light on how language changes propagate through social networks. We use a data set of Twitter users that contains all public messages for several million accounts, augmented with social network and geolocation metadata. This data set makes it possible to track, and potentially explain, every usage of a linguistic variable[1] as it spreads through social media. Overall, we make the following contributions:

- We show that non-standard words are most likely to propagate between individuals who are connected in the Twitter mutual-reply network. This validates the basic approach of using Twitter to measure language change.
- For some classes of non-standard words, we observe complex contagion—i.e., multiple exposures increase the likelihood of adoption. This is particularly true for phonetic spellings and "netspeak" abbreviations. In contrast, non-standard words that originate in speech do not display complex contagion.
- We use a parametric Hawkes process model [26,39] to test whether specific types of social network connections are more influential than others. For some words, we find that densely embedded social ties are significantly better conduits of linguistic influence. This finding suggests that individuals make social evaluations of their exposures to new linguistic forms, and then use these social evaluations to strategically govern their own language use.
- We present an efficient parameter estimation method that uses sparsity patterns in the data to scale to social networks with millions of users.

2 Data

Twitter is an online social networking platform. Users post 140-character messages, which appear in their followers' timelines. Because follower ties can be

---

[1] The basic unit of linguistic differentiation is referred to as a "variable" in the sociolinguistic and dialectological literature [50]. We maintain this terminology here.

asymmetric, Twitter serves multiple purposes: celebrities share messages with millions of followers, while lower-degree users treat Twitter as a more intimate social network for mutual communication [31]. In this paper, we use a large-scale Twitter data set, acquired via an agreement between Microsoft and Twitter. This data set contains all public messages posted between June 2013 and June 2014 by several million users, augmented with social network and geolocation metadata. We excluded retweets, which are explicitly marked with metadata, and focused on messages that were posted in English from within the United States.

2.1 Linguistic Markers

The explosive rise in popularity of social media has led to an increase in linguistic diversity and creativity [5,6,9,14,17,27], affecting written language at all levels, from spelling [18] all the way up to grammatical structure [48] and semantic meaning across the lexicon [25,30]. Here, we focus on the most easily observable and measurable level: variation and change in the use of individual words.

We take as our starting point words that are especially characteristic of eight cities in the United States. We chose these cities to represent a wide range of geographical regions, population densities, and demographics. We identified the following words as geographically distinctive markers of their associated cities, using SAGE [20]. Specifically, we followed the approach previously used by Eisenstein to identify community-specific terms in textual corpora [19].[2]

Atlanta: *ain* (phonetic spelling of *ain't*), *dese* (phonetic spelling of *these*), *yeen* (phonetic spelling of *you ain't*);

Baltimore: *ard* (phonetic spelling of *alright*), *inna* (phonetic spelling of *in a* and *in the*), *lls* (*laughing like shit*), *phony* (fake);

Charlotte: *cookout*;

Chicago: *asl* (phonetic spelling of *as hell*, typically used as an intensifier on Twitter[3]), *mfs* (*motherfuckers*);

Los Angeles: *graffiti*, *tfti* (*thanks for the information*);

Philadelphia: *ard* (phonetic spelling of *alright*), *ctfuu* (expressive lengthening of *ctfu*, an abbreviation of *cracking the fuck up*), *jawn* (generic noun);

San Francisco: *hella* (an intensifier);

Washington D.C.: *inna* (phonetic spelling of *in a* and *in the*), *lls* (*laughing like shit*), *stamp* (an exclamation indicating emphasis).[4]

Linguistically, we can divide these words into three main classes:

---

[2] After running SAGE to identify words with coefficients above 2.0, we manually removed hashtags, named entities, non-English words, and descriptions of events.

[3] Other sources, such as http://urbandictionary.com, report *asl* to be an abbreviation of *age, sex, location?* However, this definition is not compatible with typical usage on Twitter, e.g., *currently hungry asl* or *that movie was funny asl.*

[4] *ard*, *inna*, and *lls* appear on multiple cities' lists. These words are characteristic of the neighboring cities of Baltimore, Philadelphia, and Washington D.C.

Lexical words: The origins of *cookout*, *graffiti*, *hella*, *phony*, and *stamp* can almost certainly be traced back to spoken language. Some of these words (e.g., *cookout* and *graffiti*) are known to all fluent English speakers, but are preferred in certain cities simply as a matter of topic. Other words (e.g., *hella* [12] and *jawn* [3]) are dialect markers that are not widely used outside their regions of origin, even after several decades of use in spoken language.

Phonetic spellings: *ain*, *ard*, *asl*, *inna*, and *yeen* are non-standard spellings that are based on phonetic variation by region, demographics, or situation.

Abbreviations: *ctfuu*, *lls*, *mfs*, and *tfti* are phrasal abbreviations. These words are interesting because they are fundamentally textual. They are unlikely to have come from spoken language, and are intrinsic to written social media.

Several of these words were undergoing widespread growth in popularity around the time period spanned by our data set. For example, the frequencies of *ard*, *asl*, *hella*, and *tfti* more than tripled between 2012 and 2013. Our main research question is whether and how these words spread through Twitter. For example, lexical words are mainly transmitted through speech. We would expect their spread to be only weakly correlated with the Twitter social network. In contrast, abbreviations are fundamentally textual in nature, so we would expect their spread to correlate much more closely with the Twitter social network.

2.2 Social Network

To focus on communication between peers, we constructed a social network of mutual replies between Twitter users. Specifically, we created a graph in which there is a node for each user in the data set. We then placed an undirected edge between a pair of users if each replied to the other by beginning a message with their username. Our decision to use the reply network (rather than the follower network) was a pragmatic choice: the follower network is not widely available. However, the reply network is also well supported by previous research. For example, Huberman et al. argue that Twitter's mention network is more socially meaningful than its follower network: although users may follow thousands of accounts, they interact with a much more limited set of users [28], bounded by a constant known as Dunbar's number [15]. Finally, we restricted our focus to mutual replies because there are a large number of unrequited replies directed at celebrities. These replies do not indicate a meaningful social connection.

We compared our mutual-reply network with two one-directional "in" and "out" networks, in which all public replies are represented by directed edges. The degree distributions of these networks are depicted in Fig. 1. As expected, there are a few celebrities with very high in-degrees, and a maximum in-degree of 20, 345. In contrast, the maximum degree in our mutual-reply network is 248.

2.3 Geography

In order to test whether geographically local social ties are a significant conduit of linguistic influence, we obtained geolocation metadata from Twitter's location

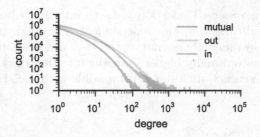

Fig. 1. Degree distributions for our mutual-reply network and "in" and "out" networks.

field. This field is populated via a combination of self reports and GPS tagging. We aggregated metadata across each user's messages, so that each user was geolocated to the city from which they most commonly post messages. Overall, our data set contains 4.35 million geolocated users, of which 589,562 were geolocated to one of the eight cities listed in Sect. 2.1. We also included the remaining users in our data set, but were not able to account for their geographical location.

Researchers have previously shown that social network connections in online social media tend to be geographically assortative [7,46]. Our data set is consistent with this finding: for 94.8 % of mutual-reply dyads in which both users were geolocated to one of the eight cities listed in Sect. 2.1, they were both geolocated to the same city. This assortativity motivates our decision to estimate separate influence parameters for local and non-local social connections (see Sect. 5.1).

3   Language Change as Social Influence

Our main goal is to test whether and how geographically distinctive linguistic markers spread through Twitter. With this goal in mind, our first question is whether the adoption of these markers can be viewed as a form of COMPLEX CONTAGION. To answer this question, we computed the fraction of users who used one of the words listed in Sect. 2.1 after being exposed to that word by one of their social network connections. Formally, we say that user i EXPOSED user j to word w at time t if and only if the following conditions hold: i used w at time t; j had not used w before time t; the social network connection $i \leftrightarrow j$ was formed before time t. We define the INFECTION RISK for word w to be the number of users who use word w after being exposed divided by the total number of users who were exposed. To consider the possibility that multiple exposures have a greater impact on the infection risk, we computed the infection risk after exposures across one, two, and three or more distinct social network connections.

The words' infection risks cannot be interpreted directly because relational autocorrelation can also be explained by homophily and external confounds. For example, geographically distinctive non-standard language is more likely to be used by young people [44], and online social network connections are assortative by age [2]. Thus, a high infection risk can also be explained by the confound of age. We therefore used the shuffle test proposed by Anagnostopoulos et al. [4], which compares the observed infection risks to infection risks under the null hypothesis that event timestamps are independent. The null hypothesis infection risks are computed by randomly permuting the word usage events. If the observed infection risks are substantially higher than the infection risks computed using the permuted usage events, then this is compatible with social influence.[5]

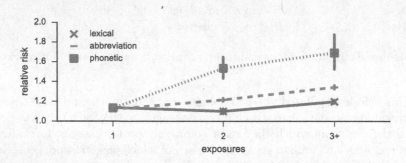

Fig. 2. Relative infection risks for words in the linguistic classes defined in Sect. 2.1. The figure depicts 95 % confidence intervals, computed using the shuffle test [4].

Figure 2 depicts the ratios between the words' observed infection risks and the words' infection risks under the null hypothesis, after exposures across one, two, and three or more distinct connections. We computed 95 % confidence intervals across the words and across the permutations used in the shuffle test. For all three linguistic classes defined in Sect. 2.1, the risk ratio for even a single exposure is significantly greater than one, suggesting the existence of social influence. The risk ratio for a single exposure is nearly identical across the three classes. For phonetic spellings and abbreviations, the risk ratio grows with the number of exposures. This pattern suggests that words in these classes exhibit COMPLEX CONTAGION—i.e., multiple exposures increase the likelihood of adoption [13]. In contrast, the risk ratio for lexical words remains the same as the number of exposures increases, suggesting that these words spread by simple contagion.

Complex contagion has been linked to a range of behaviors, from participation in collective political action to adoption of avant garde fashion [13]. A common theme among these behaviors is that they are not cost-free, particularly if the behavior is not legitimated by widespread adoption. In the case of linguistic markers intrinsic to social media, such as phonetic spellings and abbreviations, adopters risk negative social evaluations of their linguistic competency, as well as their cultural authenticity [47]. In contrast, lexical words are already well known from spoken language and are thus less socially risky. This difference may explain why we do not observe complex contagion for lexical words.

---

[5] The shuffle test assumes that the likelihood of two users forming a social network connection does not change over time. Researchers have proposed a test [32] that removes this assumption; we will scale this test to our data set in future work.

4   Social Evaluation of Language Variation

In the previous section, we showed that geographically distinctive linguistic markers spread through Twitter, with evidence of complex contagion for phonetic spellings and abbreviations. But, does each social network connection contribute equally? Our second question is therefore whether (1) strong ties and (2) geographically local ties exert greater linguistic influence than other ties. If so, users must socially evaluate the information they receive from these connections, and judge it to be meaningful to their linguistic self-presentation. In this section, we outline two hypotheses regarding their relationships to linguistic influence.

4.1   Tie Strength

Social networks are often characterized in terms of strong and weak ties [22,40], with strong ties representing more important social relationships. Strong ties are often densely embedded, meaning that the nodes in question share many mutual friends; in contrast, weak ties often bridge disconnected communities. Bakshy et al. investigated the role of weak ties in information diffusion, through resharing of URLs on Facebook [8]. They found that URLs shared across strong ties are more likely to be reshared. However, they also found that weak ties play an important role, because users tend to have more weak ties than strong ties, and because weak ties are more likely to be a source of new information. In some respects, language change is similar to traditional information diffusion scenarios, such as resharing of URLs. But, in contrast, language connects with personal identity on a much deeper level than a typical URL. As a result, strong, deeply embedded ties may play a greater role in enforcing community norms.

We quantify tie strength in terms of EMBEDDEDNESS. Specifically, we use the normalized mutual friends metric introduced by Adamic and Adar [1]:

$$s_{i,j} = \sum_{k \in \Gamma(i) \cap \Gamma(j)} \frac{1}{\log\left(\#|\Gamma(k)|\right)}, \tag{1}$$

where, in our setting, $\Gamma(i)$ is the set of users connected to i in the Twitter mutual-reply network and $\#|\Gamma(i)|$ is the size of this set. This metric rewards dyads for having many mutual friends, but counts mutual friends more if their degrees are low—a high-degree mutual friend is less informative than one with a lower-degree. Given this definition, we can form the following hypothesis:

H1 The linguistic influence exerted across ties with a high embeddedness value $s_{i,j}$ will be greater than the linguistic influence exerted across other ties.

4.2   Geographic Locality

An open question in sociolinguistics is whether and how local COVERT PRESTIGE—i.e., the positive social evaluation of non-standard dialects—affects the adoption of new linguistic forms [49]. Speakers often explain their linguistic choices in terms of their relationship with their local identity [16], but this may be a post-hoc rationalization made by people whose language is affected by factors beyond their control. Indeed, some sociolinguists have cast doubt on the role of "local games" in affecting the direction of language change [35].

The theory of covert prestige suggests that geographically local social ties are more influential than non-local ties. We do not know of any prior attempts to test this hypothesis quantitatively. Although researchers have shown that local linguistic forms are more likely to be used in messages that address geographically local friends [43], they have not attempted to measure the impact of exposure to these forms. This lack of prior work may be because it is difficult to obtain relevant data, and to make reliable inferences from such data. For example, there are several possible explanations for the observation that people often use similar language to that of their geographical neighbors. One is exposure: even online social ties tend to be geographically assortative [2], so most people are likely to be exposed to local linguistic forms through local ties. Alternatively, the causal relation may run in the reverse direction, with individuals preferring to form social ties with people whose language matches their own. In the next section, we describe a model that enables us to tease apart the roles of geographic assortativity and local influence, allowing us to test the following hypothesis:

H2 The influence toward geographically distinctive linguistic markers is greater when exerted across geographically local ties than across other ties.

We note that this hypothesis is restricted in scope to geographically distinctive words. We do not consider the more general hypothesis that geographically local ties are more influential for all types of language change, such as change involving linguistic variables that are associated with gender or socioeconomic status.

5   Language Change as a Self-exciting Point Process

To test our hypotheses about social evaluation, we require a more sophisticated modeling tool than the simple counting method described in Sect. 3. In this section, rather than asking whether a user was previously exposed to a word, we ask by whom, in order to compare the impact of exposures across different types of social network connections. We also consider temporal properties. For example, if a user adopts a new word, should we credit this to an exposure from a weak tie in the past hour, or to an exposure from a strong tie in the past day?

Following a probabilistic modeling approach, we treated our Twitter data set as a set of cascades of timestamped events, with one cascade for each of the geographically distinctive words described in Sect. 2.1. Each event in a word's cascade corresponds to a tweet containing that word. We modeled each cascade as a probabilistic process, and estimated the parameters of this process. By comparing nested models that make progressively finer distinctions between social network connections, we were able to quantitatively test our hypotheses.

Our modeling framework is based on a HAWKES PROCESS [26]—a specialization of an inhomogeneous Poisson process—which explains a cascade of timestamped events in terms of influence parameters. In a temporal setting, an inhomogeneous Poisson process says that the number of events y_{t_1,t_2} between t_1 and t_2 is drawn from a Poisson distribution, whose parameter is the area under a time-varying INTENSITY FUNCTION over the interval defined by t_1 and t_2:

$$y_{t_1,t_2} \sim \text{Poisson}\left(\Lambda(t_1, t_2)\right) \tag{2}$$

where

$$\Lambda(t_1, t_2) = \int_{t_1}^{t_2} \lambda(t) \, dt. \tag{3}$$

Since the parameter of a Poisson distribution must be non-negative, the intensity function must be constrained to be non-negative for all possible values of t.

A Hawkes process is a self-exciting inhomogeneous Poisson process, where the intensity function depends on previous events. If we have a cascade of N events $\{t_n\}_{n=1}^N$, where t_n is the timestamp of event n, then the intensity function is

$$\lambda(t) = \mu_t + \sum_{t_n < t} \alpha \, \kappa(t - t_n), \tag{4}$$

where μ_t is the base intensity at time t, α is an influence parameter that captures the influence of previous events, and $\kappa(\cdot)$ is a time-decay kernel.

We can extend this framework to vector observations $\boldsymbol{y}_{t_1,t_2} = (y_{t_1,t_2}^{(1)}, \ldots, y_{t_1,t_2}^{(M)})$ and intensity functions $\boldsymbol{\lambda}(t) = (\lambda^{(1)}(t), \ldots, \lambda^{(M)}(t))$, where, in our setting, M is the number of users in our data set. If we have a cascade of N events $\{(t_n, m_n)\}_{n=1}^N$, where t_n is the timestamp of event n and $m_n \in \{1, \ldots, M\}$ is the source of event n, then the intensity function for user $m' \in \{1, \ldots, M\}$ is

$$\lambda^{(m')}(t) = \mu_t^{(m')} + \sum_{t_n < t} \alpha_{m_n \to m'} \kappa(t - t_n), \tag{5}$$

where $\mu_t^{(m')}$ is the base intensity for user m' at time t, $\alpha_{m_n \to m'}$ is a pairwise influence parameter that captures the influence of user m_n on user m', and $\kappa(\cdot)$ is a time-decay kernel. Throughout our experiments, we used an exponential decay kernel $\kappa(\Delta t) = e^{-\gamma \Delta t}$. We set the hyperparameter γ so that $\kappa(1 \text{ hour}) = e^{-1}$.

Researchers usually estimate all M^2 influence parameters of a Hawkes process (e.g., [38,51]). However, in our setting, $M > 10^6$, so there are $O(10^{12})$ influence parameters. Estimating this many parameters is computationally and statistically intractable, given that our data set includes only $O(10^5)$ events (see the x-axis of Fig. 3 for event counts for each word). Moreover, directly estimating these parameters does not enable us to quantitatively test our hypotheses.

5.1   Parametric Hawkes Process

Instead of directly estimating all $O(M^2)$ pairwise influence parameters, we used Li and Zha's parametric Hawkes process [39]. This model defines each pairwise influence parameter in terms of a linear combination of pairwise features:

$$\alpha_{m \to m'} = \boldsymbol{\theta}^\top \boldsymbol{f}(m \to m'), \tag{6}$$

where $\boldsymbol{f}(m \rightarrow m')$ is a vector of features that describe the relationship between users m and m'. Thus, we only need to estimate the feature weights $\boldsymbol{\theta}$ and the base intensities. To ensure that the intensity functions $\lambda^{(1)}(t), \ldots, \lambda^{(M)}(t)$ are non-negative, we must assume that $\boldsymbol{\theta}$ and the base intensities are non-negative.

We chose a set of four binary features that would enable us to test our hypotheses about the roles of different types of social network connections:

F1 Self-activation: This feature fires when $m' = m$. We included this feature to capture the scenario where using a word once makes a user more likely to use it again, perhaps because they are adopting a non-standard style.

F2 Mutual reply: This feature fires if the dyad (m, m') is in the Twitter mutual-reply network described in Sect. 2.2. We also used this feature to define the remaining two features. By doing this, we ensured that features F2, F3, and F4 were (at least) as sparse as the mutual-reply network.

F3 Tie strength: This feature fires if the dyad (m, m') is in the Twitter mutual-reply network, and the Adamic-Adar value for this dyad is especially high. Specifically, we require that the Adamic-Adar value be in the 90th percentile among all dyads where at least one user has used the word in question. Thus, this feature picks out the most densely embedded ties.

F4 Local: This feature fires if the dyad (m, m') is in the Twitter mutual-reply network, and the users were geolocated to the same city, and that city is one of the eight cities listed in Sect. 2. For other dyads, this feature returns zero. Thus, this feature picks out a subset of the geographically local ties.

In Sect. 6, we describe how we used these features to construct a set of nested models that enabled us to test our hypotheses. In the remainder of this section, we provide the mathematical details of our parameter estimation method.

5.2 Objective Function

We estimated the parameters using constrained maximum likelihood. Given a cascade of events $\{(t_n, m_n)\}_{n=1}^{N}$, the log likelihood under our model is

$$\mathcal{L} = \sum_{n=1}^{N} \log \lambda^{(m_n)}(t_n) - \sum_{m=1}^{M} \int_{0}^{T} \lambda^{(m)}(t) \, dt, \qquad (7)$$

where T is the temporal endpoint of the cascade. Substituting in the complete definition of the per-user intensity functions from Eqs. 5 and 6,

$$\mathcal{L} = \sum_{n=1}^{N} \log \left(\mu_{t_n}^{(m_n)} + \sum_{t_{n'} < t_n} \boldsymbol{\theta}^{\top} \boldsymbol{f}(m_{n'} \rightarrow m_n) \, \kappa(t_n - t_{n'}) \right) -$$

$$\sum_{m'=1}^{M} \int_{0}^{T} \left(\mu_{t}^{(m')} + \sum_{t_{n'} < t} \boldsymbol{\theta}^{\top} \boldsymbol{f}(m_{n'} \rightarrow m') \, \kappa(t - t_{n'}) \right) dt. \qquad (8)$$

If the base intensities are constant with respect to time, then

$$
\mathcal{L} = \sum_{n=1}^{N} \log\left(\mu^{(m_n)} + \sum_{t_{n'} < t_n} \boldsymbol{\theta}^\top \boldsymbol{f}(m_{n'} \to m_n)\, \kappa(t_n - t_{n'}) \right) -
$$
$$
\sum_{m'=1}^{M} \left(T\mu^{(m')} + \sum_{n=1}^{N} \boldsymbol{\theta}^\top \boldsymbol{f}(m_n \to m')\, (1 - \kappa(T - t_n)) \right), \tag{9}
$$

where the second term includes a sum over all events $n = \{1, \ldots, N\}$ that contibute to the final intensity $\lambda^{(m')}(T)$. To ease computation, however, we can rearrange the second term around the source m rather than the recipient m':

$$
\mathcal{L} = \sum_{n=1}^{N} \log\left(\mu^{(m_n)} + \sum_{t_{n'} < t_n} \boldsymbol{\theta}^\top \boldsymbol{f}(m_{n'} \to m_n)\, \kappa(t_n - t_{n'}) \right) -
$$
$$
\sum_{m=1}^{M} \left(T\mu^{(m)} + \sum_{\{n : m_n = m\}} \boldsymbol{\theta}^\top \boldsymbol{f}(m \to \star)\, (1 - \kappa(T - t_n)) \right), \tag{10}
$$

where we have introduced an aggregate feature vector $\boldsymbol{f}(m \to \star) = \sum_{m'=1}^{M} \boldsymbol{f}(m \to m')$. Because the sum $\sum_{\{n : m_n = m'\}} \boldsymbol{f}(m' \to \star)\, \kappa(T - t_n)$ does not involve either $\boldsymbol{\theta}$ or $\mu^{(1)}, \ldots, \mu^{(M)}$, we can pre-compute it. Moreover, we need to do so only for users $m \in \{1, \ldots, M\}$ with at least one event in the cascade.

A Hawkes process defined in terms of Eq. 5 has a log likelihood that is convex in the pairwise influence parameters and the base intensities. For a parametric Hawkes process, $\alpha_{m \to m'}$ is an affine function of $\boldsymbol{\theta}$, so, by composition, the log likelihood is convex in $\boldsymbol{\theta}$ and remains convex in the base intensities.

5.3 Gradients

The first term in the log likelihood and its gradient contains a nested sum over events, which appears to be quadratic in the number of events. However, we can use the exponential decay of the kernel $\kappa(\cdot)$ to approximate this term by setting a threshold $\tau^\star$ such that $\kappa(t_n - t_{n'}) = 0$ if $t_n - t_{n'} \geq \tau^\star$. For example, if we set $\tau^\star = 24$ hours, then we approximate $\kappa(\tau^\star) = 3 \times 10^{-11} \approx 0$. This approximation makes the cost of computing the first term linear in the number of events.

The second term is linear in the number of social network connections and linear in the number of events. Again, we can use the exponential decay of the kernel $\kappa(\cdot)$ to approximate $\kappa(T - t_n) \approx 0$ for $T - t_n \geq \tau^\star$, where $\tau^\star = 24$hours. This approximation means that we only need to consider a small number of tweets near temporal endpoint of the cascade. For each user, we also pre-computed $\sum_{\{n : m_n = m'\}} \boldsymbol{f}(m' \to \star)\, \kappa(T - t_n)$. Finally, both terms in the log likelihood and its gradient can also be trivially parallelized over users $m = \{1, \ldots, M\}$.

For a Hawkes process defined in terms of Eq. 5, Ogata showed that additional speedups can be obtained by recursively pre-computing a set of aggregate messages for each dyad (m, m'). Each message represents the events from user m

that may influence user m' at the time $t_i^{(m')}$ of their i^{th} event [42]:

$$
R_{m \to m'}^{(i)}
= \begin{cases}
\kappa(t_i^{(m')} - t_{i-1}^{(m')}) R_{m \to m'}^{(i-1)} + \sum_{t_{i-1}^{(m')} \leq t_j^{(m)} \leq t_i^{(m')}} \kappa(t_i^{(m')} - t_j^{(m)}) & m \neq m' \\
\kappa(t_i^{(m')} - t_{i-1}^{(m')}) \times (1 + R_{m \to m'}^{(i-1)}) & m = m'.
\end{cases}
$$

These aggregate messages do not involve the feature weights θ or the base intensities, so they can be pre-computed and reused throughout parameter estimation.

For a parametric Hawkes process, it is not necessary to compute a set of aggregate messages for each dyad. It is sufficient to compute a set of aggregate messages for each possible configuration of the features. In our setting, there are only four binary features, and some combinations of features are impossible.

Because the words described in Sect. 2.1 are relatively rare, most of the users in our data set never used them. However, it is important to include these users in the model. Because they did not adopt these words, despite being exposed to them by users who did, their presence exerts a negative gradient on the feature weights. Moreover, such users impose a minimal cost on parameter estimation because they need to be considered only when pre-computing feature counts.

5.4   Coordinate Ascent

We optimized the log likelihood with respect to the feature weights θ and the base intensities. Because the log likelihood decomposes over users, each base intensity $\mu^{(m)}$ is coupled with only the feature weights and not the other intensities. Jointly estimating all parameters is inefficient because it does not exploit this structure. We therefore used a coordinate ascent procedure, alternating between updating θ and the base intensities. As explained in Sect. 5.1, θ and the base intensities must be non-negative to ensure the intensity functions are non-negative. At each stage of the coordinate ascent, we performed constrained optimization using the active set method of MATLAB's `fmincon` function.

6   Results

We used a separate set of parametric Hawkes process models for each of the geographically distinctive linguistic markers described in Sect. 2.1. Specifically, for each word, we constructed a set of nested models by first creating a baseline model using features F1 (self-activation) and F2 (mutual reply) and then adding in each of the experimental features—i.e., F3 (tie strength) and F4 (local).

We tested hypothesis H1 (strong ties are more influential) by comparing the goodness of fit for feature set F1+F2+F3 to that of feature set F1+F2. Similarly, we tested H2 (geographically local ties are more influential) by comparing the goodness of fit for feature set F1+F2+F4 to that of feature set F1+F2.

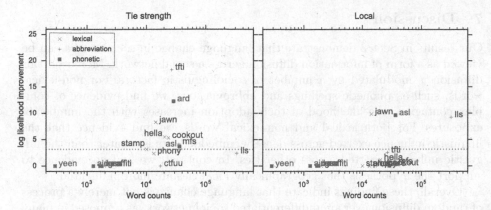

Fig. 3. Improvement in goodness of fit from adding in features F3 (tie strength) and F4 (local). The dotted line corresponds to the threshold for statistical significance at $p < 0.05$ using a likelihood ratio test with the Benjamini-Hochberg correction.

In Fig. 3, we show the improvement in goodness of fit from adding in features F3 and F4.[6] Under the null hypothesis, the log of the likelihood ratio follows a χ^2 distribution with one degree of freedom, because the models differ by one parameter. Because we performed thirty-two hypothesis tests (sixteen words, two features), we needed to adjust the significance thresholds to correct for multiple comparisons. We did this using the Benjamini-Hochberg procedure [10].

Features F3 and F4 did not improve the goodness of fit for less frequent words, such as *ain*, *graffiti*, and *yeen*, which occur fewer than 10^4 times. Below this count threshold, there is not enough data to statistically distinguish between different types of social network connections. However, above this count threshold, adding in F3 (tie strength) yielded a statistically significant increase in goodness of fit for *ard*, *asl*, *cookout*, *hella*, *jawn*, *mfs*, and *tfti*. This finding provides evidence in favor of hypothesis H1—that the linguistic influence exerted across densely embedded ties is greater than the linguistic influence exerted across other ties.

In contrast, adding in F4 (local) only improved goodness of fit for three words: *asl*, *jawn*, and *lls*. We therefore conclude that support for hypothesis H2—that the linguistic influence exerted across geographically local ties is greater than the linguistic influence across than across other ties—is limited at best.

In Sect. 3 we found that phonetic spellings and abbreviations exhibit complex contagion, while lexical words do not. Here, however, we found no such systematic differences between the three linguistic classes. Although we hypothesize that lexical words propagate mainly outside of social media, we nonetheless see that when these words do propagate across Twitter, their adoption is modulated by tie strength, as is the case for phonetic spellings and abbreviations.

---

[6] We also compared the full feature set—i.e., F1+F2+F3+F4—to feature set F1+F2+F3 and feature set F1+F2+F4. The results were almost identical, indicating that F3 (tie strength) and F4 (local) provide complementary information.

7   Discussion

Our results in Sect. 3 demonstrate that language change in social media can be viewed as a form of information diffusion across a social network. Moreover, this diffusion is modulated by a number of sociolinguistic factors. For non-lexical words, such as phonetic spellings and abbreviations, we find evidence of complex contagion: the likelihood of their adoption increases with the number of exposures. For both lexical and non-lexical words, we find evidence that the linguistic influence exerted across densely embedded ties is greater than the linguistic influence exerted across other ties. In contrast, we find no evidence to support the hypothesis that geographically local ties are more influential.

Overall, these findings indicate that language change is not merely a process of random diffusion over an undifferentiated social network, as proposed in many simulation studies [21,24,41]. Rather, some social network connections matter more than others, and social judgments have a role to play in modulating language change. In turn, this conclusion provides large-scale quantitative support for earlier findings from ethnographic studies. A logical next step would be to use these insights to design more accurate simulation models, which could be used to reveal long-term implications for language variation and change.

Extending our study beyond North America is a task for future work. Social networks vary dramatically across cultures, with traditional societies tending toward networks with fewer but stronger ties [40]. The social properties of language variation in these societies may differ as well. Another important direction for future work is to determine the impact of exogenous events, such as the appearance of new linguistic forms in mass media. Exogeneous events pose potential problems for estimating both infection risks and social influence. However, it may be possible to account for these events by incorporating additional data sources, such as search trends. Finally, we plan to use our framework to study the spread of terminology and ideas through networks of scientific research articles. Here too, authors may make socially motivated decisions to adopt specific terms and ideas [37]. The principles behind these decisions might therefore be revealed by an analysis of linguistic events propagating over a social network.

Acknowledgments. Thanks to the reviewers for their feedback, to Márton Karsai for suggesting the infection risk analysis, and to Le Song for discussing Hawkes processes. John Paparrizos is an Alexander S. Onassis Foundation Scholar. This research was supported by the National Science Foundation under awards IIS-1111142 and RI-1452443, by the National Institutes of Health under award number R01-GM112697-01, and by the Air Force Office of Scientific Research.

References

1. Adamic, L.A., Adar, E.: Friends and neighbors on the web. Soc. Netw. **25**(3), 211–230 (2003)
2. Al Zamal, F., Liu, W., Ruths, D.: Homophily and latent attribute inference: inferring latent attributes of Twitter users from neighbors. In: Proceedings of the International Conference on Web and Social Media (ICWSM), pp. 387–390 (2012)

3. Alim, H.S.: Hip hop nation language. In: Duranti, A. (ed.) Linguistic Anthropology: A Reader, pp. 272–289. Wiley-Blackwell, Malden (2009)
4. Anagnostopoulos, A., Kumar, R., Mahdian, M.: Influence and correlation in social networks. In: Proceedings of Knowledge Discovery and Data Mining (KDD), pp. 7–15 (2008)
5. Androutsopoulos, J.: Language change and digital media: a review of conceptions and evidence. In: Coupland, N., Kristiansen, T. (eds.) Standard Languages and Language Standards in a Changing Europe. Novus, Oslo (2011)
6. Anis, J.: Neography: unconventional spelling in French SMS text messages. In: Danet, B., Herring, S.C. (eds.) The Multilingual Internet: Language, Culture, and Communication Online, pp. 87–115. Oxford University Press, Oxford (2007)
7. Backstrom, L., Sun, E., Marlow, C.: Find me if you can: improving geographical prediction with social and spatial proximity. In: Proceedings of the Conference on World-Wide Web (WWW), pp. 61–70 (2010)
8. Bakshy, E., Rosenn, I., Marlow, C., Adamic, L.: The role of social networks in information diffusion. In: Proceedings of the Conference on World-Wide Web (WWW), Lyon, France, pp. 519–528 (2012)
9. Baldwin, T., Cook, P., Lui, M., MacKinlay, A., Wang, L.: How noisy social media text, how diffrnt social media sources. In: Proceedings of the 6th International Joint Conference on Natural Language Processing (IJCNLP 2013), pp. 356–364 (2013)
10. Benjamini, Y., Hochberg, Y.: Controlling the false discovery rate: a practical and powerful approach to multiple testing. J. Roy. Stat. Soc. Ser. B (Methodol.) **57**(1), 289–300 (1995)
11. Bucholtz, M., Hall, K.: Identity and interaction: a sociocultural linguistic approach. Discourse Stud. **7**(4–5), 585–614 (2005)
12. Bucholtz, M., Bermudez, N., Fung, V., Edwards, L., Vargas, R.: Hella Nor Cal or totally So Cal? The perceptual dialectology of California. J. Engl. Linguist. **35**(4), 325–352 (2007)
13. Centola, D., Macy, M.: Complex contagions and the weakness of long ties. Am. J. Sociol. **113**(3), 702–734 (2007)
14. Crystal, D.: Language and the Internet, 2nd edn. Cambridge University Press, Cambridge (2006)
15. Dunbar, R.I.: Neocortex size as a constraint on group size in primates. J. Hum. Evol. **22**(6), 469–493 (1992)
16. Eckert, P.: Linguistic Variation as Social Practice. Blackwell, Oxford (2000)
17. Eisenstein, J.: What to do about bad language on the internet. In: Proceedings of the North American Chapter of the Association for Computational Linguistics (NAACL), pp. 359–369 (2013)
18. Eisenstein, J.: Systematic patterning in phonologically-motivated orthographic variation. J. Sociolinguistics **19**, 161–188 (2015)
19. Eisenstein, J.: Written dialect variation in online social media. In: Boberg, C., Nerbonne, J., Watt, D. (eds.) Handbook of Dialectology. Wiley, Hoboken (2016)
20. Eisenstein, J., Ahmed, A., Xing, E.P.: Sparse additive generative models of text. In: Proceedings of the International Conference on Machine Learning (ICML), pp. 1041–1048 (2011)
21. Fagyal, Z., Swarup, S., Escobar, A.M., Gasser, L., Lakkaraju, K.: Centers and peripheries: network roles in language change. Lingua **120**(8), 2061–2079 (2010)
22. Granovetter, M.S.: The strength of weak ties. Am. J. Sociol. **78**(6), 1360–1380 (1973)

23. Green, L.J.: African American English: A Linguistic Introduction. Cambridge University Press, Cambridge (2002)
24. Griffiths, T.L., Kalish, M.L.: Language evolution by iterated learning with Bayesian agents. Cogn. Sci. **31**(3), 441–480 (2007)
25. Hamilton, W.L., Leskovec, J., Jurafsky, D.: Diachronic word embeddings reveal statistical laws of semantic change. In: Proceedings of the Association for Computational Linguistics (ACL), Berlin (2016)
26. Hawkes, A.G.: Spectra of some self-exciting and mutually exciting point processes. Biometrika **58**(1), 83–90 (1971)
27. Herring, S.C.: Grammar and electronic communication. In: Chapelle, C.A. (ed.) The Encyclopedia of Applied Linguistics. Wiley, Hoboken (2012)
28. Huberman, B., Romero, D.M., Wu, F.: Social networks that matter: Twitter under the microscope. First Monday **14**(1) (2008)
29. Johnstone, B., Bhasin, N., Wittkofski, D.: "Dahntahn" Pittsburgh: monophthongal /aw/ and representations of localness in Southwestern Pennsylvania. Am. Speech **77**(2), 148–176 (2002)
30. Kulkarni, V., Al-Rfou, R., Perozzi, B., Skiena, S.: Statistically significant detection of linguistic change. In: Proceedings of the Conference on World-Wide Web (WWW), pp. 625–635 (2015)
31. Kwak, H., Lee, C., Park, H., Moon, S.: What is Twitter, a social network or a news media? In: Proceedings of the Conference on World-Wide Web (WWW), pp. 591–600 (2010)
32. La Fond, T., Neville, J.: Randomization tests for distinguishing social influence and homophily effects. In: Proceedings of the Conference on World-Wide Web (WWW), pp. 601–610 (2010)
33. Labov, W.: The social motivation of a sound change. Word **19**(3), 273–309 (1963)
34. Labov, W.: Principles of Linguistic Change, vol. 2: Social Factors, vol. 2. Wiley-Blackwell, Hoboken (2001)
35. Labov, W.: Review of linguistic variation as social practice, by Penelope Eckert. Lang. Soc. **31**, 277–284 (2002)
36. Labov, W.: Principles of Linguistic Change, vol. 3: Cognitive and Cultural Factors, vol. 3. Wiley-Blackwell, Hoboken (2011)
37. Latour, B., Woolgar, S.: Laboratory Life: The Construction of Scientific Facts. Princeton University Press, Princeton (2013)
38. Li, L., Deng, H., Dong, A., Chang, Y., Zha, H.: Identifying and labeling search tasks via query-based Hawkes processes. In: Proceedings of Knowledge Discovery and Data Mining (KDD), pp. 731–740 (2014)
39. Li, L., Zha, H.: Learning parametric models for social infectivity in multi-dimensional Hawkes processes. In: Proceedings of the National Conference on Artificial Intelligence (AAAI) (2015)
40. Milroy, L., Milroy, J.: Social network and social class: toward an integrated sociolinguistic model. Lang. Soc. **21**(01), 1–26 (1992)
41. Niyogi, P., Berwick, R.C.: A dynamical systems model for language change. Complex Syst. **11**(3), 161–204 (1997)
42. Ogata, Y.: On Lewis' simulation method for point processes. IEEE Trans. Inf. Theor. **27**(1), 23–31 (1981)
43. Pavalanathan, U., Eisenstein, J.: Audience-modulated variation in online social media. Am. Speech **90**(2), 187–213 (2015)
44. Pavalanathan, U., Eisenstein, J.: Confounds and consequences in geotagged Twitter data. In: Proceedings of Empirical Methods for Natural Language Processing (EMNLP), September 2015

45. Rickford, J.R.: Geographical diversity, residential segregation, and the vitality of African American vernacular English and its speakers. Transform. Anthropol. **18**(1), 28–34 (2010)
46. Sadilek, A., Kautz, H., Bigham, J.P.: Finding your friends and following them to where you are. In: Proceedings of the Conference on Web Search and Data Mining (WSDM), pp. 723–732 (2012)
47. Squires, L.: Enregistering internet language. Lang. Soc. **39**, 457–492 (2010)
48. Tagliamonte, S.A., Denis, D.: Linguistic ruin? LOL! Instant messaging and teen language. Am. Speech **83**(1), 3–34 (2008)
49. Trudgill, P.: Sex, covert prestige and linguistic change in the urban British English of Norwich. Lang. Soc. **1**(2), 179–195 (1972)
50. Wolfram, W.: The linguistic variable: fact and fantasy. Am. Speech **66**(1), 22–32 (1991)
51. Zhao, Q., Erdogdu, M.A., He, H.Y., Rajaraman, A., Leskovec, J.: Seismic: a self-exciting point process model for predicting tweet popularity. In: Proceedings of Knowledge Discovery and Data Mining (KDD), pp. 1513–1522 (2015)

On URL Changes and Handovers in Social Media

Hossein Hamooni$^{(\boxtimes)}$, Nikan Chavoshi, and Abdullah Mueen

University of New Mexico, Albuquerque, USA
{hamooni,chavoshi,mueen}@unm.edu

Abstract. Social media sites (e.g. Twitter and Pinterest) allow users to change the name of their accounts. A change in the account name results in a change in the URL of the user's homepage. We develop an algorithm that extracts such changes from streaming data and discover that a large number of social media accounts are performing *synchronous* and *collaborative* URL changes. We identify various types of URL changes such as handover, exchange, serial handover and loop exchange. All such behaviors are likely to be automated behavior and, thus, indicate accounts that are either already involved in malicious activities or being prepared to do so.

In this paper, we focus on *URL handovers* where a URL is released by a user and claimed by another user. We find interesting association between handovers and temporal, textual and network behaviors of users. We show several anomalous behaviors from suspicious users for each of these associations. We identify that URL handovers are instantaneous automated operations. We further investigate to understand the benefits of URL handovers, and identify that handovers are strongly associated with reusable internal links and successful avoidance of suspension by the host site. Our handover detection algorithm, which makes such analysis possible, is scalable to process millions of posts (e.g. tweets, pins) and shared publicly online.

1 Introduction

Social media sites, such as Twitter, Pinterest, Tumblr and Instagram allow users to broadcast messages and content (URLs, images, videos) publicly to their followers. Many of these sites allow users to change their homepage URLs by changing their account names. Users may need to change their URLs for many reasons, such as marriage, rebranding, business acquisition and closing, and so on. Such events are relatively rare for any human or business user in social media sites. Surprisingly, we observe unusually high numbers of URL changes in some Twitter users.

For example, we identify a user changing its URL 283 times in 78 days, equivalent to roughly one change every six hours. Some of the URLs, released and claimed in the same day, are shown in Fig. 1. We identify an even more abnormal scenario where a URL, twitter.com/MalumaOficiaI, belonged to ten users in

© Springer International Publishing AG 2016
E. Spiro and Y.-Y. Ahn (Eds.): SocInfo 2016, Part I, LNCS 10046, pp. 58–74, 2016.
DOI: 10.1007/978-3-319-47880-7_4

three months. Each user *handed over* the URL to another user collaboratively. In Fig. 2, we show the sequence of handovers where nodes are user accounts and an arrow represents the direction of a handover. Such abnormal URL *handovers* are highly unlikely to be performed by a group of normal users, and most likely are generated by automated bots. As Twitter is one of the most popular social media site, we set to study such URL manipulating bots in Twitter.

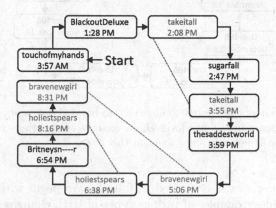

Fig. 1. A user (Twitter id: 2664619086) with ten URL changes on 7 November 2015. Some URLs are used more than once which form a loop of URLs. Repetitive URLs are connected by dotted lines.

There have been dozens of papers on mining Twitter data [5,11,12,14,16]. However, URL changes have not been studied with due diligence. An estimated 8.5 % accounts in Twitter are bot accounts [15]. Our work shows that bot accounts carry out automated URL changes on a regular basis. Irregular URL changes waste resources on Twitter, create many broken URLs, and mislead Twitter users to spam account pages. These negative consequences of URL manipulation motivate this work.

In this paper, we investigate to discover *why* and *how* users make such abnormal changes. We develop a parallel algorithm using the map-reduce framework to identify URL changes in streaming data. Our algorithm is incremental and scalable to support social media similar to Twitter in size and traffic. We extract a set of 231K URL changes in Twitter over a period of three months (10/15-01/16). Note that we use only 1 % of the data that Twitter publicly shares. We perform temporal, textual, and graph-based analyses on this data and discover several interesting facts about URL changes. Our findings are summarized below.

- Both URL changes and URL handovers are atomic operations.
- URLs that are handed over are more frequently mentioned by other users.
- URL handovers are associated with changes in content after the handover.
- URL handovers can be temporally correlated.
- URL changes are done in an organized and collaborative way by large groups of users.

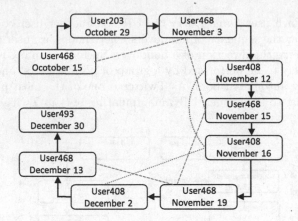

Fig. 2. The URL twitter.com/paradisecameron was handed over among four users nine times. User468 appears in the handover chain exactly every other time. The dashed lines connecting the same users show *loops* in this chain.

The rest of the paper is organized as follows. We begin with a background section that provides examples of various types of URL changes and handovers. We describe our algorithm to discover URL changes and handovers in Sect. 3. We provide association analysis with temporal, textual and graph-based features in Sect. 4. We investigate *why* and *how* frequent handovers are performed in Sect. 5. We discuss related work in Sect. 6, and conclusion in Sect. 7.

2   Background

We start with sufficient background information so readers are more familiar with URL changes and handovers.

URL Changes: Imagine a Twitter user with the name tom_hanks. The URL to the profile page of this user would be twitter.com/tom_hanks. If this user changes the screen name to thanks, the URL of its profile page will change to twitter.com/thanks. Such a change in the URL does not affect the social connections of tom_hanks in the Twitter network. All of the followers and followings of the account before and after the change remain the same. However, the URL change invalidates the old URL, which will no longer be accessible from other places on the Internet. URL changes also invalidate all of the old *mentions*[1] within Twitter, since mentions are the short form of URLs. In some social media, such as Pinterest, the old URL still functions because the site automatically redirects visitors to the new URL unless the old URL is taken by some other account.

---

[1] Twitter users can mention other users by using the '@' symbol which creates a link to the profile page of the mentioned user. For example @thanks is a link to the address twitter.com/thanks.

URL Handovers: A URL handover consists of two URL changes, in which one user releases a URL and another user claims that URL. Let us consider the example in Fig. 3 to describe URL handovers in reality. A user (user1) changes its screen name (URL) from Tom to John. The name Tom is then free on the network and can be claimed by any other user. If another user (user2) claims the name Tom by releasing its previous name Bill, a handover happens. We say the URL twitter.com/Tom has been handed over from user1 to user2. Here the user1 is the *from-account* and the user2 is the *to-account*. We also define the *handover lag* as the time duration between user1 releasing the URL twitter.com/Tom and the user2 claiming it.

In sites like Pinterest, the old URLs are redirected to the new ones. When a user changes URL, he does not need to worry about his followers who can still visit him via the old URL. However, in Twitter, old URLs are not redirected automatically. Therefore, a user often creates a new account to keep the old URL and leaves a pointer to the new URL. Thus, human users can do valid and legitimate handovers. However, in such handovers, one of the from-account or to-account should be inactive (e.g. no tweeting) after the handover under the assumption that no user wants to divide his followers among many accounts. We identify a *suspicious handover* if either of the following statements are true for a handover.

- Both of the from-account and to-account continue posting after the handover.
  OR
- Both of the from-account and to-account were active before the handover.

In the remainder of the paper, we will refer to a suspicious handover as simply a handover unless otherwise specified.

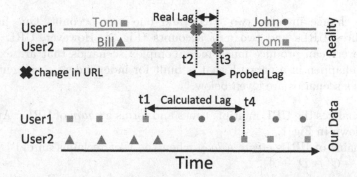

Fig. 3. $User_1$ handed over the URL Tom to $User_2$. The release time is t_2 and the claim time is t_3, and the real handover lag is $t_3 - t_2$. We calculate an upper bound, $t_4 - t_1$, for the handover lag based on the last tweet of $User_1$ at t_1 before the handover, and the first tweet from $User_2$ at t_4 after the handover.

2.1 Data Collection

We use the Twitter streaming API to collect data and produce a set of suspicious handovers. The Twitter streaming API caps the number of tweets sent to the client to a small fraction of the total volume of Tweets at any given moment [3]. We have never exceeded 48 Hz in practice. Our data collection module receives the tweets which contain the timestamp of the tweet, the URL, the user ID, the follower count, and some other information about the author of the tweet. The Twitter API provides tweets that satisfy a given condition, such as the tweet matches a given keyword, the tweet has a given topic, the tweet is made from a geo-location, or the tweet is authored by a specific set of users. We consider each tweet as a singleton object with a set of predefined features including timestamp, user ID, URL, geo-location, number of followers, number of accounts the user is following, and tweet content. In order to detect handovers, we consider three relevant features: the timestamp, the user ID, and the URL. Although a user can change the URL of the account, the user ID is fixed for an account throughout the lifetime of the account.

We use different keyword filters to collect tweets for a week. We sort users based on their number of tweets and pick the top 40,000 as the seed for the rest of the data collection from the Twitter streaming API. We make the data collection process parallel on eight computers, each of which listens to 5,000 users continuously. This parallelization maximizes the number of tweets that can be collected from the streaming source.

We have started collecting data from Twitter on 15 October 2015 and continued until 31 December 2015. We have collected 130 million tweets with 5.7 million unique users[2] and 6 million unique URLs.

2.2 Complex Handovers

One URL change involves two URLs and one user account. One handover involves three URLs and two user accounts (Fig. 3). However, URL changes and handovers can produce much more complex scenarios that are extremely unlikely to happen in a network that is built for independent social entities. A few complex scenarios are given below.

- A user changes the URL multiple times and forms a *chain* of URLs. An example is shown in Fig. 1.
- Some chains of URLs create a *loop* when the user reclaims an old URL (i.e. $A \to B \to C \to D \to A$).
- A URL can be handed over in a chain from user A to user B, and then from user B to user C. This is a suspicious behavior because it shows that multiple accounts are interested in having the same URL. It gets more suspicious if each of these accounts own the URL for a short time.

---

[2] Although our data collection seed contained 40,000 users, in total we collected tweets from 5.7 million different users.

– The handovers on a URL can also create a loop of users. This indicates that they either have a signaling mechanism to let each other know when the URL is free and ready to claim, or they are controlled by the same entity (Fig. 2).

Although a single URL change may not be an abnormal behavior, the chance of all of the abnormal scenarios described above happening inadvertently is very low. We find a multitude of evidence showing that users are performing such changes and handovers automatically using computer programs.

3   Detecting URL Handovers

Since Twitter does not provide an event flag representing a URL change, we devise an algorithm to identify handovers based on the last tweet from the from-account and the first tweet from the to-account before and after the handover respectively. Figure 3 shows a toy example of handover detection using the streaming data provided by Twitter. Note that the handover lag can be calculated as the time between the last and the first tweets from the from-account and to-account, respectively.

Computationally, handover detection is very similar to the *group-by order-by* queries for relational databases. We require grouping the tweets from the same URL and sorting the tweets for the same URL based in order of timestamps. We need to compare successive pairs of tweets from the same URL to detect change in their user IDs. Each such change in user ID corresponds to a handover. The process is further complicated by the scale of the data. A single processor cannot manage millions of tweets in reasonable time, guiding us to develop parallel solutions. We adopt map-reduce framework to distribute the computation and discuss our algorithm below.

Every map-reduce algorithm has two key components: a map function (mapper), and a reduce function (reducer). There can be other useful functions such as *filters* in a map-reduce framework. We discuss each of these components in this section. For clarity we define the input and the output of our map-reduce framework. The input is a set of tweets $T = \{tw_1, tw_2, \ldots, tw_n\}$ and the output is a set of URLs $U = \{url_1, url_2, \ldots, url_m\}$ where $url_i = \{(user_1, t_1, t_2), (user_2, t_3, t_4), \ldots, (user_k, t_{2k-1}, t_{2k})\}$, $k \geq 2$ and $t_j \leq t_{j+1}$, $\forall j$ $1 \leq j \leq 2k - 1$.

Mapper: The map function in our framework converts a tweet object to an object that can be used by the reducers. It produces a set of key-value tuples where the key is the URL of the tweet and the value is the user ID plus two timestamps. Initially both timestamps are equal to the tweet timestamp, but they will be converted to a start timestamp and an end timestamp in the next steps, which reflect the period of time in which the URL was associated with each account. In other words, initially: $mapper(tweet_i) = <url_i, \{(user_i, t_i, t_i)\} >$

Reducer: The reduce function plays the key role in our map-reduce framework. The map-reduce framework guarantees that all objects with the same URL will

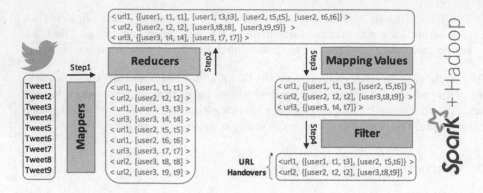

Fig. 4. The process of detecting URL handovers in the Twitter network. 2 URL handovers are detected from 9 tweet objects in this example.

be reduced together and that they produce one last merged object. A merged object in our case is $< url_i, \{(user_a, t_1, t_1), (user_b, t_2, t_2), \ldots, (user_z, t_k, t_k)\} >$ where $t_i \leq t_{i+1}$. The reducer function takes two key-value tuples as the input and produces one merged tuple as the output. As mentioned earlier, the value part is made up of a set of user IDs, each of which has a starting and an end time. The reducer function takes these two lists and creates a sorted output based on the start times of each object[3]. Since all the lists have just one element at the beginning of the reduce task, they are trivially sorted. As the reducer combines them, the merged lists are also sorted. In other words, the input to the reducer function is two sorted list of length m and n, and it just takes $O(m + n)$ steps for the reducer to sort them by using the merge sort approach.

Mergers: After the reducer produces a sorted list of user IDs with timestamps for each URL, we need to merge all the consecutive tweets with the same user ID to create a shorter list for each URL where there is a start and an end time for each user ID. For example, if the output of the reducer for a URL is $\{(A, 1, 1), (A, 4, 4), (B, 7, 7), (B, 9, 9), (A, 15, 15), (A, 20, 20)\}$ then the output of running the map value function would be: $\{(A, 1, 4), (B, 7, 9), (A, 15, 20)\}$

Filter: At this stage, we have lists of users associated with every URL in our map-reduce framework. However, we are not interested in detecting URLs that are only associated with one user ID. To filter out these URLs from the output of the map-reduce framework, we use a simple *filter* function. This function checks the length of the list of the users and outputs only the lists that have more than one user ID.

An overview of our system is shown in Fig. 4. Our algorithms have detected a total of 13,831 URL handovers involving 12,326 unique URLs and 21,257 unique users in the 78 days of data collection. We also detect 231,800 users who changed

---

[3] At this stage of the algorithm, the start time and the end time of the objects are still the same.

their URL at least once in this time period. We share the entire set of handovers and the source code to detect them in [1].

4 Handover Analysis

In this section we analyze the handovers detected by our method to observe several suspicious behaviors related to the user's temporal profile, tweet content, and the frequency of URL changes. We also discuss how multiple users can be connected through handovers. We finally analyze the handover lags to show that the handovers are automated.

4.1 Temporal Profile

We investigate questions related to the temporal profile of a user involved in a handover. We extract hourly time series of every user in every handover for 78 days. As mentioned in Sect. 2.1, each tweet object contains the timestamp of that tweet in millisecond resolution, and the number of followers of the user at that time. We construct hourly *activity time series* of every user by aggregating the total number of activities the user performs in each hour. Note that Twitter does not guarantee to provide all of the tweets of a user; therefore we achieve a lower bound time series on user activity. As we shall see, such partial data is enough to reveal abusive behaviors on Twitter.

Similarly, we create the follower time series of a user which shows the changes in the popularity of that user. We receive follower information embedded in the tweets, yielding some unevenly spaced measurements of the follower counts of a user. We interpolate the in-between follower counts by the last received count with an assumption that follower counts change very slowly (particularly for non-popular and old accounts).

Activity Association: We first consider the distribution of handovers over 11 weeks. We only consider handovers that have less than a day of calculated lag. This ensures that the real lag is at most 24 h, a reasonable value. In Fig. 5(top)

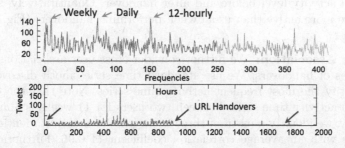

Fig. 5. (top) Frequency distribution of hourly count of handovers. (bottom) An example user with daily periodicity and a strong activity association with handover.

we show the frequency distribution of the hourly aggregates of handover counts over 1890 h. We use the method in [19] and identify three sharp peaks pointing to weekly, daily and 12-hourly periodicity. Figure 5(bottom) shows an example activity sequence of a user with daily and weekly periodicity.

We investigate if the handovers are related to a change in activity patterns. We check if the average activity levels of a user in the 6-hour windows before and after a handover are significantly different. 91 % of the times the difference is less than 1 tweet an hour. Therefore, we conclude there is no significant change in the activity level before and after the handovers. However, exceptions are possible. Figure 5(bottom) shows an example where the activity starts and stops with handovers.

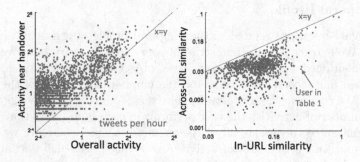

Fig. 6. Comparing the normal activity of a user with its activity near the handovers. 97.4 % of the users have higher activity-per-hour around handovers. Both x-axis and y-axis are in log scale. (right) The plot shows how the content of the tweets changes with URL changes.

Next, we consider the association between handovers and the activity level around them. We calculate the average activity-per-hour for every from-account in the 6-hour window just before releasing the URL, and the same for every to-account in the 6-hour window just after claiming the URL. We compare such pre- and post-handover activities with the average activity-per-hour of the user, calculated over the entire duration of data collection. We identify a significant difference in activity level before and after handover. Quantitatively, 97.4 % of the users are more active than usual when performing handovers (Fig. 6 (left)).

Cross-user Association: We further consider cross-user associations in temporal profiles of handovers. We use standard time series motif discovery tools [13] to identify the most frequent activity time series. Note that the expected similarity in activity time series between two users for 11 weeks is almost zero. Interestingly, we identify a motif of three users who have almost identical activity patterns with an average correlation coefficient of 0.96. Furthermore, the accounts perform URL handovers within the same hour in the same manner (e.g. to-from-to). The motifs are shown in Fig. 7. The URLs that were handed over by these accounts are all related to celebrities such as MacMiller, Rihanna, Drake, Megan Fox and Lil Wayne.

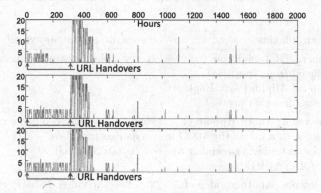

Fig. 7. Three accounts with almost identical activity profile and correlated handovers. Handovers initiate change in activity patterns.

We consider the motif as a significant discovery because it reveals that offenders work in correlation, possibly using the same codebase, and that they hand over at the same time to swap or pass URLs that they do not want to lose. In the future, we will investigate how to scale handover detection in real time so we can track the interest areas of the offenders to take countermeasures.

4.2   Content Profile

Users tweet about various topics. The topic of a tweet can be determined by analyzing the keywords in it. We first remove all useless words like *is, are, the, to, from, RT*[4], ..., and then process the tweet content. We use the content of the tweets to determine the similarity between two tweets, and also two sets of tweets. We use the Jaccard similarity coefficient [20] as our metric. The similarity between two sets of tweets X and Y can be defined as the average similarity of all pairs of tweets in them. We use this measure to calculate the similarity between two Twitter users, or between two different periods of time (before and after the handover) for the same user to profile content changes around handovers.

Content Association: We consider the content of tweets for a user before and after the URL change[5] to see *if the content changes with the change in URL*. Let T_1 and T_2 be the sets of tweets of a user from its first and second URL respectively. We calculate the in-URL similarity as the weighted average of $Sim(T_1, T_1)$ and $Sim(T_2, T_2)$, and the across-URL similarity as $Sim(T_1, T_2)$. For

---

[4] The word "RT" appears at the beginning of all retweets and has nothing to do with the content.

[5] We specifically are interested in URL changes that were part of a handover.

Table 1. The tweets from the same user with 2 URLs. The user change its URL from `zflexins` to `loveyorslf` on 11 December 2015. All the tweets of the left column are about Justin Bieber, and the ones in the right column are about Harry Styles (both are famous singers). The average in-URL similarity is 0.35, while the across-URL similarity is 0.03.

www.twitter.com/zflexins	www.twitter.com/loveyorslf
RT @justinbieber:UK! Tonight on @CapitalOfficial from 7pm 'Justin Bieber's Capital Album Party Replay'. Hear the tracks from #Purpose	Harry styles coisa mais linda gente!!!
RT @JBCrewdotcom: Another photo of Justin Bieber with a fan at the M&G in Tokyo, Japan yesterday. (December 4) https://t.co/ofAYAjzP1M	Harry s, tao precioso gente como vcs nao gostam dele???????? https://t.co/o0x2DG38JI
RT @JBCrewdotcom: Another video of Justin Bieber singing at a restaurant in Japan today. (December 5) https://t.co/jZqaMaezrO	vou tweetar video de harry stylesN
RT @favjarbara: interviewer: what do you think about justin bieber's relationships?bp: hahaha he's mine	harry w kendall eu to gRITANDO AQUI, OPSSS https://t.co/MURzVWnc0Q
RT @NME: Justin Bieber announces UK Arena tour dates for 2016 https://t.co/ECsRUqEPxk	@KendallJBrasil: 31/12- Mais fotos de Kendall e Harry Styles em St. Barts, Frana. https://t.co/CytM8Hixk

example, Table 1 shows the tweets of a user with two different URLs: `zflexins` and `loveyorslf`. The user tweeted 98 times with the first URL about Justin Bieber, and 94 times with the second URL about Harry Styles. These are two of the most popular celebrities in Twitter with millions of followers. There is a clear change in the topic of the tweets after the URL change. The average in-URL similarity for this user is 0.35 while the across-URL similarity is 0.03. It is humorous that the content of the first tweet after URL change is: `RIP zflexins`. Both of these URLs are now associated with some other accounts.

In order to check this hypothesis for other users, we select a random set of handover users that have exactly two URLs associated with them in our dataset. We filter out the users for which $|T_1| < 5$ or $|T_2| < 5$, and finally come up with 1,051 users. Figure 6(right) shows the comparison of in-URL with across-URL similarity for each of these users. We have 100 % of the users with higher in-URL similarity than the across-URL similarity. It means that the overall topic of the tweets changes when a user changes its URL, especially if that URL change is a part of URL handover.

4.3 URL Change Analysis

In this experiment, we check if the frequency of URL changes (average number of URL changes per day) of a user has any relation with the probability of that user being involved in a URL handover, since we believe both a high number of URL changes and being involved in a handover are suspicious behaviors. There are 231,800 users in our dataset which changed their URL at least once during our data collection. Figure 8 (I) shows the percentage of these users for different frequencies of URL changing. The probability of a user being involved in a handover given the frequency it has changed its URL is shown in Fig. 8 (II). The higher the change frequency, the larger the probability of performing handovers. The reason why we do not show users with more than 9 URL change frequency on the left side is that they comprise less than 1 % of our dataset. However, we can say almost all of this 1 % have done a URL handover by looking at the right hand side of the figure.

4.4 Connectivity Profile

We create a bipartite graph where the left side is the set of all users and the right side is the set of all URLs, and a link between a user u_i and a URL v_j exists if u_i owned v_j at some point in our dataset, and the URL was used for a handover. A handover is defined as a subgraph with three nodes and two edges in which two nodes from the user side have a link to the same node on the URL side.

We use the classic co-clustering approach to identify clusters in the user-URL bipartite graph [9]. Any balanced cluster with more than three members points to organized teamwork by the accounts. It is very unlikely that a large balanced cluster was created in this bipartite graph by accident.

We find a cluster of size 2,273 (1,205 users + 1,068 URLs) which has 2,399 edges. The average degree of each node in this cluster is 2.11. It is highly unlikely that such a cluster is formed randomly, and thus this cluster supports our original hypothesis that correlated and frequent handovers are signatures of automated accounts managed by the same entity. If we had more data, we could have identified more handovers, and the cluster could have been much larger. About 6 % of the all users that has performed suspicious behavior (URL handover) are in this particular cluster. This again proves that our suspicion is correct beyond a doubt since such a huge cluster can not be formed randomly.

If we consider all of the clusters with more than three members (non-trivial handover clusters), they cover 31 % of all users who have been involved in handovers. Although any URL handover is a suspicious behavior, this provides us with additional evidence of misbehavior from this 31 %. We believe that the majority of the other 69 % also belongs to a non-trivial cluster, but we are not able to catch them due to lack of data. As we show in the next section, social media sites are slow in suspending such offenders. We have detected thousands of automated spammers, even without the complete dataset, and yet Twitter has suspended only a fourth of them in six weeks.

These users who are doing URL handovers usually change their URLs more than once. Not all of their URLs are included in the discussed bipartite graph since we just add the URLs which have been handed over. If we include all of the URLs of the users who have done a handovers (even the URLs that have not been used in any handover so far) in our graph and re-cluster, the biggest cluster would have 1,205 users and 6,040 URLs. These newly added URLs are good candidates for our active probing technique (future work) since they belonged to a suspicious user at some point in the past.

4.5   Lag Profile

To examine whether these handovers are organized from a central source as opposed to independent actions, we perform an analysis on handover time-lag. The real lag between releasing a URL and claiming it back is not detectable from the publicly available tweets. Our active probing tool, which is not scalable because of a capacity limit set by Twitter, estimates handover lag at most an hour longer than the real lag.

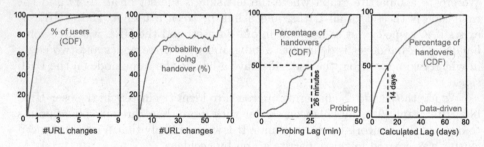

Fig. 8. (I) Percentage of users (out of 231,800 users) based on the frequency of changing URL. (II) The probability of a user doing a URL handover given the URL change frequency. The probability reaches 1 for a user with frequency higher than 68 (almost one URL change every day). (III) The distribution of handovers based on their lags calculated using our probing technique. (IV) The distribution of handovers based on their lags calculated using the data-driven approach. 50th percentiles are shown in both III and IV

We can only probe 180 users and/or URLs every 15 min. We start probing every day with a list of 100 users who we know have high numbers of URL changes (based on our dataset). We have done this experiment for 8 consecutive days and observed 210 handovers. Figure 8 (III) shows the CDF of the percentage of handovers for different lags. The sharp increases at minutes 15, 30, and 45 are the result of the discontinuities in the probing algorithm caused by Twitter API limitations imposed on our algorithm. The approximately linear CDF illustrates the remarkable fact that handovers are instantaneous operations. We can verify this claim by simulating a set of instantaneous handovers spread uniformly over time and applying our probing algorithm to calculate an estimated CDF. The estimated CDF is, indeed, a line and the slope of the line is very similar to what we have observed.

This analysis formed the basis of our data-driven detection process. In the data-driven detection process, we can only detect a handover if the pair of accounts tweet something before and after the handover, and Twitter provides us the tweets. Under such stringent condition, the lags we calculate are weak upper bounds of the real lags.

We show the CDF of the handover lags detected by the data-driven technique in Fig. 8 (IV). Although the data-driven process detects larger lags compared to the real lag, since we know from this analysis that the handovers are mostly instantaneous operations performed by automated programs, we trust that the handovers detected using the data-driven technique are highly suspicious. Also note that the lag for half of the handovers we find is less than 14 days. Therefore, if a URL is not claimed after few days of releasing, it (probably) will not ever be claimed.

5   Why Handovers?

Such a magnitude of automated URL changes must have good reasons behind. In this section, we discuss association of handovers with potential benefits such as obtaining human followers and avoiding suspension, and thus attempt to answer the question *why are handovers so frequent?*.

Mentions and External URLs: Although URL changes do not impact the internal connectivity among users (who follows whom), they have a direct impact on URLs linked from external web pages. It also affects the links created by *mentions* within Twitter. For example, when user1 mentions user2 as @DavidW (whose screen name is *DavidW*) in a tweet, Twitter creates the URL twitter.com/DavidW and embeds it in the tweet content. If user2 hands over this URL to user3, the mention DavidW would point to user3's profile page. Thus, thousands of mentions within Twitter are being abused by URL changes.

Our hypothesis is that the miscreants change URLs frequently to fool users in visiting different pages every time they follow the same mention. The motivation is to increase the chance of getting a human visitor or follower in the process.

To test this hypothesis, we use the Twitter Advanced Search page in which one can search for the mentions of a certain screen name (i.e. URL). We count the number of mentions a URL receives in the first fifteen pages of the search result. Figure 9 (left) shows the percentage of URLs based on the number of mentions for 1000 random URLs and 1000 handover URLs. The URLs that have been handed over have a higher number of mentions compared to random URLs. The average number of mentions for handover URLs is 80 compared to 22 for random URLs.

Suspension: Twitter suspends accounts that violate some of its rules [2]. Twitter rules says, *Creating multiple accounts with overlapping uses or in order to evade the temporary or permanent suspension of a separate account is not*

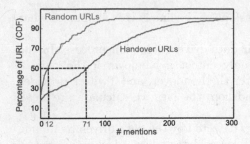 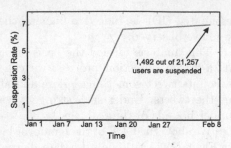

Fig. 9. (left) The percentage of URLs based on the number of mentions for random URLs and handover URLs. (right) Twitter suspension rate of handover users.

allowed. Handovers are strong signals of overlapping uses, hence, handover accounts are violating the Twitter rules.

We have detected URL changes and handovers until 31 December 2015. In order to see whether or not doing the handover has any impact on the suspension of the involved users, we check the status of all handover users almost every week from 1 January 2016 to 8 February 8 2016. Each point in Fig. 9 (right) shows the percentage of the handover accounts being suspended by Twitter until that day. The interesting point is that although these users had done many URL handovers in our data collection period, just 0.6 % of them were suspended by 1 January 2016. In Sect. 4.4, we mentioned that we have additional strong evidence of 31 % of the handover users being suspicious, and still just 7 % of these accounts are suspended by Twitter at the time of writing, while we had found these suspicious users weeks earlier.

6   Related Work

URL changes and handovers have been actively performed by users in social media. To the best of our knowledge, our work is the first to investigate the association between these activities and abuse in social media. Research has been done on other various aspects of abuse in social media including account hijacking [17], trolling [6], faking [4] and trafficking fraudulent accounts [18]. All of these works provide an important perspective on how fraudsters, merchants and abusers are manipulating social media for their own benefit. Our work considers URL handovers in the same manner. There are several works on bot and automated user account detection in social media using data mining techniques. In [8], the authors have modeled the inter-arrival time between tweets to understand bot behavior. In [7,10], supervised techniques are used to detect bots at registration time.

7   Conclusion

We develop methods to detect URL handovers between accounts in social media using publicly available data. We perform an in-depth analysis on the users who perform URL changes and handovers and identify several interesting characteristics. Collaborative abusers exploit this ability to change their URLs in social media to trick regular human users into following spam accounts. Our data analysis discovers automated and collaborative handovers in temporal and connectivity profiles of these users, and provides useful insights into how the abusers are operating. In future work we will develop active prevention based on these insights by predicting which users are going to do a URL handover.

References

1. Project repository. http://cs.unm.edu/~hamooni/papers/handover
2. The Twitter rules. https://support.twitter.com/articles/18311
3. Twitter streaming API. https://dev.twitter.com/tags/streaming-api
4. Akoglu, L., Chandy, R., Faloutsos, C.: Opinion fraud detection in online reviews by network effects. In: ICWSM, pp. 2–11 (2013)
5. Asur, S., Huberman, B.A.: Predicting the future with social media. In: IEEE/WIC/ACM International Conference on Web Intelligence and Intelligent Agent Technology, vol. 1, pp. 492–499. IEEE (2010)
6. Cheng, J., Danescu-Niculescu-Mizil, C., Leskovec, J.: Antisocial behavior in online discussion communities. In: Proceedings of ICWSM (2015)
7. Chu, Z., Gianvecchio, S., Wang, H., Jajodia, S.: Detecting automation of Twitter accounts: are you a human, bot, or cyborg? IEEE Trans. Dependable Secure Comput. **9**(6), 811–824 (2012)
8. Costa, A.F., Yamaguchi, Y., Traina, A.J.M., Traina Jr., C., Faloutsos, C.: RSC: mining and modeling temporal activity in social media. In: Proceedings of the 21st ACM SIGKDD International Conference on Knowledge Discovery and Data Mining, KDD 2015. ACM (2015)
9. Dhillon, I.S.: Co-clustering documents and words using bipartite co-clustering documents and words using bipartite spectral graph partitioning. In: Proceedings of 7th ACM SIGKDD Conference, pp. 269–274 (2001)
10. Lee, K., Caverlee, J., Webb, S.: Uncovering social spammers. In: Proceeding of the 33rd International ACM SIGIR Conference on Research and Development in Information Retrieval - SIGIR 2010, p. 435. ACM Press (2010)
11. Li, H., Mukherjee, A., Liu, B., Kornfield, R., Emery, S.: Detecting campaign promoters on Twitter using Markov random fields. In: IEEE International Conference on Data Mining (ICDM), pp. 290–299 (2014)
12. Matsubara, Y., Sakurai, Y., Ueda, N., Yoshikawa, M.: Fast and exact monitoring of co-evolving data streams. In: IEEE International Conference on Data Mining, pp. 390–399. IEEE (2014)
13. Mueen, A.: Enumeration of time series motifs of all lengths. In: Proceedings - IEEE International Conference on Data Mining, ICDM, ICDM, pp. 547–556 (2013)
14. Ratkiewicz, J., Conover, M., Meiss, M., Goncalves, B., Flammini, A., Menczer, F.: Detecting and tracking political abuse in social media (2011)
15. Subrahmanian, V., Azaria, A., Durst, S., Kagan, V., Galstyan, A., Lerman, K., Zhu, L., Ferrara, E., Flammini, A., Menczer, F., Waltzman, R., Stevens, A., Dekhtyar, A., Gao, S., Hogg, T., Kooti, F., Liu, Y., Varol, O., Shiralkar, P., Vydiswaran, V., Mei, Q., Huang, T.: The DARPA Twitter bot challenge. IEEE Computer (2016, in press)

16. Thomas, K., Grier, C., Song, D., Paxson, V.: Suspended accounts in retrospect: an analysis of Twitter spam. In: Proceedings of the ACM, IMC 2011, pp. 243–258 (2011)
17. Thomas, K., Li, F., Grier, C., Paxson, V.: Consequences of connectivity: characterizing account hijacking on Twitter. In: Proceedings of the ACM SIGSAC Conference on Computer and Communications Security - CCS 2014, pp. 489–500. ACM Press (2014)
18. Thomas, K., Paxson, V., Mccoy, D., Grier, C.: Trafficking fraudulent accounts: the role of the underground market in Twitter spam and abuse trafficking fraudulent accounts. In: USENIX Security Symposium, SEC 2013, pp. 195–210 (2013)
19. Vlachos, M., Gunopulos, D., Das, G.: Rotation invariant distance measures for trajectories. In: Proceedings of the ACM SIGKDD International Conference on Knowledge Discovery and Data Mining - KDD 2004, p. 707. ACM Press (2004)
20. Wikipedia. Jaccard index – wikipedia, the free encyclopedia. https://en.wikipedia.org/w/index.php?title=Jaccard_index&oldid=688763411

Comment-Profiler: Detecting Trends and Parasitic Behaviors in Online Comments

Tai-Ching Li[1(✉)], Abdullah Mueen[2], Michalis Faloutsos[1], and Huy Hang[1]

[1] University of California, Riverside, CA, USA
{tli010,michalis,hangh}@cs.ucr.edu
[2] University of New Mexico, Albuquerque, NM, USA
mueen@cs.unm.edu

Abstract. Can we detect anomalies and abuse among users of commenting platforms? Commenting has become a significant activity and specialized platforms provide commenting capability to many popular websites, such as Huffington Post. These platforms have become a new type of online social interaction, but have received very little attention. We conduct an extensive study on 19M comments from Disqus, one of the largest commenting platforms. Our work consists of two thrusts: (a) we identify features and patterns of commenting behavior, and (b) we detect peculiar and parasitic users. First, we study and evaluate features of user behavior that capture different aspects: user-user interaction ("social"), user-article interaction ("engagement"), and temporal properties. We also develop a method which we call, DownTimeFinder, to determine users' downtime (think night-time) in their daily behavior, which helps identify three major groups of users based on their utilization (3, 9, 15 h of up-time). Second, we identify surprising and abnormal behaviors using our features. Interestingly, we find: (a) two tightly collaborative groups of size at least 29 users that seem to be promoting the same ideas, (b) 38 users with behavior that points to spamming and trolling activities, and (c) 19 different instances where Disqus is used as a chat room. The goal of our work is to highlight commenting platforms as an ignored, but information-rich, online activity.

1 Introduction

User comments on news articles has emerged as a platform for social interaction over the last 10 years or so. Typically, one thinks of commenting on an article as an isolated activity. However, two interesting phenomena have changed this. First, there are a few companies that facilitate the backend management of comments for **a wide range of websites**. We use the term **commenting platform** to refer to such platforms, which include Disqus [9], LiveFyre [16], and IntenseDebate [13]. Second, many users exhibit intense commenting activity, such as spending many hours daily leaving comments. As a result of these two phenomena, one can obtain a comprehensive view of the commenting behavior of people across a large number of sites. These platforms are becoming more and more like social networks, since users can "follow" other users, and often

© Springer International Publishing AG 2016
E. Spiro and Y.-Y. Ahn (Eds.): SocInfo 2016, Part I, LNCS 10046, pp. 75–91, 2016.
DOI: 10.1007/978-3-319-47880-7_5

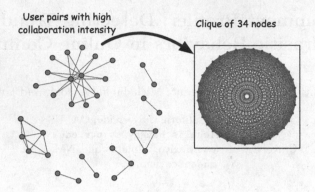

Fig. 1. Detecting collusion: Groups of real users in Disqus whose "social" behavior suggests collusion: the edges indicate collaboration intensity of more than 2^{10} distinct articles (referred to as SG_{10} graph in Sect. 3.1). The largest component is part of the clique of colluding nodes in the inset.

comments are addressed to other users (e.g. "Where have you been john123? Long time no see.").

Here are some terms and definitions that we use in this paper. A **user** is defined by a platform account, which enables her to leave comments to articles on a website that uses the commenting platform. A user may leave more than one comment for an article, which leads us to define the **engagement** of a user for that article. An engagement has a time duration and intensity in terms of number of comments, with both metrics being an indirect indication of the user's interest in that article or topic in general. Due to the lack of a better term, when two users comment on the same article, we say that they **collaborate** and we use the term **collaboration** to describe this joint activity. Users can leave comments for an article or respond to a previous comment, but we do not distinguish between these two types of comments in this work, despite some initial exploration. We use the term **collaboration intensity** to refer to the number of articles for which two users collaborate.

How can we detect unusual and parasitic users in these commenting platform? This is the overarching question in our work. We want to identify patterns and anomalies focusing mostly on the behaviors of the users, and we use the content of the posts very lightly or as a validation of suspicious activity. Specifically, the input to the problem is the commenting information of the users. This includes: the author of the comment, the time it was posted, and information on the article it was posted for. The goal is to determine patterns of behavior, per user, and per user-article pair and then identify and explain surprising and anomalous phenomena.

Detecting parasitic and abusive behaviors is a critical building block for ensuring that these platforms serve their primary purpose, which is the honest exchange of opinions among readers. We want to empower commenting platforms with techniques and tools to identify parasitic behaviors. The goal is to shield users from malicious actors, who could try to push their opinions and agendas, and use it as a mechanism to bias the public opinion, and even intimidate and propagate hostility and hate in society. For example, an organized group could systematically and aggressively attack a product, an idea, a person, or an institution.

Commenting platforms have attracted very little attention so far, as most work focuses on analyzing the posts of Online Social Networks (OSNs) like Facebook, Twitter and blogs. We identify three related areas to our work: (a) inferring the geographic locations of users based usually on the textual content; (b) detecting spam, abusive behaviors and malware in OSNs; and (c) inferring users' psychological state and personality traits. We expand on related work in Sect. 5.

As our key contribution, we identify patterns and anomalies among users through an extensive study. We collect our data by "crawling" Disqus, arguably the largest commenting platform. In our effort to develop a systematic approach to behavior profiling, we identify three dimensions in the user behavior: (a) social interaction or user-user interaction, (b) engagement or user-article interaction, and (c) temporal features of the user behavior. Our work focuses on two thrusts: (a) we identify patterns and common behaviors, (b) we identify surprising behaviors. Crawling Disqus, we collect 19 million comments between Nov-2007 to Jan-2015, and study 109K unique user accounts who have more than 10 comments. We summarize our key results below.

a. Identifying behavioral patterns. We study the behavior of users along the aforementioned three dimensions, and we make the following observations.

(i) Social behavior: We find that users have many ephemeral collaborations, but very few intense collaborations: 82 % of users have more than 10 collaborators, but, 99 % of these pair-wise collaborations have an intensity lower than 13, while we find only 0.03 % collaborations with intensity larger than 100.

(ii) User-article engagement: The engagement of users per article exhibit a skewed distribution with most engagements being short lived, and with few exceptionally long ones. Specifically, we find that 78.9 % of engagements last less than ten minutes, while 4.2 % engagements last more than one day and 0.54 % of the engagements last more than one month.

(iii) Temporal behavior. Going beyond the expected periodic behaviors, we focus on the downtime (think night-time) in the users' daily behavior and identify three major groups with 3, 9, 15 h a day of engagement time. In other words, there are users that spend 15 h a day writing comments, which is very helpful in detecting anomalous behaviors.

b. Identifying surprising behaviors. We identify abnormal behaviors and potentially malicious users by leveraging our metrics and observations. Our goal here is to show that: (a) our metrics are informative, and (b) some very surprising behaviors indeed exist.

(i) Collusion: We find two large cliques of sizes 29 and 34 with unusual social behavior that suggests collusion: all clique members collaborate with each other with collaboration intensity of 100 or more articles, and we also find that they support each other's opinions. Figure 1 depicts our detection approach: we start from groups of high collaboration intensity (>1024) and then upon closer investigation, the largest such group leads us to the colluding clique of size 34.

(ii) Spamming and Trolling: We find 38 users that exhibit trolling and spamming behavior. Their engagement behavior and temporal patterns are unusual: their engagement intensity is very low, mostly one comment per article, and they do not show the periodic and predictable temporal pattern.

(iii) Chatroom use: We find 24 users on 19 different websites where Disqus is used as a chat room, anchored by a fake empty article. We identify these chatrooms by finding the chatroom users, who are characterized by unusually intense engagements: more than 500 comments and engagement durations that exceed one month.

This work simply demonstrates the richness of information and interesting user behaviors that can be found in these much-less-studied online platforms, which we plan to fully explore in future work.

2   Data Sets

Disqus is arguably the most widespread commenting platform as it reached one billion unique visitors a month and about roughly two and a half million site installs by May 2013 [8]. There are two main reasons for the success of commenting platform. First, it enables users to comment on multiple sites with a single sign-in, thus eliminating the need for multiple registration, passwords and logins. Second, the host website does not have to develop and manage its own commenting infrastructure, which save both resources and bandwidth.

Disqus Data Collection: We explain how we collect data from Disqus. Disqus exposes its database to its clients (the blogs and news services) through an Application Programming Interface, or API. In fact, if a *user ID* associated with a user profile is known, anyone can retrieve every comment that this user has made on any website that uses Disqus. Furthermore, the following pieces of information are also available:

- The timestamp of the comment (in UTC), and any URL of attached media (video or image).
- Basic metadata about the article which includes: (a) the title of the article, and (b) the URL of the article.

As mentioned above, we only need a set of user ID to retrieve all their comments from the user. The user ID happens to be number, which seems to correspond to the order with which the person joined Disqus. We decided to focus on ID c within $1,000,000 \leq c \leq 1,999,999$. This seemed to be the more densely populated range compared to the zero to 1M range of IDs that we tried initially. Given the throttling mechanisms of the site, it took us roughly three months to retrieve comments from every user in that range. We collected 19M comments spanning Nov-2007 to Jan-2015.

Our User-centric data set: D_{Comm}. To create meaningful user profile, we need them to have non-trivial posting behavior. Thus, for this kind of analysis, we filter out users with fewer than 10 comments each. The dataset has 109,564 users, 19,121,250 comments, 3,474,360 articles, and 91,878 *forums* as defined by Disqus. Disqus uses the term forum to indicate a website that provides articles, and we adopt their definition of a website in our work. For the rest of this paper, we will use the term *forum* and *website* interchangeably.

The daily behavior dataset: D^t_{Comm}. To study the daily pattern of users, we create a new data set, D^t_{Comm}. For each user, we identify their 90-day window that has the maximum number of comments. We select users that have more than 100 comments in the window, and this leads to 26,009 unique users. The starting day of that window varies among users. This was necessary as some users are active for a while and then disappear for a long-time or forever. For each user, we aggregate their comments from that 90-day interval per hour of the day. These bins form a time series for each user as the one shown later in Fig. 4(c).

3   Modeling User Behavior

We study the behavior of users on Disqus to identify major patterns and develop a frame of reference, which we leverage in detecting surprising behaviors in Sect. 4.

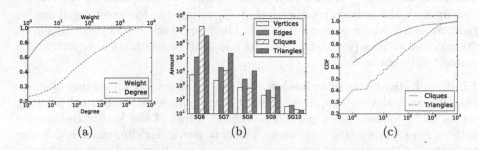

(a)   (b)   (c)

Fig. 2. Social properties. (a) CDF of the node degrees and the edge weight distributions of the SG_0 graph. (b) The number of vertices, edges, cliques and triangles for various SG_n. (c) Distribution of cliques and triangles of users in SG_7.

3.1   The Social Properties of the User Interactions

We model and study how users relate to each other in terms of commenting on the same articles, and we use the term collaboration to capture this relationship as we defined earlier. We define the **Social Graph** $SG = \langle V, E \rangle$ as an undirected weighted graph as follows:

1. V is the set of vertices and each vertex $v \in V$ is a user.
2. E is the set of edges, where the edge e_{ij} between nodes v_i and v_j exists, if and only if the users i and j collaborated on at least one article.
3. The weight $W(e_{ij})$ of the edge is the collaboration intensity (the number of articles that i and j have collaborated on).

The SG graph is quite dense with 102,028 vertices, 87,667,686 edges, an average degree of 1718.5 and a median degree of 130. We are interested in both the connectivity and the weight distribution of the graph. Figure 2(a) that 95 % of the edges have weights lower than 10 and there are extreme cases: 19,018 edges (2.1 %) have weights higher than 128, 485 edges higher than 512, and 40 higher than 1024. The highest value of collaboration intensity is 2,198 articles. Although 82 % of users have more than 10 collaborators in SG, there are few intense collaborations as we show below.

Studying the collaboration intensity: the SG_n graph sequence. To focus on highly collaborative user-pairs, we define a sequence of the subgraphs SG_n from SG by filtering out edges with low collaboration intensity. A subgraph SG_n is a subset $\langle V_n, E_n \rangle$ of $\langle V, E \rangle$ and $E_n = \{e_{ij} \in E \mid W(e_{ij}) \geq 2^n\}$. In other words, subgraph SG_n focus on edges with higher collaboration intensity as n increases. It is easy to see that SG_0 and SG are the same graph. Also, the lower n subgraphs include the higher n subgraphs: $SG_k \subseteq SG_l$ for $k > l$.

Focusing on highly collaborative user pairs. We find that for $n > 10$, we end up with the null graph, therefore SG_{10} represents the most collaborative pairs of nodes, which is shown in Fig. 1. The graph includes 33 users, 40 edges, 20 maximal cliques (we always refer to maximal cliques in this paper), and 8 connected components.

Finding indications of collusion. We can make several interesting observations by analyzing subgraph SG_{10}. First, each connected component in SG_{10} seems to be focusing on a particular website. Most of the jointly commented articles in each connected component belong to particular site, which is different than that of the other components. These sites are Newser, Foxnews, Milenio, RawStory, Fotbollskanalen, EscantonBlog, nonamedufus.blogspot, and stfueverybody.tumblr.com (which recently was shut down by its owner). Second, all the usernames of the users in the largest component of SG_{10} start with the prefix **newser-**. This shows that the users registered in Disqus through a link from Newser, as this prefix is automatically given by Disqus. Looking more closely, we also find that all of these users registered on the same day. In fact, two of these users have collaborated on 2,198 articles, which is the highest collaboration intensity, and their registration times are only forty minutes apart. We continue the exploration of collusion in Sect. 4.

Social density: studying the local connectivity. We study the local connectivity of the nodes using the number of maximal cliques and triangles for which a node is part of in a given SG_n graph.

In Fig. 2(b), we show the total number of nodes, vertices, maximal cliques and triangles for subgraphs SG_n, for $6 \leq n \leq 10$. There are roughly three orders of magnitude in the drop of the number of maximal cliques from SG_6 to SG_7: there are more than 16 million in SG_6 and 11,546 maximal cliques in SG_7. Thus, we select SG_n for $n \geq 7$ as good places to look for social structure, since SG_6 is very large and dense. Note that finding the maximal cliques in subgraphs with $n \leq 5$ was computational expensive due to the size of these graphs. In addition, we are more interested in node pairs with high collaboration intensity anyway. Recall that there are no collaboration edges in the graphs with $n > 10$.

Most users (82 %) participate in less than 10 cliques in SG_7. In Fig. 2(c), we plot the Cumulative Distribution Function of cliques and triangles of user in SG_7. We see that only 18 % of users have more than 10 cliques and there are 21 users that participate in more than 5,773 cliques, which is 50 % of total cliques in SG_7. Note that the distribution of cliques starts from 1, but the distribution of triangles from 0: a pair of users each with a degree of one constitutes a trivial clique, but has zero triangles.

We leverage these observations to profile users in Sect. 4.

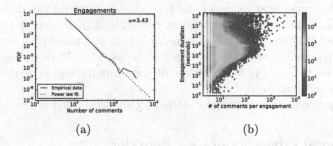

(a)	(b)

Fig. 3. Engagement properties. (a) Distribution of comments per engagement. (b) The heatmap of the engagement duration versus intensity.

3.2   User-Article Engagement Properties

An engagement refers to the interaction of a user with a given article for which she has left at least one comment. We consider two attributes for an engagement: (a) **engagement duration**, which is the amount of time interval between the first and last comments of the user, and (b) **engagement intensity**, which is the number of comments that the user left for that article.

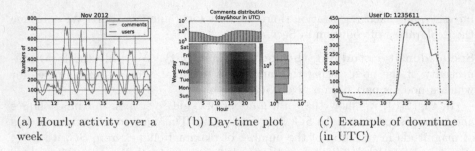

(a) Hourly activity over a week   (b) Day-time plot   (c) Example of downtime (in UTC)

Fig. 4. Temporal properties. (a) The hourly number of comments of Disqus starting on Sunday Nov 11-th. (b) Date-time plot for comments of Disqus shown as a heatmap. The color of each square represents the number of comments for a given time (x-axis) fora given day of the week (y-axis). (c) The aggregate number of comments per time of day for a single user and showing the downtime as calculated by our algorithm.

Figure 3(a) shows the skewed distribution of comments per engagement, which seems to follow a powerlaw with high accuracy and a slope of 3.43. We observe that: (a) 94 % of the engagements contain fewer than 10 comments, (b) only 0.2 % of engagements contain more than 50 comments, and (c) there are 20 engagements with more than *1000* comments.

Duration and intensity of engagement are not correlated. Intuitively, one would have expected that the duration and the number of comments per engagement should be correlated, but we find that this is not the case. Figure 3(b) depicts the heatmap that captures the relationship between the duration of an engagement and the number of comments posted by a single user in that engagement. We find that only 1.4 % engagements which last more than an year have more than 100 comments while 74 % have only less than 10 comments. Engagements are tend to be short lived, 78.9 % of engagements end in 10 min and only 4.2 % last more than one day.

We leverage our observations for detecting surprising behaviors in Sect. 4.

3.3   Temporal Properties of User Behavior

Users exhibit persistent behavior in their daily and weekly behavior. We observe a strong periodicity in the temporal distribution of the number of comments as well as the number of users that make comments on Disqus. This suggests that the behaviors of a user exhibit reasonably stable temporal patterns. From the D_{Comm} data set, we plot the number of comments and the number of unique users (identified by their IDs) at two different levels: (a) hourly activity over a week, and (b) the heatmap of daily and hourly activity in Fig. 4. We observe a strong periodicity in both plot. Due to the space limitation, we do not show the daily activity over a month which also have a strong periodicity could be imagine as a repetition of Fig. 4(a). Although we only show plots generated from slices of 2012 data, we have observed the same patterns in the data for other years (2009 to 2014) in our D_{Comm} data set.

Interestingly, Fig. 4(a) shows that the commenting activities begin to rise at the start of the week, peak in the middle, and drop sharply afterwards. This suggests that for a typical Disqus user, they post the most content toward the middle of a typical *workweek*. In Fig. 4(b), we show the histogram of all comments in the D_{Comm} data set in terms of hour-of-day (X-axis) and day-of-week (Y-axis) as a heatmap. We see that daily commenting activity is more intense on weekdays, compared to weekends. Interestingly, the drop in weekends is also observed in the posting on **Facebook** [26] and **Twitter** [24].

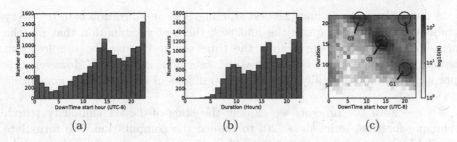

(a)   (b)   (c)

Fig. 5. Distributions of (a) Downtime start hours, (b) Downtime durations, and (c) both start housr and durations expressed as a heatmap (in grayscale to show the distinct "peaks")

Detecting the "downtimes" of users. A key temporal property is the time that the user spends at the platform, namely the uptime and downtime of the user. Knowing the downtime helps us develop user profiles in Sect. 4.

We describe, DownTimeFinder, the algorithm we use to detect a user's downtime period. Defining downtime for a user is a challenge as there is noise and variations among users. We are exploiting two facts from our previous analysis: (a) some users interact heavily with the platform, and (b) most users have periodic and predictable behaviors. Intuitively, we define a person's downtime as the *largest* period of time (a few hours in length at least) during the day when the person's commenting activity is significantly lower compared to the rest of the activities on that day. An example can be seen in Fig. 4(c), which shows the commenting activity of one user expressed as a time series. In this figure, the time interval from hour 0 to hour 13 can potentially be the user's downtime.

DownTimeFinder Algorithm: The DownTimeFinder algorithm takes in a time series for a user as input and outputs a time range when the user is the least active. The algorithm assumes that the downtime of a user is a straight sequence of hours without any gap. With this assumption, activity time series of user can be modeled by a square wave with exactly one *high* and *one* low segment. However, even after this simplifying assumption, the algorithm needs to find the starting time and the duration of the low segment of the square wave.

Algorithm 1. *DownTimeFinder(TS)*

Require: $TS \leftarrow$ an aggregate time series of daily activity
Ensure: Output downtime, a range of ours of inactivity
1: $TS \leftarrow$ append TS with TS
2: **for** $d \leftarrow 1$ to 20 hours **do**
3:    $q \leftarrow$ a square wave of d hours of low segment and 24-d hours of high segment
4:    Search the best fit of q in TS
5:    **if** error of the best fit < global-best **then**
6:       Update the global-best
7: **return** the global best

A naive algorithm to find the best starting time and duration is to try every combination of the two quantities and pick the best combination that fits the corresponding square wave against the time series. We use the simplest and the most effective goodness-of-fit measure: *sum of squared error*. However, this approach is computationally expensive for millions of users that a commenting service hosts.

To speed-up the process, we exploit the state-of-the-art similarity search technique for time series data [20] to reduce the computation. We formulate the problem as a similarity search problem as shown in the Algorithm 1. The algorithm start with self appending the time series. It then loops over possible durations. For each duration, the algorithm the similarity search function [20] as a subroutine to identify the best starting time. The similarity search algorithm exploits the overlap between successive slides of the square wave and finds the best match in only $O(n \log n)$ time where n is the length of the time series. The Algorithm 1 is fast enough to process 26K users in just few minutes. By contrast, the naive approach would take roughly six hours, several orders of magnitude slower.

Observations. Using the DownTimeFinder algorithm, we extract: (a) downtime start hour and (b) downtime duration from each of the time series in D^t_{Comm}. Figures 5(a) and (b) show the distribution for (a) and (b) respectively while Fig. 5(c) represents the joint distribution of both (a) and (b) in the form of a heatmap. Even though the original data is in UTC, we shifted the data to UTC-8 (Pacific Time Zone) for this section.

From Fig. 5(b), we see there are three dominant peaks, so we identify three groups of users, and we attempt to interpret their behavior assuming that they are US-based. In our analysis, not reported here, we actually verify that this is the case for the majority of these users. Clearly for non US-based users, our interpretation will need to be corrected.

(a) One group with roughly 3 h of "up time" (21 h of downtime). These users seem to post their comments during their breaks at work.
(b) One group with roughly 9 h of "up time" (15 h of downtime). These users seem to post their comments throughout their day at work and stop afterwards.

(c) One group with roughly 15 h of "up time" (9 h of downtime). These users are highly active and therefore also highly suspicious, as they seem to spend the majority of their time making comments.

Furthermore, Fig. 5(a) shows that there are two dominant peaks for downtime start hours at hours 14 and 23 (UTC-8), which would translates respectively to 2pm and 11pm on the United States West Coast. When we cross-reference the two peaks with the heatmap in Fig. 5(c) (the dark spots labeled G2 and G1 in the Figure), we find that they are associated with the durations of 15 and 9 h of downtime (or 9 and 15 h of "up time") respectively.

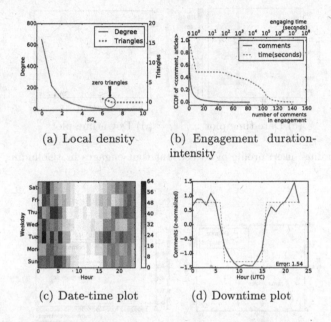

(a) Local density

(b) Engagement duration-intensity

(c) Date-time plot

(d) Downtime plot

Fig. 6. An example of a typical user profile. (ID:1164515)

4   Identifying Parasitic Behaviors

We develop user profiles using the features and observations in the previous section. We focus on features that capture social, engagement, and temporal properties of users as shown in Fig. 6. Our work here is the first step towards techniques that can automatically search for suspicious users, which is in our future plans. Here we provide a useful set of features and the intuition that a detection mechanism could leverage.

Profile of a benign user: features and plots. We selected these four plots to visually capture key properties and we show how we can identify unusual

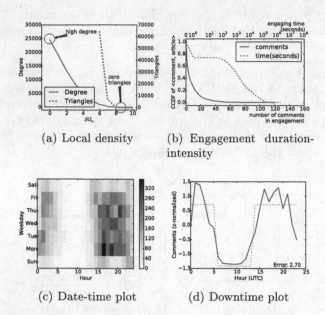

(a) Local density   (b) Engagement duration-intensity

(c) Date-time plot   (d) Downtime plot

Fig. 7. "Colluding" user: profile of an account that engages in "Colluding" activities. (ID:1742861)

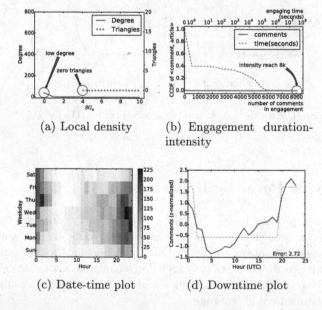

(a) Local density   (b) Engagement duration-intensity

(c) Date-time plot   (d) Downtime plot

Fig. 8. "Chat room" user: profile of an account that engages in "chat room" activities. (ID:1204188)

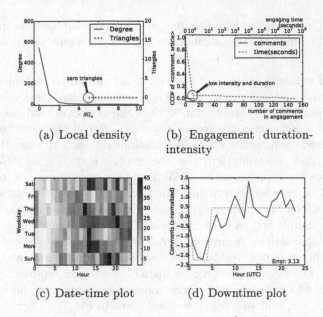

(a) Local density

(b) Engagement duration-intensity

(c) Date-time plot

(d) Downtime plot

Fig. 9. "Hit-n-away" user: profile of a user that engages in trolling and spamming. (ID:1026110)

users, as explained below. Due to space limitations, we were forced to overload the plots with multiple features per plot.

First, we capture the user-user interactions with the **local density plot**, shown in Fig. 6(a). The plot shows the number of collaborators of a user and the number of triangles it forms with collaborators in each SG_n. The higher values of n, the higher the collaboration intensity between a user and its neighbors. A typical user exhibits moderate local density, and SG_7 is the first index for which SG_n has zero triangles, as highlighted in the figure. Here, we study the social density of a node for each of the subgraphs SG_n starting from $n = 10$ and decrease n until SG_n becomes too large to analyze.

Second, the **the engagement duration-intensity plot**, shown in Fig. 6(b), depicts the distribution of the number of comments (x-axis bottom scale and solid line) and the distribution of the duration times (x-axis top scale and dotted line). For a typical user, the engagements tend to involve few comments per engagement, whose distribution follows a smooth exponential distribution, as seen in Fig. 6(b).

Finally, we capture temporal properties using the Date-time plot, shown in Fig. 6(c) and the Downtime plot of the user, shown in Fig. 6(d), as discussed earlier. A typical user has a well-defined downtime shape, and pronounced regularity of commenting per day-of-week and hour-of-day.

a. "Colluding" users: (Profile: high local density). In Fig. 7(a), we plot a colluding user: the user is tightly collaborating with a group with the most likely goal of promoting a particular idea. The tell-tale signs are captured in the social density: high degree of connections, and high local density as indicated by the number of triangles in its neighborhood. We discuss how we select this node below.

Detecting and validating collusion. In Sect. 3, we presented circumstantial evidence that implies that some users in Fig. 1 are organized and work together for a shared purpose. E.g. users joined the Disqus network on the same day, and comment mostly on a particular news site. Recall that in the SG_{10}, we have 8 connected components with collaborations intensity of more than 2^{10}. To investigate further, we study how the size of their cliques from SG_{10} to SG_7, which increases for lower values of n. We find two large cliques of interest: one consists of 34 users whose usernames start with **cnn-**, which was not present in SG_{10}, and second one consists of 29 users whose usernames start **newser-**. We randomly sample 500 pairs of comments from each clique which their users made on the same article. We manually validate that *at least* 86 % of pairs from the **newser** clique, and *at least* 73 % of the pairs in **cnn** clique are supportive of each other. The percentage could be higher, as we only report the pairs that the agreement was beyond doubt. Since most of users in **cnn** clique have very similar profiles, we randomly select one and show in Fig. 7.

"Hit-n-away" users: spamming and trolling. (Profile: low intensity of engagement, less regular temporal patterns). We identify users who post a large number of comments, but each comment is at a *different* article. We discover these users by focusing on engagements which have low intensity and then we examine the top 100 users with the highest number of such engagements. We find 27 users that are engaged in spamming: they keep posting the same comment or URL. We also find 11 users that exhibit behavior consistent with "trolling". Troller attempts to incite others through insults, nonsensical, or outrageous comments. These users do not have any lengthy engagements (Fig. 9(d)) and tend to have less regular day and time patterns, as shown in Fig. 9(c). Interestingly, two of those users posted thousands of (offensive) comments and each one was never more than two or three words long.

"Chat room" users: (Profile: extremely intense and long lived engagement). In Fig. 3(b), we see that there are outliers with engagements that can last for months and containing upwards of 500 comments. This is unusual, since 90 % of all engagements last less than one day and contain fewer than 10 comments. Upon further investigation, we find a total of 53 engagements that exhibit extremely high intensity (at least 500 comments each) and longevity (lasting at least one month). These engagements are conducted by 24 unique user accounts on 19 different websites. It turns out that these 19 websites seem to be using Disqus as their chat room. These websites create an "empty" article with no content and its sole purpose is to facilitate the discussion between a group of people, say an organization. Though not illegal, this is a surprising use of Disqus. For example, the user in Fig. 8 spends almost all their nearly 8,000 comments discussing video games.

The profile of one of the 24 suspicious users is shown in Fig. 8. We see in Fig. 8(a) that their local density is low and the degree has a sharp decrease from SG_0 to SG_1 followed by a straight line to SG_{10}. Besides, their engagement intensity is not typical: it visually looks like a straight line and it extends to 8,000 comments over a period of slightly more than one year.

5   Related Work

Due to space limitations, we highlight indicative studies in five related areas.

Detecting abusive behavior. Many studies detect misbehavor [2,4,22] but they focus on OSNs. E.g. the first work [2] analyzes the comments on three websites which adopt Disqus but the focus is to predict if a user will be banned due to misbehavior, which is different from ours.

Geolocation. Many works focus on inferring the geographic locations of users of OSNs by analyzing their posts [3,11,12], their social/spatial proximity to other users [1,5,14] or their behavioral activity over time [17,18].

Content analysis. Many studies leverage textual analysis to detect/block spammers on OSNs [21] and blogs [19] or how to achieve the same objective by analyzing the users's behavioral patterns [23]. We focus on behavioral properties for discovering unusual user.

Inferring users' state. Several recent studies use mobile phones and OSNs to profile users' psychological states [6,25], behavioral pattern and personality traits [7,15], but no work has used commenting platforms.

Identifying users' activities pattern. Some other works have developed techniques to estimate the downtime and uptime of users [10], and we intend to compare our approach with those in the future.

6   Conclusion

In our effort to detect surprising and anomalous behaviors, our work shows the interesting user behaviors that emerge in commenting platforms, which have receive limited attention. A key novelty of our approach is that we analyze user behavior along three dimensions: (a) The social behavior, (b) that user-article engagement, (c) the temporal properties of users. We ultimately propose a way to profile users relying on features along these dimensions. We have collected information from 1 million users between Nov-2007 to Jan- 2015, and study 109K users, who have more than 10 comments in the platform. Using our profiles, we show how we can detect anomalous users and how the profiles gives us a glimpse of their unusual behavior. In other words, our profiles provide a first step of forensics analysis: we identify profiles for colluding users, spammers and trollers. We are forced to limit the discussion here due to space limitations.

In the future, we intend to: (a) consider more features, (b) develop automated techniques to detect and classify misbehaving users, and (c) mine comment information to extract trends in terms of public opinions and preferences.

References

1. Backstrom, L., Sun, E., Marlow, C.: Find me if you can: improving geographical prediction with social and spatial proximity. In: WWW, pp. 61–70. ACM (2010)
2. Cheng, J., Danescu-Niculescu-Mizil, C., Leskovec, J.: Antisocial behavior in online discussion communities. arXiv preprint arXiv:1504.00680 (2015)
3. Cheng, Z., Caverlee, J., Lee, K.: You are where you tweet: a content-based approach to geo-locating Twitter users. In: Proceedings of the 19th ACM International Conference on Information and Knowledge Management, pp. 759–768. ACM (2010)
4. Chu, Z., Gianvecchio, S., Wang, H., Jajodia, S.: Detecting automation of Twitter accounts: are you a human, bot, or cyborg? IEEE Trans. Dependable Secure Comput. **9**(6), 811–824 (2012)
5. Davis Jr., C.A., Pappa, G.L., de Oliveira, D.R.R., de L Arcanjo, F.: Inferring the location of Twitter messages based on user relationships. Trans. GIS **15**(6), 735–751 (2011)
6. De Choudhury, M., Counts, S., Horvitz, E.J., Hoff, A.: Characterizing and predicting postpartum depression from shared Facebook data. In: Proceedings of the 17th ACM Conference on Computer Supported Cooperative Work & Social Computing, pp. 626–638. ACM (2014)
7. Devineni, P., Koutra, D., Faloutsos, M., Faloutsos, C.: If walls could talk: patterns and anomalies in Facebook wallposts. In: ASONAM, pp. 367–374. ACM (2015)
8. Disqus: What's Cooler Than a Billion Monthly Uniques? May 2013. http://blog.disqus.com/post/50374065365/whats-cooler-than-a-billion-monthly-uniques
9. Disqus: Disqus: Blog-comment hosting service (2015). https://disqus.com
10. Ferraz Costa, A., Yamaguchi, Y., Juci Machado Traina, A., Traina Jr., C., Falout-sos, C.: Rsc: mining and modeling temporal activity in social media. In: Proceedings of the 21th ACM SIGKDD International Conference on Knowledge Discoveryand Data Mining, pp. 269–278. ACM (2015)
11. Gelernter, J., Balaji, S.: An algorithm for local geoparsing of microtext. GeoInformatica **17**(4), 635–667 (2013)
12. Gelernter, J., Zhang, W.: Cross-lingual geo-parsing for non-structured data. In: Proceedings of the 7th Workshop on Geographic Information Retrieval, pp. 64–71. ACM (2013)
13. Intense Debate: Intense Debate: Imagine better comments. http://intensedebate.com
14. Jurgens, D.: That's what friends are for: inferring location in online social media platforms based on social relationships. ICWSM **13**, 273–282 (2013)
15. Kosinski, M., Stillwell, D., Graepel, T.: Private traits and attributes are predictable from digital records of human behavior. Proc. Natl. Acad. Sci. **110**(15), 5802–5805 (2013)
16. LiveFyre: LiveFyre: Real-time Content Marketing and Engagement. http://web.livefyre.com
17. Mahmud, J., Nichols, J., Drews, C.: Where is this tweet from? Inferring home locations of Twitter users. ICWSM **12**, 511–514 (2012)
18. Mahmud, J., Nichols, J., Drews, C.: Home location identification of Twitter users. ACM Trans. Intell. Syst. Technol. (TIST) **5**(3), 47 (2014)
19. Mishne, G., Carmel, D., Lempel, R.: Blocking blog spam with language model disagreement. AIRWeb. **5**, 1–6 (2005)
20. Mueen, A., Viswanathan, K., Gupta, C., Keogh, E.: The fastest similarity search algorithm for time series subsequences under euclidean distance, August 2015. http://www.cs.unm.edu/mueen/FastestSimilaritySearch.html

21. Sculley, D., Wachman, G.M.: Relaxed online svms for spam filtering. In: Proceedings of the 30th Annual International ACM SIGIR Conference on Research and Development in Information Retrieval, pp. 415–422. ACM (2007)
22. Sood, S.O., Churchill, E.F., Antin, J.: Automatic identification of personal insults on social news sites. J. Am. Soc. Inf. Sci. Technol. **63**(2), 270–285 (2012)
23. Sureka, A.: Mining user comment activity for detecting forum spammers in youtube. arXiv preprint arXiv:1103.5044 (2011)
24. Sysomos: Inside Twitter: An in-depth look inside the twitter world, April 2014. http://sysomos.com/sites/default/files/Inside-Twitter-BySysomos.pdf
25. Wang, R., Chen, F., Chen, Z., Li, T., Harari, G., Tignor, S., Zhou, X., Ben-Zeev, D., Campbell, A.T.: Studentlife: assessing mental health, academic performance and behavioral trends of college students using smartphones. In: UbiComp, pp. 3–14. ACM (2014)
26. Warren, C.: When are Facebook users most active? [study] (2010). http://mashable.com/2010/10/28/facebook-activity-study/

On Profiling Bots in Social Media

Richard J. Oentaryo[✉], Arinto Murdopo, Philips K. Prasetyo,
and Ee-Peng Lim[✉]

Living Analytics Research Centre, Singapore Management University,
Singapore, Singapore
{roentaryo,arintom,pprasetyo,eplim}@smu.edu.sg

Abstract. The popularity of social media platforms such as Twitter has led to the proliferation of automated bots, creating both opportunities and challenges in information dissemination, user engagements, and quality of services. Past works on profiling bots had been focused largely on malicious bots, with the assumption that these bots should be removed. In this work, however, we find many bots that are benign, and propose a new, broader categorization of bots based on their behaviors. This includes *broadcast*, *consumption*, and *spam* bots. To facilitate comprehensive analyses of bots and how they compare to human accounts, we develop a systematic profiling framework that includes a rich set of features and classifier bank. We conduct extensive experiments to evaluate the performances of different classifiers under varying time windows, identify the key features of bots, and infer about bots in a larger Twitter population. Our analysis encompasses more than 159K bot and human (non-bot) accounts in Twitter. The results provide interesting insights on the behavioral traits of both benign and malicious bots.

Keywords: Bot profiling · Classification · Feature extraction · Social media

1 Introduction

In recent years, we have seen a dramatic growth of people's activities taking place in social media. Twitter, for example, has evolved from a personal microblogging site to a news and information dissemination platform. The openness of the Twitter platform, however, has made it easy for a user to set up an automated social program called *bot*, to post tweets on his/her behalf.

The proliferation of bots has both good and bad consequences [4,8]. On the one hand, bots can generate benign, informative tweets (e.g., news and blog updates), which enhance information dissemination. Bots can also be helpful for the account owners, e.g., bots that aggregate contents from various sources based on the owners' interests. On the other hand, spammers may exploit bots to attract regular accounts as their followers, enabling them to hijack search engine results or trending topics, disseminate unsolicited messages, and entice users to visit malicious sites [8,10,11]. In addition to deteriorating user experience and

© Springer International Publishing AG 2016
E. Spiro and Y.-Y. Ahn (Eds.): SocInfo 2016, Part I, LNCS 10046, pp. 92–109, 2016.
DOI: 10.1007/978-3-319-47880-7_6

trust, malicious bots may cause more severe impacts, e.g., creating panic during emergencies, biasing political views, or damaging corporate reputation [8,21].

It is thus important to characterize different types of bots and understand how they compare with human users. Recent studies have shown the importance of profiling bots in social media [1,2,4,8,10–13,17,18,20,21], but these works have focused mainly on malicious (e.g., spam) bots, failing to account for other types of benign bots. With the rise of new services and intelligent apps in Twitter, benign bots are increasingly becoming prominent as well.

Comprehensive profiling of both malicious and benign bots would offer several major benefits. In information dissemination and retrieval, knowing the activity traits of both bot types and the nature of their tweet contents can improve search and recommendation services by separating tweets of bots from those of humans, returning more relevant, personalized search results, and promoting certain products/services more effectively. For social science research, a more accurate understanding of human interactions and information diffusion patterns [8,9] can also be obtained by filtering out activity biases generated by bots. In turn, these would benefit the overall user community as well.

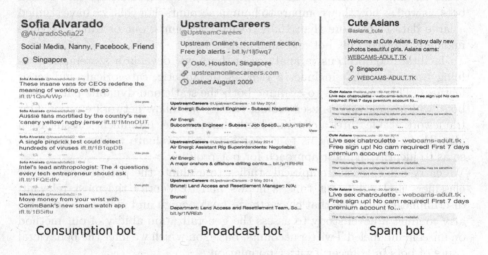

Fig. 1. Examples of broadcast, consumption and spam bots in Twitter

To illustrate the usefulness of profiling bots, consider the examples in Fig. 1, of different types of benign and malicious bots (which we further describe in Sect. 3). The first example is a user who utilizes the IFTTT service[1] to gather contents from diverse sources for her own consumption. Knowing that she uses a consumption bot, Twitter can provide a new service to organize the unstructured contents, or recommend new contents that match her interest. The second example involves a broadcast bot managed by a job agency to advertise job openings.

---

[1] https://ifttt.com.

Twitter recently introduced a new feature called *promoted tweets*[2] and, knowing it is a (benign) broadcast bot, Twitter can recommend the feature to help the agency reach a wider audience. The last example shows a malicious, spam bot that lures users to visit adult websites, posssibly containing harmful malware. For such a bot, Twitter may develop a strategy to demote—or even filter out—its posts, so that the followers do not see them on their tweet streams.

Contributions. In this paper, we present a new categorization of bots based on long-term observations on the behaviors of various automated accounts in Twitter. To our best knowledge, this work is the first extensive study on both *benign* and *malicious* Twitter bots, with detailed analyses on both their static and dynamic patterns of activity. In recent years, Twitter bots have evolved rapidly, and so our work also provides a more timely study that offers updated insights on the bot characteristics. Our findings should also benefit social science and network mining researches. We summarize our key contributions below:

- We propose a new categorization of Twitter bots based on their behavioral traits. In contrast to past studies that focus largely on malicious bots, our study encompasses more detailed examinations of both malicious and benign bots, as well as how they compare to human accounts. For this, we have studied a large dataset of more than 159K Twitter accounts, out of which we have manually labeled 1.6K bot and human accounts.
- To facilitate comprehensive analyses on bots, we develop a systematic profiling framework that includes a rich set of numeric, categorical, and series features. This enables us to examine both the static and dynamic patterns of bots, which span various user profile, tweet, and follow network entities. Our framework also features a classifier bank that includes prominent classification algorithms, thus allowing us to comprehensively evaluate various algorithms so as to identify the best approach for bot profiling.
- We carry out extensive empirical studies to evaluate the performance of our classifiers under different time windows and to identify the most relevant, discriminating features that characterize both benign and malicious bots. We also conduct a novel study to assess the generalization ability of our method on unseen, unlabeled Twitter accounts, based on which we infer the behavioral traits of bots in a larger Twitter population.

2   Background and Related Work

A number of studies have been conducted to identify and profile bots in social media. To detect spam bots, Wang [21] utilized content- and graph-based features, derived from the tweet posts and follow network connectivity respectively. Chu *et al.* [4] investigated whether a Twitter account is a human, bot, or cyborg. Here a bot was defined as an aggresive or spammy automated account, while cyborg refers to a bot-assisted human or human-assisted bot. Different from our

---

[2] https://business.twitter.com/solutions/promoted-tweets.

work, the bots defined in [4] are more of malicious nature, and the study did not provide further categorization/analysis of benign and malicious bots in Twitter.

To investigate on spam bots, Stringhini *et al.* [17] created honey profiles on Facebook, Twitter and MySpace. By analyzing the collected data, they identified anomalous accounts who contacted the honey profiles and devised features for detecting spam bots. Going further, Lee *et al.* [13] conducted a 7-month study on Twitter by creating 60 social honeypots that try to lure "content polluters" (a.k.a. spam bots). Users who follow or message two or more honeypot accounts are automatically assumed to be content polluters. There are also related works on spam bot detection based on social proximity [10] or both social and content proximities [11]. Tavares and Faisal [19] distinguished between personal, managed, and bot accounts in Twitter, according to their tweet time intervals.

Ferrara *et al.* [8] built a web application to test if a Twitter account behaves like a bot or human. They used the list of bots and human accounts identified by [13], and collected their tweets and follow network information. This study, however, covers only malicious bots. Dickerson *et al.* [5] used network, linguistic, and application-oriented features to distinguish between bots and humans in the 2014 Indian election. Abokhodair *et al.* [1] studied on a network of bots that collectively tweet about the 2012 Syrian civil war. This study covers both malicious (e.g., phishing) and benign (e.g., testimonial) bots. In contrast to our work, however, their findings are tailored to a specific event (i.e., the civil war) and may not be applicable to other bot types in a larger Twitter population.

There are also studies aiming to quantify the susceptibility of social media users to the influence of bots [2,12,20]. By embedding their bots into the Facebook network, Boshmaf *et al.* [2] demonstrated that users are vulnerable to phishing (e.g., exposing their phone number or address). The susceptibility of users is also evident in Twitter [12,20]. Freitas *et al.* [9] tried to reverse-engineer the infiltration strategies of malicious Twitter bots in order to understand their functioning. Most recently, Subrahmanian *et al.* [18] reported the winning solutions of the DARPA Twitter Bot Detection Challenge. Again, however, all these studies deal mainly with malicious bots and ignore benign bots.

3   New Categorization of Bots

We define a bot as a Twitter account that generates contents and interacts with other users automatically—at least according to human judgment. Our definition thus includes *both* benign and malicious bots. Based on long-term observations on Twitter data, we propose to categorize Twitter bots into three main types:

- **Broadcast bot**. This bot aims at disseminating information to general audience by providing, e.g., benign links to news, blogs or sites. Such bot is often managed by an organization or a group of people (e.g., bloggers).
- **Consumption bot**. The main purpose of this bot is to aggregate contents from various sources and/or provide update services (e.g., horoscope reading, weather update) for personal consumption or use.

Fig. 2. Bot and human accounts in Twitter

– **Spam bot**. This type of bots posts malicious contents (e.g., to trick people by hijacking certain account or redirecting them to malicious sites), or promotes harmless but invalid/irrelevant contents aggressively.

Figure 2 illustrates the three bot types, where the arrow direction represents the flow of information. It is worth noting that our proposed categorization is more general than the taxonomy put forward in [15], which covers mainly malicious bots. Our categorization is also general enough to cater for new, emerging types of bot (e.g., chatbots can be viewed as a special type of broadcast bots).

4   Dataset

Data collection. Our study involves a Twitter dataset generated by users in Singapore and collected from 1 January to 30 April 2014 via the Twitter REST and streaming APIs[3]. Starting from popular seed users (i.e., users having many followers), we crawled their follow, retweet, and user mention links. We then added those followers/followees, retweet sources, and mentioned users who state Singapore in their profile location. With this, we have a total of 159,724 accounts.

Table 1. Distribution of our Twitter dataset

Labeled data				Unlabeled data
Consumption bot	Broadcast bot	Spam bot	Human account	
313	171	105	1,024	158,111

Total no. of labeled data = 1,613; Total no. of data = 159,724

To identify bots, we first checked active accounts who tweeted at least 15 times within the month of April 2014. We then manually labeled these accounts and found 589 bots. As many more human users are expected in the Twitter population, we randomly sampled the remaining accounts, manually checked them, and identified 1,024 human accounts. In total, we have 1,613 labeled accounts,

---

[3] https://dev.twitter.com/overview/.

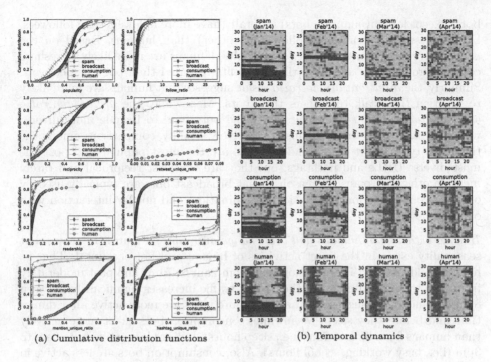

(a) Cumulative distribution functions   (b) Temporal dynamics

Fig. 3. Statistics of humans and bots in our labeled Twitter data

as summarized in Table 1. The labeling was done by four volunteers, who were carefully instructed on the definitions in Sect. 3. The volunteers agree on more than 90 % of the labels, and any labeling differences in the remaining accounts are resolved by consensus. Also, if an account exhibits both human and bot characteristics, we determine the label based on the majority posting patterns.

Exploratory analysis. We conducted a preliminary study on our 1,613 labeled data to get a glimpse of the activity patterns of bots as well as human accounts. Figure 3(a) shows the cumulative distribution functions (CDF) of several key attributes. An early increase in CDF value means a more skewed distribution. We focus on key attributes that reflect a user's social and posting patterns: $popularity = \frac{|F|}{|E|+|F|}$, $follow\_ratio = \frac{|E|}{|F|}$, $reciprocity = \frac{|E \cap F|}{|E \cup F|}$, $retweet\_unique\_ratio = \frac{|R|}{|T|}$, $url\_unique\_ratio = \frac{|U|}{|T|}$, $mention\_unique\_ratio = \frac{|M|}{|T|}$, $hashtag\_unique\_ratio = \frac{|H|}{|T|}$, where E, F, R, T, U, M, H are the set of followees, followers, retweets, tweets, URLs, user mentions, and hashtags for a given account, respectively. We also define $readership = \frac{retweeted}{|T|}$, where $retweeted$ is the number of times a user's tweets get retweeted (by others). Figure 3(b) shows heatmaps of tweet counts $|T|$ for different days and hours over 4 months.

How do humans compare with bots and how do bots differ from one another? The *popularity*, *follow_ratio*, and *reciprocity* results in Fig. 3(a) suggest that bots (except for consumption bots) generally have more followers than followees, but are less reciprocal (i.e., follow each other) than humans. Based on the *retweet_unique_ratio* and *readership* results, humans are more likely to reshare contents from others and have their contents reshared than bots, respectively. Similarly, the *mention_unique_ratio* result suggests that humans are more likely to mention (i.e., talk to) others than bots. Meanwhile, the *url_unique_ratio* and *hashtag_unique_ratio* results show the bots tend to include more diverse web links and topics than humans, respectively. Finally, comparisons among the three bot types show that broadcast bots are the most popular and post the most diverse URLs and hashtags, but they are the least reciprocal and rarely mention others. A plausible reason is that broadcast bots are typically used by organizations solely for information dissemination, and not for interaction with others.

How do activities of humans and bots change over time? Figure 3(b) shows that seasonality exists in the tweet activities of human and bot accounts[4]. That is, humans seldom tweet in early morning (from 2am to 7am) and post moderately from 7am to 8pm. Afterwards, their tweet traffic increases significantly between 8pm and midnight, suggesting that Singapore users are more active after dinner time and before they sleep. Meanwhile, consumption bots tweet more actively than humans from 3am to 7am (i.e., sleep hours), but are less active from 9am to 3pm (i.e., busy working/school hours). Also, consumption bots are less active in the weekends than in the weekdays. While broadcast bots have generally similar patterns to consumption bots, the former is less active during sleep hours (3am–7am) whereas the latter during busy hours (9am–3pm). We can attribute this to the intuition that broadcast bots aim to reach a wider audience during their non-sleep hours. Lastly, unlike broadcast and consumption bots, spam bots are active all days/hours, and they exhibit very random timings. In summary, different bots serve different purposes and their temporal signatures reflect these.

5   Profiling Framework

We develop a systematic profiling framework to facilitate comprehensive analyses of bots. Below we describe each component of the framework in turn.

Database. Our framework takes as input three types of database: *profile*, *tweet*, and *follow* databases. The profile database contains user information such as the Twitter user id, screenname, location, and profile description. The tweet database contains all the tweets posted by different users, which may include various entities such as hashtags, URLs, user mentions, videos/images, retweet information, and tweet sources/devices. We collectively refer to these as *tweet entities*. Finally, the follow database contains the snapshots of users' relationship network over time, which include both followers and followees of the users at different time periods. We collectively call these *follow entities*.

---

[4] The exceptionally low tweet frequencies in the first week of January and 12-14 February are due to major downtime of our servers.

Feature extraction. This component serves to construct a *feature vector* that represents a Twitter account. It takes three types of feature: *numeric*, *categorical*, and *series*. We describe the extraction steps for each type below:

- For **numeric features**, we perform *standarization* by scaling each feature to a unit range $[0, 1]$. This would allow us to mitigate feature scaling issues, particularly for classification methods that rely on some distance metric. Examples of numeric features are count and ratio attributes (see Table 2).
- For **categorical features**, we first select the top K categories based on their frequencies in each data point, and then filter out the remaining categories. Next, we perform *one-hot encoding* by transforming the top K categories into a binary vector with K elements. For example, a categorical attribute with four possible values: "A", "B", "C", and "D" is encoded as $[1, 0, 0, 0]$, $[0, 1, 0, 0]$, $[0, 0, 1, 0]$, and $[0, 0, 0, 1]$, respectively.
- For **series features**, we first count the frequency of every (discrete) number in the series. For instance, given a series $[a, a, b, a, c, b, c, a, b]$, we can compute the histogram bins: $(a, 4), (b, 3), (c, 2)$. To ensure a moderate feature size, we keep only top 100 bins with the highest count frequencies. Subsequently, we normalize the frequencies such that they sum to 1, thus forming a probability distribution. For the previous histogram bins $(a, 4)$, $(b, 3)$, $(c, 2)$, the normalization will result in $(a, \frac{4}{9})$, $(b, \frac{3}{9})$, $(c, \frac{2}{9})$.

Classifier bank. Finaly, to learn the association between the extracted features and different bot types (or human), our framework includes a classifier bank that comprises a rich collection of classification algorithms. In our study, we employ four prominent classifiers: *naïve Bayes* (NB) [6], *random forest* (RF) [3], and two instances of generalized linear model, i.e., *support vector machine* (SVM) and *logistic regression* (LR) [7]. These algorithms represent the state-of-the-art methods previously used for (malicious) bot classification. For instance, RF was utilized in [4,5,8,13], while SVM and NB were used in [5,21].

6 Feature Engineering

We have crafted a rich set of features based on the feature extraction component in our bot profiling framework. Our feature set consists of three groups: *tweet*, *follow* and *profile* features. For tweet features, we also distinguish between *static* (i.e., time-independent) and *dynamic* (i.e., time-dependent) tweet features. Table 2 provides a listing of all the features used in our empirical study.

Static tweet features. We generate static tweet features based on the combination of entities and statistical metrics, as shown in Table 2. For instance, to generate the hashtag features of a user, we treat each hashtag as a "bag" and count how many times the word occurs in all of x's tweets. This yields a bag-of-hashtag vector, from which we can compute first-order statistics (i.e., *count*, *unique_count*, *mean*, *median*, *min*, and *max*) as well as second-order metrics (i.e., standard deviation (*std*) and Shannon entropy [16] (*entropy*)). We note

Table 2. List of features used in our bot classification task

Group	Entity	Features
Static tweet features	Tweet_word	Count (N), unique_count (N), unique_ratio (N), basic_stats (N)
	Retweet	Retweeted (N), readership (N), count (N), unique_count (N), ratio (N), unique_ratio (N), basic_stats (N)
	Hashtag	Count (N), unique_count (N), ratio (N), unique_ratio (N), basic_stats (N)
	Mention	Count (N), unique_count (N), ratio (N), unique_ratio (N), basic_stats (N)
	Url	Count (N), unique_count (N), ratio (N), unique_ratio (N), basic_stats (N)
	Media	Count (N), unique_count (N), ratio (N), unique_ratio (N), basic_stats (N)
	Source	Sources (S)
Dynamic tweet features	Tweet	Hours (S), days (S), weekdays (S), timeofdays (S), extended_stats (N)
	Retweet	Hours (S), days (S), weekdays (S), timeofdays (S), extended_stats (N)
	Hashtag	Hours (S), days (S), weekdays (S), timeofdays (S), extended_stats (N)
	Mention	Hours (S), days (S), weekdays (S), timeofdays (S), extended_stats (N)
	Url	Hours (S), days (S), weekdays (S), timeofdays (S), extended_stats (N)
	Media	Hours (S), days (S), weekdays (S), timeofdays (S), extended_stats (N)
Follow features	Followees_count	basic_stats (N)
	Followers_count	Basic_stats (N)
	Mutual_count	Basic_stats (N)
	Reciprocity	Basic_stats (N)
	In_reciprocity	Basic_stats (N)
	Out_reciprocity	Basic_stats (N)
	Popularity	Basic_stats (N)
	Follow_ratio	Basic_stats (N)
Profile features	Profile	Is_geo_enabled (C), lang (C), time_zone (C), account_age (N), favourites_count (N), listed_count (N), statuses_count (N), utc_offset (N)

Basic_stats: set of statistical metrics {mean, median, min, max, std, entropy}
Extended_stats: Cartesian product of {timegap, hour, day, weekday, timeofday} and basic_stats
N: numeric feature, C: categorical feature, S: series feature

that the second-order metrics serve to quantify the *diversity* of the entities. We also compute the $ratio = \frac{count}{|T|}$ and $unique\_ratio = \frac{unique\_count}{|T|}$, where $|T|$ is the total number of tweets posted by a user. For the retweet entity, we additionally consider *retweeted* and *readership* features, as described in Sect. 4. Finally, we consider a series feature to represent the source entity, whereby each source maps to a histogram bin containing the normalized frequency of the source.

Dynamic tweet features. For these features (cf. Table 2), we introduce additional time dimensions that capture the dynamics of tweet activities, namely: $hours \in \{0, \ldots, 23\}$, $days \in \{1, \ldots, 31\}$, $weekdays \in \{Monday, \ldots, Sunday\}$, $timeofdays \in \{morning$ (4am–12pm), $afternoon$ (12pm–5pm), $evening$ (5pm–8pm), $night$ (8pm–4am)$\}$, and *timegaps*. The timegap dimension refers to the gap (in milliseconds) between two *consecutive* entity timestamps, e.g., for N tweets posted by a user x, we can compute a timegap vector with length $(N - 1)$. For each time dimension, we can then generate the series features based on the histogram binning described in Sect. 5, as well as compute the statistical metrics such as *mean, median, min, max, std* and *entropy*.

Follow features. These features are derived by computing metrics that summarize snapshots of the follow network at different time points (cf. Table 2). Let E and F be the set of followees and followers of a given user. In turn, we compute the $followees\_count = |E|$, $followers\_count = |F|$, $mutual\_count = |E \cap F|$. as well as ratio metrics such as $reciprocity = \frac{|E \cap F|}{|E \cup F|}$, $in\_reciprocity = \frac{|E \cap F|}{|F|}$, $out\_reciprocity = \frac{|E \cap F|}{|E|}$, $popularity = \frac{|F|}{|E| + |F|}$, and $follow\_ratio = \frac{|E|}{|F|}$. We calculate these metrics for every snapshot of the follow network at a given time point, and then compute the statistics *mean, median, min, max, std* and *entropy* to summarize the metrics over all time points.

Profile features. Finally, we also consider several basic user profile features, as per Table 2. Here, *account_age* refers to the lapse between the time a user first joined Twitter and the current reference time. Further details on the definitions of the other profile features can be found in https://dev.twitter.com/.

7   Results and Findings

This section elaborates our empirical study on bots. We first describe our experiment setup, and then address several research questions in Sects. 7.1–7.3.

Evaluation metrics. To evaluate our classifiers, we utilize three metrics popularly used in information retrieval [14]: *Precision, Recall* and $F1$. We report, for each class $c \in \{broadcast, consumption, spam, human\}$, the $Precision(c) = \frac{TP(c)}{TP(c)+FP(c)}$, $Recall = \frac{TP(c)}{TP(c)+FN(c)}$, and $F1(c) = \frac{2Precision(c)Recall(c)}{Precision(c)+Recall(c)}$, where $TP(c)$, $FP(c)$ and $FN(c)$ are the true positives, false positives, and false negatives respectively. Based on these, we also report the macro-averaged $Precision = \frac{1}{4}\sum_{c=1}^{4} Precision(c)$, $Recall = \frac{1}{4}\sum_{c=1}^{4} Recall(c)$, and $F1 = \frac{1}{4}\sum_{c=1}^{4} F1(c)$.

Experiment protocols. In this work, we consider two sets of experiment:

- **Experiment E_1:** This set of experiment involves evaluation on our **1,613** *labeled data* (see Table 1). For this evaluation, we use a *stratified* 10-fold cross-validation (CV), whereby we split the labeled data into 10 mutually exclusive groups, each retaining the class proportion as per the original data. This stratification serves to ensure that each fold is a good representative of the whole, i.e., it retains the (unbalanced) class distribution as in the original data. For each CV iteration f, we then use group f (10 %) for testing and the remaining groups $f' \neq f$ (90 %) for training. We report the results averaged over 10 iterations, which include $Precision(c)$, $Recall(c)$ and $F1(c)$ for each class c, as well as the macro-averaged $Precision, Recall$ and $F1$.

- **Experiment** E_2: This set of experiment serves to evaluate predictions on the remaining **158,111** *unlabeled data* (see again Table 1). Based on this, we can infer the behavioral traits of bots in a larger Twitter population. For this experiment, we are unable to compute *Recall*, as we would have to manually verify one by one a large number of unlabeled data. Instead, we evaluate based on *Precision* at top K for each class ($K \ll 158, 111$).

Model parameters. We configured our classifier bank as follows: For the NB classifier, we use the smoothing parameter $\alpha = 1$. For RF, we use $N = 100$ decision trees. Finally, for SVM and LR, we set the cost parameter $C = 1$ and `class_weight` = "balanced"; the latter is for automatically handling the imbalanced class distribution. We performed grid search to determine all these parameters, which give the optimal performances for each classifier. In particular, we varied the NB parameter from the range $\alpha \in \{0.1, 1, 10\}$. For RF, we tried $N \in \{10, 20, \ldots, 100\}$, and for SVM and LR, we tried $C \in \{0.01, 0.1, 1, 10, 100\}$.

Significance test. Finally, we use *Wilcoxon signed-rank test* [22] to test for the statistical significance of our results. When comparing between two performance vectors, we look at the p-value at a significance level of 0.01. If the p-value is less than 0.01, we say that the performance difference is indeed significant.

7.1   How Well Can the Classifiers Predict for Bots?

To answer this research question, we first conduct a sensitivity study by varying the time duration for which features (cf. Table 2) are generated. For this study, we use the CV procedure on our labeled data (i.e., Experiment E_1), whereby the classifiers were trained using all features listed in Table 2. Figure 4 shows the macro-averaged *Precision*, *Recall*, and *F*1 over 10 CV folds, with the duration varied from 1 week, 2 weeks and 1 month to 2 months and 4 months (up to 30 April 2014). Based on the *F*1 results, we can conclude that 2 weeks is the best duration and that LR outperforms the other classifiers. In this case, RF gives higher *Precision* than LR, but its *Recall* is much lower, and so is its *F*1. It is also shown that a tradeoff exists in choosing the duration; an overly short duration degrades the performance, which can be attributed to data scarcity. The same goes for an overly long duration, due to inclusion of outdated data.

Table 3 shows further breakdown of the CV results for the best time duration (i.e., 2 weeks). Overall, LR and SVM give the best results, and outperform the more complex RF and simpler NB methods (except for *Precision* of the "spam" class). For spam bots, RF yields higher *Precision*, but much lower *Recall* and *F*1 than LR and SVM. While SVM and LR perform very similarly, we decided to use LR as our main classifier for two reasons: (i) LR outputs more meaningful probabilitic scores than the unbounded decision scores in SVM; and (ii) LR is more robust than SVM against variation in time duration, as we saw in Fig. 4.

Based on the individual *Precision*(c), *Recall*(c) and *F*1(c) of each class c, we can conclude that, among the bots, consumption bots are the easiest to detect, followed by broadcast and spam bots. This is expected, owing to the imbalanced

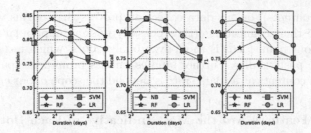

Fig. 4. Classification results for varying durations

Table 3. Breakdown of 10-fold cross-validation results using 2-week training data

Metric	Method	Class label				Macro average
		Broadcast	Consumption	Spam	Human	
Precision	NB	0.6519 (−)	0.7206 (−)	0.7069 (+)	0.9929	0.7681 (−)
	RF	0.5880 (−)	**0.9462**	**0.8636** (+)	0.9750 (−)	**0.8432** (+)
	SVM	**0.6952**	0.9278	0.6574 (−)	**0.9961**	0.8191
	LR	0.6798	0.9366	0.6869	0.9942	0.8244
Recall	NB	0.6901 (−)	**0.8818** (+)	0.3905 (−)	0.9609(−)	0.7308 (−)
	RF	**0.8596** (+)	0.8435	0.3619 (−)	0.9902	0.7638 (−)
	SVM	0.7602(−)	0.8626	**0.6762** (+)	**0.9990**	0.8245
	LR	0.8070	0.8498	0.6476	0.9971	**0.8254**
F1-score	NB	0.6705 (−)	0.7931 (−)	0.5031 (−)	0.9767 (−)	0.7358 (−)
	RF	0.6983 (−)	0.8919	0.5101 (−)	0.9826 (−)	0.7707 (−)
	SVM	0.7263	**0.8940**	**0.6667**	**0.9976**	0.8211
	LR	**0.7380**	0.8911	**0.6667**	0.9956	**0.8228**

NB: näive Bayes, SVM: support vector machine, LR: logistic regression, RF: random forest
(−): significantly worse than LR at 0.01, (+): significantly better than LR at 0.01

class distribution as per Table 1. We can also compare the results of our classifiers with that of a random guess[5]. Based on the statistics in Table 1, the expected $F1$ scores of a random guess for broadcast bot, consumption bot, spam bot, and human classes are 10.6 %, 19.40 %, 6.51 % and 63.49 %, respectively. Our four classifiers thus outperform the random guess baseline by a large margin.

For spam bots, several studies [4,8,13] have reported high classification accuracies, while our results are modest by comparison, largely due to the lack of spam bot accounts in our data. However, it must be noted that these works focused largely on distinguishing between (malicious) bots vs. other accounts, whereas our study deals with a much more challenging and fine-grained categorization of broadcast, consumption and spam bots. Also, the lack of spam bots in our data can be attributed to several factors, such as our relatively strict definition of spam bot (whereby the majority of its postings need to have malicious

---

[5] Random guess w.r.t. a class c refers to a classifier that assigns a proportion p_c% of the instances to class c, and $(1-p_c)$% to classes other than c. In this case, $Precision(c) = Recall(c) = F1(c) = p_c$, where $p_c = \frac{P(c)}{P(c)+N(c)} = \frac{TP(c)+FN(c)}{TP(c)+FN(c)+TN(c)+FP(c)}$.

or irrelevant contents), or our data collection process that begins with popular seed users and their connections (thus possibly missing unpopular spam bots). Nevertheless, our main focus is to analyze benign bots, which has been largely ignored in the past studies. Further studies on less prominent spam bots that post malicious contents at a sparse rate is beyond the scope of our current study.

7.2   Which Features Are the Most Indicative of Each Bot Type?

In light of this research question, we trained our best classifier (i.e., LR) using all 1,613 labeled data, and look at the weight coefficients $w_{i,c}$ of each class in the trained LR. Here we use the raw weights $w_{i,c}$ instead of the absolute values $|w_{i,c}|$ or squared values $w_{i,c}^2$, as the raw weights allow us to distingush between features that correlate positively with a class label (which are our main interest) and those that correlate negatively. Figure 5 shows the top 15 positively-correlated features for each class. In general, we find that the top features are dominated by the *source* (i.e., where the tweets come from) and *entropy-based dynamic tweet* features. Below we elaborate our feature analysis for each class further.

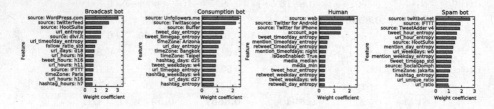

Fig. 5. Top discriminative features for each label in bot classification task

Broadcast bots. Among the top features for broadcast bots, certain sources that are popularly used for blogging (such as WordPress and Twitterfeed) or brand management (such as HootSuite) are found to be highly indicative. It is also shown that the entropy-based features for the url entity correlate strongly with broadcast bots. Recall from Sect. 6 that entropy is a second-order metric that quantifies how diverse a distribution is. Accordingly, as broadcast bots generally aim to disseminate information about certain sites/brands, we can expect that they would have more concentrated url distribution (i.e., low entropy). We will further verify this in Sect. 7.3. Figure 5 also suggests that certain critical timings of the url postings are highly indicative of broadcast bots.

Consumption bots. From Fig. 5, we firstly find that the top three sources for consumption bots (i.e., Unfollowers, Twittascope, and Buffer) are service apps that allow users to track their followers/followees status, horoscope readings, and scheduled postings, respectively. Secondly, we discover that the diversity (entropy) of tweet postings is a strong indicator for consumption bots. Lastly, Fig. 5 shows that certain timezones and timings (weekday and day) of the hashtag and url activities constitute yet another important set of indicators. All these led us to conclude that consumption bots post tweets in a way that follows certain timings/schedules. We will further analyze this in Sect. 7.3.

Spam bots. The result in Fig. 5 suggests that there are certain sources that can be exploited by spammers to post irrelevant or unsolicited tweets. For example, TwittBot is an application that allows multiple users (and thus spammers) to post to a single Twitter account. In addition, the timing diversities of the url, mention, tweet and hashtag activities are found to be the key signatures of spam bots. As also shown in Fig. 3(b) (of Sect. 4), the temporal patterns of spam bots are highly irregular. Altogether, these suggest that spam bots have highly diverse timings (i.e., high entropy), which we will again verify in Sect. 7.3.

Humans. The top three features in Fig. 5 suggest that human accounts typically use credible sources such as "web" (i.e., Twitter website) and the official Twitter mobile apps. Next, the *account_age* and *isGeoEnabled* features suggest that human accounts have lived relatively long in Twitter and usually have his/her tweets' location enabled, respectively. Also, high timing diversity (entropy) of the tweet, retweet and mention activities are indicative of human accounts, although it is not as high as that of spam bots. Again, Sect. 7.3 analyzes this further. Lastly, the *media_median* and *media_mean* features suggest that human accounts like to attach media files (e.g., photos) in their tweets.

7.3   What Can We Tell About Bots in a Larger Twitter Population?

To address this question, we performed Experiment E_2 by deploying our trained LR classifier to predict for the unlabeled 158, 111 accounts. We then picked the top K accounts with the highest probability scores for each class, and manually assessed the class assignments of these accounts. The assessment results can be found in Appendix A (Table 4). We found that the prediction results generally match well with our manual judgments. Based on this, we can make inference on the behavior of bots in a larger Twitter population, i.e., the entire population of Singapore Twitter users. We focus our analyses on the entropy-based dynamic tweet features, which dominate the top features as shown in Fig. 5. That is, we analyze the entropy distributions of the tweet, retweet, mention, hashtag and url activities. The complete distributions can be found in Appendix A (Fig. 6), which reveals several interesting insights as elaborated below.

Tweet patterns. We first compared the distributions of the tweet timings, and discovered that consumption and spam bots exhibit higher diversity (entropy) than that of humans. In contrast, broadcast bots were found to have more concentrated timings. These suggest that broadcast bots post tweets at more specific timings than humans and other types of bots. We also found that consumption and spam bots are very similar in terms of daily timings (i.e., weekday and day entropies), but the former is less diverse than the latter in terms of hourly timings. We can thus conclude that consumption and spam bots tweet equally regularly on a daily basis, but the latter tend to post at random hours.

Retweet and mention patterns. Retweet and mention activities can be used to gauge how much a bot (or human) cares about other accounts. Comparing the distributions of the retweet and mention timings in Fig. 6, we can see again that spam bots have the most random patterns compared to humans and other bot types. But unlike the results for tweet timings, consumption bots have the lowest diversity in terms of daily and hourly timings for the retweet and mention activities. This suggests that consumption bots reshare contents and mention other users at more specific timings, respectively. Such regularity makes sense, especially for consumption bots that provide update services to their users, e.g., Unfollowers and Twittascope (cf. Sect. 7.2).

Hashtag patterns. In Twitter, a hashtag can be viewed as representing a topic of interest. As shown in Fig. 6, humans and consumptions bots have very similar diversities of hashtag timings. It is also shown that spam bots have the most diverse hashtag timings (as expected), whereas broadcast bots exhibit very focused hashtag timings. The latter suggests that broadcast bots tend to talk about different topics at more regular time intervals. This is intuitive, especially if we consider the nature of the account owners of broadcast bots (e.g., news/blogger sites), which aim to disseminate various information on a regular basis.

URL patterns. For the URL timings, we find that in general humans and broadcast bots use URLs at more specific timings than consumption and spam bots. Interestingly, however, we observe that consumption bots exhibit higher diversity in daily timings than spam bots, but the reverse is true for hourly timings. This suggests that consumption bots use URLs on a more regular daily basis than spam bots, but the latter post URLs at more random hours.

Comparisons. It is also interesting to see how our results in Figs. 5 and 6 put little emphasis on the importance of the follow network features in the classification task. This is different from previous studies on (malicious) bots [4,5,13,17,20], whereby the follow features play a key role. We can attribute this to the evolution of bot activities as well as stricter regulations imposed by Twitter (especially for spam bots). Also, to our best knowledge, no attempt has been made in the previous works to infer on a larger population. Thus, our work offers more comprehensive insights on the behavioral traits of bots.

8   Conclusion

In this paper, we present a new categorization of bots, and develop a systematic bot profiling framework with a rich set of features and classification methods. We have carried out extensive empirical studies to analyze on broadcast, consumption and spam bots, as well as how they compare with regular human accounts. We discovered that the diversities of timing patterns for posting activities (i.e., tweet, retweet, mention, hashtag and url) constitute the key features to effectively identify the behavioral traits of different bot types.

This study hopefully will benefit social science studies and help create better user services. In the future, we plan to examine the prevalence of our findings across multiple countries, beyond our current Singapore data. We also wish to study information diffusion and user interaction in Twitter with the aid of bots.

Acknowledgments. This research is supported by the National Research Foundation, Prime Ministers Office, Singapore under its International Research Centres in Singapore Funding Initiative.

A   Predictions on Unlabeled Twitter Accounts

To facilitate our study on a larger Twitter population, we first examined how well our best classfier (i.e., LR) can predict for unlabeled data that it never sees in the (labeled) CV data. Table 4 summarizes the top K prediction results, whereby we varied K from 10 to 50 to verify the robustness of the predictions. For each class, we computed the number of correctly predicted instances (TP) as well as precision at top K, i.e., $Precision = \frac{TP}{K}$.

Table 4. Top K predictions on unlabeled 158,111 Twitter accounts

Label	$K = 10$		$K = 20$		$K = 30$		$K = 40$		$K = 50$	
	TP	Precision	TP	Precision	TP	Precision	TP	Precision	TP	Precision
Broadcast bot	9	0.80	18	0.90	27	0.90	34	0.85	38	0.76
Consumption bot	10	1.00	20	1.00	30	1.00	38	0.95	48	0.96
Spam bot	4	0.40	9	0.45	12	0.43	19	0.475	23	0.48
Human	10	1.00	20	1.00	30	1.00	40	1.00	40	1.00

TP: number of true positives

As shown in Table 4, our LR classifier produces fairly accurate and consistent predictions across different K values. With respect to human accounts, our LR classifier achieved perfect *Precision* for all K values. Unsurprisingly, we can expect that human accounts constitute the largest proportion of the Twitter population, and thus they should be the easiest to classify. We also obtained good results for the broadcast and consumption bots, with precision scores greater than 75 % and 95 % respectively. On the other hand, we observe rather modest *Precision* scores for spam bots (i.e., 40–47.5 %). We can attribute this to the insufficient number of instances for spam bots, which form only $\frac{105}{1,613} = 6.51\,\%$ of our labeled data (cf. Table 1). This may (again) be due to our data collection procedure that involved popular users as seeds and/or due to our relatively strict criteria for the characterization of spam bot accounts (cf. Sect. 7.1). Nevertheless, the *Precision* scores of 40–47.5 % remain relatively good, if we compare with that of a random guess for our labeled data (i.e., 6.51 %).

All in all, we find our top K predictions on unlabeled data to be satisfactory. Based on this, we can use our predictions to infer the behavioral profiles of bots in a larger Twitter population, which in this case spans the overall Singapore users.

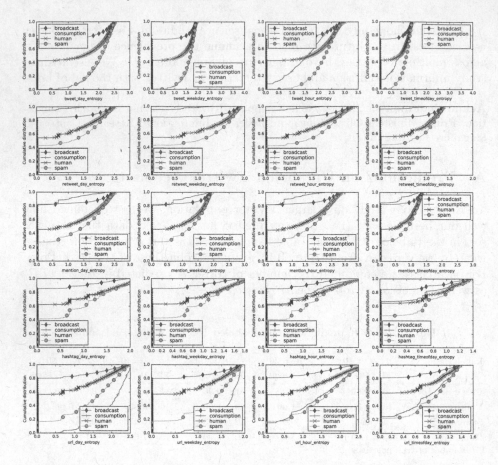

Fig. 6. Distribution of entropy-based features for 158,111 Twitter accounts

In particular, we analyze the entropy-based dynamic tweet features, namely the entropy distributions of the tweet, retweet, mention, hashtag and url activities, which constitute the majority group of the top discriminative features in Fig. 5. Figure 6 presents the cumulative distribution functions of these features. The detailed analysis of the distributions can be found in Sect. 7.3.

References

1. Abokhodair, N., Yoo, D., McDonald, D.W.: Dissecting a social botnet: growth, content and influence in Twitter. In: CSCW (2015)
2. Boshmaf, Y., Muslukhov, I., Beznosov, K., Ripeanu, M.: Design and analysis of a social botnet. Comput. Netw. **57**(2), 556–578 (2013)
3. Breiman, L.: Random forests. Mach. Learn. **45**(1), 5–32 (2001)
4. Chu, Z., Gianvecchio, S., Wang, H., Jajodia, S.: Detecting automation of Twitter accounts: are you a human, bot, or cyborg? IEEE Trans. Dependable Secure Comput. **9**(6), 811–824 (2012)

5. Dickerson, J.P., Kagan, V., Subrahmanian, V.: Using sentiment to detect bots on Twitter: are humans more opinionated than bots? In: ASONAM (2014)
6. Domingos, P., Pazzani, M.: On the optimality of the simple Bayesian classifier under zero-one loss. Mach. Learn. **29**(2–3), 103–130 (1997)
7. Fan, R.-E., Chang, K.-W., Hsieh, C.-J., Wang, X.-R., Lin, C.-J.: LIBLINEAR: a library for large linear classification. JMLR **9**, 1871–1874 (2008)
8. Ferrara, E., Varol, O., Davis, C., Menczer, F., Flammini, A.: The rise of social bots. Commun. ACM **59**(7), 96–104 (2016)
9. Freitas, C., Benevenuto, F., Ghosh, S., Veloso, A.: Reverse engineering socialbot infiltration strategies in Twitter. In: ASONAM, pp. 25–32 (2015)
10. Ghosh, S., Viswanath, B., Kooti, F., Sharma, N.K., Korlam, G., Benevenuto, F., Ganguly, N., Gummadi, K.P.: Understanding and combating link farming in the Twitter social network. In: WWW, pp. 61–70 (2012)
11. Hu, X., Tang, J., Zhang, Y., Liu, H.: Social spammer detection in microblogging. In: IJCAI, pp. 2633–2639 (2013)
12. Hwang, T., Pearce, I., Nanis, M.: Socialbots: voices from the fronts. Interactions **19**(2), 38–45 (2012)
13. Lee, K., Eoff, B.D., Caverlee, J.: Seven months with the devils: a long-term study of content polluters on Twitter. In: ICWSM, pp. 185–192 (2011)
14. Manning, C., Raghavan, P., Schütze, H.: Introduction to Information Retrieval. Cambridge University Press, Cambridge (2008)
15. Mitter, S., Wagner, C., Strohmaier, M.: A categorization scheme for socialbot attacks in online social networks. In: ACM Web Science (2013)
16. Shannon, C.E.: A mathematical theory of communication. Bell Syst. Tech. J. **27**(3), 379–423 (1948)
17. Stringhini, G., Kruegel, C., Vigna, G.: Detecting spammers on social networks. In: ACSAC (2010)
18. Subrahmanian, V., Azaria, A., Durst, S., Kagan, V., Galstyan, A., Lerman, K., Zhu, L., Ferrara, E., Flammini, A., Menczer, F., Waltzman, R., Stevens, A., Dekhtyar, A., Gao, S., Hogg, T., Kooti, F., Liu, Y., Varol, O., Shiralkar, P., Vydiswaran, V., Mei, Q., Huang, T.: The DARPA Twitter bot challenge. IEEE Comput. **49**(16), 38–46 (2016)
19. Tavares, G., Faisal, A.A.: Scaling-laws of human broadcast communication enable distinction between human, corporate and robot Twitter users. PloS One **8**(7), e65774 (2013)
20. Wagner, C., Mitter, S., Körner, C., Strohmaier, M.: When social bots attack: modeling susceptibility of users in online social networks. In: MSM (2012)
21. Wang, A.H.: Detecting spam bots in online social networking sites: a machine learning approach. In: DBSec, pp. 335–342 (2010)
22. Wilcoxon, F.: Individual comparisons by ranking methods. Biometrics Bull. **1**(6), 80–83 (1945)

A Diffusion Model for Maximizing Influence Spread in Large Networks

Tu-Thach Quach$^{(\boxtimes)}$ and Jeremy D. Wendt$^{(\boxtimes)}$

Sandia National Laboratories, Albuquerque, NM, USA
{tong,jdwendt}@sandia.gov

Abstract. Influence spread is an important phenomenon that occurs in many social networks. Influence maximization is the corresponding problem of finding the most influential nodes in these networks. In this paper, we present a new influence diffusion model, based on pairwise factor graphs, that captures dependencies and directions of influence among neighboring nodes. We use an augmented belief propagation algorithm to efficiently compute influence spread on this model so that the direction of influence is preserved. Due to its simplicity, the model can be used on large graphs with high-degree nodes, making the influence maximization problem practical on large, real-world graphs. Using large Flixster and Epinions datasets, we provide experimental results showing that our model predictions match well with ground-truth influence spreads, far better than other techniques. Furthermore, we show that the influential nodes identified by our model achieve significantly higher influence spread compared to other popular models. The model parameters can easily be learned from basic, readily available training data. In the absence of training, our approach can still be used to identify influential seed nodes.

1 Introduction

Social networks often show that different users have varying levels of influence. As an example, tweets from some users are more likely to spread than from others. In a network of friends, an individual adopting a product may cause others to do the same. Identifying these influential nodes has important applications. For instance, in marketing, an organization wants to identify which small set of nodes will return the highest influence spread given a limited budget. Finding the seed nodes that maximize influence spread is called *influence maximization*.

Influence maximization requires two inputs: a graph (with nodes representing individuals and edges representing relationships between any two individuals), and a diffusion model. Given the graph and the diffusion model, influence maximization finds k seed nodes such that the expected number of nodes influenced is maximized [11].

A variety of diffusion models have been proposed and analyzed. Two popular diffusion models are the independent cascade (IC) model and linear threshold (LT) model [11]. In the IC model, each active node i has one opportunity to

© Springer International Publishing AG 2016
E. Spiro and Y.-Y. Ahn (Eds.): SocInfo 2016, Part I, LNCS 10046, pp. 110–124, 2016.
DOI: 10.1007/978-3-319-47880-7_7

activate a neighboring node j with probability p_{ij}. In the LT model, each node j is influenced jointly by all neighboring nodes $i \in N(j)$ ($N(j)$ is the set of neighbors of node j). Each node j is influenced by each neighbor i with weight p_{ij} such that the sum of all incoming weights to j is at most 1. Each j determines a threshold t_j. If the sum of the incoming weights exceeds t_j, then j is activated.

A major drawback of these models is the computation: to get reasonable estimates of influence spread for a single node, these diffusion models require running Monte-Carlo simulations on the network many times (typically 10,000). This is clearly feasible only on small networks. In an effort to minimize this problem, several heuristics have been proposed to estimate the spread without resorting to Monte-Carlo simulations [2,3,14]. Others have proposed entirely new diffusion models. A probabilistic voter model found that the optimal seed nodes are those with the highest degree [6]. Markov models have also been proposed [5, 17]. Unlike cascade models, which capture the evolution of influence over time, Markov models capture the interactions of nodes as a set of interdependent random variables. Abandoning diffusion models altogether, the credit assignment approach uses historical logs to directly compute the influence of a node [8]. Many of these methods have parameters that need to be defined as well. When training data is available, a model's parameters can be learned [7,18]. In the absence of training data, constants and heuristics, such as weighted cascade where p_{ij} is inversely proportional to the in-degree of node j, are often used instead.

Given the existence of several proposed diffusion methods, practitioners must determine which diffusion model is the right one to use in any given situation. We propose three considerations for identifying which model to use:

1. A diffusion model should match well with ground-truth data when available. This validates the model and justifies its use for influence maximization.
2. A diffusion model's parameters can be learned from readily available data. In other words, the training data required should be practical to obtain. Furthermore, the model should still be usable when no training data is available.
3. A diffusion model should be computationally efficient so that it scales to large networks. Thus, practical diffusion models cannot rely on costly Monte Carlo simulations.

With these considerations in mind, we present a new diffusion model based on pairwise factor graphs that predicts influence spread for a given seed set. An efficient belief propagation algorithm is used to compute influence spread; it can be used on large real-world graphs with high-degree nodes. We provide experimental results showing that our model predictions match ground-truth spreads using a large Flixster dataset [10] and an Epinions dataset [19]. We then investigate the influence maximization problem under our model and show that the influential nodes identified by our model achieve higher influence spread compared to other popular models. The model parameters can easily be learned entirely from data. Moreover, the type of training data required is simple and practical. In the absence of training data, our model can still identify influential seeds.

This paper is organized as follows. Details of our diffusion model and the associated algorithms are presented in Sect. 2. Experimental results are provided in Sect. 3. Concluding thoughts are provided in Sect. 4.

2   Influence Spread

Given a graph $G = (\mathcal{V}, \mathcal{E})$ of n nodes, a set of seed nodes $\mathcal{S} \subseteq \mathcal{V}$, and a diffusion model Ω, the influence spread, $\sigma_\Omega(\mathcal{S})$, is the expected number of influenced or activated nodes. Here we adopt a factor graph, which can represent general graphical models including Markov networks and Bayesian networks. Our diffusion model consists of unary (ϕ) and pairwise (ψ) factors (potential functions). Specifically, each node $i \in \mathcal{V}$ has a corresponding state $x_i \in \{0, 1\}$ that indicates whether or not node i adopts the product (e.g., $x_i = 1$ means node i adopts the product). The adoption probability, $p_i(x_i)$, depends on not only i's preference, but also the states of its neighbors. The joint probability distribution of the states of the network is

$$p(x_1, \ldots, x_n) \propto \prod_{i \in \mathcal{V}} \phi_i(x_i) \prod_{(i,j) \in \mathcal{E}} \psi_{ij}(x_i, x_j). \tag{1}$$

Note that the unary potential function expresses the state preference of node i independent of its neighbors. The pairwise potential function expresses the dependency between neighboring nodes i and j whenever an edge between i and j exists in $\mathcal{E}$. The pairwise potential function depends on only two nodes and allows the model to deal with high-degree nodes directly (instead of pruning excess edges on high-degree nodes as in [14]). With this model, the marginal probability of each node i is

$$p_i(x_i) = \sum_{x_1, \ldots, x_{i-1}, x_{i+1}, \ldots, x_n} p(x_1, \ldots, x_n). \tag{2}$$

Computing the marginal probabilities can be done efficiently using belief propagation [21].

For undirected graphs, the above model can be used to compute the marginal probability of each node, which corresponds to its adoption probability. For directed graphs, an edge's direction indicates the *direction* of influence. Therefore, we propose to adapt the above model for directed influence. Consider the graph shown in Fig. 1. In this case, the state of node 2 depends on node 1's state, but not on node 3's. This is not true of an undirected model. To capture directionality, we compute the forward probabilities [16] instead of the marginal probabilities. For a chain (such as the one shown in Fig. 1) the forward probability of the state of node i is

$$f_i(x_i) = \sum_{x_1, \ldots, x_{i-1}} p(x_1, \ldots, x_{i-1}, x_i), \tag{3}$$

where the joint probability is

$$p(x_1, \ldots, x_i) \propto \prod_{j \in \mathcal{V}: j \leq i} \phi_j(x_j) \prod_{(j,k) \in \mathcal{E}: j \leq i, k \leq i} \psi_{jk}(x_j, x_k). \qquad (4)$$

It is clear that the forward probability of node i considers only those nodes that can influence it (e.g., the previous nodes in the chain), which is consistent with the meaning of directed edges. To compute the forward probabilities, we augment the belief propagation algorithm so that messages are only sent from node i to node j if there is an edge $(i, j) \in \mathcal{E}$. We provide further implementation details in Subsect. 2.4.

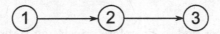

Fig. 1. A directed graph consisting of three nodes. An edge's direction determines the influencer-influencee relationship. In this case, node 2 is influenced by node 1, but not node 3.

We emphasize that an advantage of our approach is that efficient inference algorithms, such as belief propagation, can be used to approximate these forward probabilities [21]. Although exact computation of the forward probabilities is feasible only on graphs without loops, belief propagation is widely used on graphs with loops and generally provides good results [13,15,20].

2.1 Learning

The unary (ϕ_i) and pairwise (ψ_{ij}) potential functions are our model's parameters. The pairwise potential function is shown in Table 1, where p_{ij} is an edge-specific influence probability. If causal information between nodes were available, these could be learned individually. In the absence of good causal information, p_{ij} can be set using a heuristic, such as those based on the degree of a node, or some constant [2,3,11,18]. In our model, we set $p_{ij} = 0.5 + 0.5/\text{in-degree}(j)$. As the in-degree of a node increases, $p_{ij} \rightarrow 0.5$. This essentially implies that for any high in-degree node, each influencer exerts less influence on it. The rationale for this function is that when a node has many influencers, each influencer, individually, has a smaller impact on the decision of that node, allowing the node to make a decision based on the aggregate of the states of the influencers. Note that the potential functions express the fact that when $x_i = 0$, node i does not influence node j because both states are equally likely; the lack of adoption does not spread influence. Note that pairwise potential functions are general enough to accommodate other situations, including the case where the lack of adoption could spread influence. For completeness, we also consider the case when p_{ij} is a constant in our experiments.

Each node's unary potential function can be learned from training data, if available. Specifically, the unary potential is any node's adoption probability

Table 1. Pairwise potential functions.

$\psi_{ij}(x_i, x_j)$	$x_j = 0$	$x_j = 1$
$x_i = 0$	0.5	0.5
$x_i = 1$	$1 - p_{ij}$	p_{ij}

Table 2. Unary potential functions.

	$\phi_i(x_i)$
$x_i = 0$	$1 - \rho_i$
$x_i = 1$	ρ_i

independent of all other nodes. Therefore, given any training dataset, we can compute a node's adoption probability as simply the number of historical node adoptions divided by the total number of possible adoptions. Let ρ_i be this probability for node i. If ρ_i is too small, we set it to a minimum value (that is, $\rho_i \geq 10^{-5}$). This allows nodes that are not activated in the training set to still participate in influence propagations in the test set. The unary potential function is shown in Table 2.

The type of data required for training our model is minimal and generally available. In particular, it is far more realistic to assume that we can obtain historical states of the nodes in a network than to capture other higher-level information, such as the causal spread of information from one node to another, as required by the credit assignment model [8]. That is, it is easier to capture *which* nodes are activated than *how* nodes are activated. Nonetheless, in our experiments, we also consider the situation when no training data is available and set ρ_i to a small positive constant.

2.2 Computing Influence Spread

Using our diffusion model, we can compute the influence spread of seed set $\mathcal{S}$. For each seed node $i \in \mathcal{S}$, we set $\phi_i(1) = 1$ and $\phi_i(0) = 0$. We then run our forward belief propagation algorithm to compute $f_i(x_i)$ as defined by (3). Since social networks tend to have loops, belief propagation requires several iterations to converge (we use a maximum of 20 iterations in our experiments). Once converged, or the maximum number of iterations is reached, the influence spread of seed set $\mathcal{S}$ is quantified by

$$\sigma_\Omega(\mathcal{S}) = \sum_{i \in \mathcal{V}} f_i(x_i = 1). \tag{5}$$

2.3 Influence Maximization

The influence maximization problem is to find seed set $\mathcal{S}$ of specified size k that maximizes the influence spread [11]. A greedy approach can be used to

---

Algorithm 1. Greedy Influence Maximization

Input: $G = (\mathcal{V}, \mathcal{E})$, k, σ_Ω
Output: $\mathcal{S}$
$\mathcal{S} \leftarrow \emptyset$
while $|\mathcal{S}| < k$ **do**
    $u \leftarrow \arg\max_{v \in \mathcal{V} \setminus \mathcal{S}} \sigma_\Omega(\mathcal{S} \cup v) - \sigma_\Omega(\mathcal{S})$
    $\mathcal{S} \leftarrow \mathcal{S} \cup u$

Algorithm 2. CELF

Input: $G = (\mathcal{V}, \mathcal{E})$, k, σ_Ω
Output: $\mathcal{S}$
$\mathcal{S} \leftarrow \emptyset$
for $u \in \mathcal{V}$ **do**
    $u.priority \leftarrow \sigma_\Omega(\{u\})$
    $u.n \leftarrow 0$
    enqueue(u)
while $|\mathcal{S}| < k$ **do**
    $u \leftarrow$ dequeue()
    if $u.n = |\mathcal{S}|$ **then**
        $\mathcal{S} \leftarrow \mathcal{S} \cup u$
    else
        $u.priority \leftarrow \sigma_\Omega(\mathcal{S} \cup u) - \sigma_\Omega(\mathcal{S})$
        $u.n \leftarrow |\mathcal{S}|$
        enqueue(u)

---

approximate the influence maximization problem, which is NP-hard in general. The greedy algorithm, taken from [8], is shown in Algorithm 1.

The problem with the greedy approach is that it searches all nodes in the network at each iteration to find the best node. This can be prohibitively expensive, especially if the diffusion model uses Monte Carlo simulations. Several methods have been proposed to improve the greedy algorithm so as to reduce the number of nodes evaluated [3,9,12,22]. In particular, CELF (Cost-Effective Lazy Forward) significantly reduces the number of nodes to evaluate, resulting in 700 times speedup [12]. It uses a priority queue to greedily select the node that has the largest gain in influence spread at each iteration so as to minimize the number of nodes evaluated. In our experiments, we use CELF. For completeness, the CELF algorithm is shown in Algorithm 2. We note that CELF is not the only algorithm that can be used for seed selection. Other algorithms, such as CELF++ [9], can serve as alternatives. We prefer CELF due to its simplicity and find it sufficient, as computing spread on our model via belief propagation is fast.

2.4   Implementation Details

We now briefly describe an implementation of belief propagation and our modification for directed graphs so that the direction of influence is preserved. For a
more thorough treatment, see [21].

To solve the marginal probabilities from (2), belief propagation defines a per-
edge message from i to j about the likelihood of node j being in state x_j from
the perspective of node i:

$$m_{ij}(x_j) = \sum_{x_i} \phi_i(x_i)\, \psi_{ij}(x_i, x_j) \prod_{k \in N(i)\backslash j} m_{ki}(x_i) \qquad (6)$$

where ϕ_i and ψ_{ij} are potentials as defined before, and the final term is the
product of all messages sent to i by its neighbors (excluding j). These messages
are initialized to a fixed value at all nodes (usually 1). At each iteration, new
messages are computed from the previous iteration's messages in both directions
along each edge. Iterations continue until either all messages converge to a steady
value or a maximum number of iterations is reached.

Once converged, the belief over the states of node i is

$$b_i(x_i) \propto \phi_i(x_i) \prod_{j \in N(i)} m_{ji}(x_i). \qquad (7)$$

The normalized belief $b_i(x_i)$ corresponds to the marginal probability $p_i(x_i)$.

The solution to both (6) and (7) may have numerical issues if any node
involved has high degree. That is, the product of hundreds or thousands of
messages with values between $[0,1]$ leads to products that are unrepresentable
by finite-precision machines. The solution to this problem is to use the well-
known log trick.

Since $m_{ij}(x_j)$ can be normalized by an arbitrary positive constant c_{ij} (that
is fixed for all values of x_i and x_j on an edge), we can reformulate (6) as

$$m_{ij}(x_j) = \sum_{x_i} c_{ij}\, \phi_i(x_i)\, \psi_{ij}(x_i, x_j) \prod_{k \in N(i)\backslash j} m_{ki}(x_i)$$

$$= \sum_{x_i} \exp\left[\ln(c_{ij}) + \ln(\phi_i(x_i)) + \ln(\psi_{ij}(x_i, x_j)) + \sum_{k \in N(i)\backslash j} \ln(m_{ki}(x_i)) \right].$$

$$(8)$$

By exploiting log space, the products become sums, and we avoid numerical
underflow during message computation. We can ensure the result is within the
representable double-precision range before exponentiation by setting $\ln(c_{ij})$ to
an appropriate value. Specifically, we use

$$\ln(c_{ij}) = -\max_{x_i, x_j}\{\ln(\phi_i(x_i)) + \ln(\psi_{ij}(x_i, x_j)) + \sum_{k \in N(i)\backslash j} \ln(m_{ki}(x_i))\}. \qquad (9)$$

A similar trick is used for computations involving (7).

We found that a graph containing a node with degree greater than 750 would underflow with the default implementation. With the exponentiated version, we have tested up to degree 20,000 with no numerical issues.

Finally, the above belief propagation works on undirected graphs. However, as already described, the influence problem can be directed or asymmetric along edges – that is, i may influence j more than j does i. The only alteration required in the implementation to solve for (3) instead of (2) is to send messages downstream only. The beliefs now correspond to the forward probabilities.

We have implemented the above algorithms and models in Java and integrated them into the open-source Algorithm Foundry package.[1] For a commented example of how to run our code, see the class InfluenceSpread in the GraphExamples Component.

3   Experiments

We demonstrate the utility of our model using two datasets: Flixster [10] and Epinions [19]. The Flixster dataset contains movie reviews with timestamps and a network of friends. The edges are undirected, but to allow asymmetric influence along edges, we convert each edge into two opposite directed edges. The Epinions dataset contains product reviews with timestamps and a directed network of trust among reviewers. The basic statistics of the two datasets are summarized in Table 3.

Table 3. Statistics of the two datasets.

	Flixster	Epinions
# Nodes	800K	18K
# Directed Edges	12M	1.2M
# Products/Movies	49K	262K
Avg. Degree	30	64
Max. Degree	2K	4K

In the following, we use the word propagations to refer to movies in the Flixster dataset and products in the Epinions dataset. For each dataset, we split the propagations into two sets: training (80 %) and test (20 %). As in [8], to ensure a fair distribution of the propagation sizes across the training and test sets, we order all propagations by size and assign every fifth propagation into the test set. The training set is used to learn the unary potential functions, ρ_i.

---

[1] https://github.com/algorithmfoundry/Foundry/.

3.1   Diffusion Model Validation

We use the test sets to quantify how well our diffusion model predicts actual influence spreads using the method proposed in [8]. Specifically, for a given seed set $\mathcal{S}$, we calculate the predicted spread, $\sigma_\Omega(\mathcal{S})$, using our diffusion model. We can compare our predictions against ground-truth spreads. As in [8], for each propagation in the test set, the seed set is the set of users who are first to review among their immediate friends. The ground-truth spread is the actual number of users who reviewed the propagation.

We consider two strategies for choosing the pairwise potential functions (ψ_{ij}, Table 1):

- **DW**: Degree Weighted – $p_{ij} = 0.5 + 0.5/\text{in-degree}(j)$.
- **CW**: Constant Weight – p_{ij} is set to a constant of 10^{-3}. Note that for the Flixster dataset the weights are symmetric on all edges and the resulting model is undirected.

We consider two strategies for the unary potential functions (ϕ_i, Table 2):

- **LU**: Learned Unary – ρ_i is number of reviewed propagations by node i divided by the total number of propagations in the training set; must be at least 10^{-5}.
- **CU**: Constant Unary – ρ_i is set to a constant (5×10^{-3}).

Thus, for any experiment, we must select both a unary and a pairwise strategy. Hereafter, we refer to a combined strategy as a *pairwise-unary* strategy. As an example, the DW-LU strategy uses the degree weighted pairwise and learned unary strategies. For completeness, we also consider the weighted cascade IC model, a first in itself for the large Flixster graph. For the IC model, we use 10,000 Monte Carlo simulations to compute the spread of each propagation.

We show the scatter plots of the predicted and actual spread of each of these strategies on the test sets in Fig. 2. To improve the readability of the scatter plots, if there are several propagations that have the same actual spread, we report the average predicted spread. The ideal spread is shown as the green dashed line. The CW-LU strategy consistently underestimates the actual spreads. The DW-CU strategy overestimates (or underestimates, depending on the constant ρ_i) the actual spreads. The IC model significantly overestimates the spreads. This is consistent with past observations on smaller networks [8]. The DW-LU strategy performs the best – surrounding the actual spread.

In Fig. 3, we show the same scatter plots of DW-LU along with the corresponding seed sizes. The plots show that this model is able to take the initial seeds and spread their influence to other nodes in the network.

We believe DW-LU performs well on both datasets for two reasons. First, the learned unary strategy is able to incorporate node-specific data. For instance, some social media users are much more likely to produce content than others: incorporating this into the model improves results. This also implies that nodes that tend to adopt products on their own should not be targeted, as resources are better spent on other nodes. Second, as mentioned earlier, the choice of DW for p_{ij} implies that influencers of high in-degree nodes have small impact on their

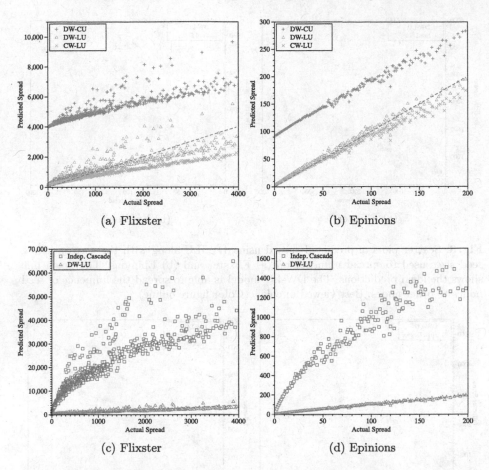

Fig. 2. Scatter plots of predicted spread vs. actual spread for different choices of potential functions: DW-CU (degree weighted and constant unary), CW-LU (constant weight and learned unary), and DW-LU (degree weighted and learned unary), as well as IC for the Flixster and Epinions datasets. The green line shows the ideal predictions. The DW-LU model best predicts the actual spread. Best viewed in color. (Color figure online)

influencees individually, allowing the influencees to make their decisions based on the aggregate of the states of their influencers.

3.2   Influence Maximization

Since our results establish that the DW-LU model is the best in predicting the spread of a seed set, we now investigate the influence maximization problem to determine how much spread is achieved under the DW-LU model on seeds selected by various models, obtained by running CELF on each model as appropriate. In addition, we also investigate the similarity between the seeds selected

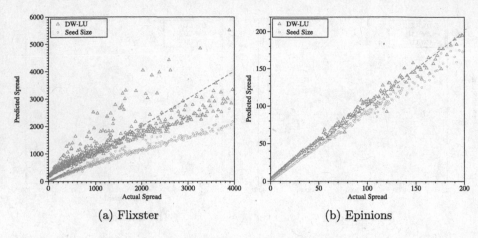

Fig. 3. Scatter plots of predicted spread using DW-LU along with the corresponding seed sizes used to spread influence on (a) Flixster and (b) Epinions. The green line shows the ideal predictions. The DW-LU model is able to spread the influence of seed nodes to other nodes. Best viewed in color. (Color figure online)

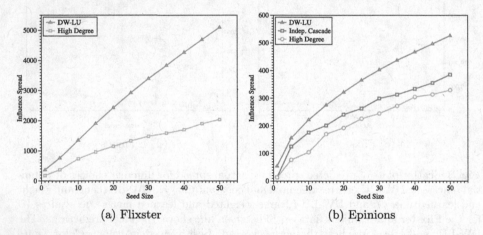

Fig. 4. Influence spread achieved under the DW-LU model by seed sets selected by various models: DW-LU, IC, and High Degree on (a) Flixster and (b) Epinions.

by DW-LU and other models. We consider several models: DW-LU, DW-CU, IC, and High Degree which selects the top k nodes as seeds based on degree (CELF is not needed). Since the IC model is computationally demanding, we run it only on the Epinions dataset, which is the smaller of the two datasets. Even then, it takes 22 days to find 50 seeds.

The plot of the influence spreads of seeds selected by DW-LU, IC, and High Degree are shown in Fig. 4. The results show that the seeds identified by our DW-LU model achieve significantly higher influence spread than High Degree and IC.

Table 4. Number of overlapping seeds between DW-LU and other models for various seed sizes.

		10	20	30	40	50
Flixster	High Degree	0	1	3	4	5
	DW-CU	10	20	30	40	49
Epinions	High Degree	3	6	10	16	18
	DW-CU	10	18	29	36	45
	IC	6	9	14	18	24

Table 4 shows the number of overlapping seeds between DW-LU and the other models. It is clear that the seeds selected by High Degree have low overlap with DW-LU. Even with $k = 50$ seeds, the overlap between High Degree and DW-LU is only 5 for the Flixster dataset and only 18 for the Epinions dataset. For the Epinions dataset, with 50 seeds, the number of overlapping seeds between DW-LU and IC is 24. The DW-CU and DW-LU models have almost identical seeds, which is why we did not plot DW-CU in Fig. 4 as those two curves are on top of each other. An important consequence of this result is that the social network analyst can leverage the DW-CU model to identify influential seed nodes in the absence of any training data. This is significant as training data may not be available in some applications.

We examine various graph metrics for the seeds selected by DW-LU, IC, and High Degree.

- **Community Overlap**: We run Louvain community detection [1] on the Flixster and Epinions graphs to identify the community assignment for each of the identified seeds. Since we have 50 seeds and between 15 and 25 communities identified on each graph, there is some overlap in community assignment for seed nodes. However, the High Degree technique selects far more nodes from its two most common communities (24 and 10 on Flixster; 30 and 9 on Epinions) than our DW-LU model (12 and 9 on Flixster; 20 and 14 on Epinions). On Epinions, IC is approximately the same as DW-LU (19 and 14).
- **Average Distance**: We compute each seed's average distance to all other seeds using Dijkstra's Algorithm [4]. In both graphs, our DW-LU model selects nodes that are farther apart on average (2.38 vs. 2.14 edges apart on Flixster; 1.82 vs. 1.39 edges apart on Epinions). On Epinions, IC averages 1.58–further than degree, but closer than DW-LU.
- **Node Degree**: We investigate the degrees of seed nodes in the order selected by CELF. Although the degree of each seed selected by IC and DW-LU varies from the degrees of the seeds selected before or after it, when we fit a line to the seeds' degrees, there is a clear negative trend. The degrees of the seeds are mostly well above the average degree on both graphs (one of the seeds selected by DW-LU for Epinions is just below the graph-wide average degree). IC consistently selects higher degree nodes than DW-LU.

These results indicate several interesting features for effective seed nodes. First, while high degree seems to be a useful feature for a seed node (nearly all DW-LU seeds have high degree), it is not sufficient (High Degree achieves lower influence spread and IC's higher degree nodes achieve lower influence as shown in Fig. 4). The community overlap and average distance measures indicate a second critical feature: the best seeds spread out from each other. Note that the most spread out set of nodes are among the leaves, but those nodes do not have a high enough degree to spread influence. Thus, there must be a balance between spread and high degree. Both IC and DW-LU balance these two features, although DW-LU balances them better.

3.3   Computing Resources

Our model uses an augmented belief propagation algorithm to compute influence spread. The runtime and memory requirement are both bounded by $O(|\mathcal{E}| + |\mathcal{V}|)$. The quantities provided here are relevant to the Flixster dataset, which is the larger of the two. Our Java implementation uses 3.4 GB of memory. As for computing time, the most expensive operation in influence maximization is the computation of influence spread of each node as a seed node, which is required by CELF. For this, we use a compute cluster of 60 compute nodes to run our model, which took 12 h to complete. Note that we do this only once. Once done, we select the top 50 nodes using CELF on a single workstation. The total time to find the top 50 nodes is approximately 16 min. On average, propagating each seed set takes 4 s. In contrast, using 10,000 Monte Carlo simulations to compute the spread of each seed set under the IC model takes 6 min on average.

4   Discussion and Conclusion

Influence maximization is a relevant and important problem in social network analysis. As such, it is important to have models that are efficient, provide a certain level of validation against ground-truth data, and can be learned from readily available data. To this end, we have presented a model that addresses these concerns. Our model uses belief propagation instead of Monte Carlo simulations to compute influence spread. Our model parameters can be learned from basic training data, such as frequency of adoptions, which we believe is more readily available in practical applications than other models that require causal relationships. In the absence of training data, our model can still identify influential seeds. As mentioned earlier, we use a heuristic based on in-degree to set the pairwise potentials. Our model, however, is general enough that these pairwise functions can be set to arbitrary values, including those learned from a training dataset, if available.

The results of this work raise an important question: what intrinsic graph properties are important in identifying influential seeds? As we have seen, high degree alone is not sufficient. Yet, our model, using pairwise functions that are based on in-degree, identifies seed nodes that achieve high influence spread. Are seed nodes intrinsic to graph structures? We hope to provide further insight into these questions in our future work.

Acknowledgment. We are grateful to Cristopher Moore for discussions on belief propagation and implementation considerations, Rich Field for improving the quality of the paper, and Dave Zage for discussions on implementation considerations. This work was supported by the Laboratory Directed Research and Development program at Sandia National Laboratories, a multi-program laboratory managed and operated by Sandia Corporation, a wholly owned subsidiary of Lockheed Martin Corporation, for the U.S. Department of Energy's National Nuclear Security Administration under contract DE-AC04-94AL85000.

References

1. Blondel, V.D., Guillaume, J.L., Lambiotte, R., Lefebvre, E.: Fast unfolding of communities in large networks. J. Stat. Mech.: Theory Exp. **10**, P10008 (2008)
2. Chen, W., Wang, C., Wang, Y.: Scalable influence maximization for prevalent viral marketing in large-scale social networks. In: Proceedings of the 16th ACM SIGKDD International Conference on Knowledge Discovery and Data Mining, pp. 1029–1038. ACM (2010)
3. Chen, W., Wang, Y., Yang, S.: Efficient influence maximization in social networks. In: Proceedings of the 15th ACM SIGKDD International Conference on Knowledge Discovery and Data Mining, pp. 199–208 (2009)
4. Dijkstra, E.W.: A note on two problems in connection with graphs. Numer. Math. **1**, 269–271 (1959)
5. Domingos, P., Richardson, M.: Mining the network value of customers. In: Proceedings of the Seventh ACM SIGKDD International Conference on Knowledge Discovery and Data Mining, pp. 57–66. ACM (2001)
6. Even-Dar, E., Shapira, A.: A note on maximizing the spread of influence in social networks. Inf. Proces. Lett. **111**(4), 184–187 (2011)
7. Goyal, A., Bonchi, F., Lakshmanan, L.V.: Learning influence probabilities in social networks. In: Proceedings of the Third ACM International Conference on Web Search and Data Mining, pp. 241–250. ACM (2010)
8. Goyal, A., Bonchi, F., Lakshmanan, L.V.: A data-based approach to social influence maximization. Proc. VLDB Endow. **5**, 73–84 (2011)
9. Goyal, A., Lu, W., Lakshmanan, L.V.: CELF++: Optimizing the greedy algorithm for influence maximization in social networks. In: Proceedings of the 20th International Conference Companion on World Wide Web, pp. 47–48. ACM (2011)
10. Jamali, M., Ester, M.: A matrix factorization technique with trust propagation for recommendation in social networks. In: Proceedings of the Fourth ACM Conference on Recommender Systems, pp. 135–142. ACM (2010)
11. Kempe, D., Kleinberg, J., Tardos, É.: Maximizing the spread of influence through a social network. In: Proceedings of the Ninth ACM SIGKDD International Conference on Knowledge Discovery and Data Mining, pp. 137–146. ACM (2003)
12. Leskovec, J., Krause, A., Guestrin, C., Faloutsos, C., VanBriesen, J., Glance, N.: Cost-effective outbreak detection in networks. In: Proceedings of the 13th ACM SIGKDD International Conference on Knowledge Discovery and Data Mining, pp. 420–429. ACM (2007)

13. Mooij, J.M., Kappen, H.J.: Sufficient conditions for convergence of the sum-product algorithm. IEEE Trans. Inf. Theory **53**(12), 4422–4437 (2007)
14. Nguyen, H., Zheng, R.: Influence spread in large-scale social networks – a belief propagation approach. In: Flach, P.A., De Bie, T., Cristianini, N. (eds.) ECML PKDD 2012, Part II. LNCS, vol. 7524, pp. 515–530. Springer, Heidelberg (2012)
15. Pearl, J.: Probabilistic Reasoning in Intelligent Systems: Networks of Plausible Inference. Morgan Kaufmann, Burlington (2014)
16. Rao, V., Teh, Y.W.: Fast MCMC sampling for Markov jump processes and extensions. J. Mach. Learn. Res. **14**(1), 3295–3320 (2013)
17. Richardson, M., Domingos, P.: Mining knowledge-sharing sites for viral marketing. In: Proceedings of the Eighth ACM SIGKDD International Conference on Knowledge Discovery and Data Mining, pp. 61–70. ACM (2002)
18. Saito, K., Nakano, R., Kimura, M.: Prediction of information diffusion probabilities for independent cascade model. In: Lovrek, I., Howlett, R.J., Jain, L.C. (eds.) KES 2008, Part III. LNCS (LNAI), vol. 5179, pp. 67–75. Springer, Heidelberg (2008)
19. Tang, J., Gao, H., Liu, H., Sarma, A.D.: eTrust: understanding trust evolution in an online world. In: Proceedings of the 18th ACM SIGKDD International Conference on Knowledge Discovery and Data Mining, pp. 253–261. ACM (2012)
20. Weiss, Y.: Correctness of local probability propagation in graphical models with loops. Neural Comput. **12**(1), 1–41 (2000). http://dx.doi.org/10.1162/089976600300015880
21. Yedidia, J.S., Freeman, W.T., Weiss, Y.: Understanding belief propagation and its generalizations. Technical report. TR2001-22, Mitsubishi Electric Research Laboratories, November 2001
22. Zhou, C., Zhang, P., Zang, W., Guo, L.: On the upper bounds of spread for greedy algorithms in social network influence maximization. IEEE Trans. Knowl. Data Eng. **27**(10), 2770–2783 (2015)

Lightweight Interactions for Reciprocal Cooperation in a Social Network Game

Masanori Takano[✉], Kazuya Wada, and Ichiro Fukuda

CyberAgent, Inc., Chiyoda-ku, Tokyo, Japan
takano_masanori@cyberagent.co.jp

Abstract. The construction of reciprocal relationships requires cooperative interactions during the initial meetings. However, cooperative behavior with strangers is risky because the strangers may be exploiters. In this study, we show that people increase the likelihood of cooperativeness of strangers by using lightweight non-risky interactions in risky situations based on the analysis of a social network game (SNG). They can construct reciprocal relationships in this manner. The interactions involve low-cost signaling because they are not generated at any cost to the senders and recipients. Theoretical studies show that low-cost signals are not guaranteed to be reliable because the low-cost signals from senders can lie at any time. However, people used low-cost signals to construct reciprocal relationships in an SNG, which suggests the existence of mechanisms for generating reliable, low-cost signals in human evolution.

Keywords: Data mining · Human cooperation · Reciprocal altruism · Signaling · Social network game

1 Introduction

Evolutionary game theory research has shown that reciprocal altruism drives the evolution of cooperation [1,2,12,14,15,20,31]. In this behavior, an individual acts in a manner that temporarily reduces its fitness, while increasing another individual's fitness, with the expectation that the other individual will behave in a similar manner at a later time. This behavior has been observed in humans [8,9,19] and other primates [17]. In addition, the possibility of this behavior has even been suggested in vampire bats [33] and fishes [4].

Axelrod [2] showed that cooperation based on reciprocity requires friendly interactions during the initial meeting in simulations of the iterated Prisoner's Dilemma game. Because reciprocal cooperators cooperate with individuals who cooperated with them previously. Indeed, experimental studies using game theory have shown that humans tend to be cooperative in their first meetings without prior interactions [9,18,19,32].

However, an interaction with strangers can be risky because it is difficult to know each other's levels of cooperativeness. Therefore, mechanisms for cooperation (kin selection [10], direct reciprocity [1,2,12,15,31], indirect reciprocity [16],

© Springer International Publishing AG 2016
E. Spiro and Y.-Y. Ahn (Eds.): SocInfo 2016, Part I, LNCS 10046, pp. 125–137, 2016.
DOI: 10.1007/978-3-319-47880-7_8

and tags [21]) generate a structured interaction where individuals interact more frequently with acquaintances because strangers may be exploiters. Nonetheless, humans tend to be cooperative during their first meetings without prior interactions [9,18,19,32]. The evidences [2,9,18,19,32] has been acquired in modeled environments based on the constrained behaviors of humans, or agents, to explicitly analyze their social behavior, e.g., they had to select their strategies without prior interactions. However, in the real world, we engage in lightweight preliminary interactions, such as observing, eye contact, bowing, and greeting each other. Therefore, it is important to study these preliminary interactions in a less restrictive environment than that imposed in experimental studies.

In this study, we analyzed the interactions during initial meetings to understand risk reduction behavior in the construction of reciprocal relationships in a social network game (SNG). In the game, numerous players can behave more freely than possible in the environments used in previous theoretical and experimental studies [14,20], i.e., they did not need to select from a sequence of several alternatives because they always had multiple alternatives and the actions of all the players can be recorded. In addition, the following features of the SNG make it easier to analyze reciprocal relationships. The game allows real players to cooperate and compete with others in situations where the player's benefit is represented by a quantitative value, such as a payoff in game theory. A previous study [29] demonstrated the existence of reciprocal relationships where cooperators had more advantages than non-cooperators in this SNG.

Many previous studies have used data obtained from interactive online games, particularly in social science [3,5,24–29], e.g., the dynamics of virtual world economics [3,5], human migration behavior [24,27], gender differences in social behavior [26], and reciprocal cooperation [29].

2   Materials and Methods

In this section, we provide the minimal SNG information and we define cooperative behavior in the SNG (see Appendices A.1, A.2, and A.3 for the game information, rules, and definition, respectively).

We analyzed cooperative behavior in the SNG, "Girl Friend BETA," where players acquired "event points" and competed in the rankings based on these points because the players received better awards as their rankings increased (Fig. 1). This SNG was released on 10/29/2012. The player's ranking order was determined by the sum of event points obtained in the period from 3/25/2013 to 4/8/2013.

Players must use their energy to obtain event points; therefore, the number of actions by player is finite. There are two methods for replenishing these points: waiting for the points to replenish over time and using a paid item. Players must use their resources (items and time) in an effective manner to progress to a higher ranking because their time and money are finite.

Players belong to groups and they must cooperate with each other to play the game efficiently. The groups are limited to 1–50 players. The SNG was designed

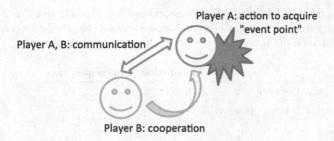

Fig. 1. Overview of players' interaction. Player A acts to acquire "event point", then player B belonging to a same group can cooperate for A. They can communicate each other at any time by using three types of simple text messaging.

to ensure that cooperation with group members results in more effective game play. We filtered out players who did not belong to groups because most of the active players must belong to groups to play effectively. Players can create groups on their own. Others can apply to join groups at any time and then join a group after the acceptance of their application by an administrator, who was typically a group founder. Players can leave a group at any time and apply to join a different group. We regarded this behavior as migration. The migrants were newcomers for the existing group members. We regarded interactions between migrants (newcomers) and the existing group members within 48 h of migration as initial meetings. Players can observe the behavior of members of their group (e.g., attack on common enemies; the details are provided later) because the game system showed their behavior on the game screen. We targeted groups of five or more active players who logged in at least one or more times to analyze their social interactions.

Players can communicate at any time using three types of simple text messaging. The first type is a message from one player to another (direct messaging). The second type is a message from a player to their group members (group messaging). The third type is a posting on the forum for their group (forum posting). These messages have no negative effects on either the senders or receivers, but they also have few or no positive effects[1]. We limited the data to intragroup communication and cooperation.

We analyzed cooperative behavior in the environment described above. It was difficult to track all of the cooperative behaviors because the players can perform various actions in the SNG. Thus, we selected a specific cooperative behavior and regarded the frequency of that behavior as a measure of a player's cooperativeness.

We focused on a game scenario where the relationship between players was similar to that in the Leader game (Table 1), but it was not possible for both

---

[1] Players can acquire a few points for a lottery, which provides a card when the players sent messages to each other at the beginning of each day. However, the players had to pay 200 points to enter the lottery and the effect of the card is small, i.e., the points do not increase the players' abilities.

Table 1. Payoff matrix for the leader game, where $S + T > 2R$ and $T > S > R > P$, i.e., Pareto efficiency is achieved when one cooperates and the other does not cooperate. The cooperator then obtains S and the noncooperator receives T.

	Cooperation	Noncooperation
Cooperation	R, R	S, T
Noncooperation	T, S	P, P

players to cooperate at the same time in this scenario (see Appendix A.3). Pareto efficiency is achieved in the Leader game when one player cooperates and the other do not. The cooperator then receives S and the noncooperator receives T. However, both try to avoid the worst situation (i.e., they receive P), but they also do not want to pay the cost for avoiding the worst situation (i.e., they do not want to receive S), i.e., the players receive a high payoff by sharing S and T in repeated plays of the game in a process known as ST reciprocity [30]. We recognized this cooperative behavior, which provided a payoff T from one to the other, as a cooperative behavior in this scenario.

3   Results

First, we evaluated the effects of social behavior on the number of cooperation behaviors by others. We compared the social interactions by migrants (newcomers) within 48 h of migration and those by existing group members within 48 h from a random time. We employed the following generalized linear model (GLM) to analyze these data:

$$C_i' \sim \mathrm{NB}(\lambda_i), \tag{1}$$
$$\ln \lambda_i = \beta_1 \ln \overline{a}_i H_i + \beta_2 f_i + \beta_3 f_i C_i + \beta_4 (1 - f_i) C_i$$
$$+ \beta_5 f_i g_i + \beta_6 (1 - f_i) g_i + \beta_7 f_i G_i + \beta_8 (1 - f_i) G_i$$
$$+ \beta_9 f_i b_i + \beta_{10} (1 - f_i) b_i + \beta_{11t} d_{ti} + \beta_{12}.$$

This model was used to explain the number of cooperative behaviors from group members to player i (C_i') based on a migrant flag f_i (if i migrated, then $f_i = 1$; else $f_i = 0$), an interaction between f_i and their social behaviors (the number of cooperative behaviors by i (C_i), the number of direct messages by i (g_i), the number of group messages by i (G_i), and the number of forum posts by i (b_i)), and trends in the cooperative behavior on day t (d_{ti}) as dummy variables. In addition, we used the log of the product of the number of attacks by the group members $\overline{a}_i$ and the number of help requests from i to their group members (H_i) because this value was expected to increase C_i' proportionally if group members cooperated at random (see Appendix A.3), i.e. this controls i's group effect. d_{ti} was entered as covariates to control for the influence of each day. $\mathrm{NB}(x)$ shows that x follows a negative binomial distribution. We estimated its parameters with 80, 880 relationships between players, sampled at random. We considered

Table 2. Results of the regression analysis based on the effects of social behavior relative to the number of cooperative behaviors by others (Eq. 1). $***$, $**$, and $*$ indicate that the signs of the regression coefficients did not change in Wald-type 99.9 %, 99 %, and 95 % confidence intervals, respectively (the symbols have the same meaning in the following tables). The regression coefficient of f_i, $f_i C_i$, $(1 - f_i)C_i$, $f_i g_i$, $(1 - f_i)g_i$, $f_i G_i$, $(1 - f_i)G_i$, and $f_i b_i$ were positive and significant, even after controlling for the other explanatory variables. The positive coefficient of f_i shows that newcomers tended to cooperate more than existing group members. The regression coefficients of C_i, g_i, G_i, and b_i were positive regardless of whether $f_i = 1$ was or not, and those for $f_i = 1$ were larger than those for $f_i = 0$ (excluding the forum posts by existing group members, which was not significant).

Explanatory variable	Regression coefficient	Standard error
$\ln \overline{a}_i H_i$	0.7836054	$(0.0052732)^{***}$
f_i	4.9450949	$(0.0363335)^{***}$
$f_i C_i$	0.1642087	$(0.0068994)^{***}$
$(1 - f_i)C_i$	0.1018723	$(0.0019206)^{***}$
$f_i g_i$	0.0079778	$(0.0008424)^{***}$
$(1 - f_i)g_i$	0.0004976	$(0.0002189)^{*}$
$f_i G_i$	0.0941003	$(0.0111839)^{***}$
$(1 - f_i)G_i$	0.0494162	$(0.0069425)^{***}$
$f_i b_i$	0.0170395	$(0.0046058)^{***}$
$(1 - f_i)b_i$	−0.0002771	(0.0015286)
d_1	0.3258377	$(0.0507144)^{***}$
d_2	0.5307506	$(0.0508597)^{***}$
d_3	0.7443556	$(0.0513828)^{***}$
d_4	0.7133664	$(0.0506845)^{***}$
d_5	0.8200644	$(0.0500531)^{***}$
d_6	0.9167403	$(0.0502217)^{***}$
d_7	0.9726432	$(0.0511331)^{***}$
d_8	0.9641601	$(0.0520516)^{***}$
d_9	1.0840990	$(0.0519211)^{***}$
d_{10}	0.9394478	$(0.0529310)^{***}$
d_{11}	0.8924624	$(0.0513280)^{***}$
d_{12}	1.0503286	$(0.0503736)^{***}$
d_{13}	1.3767783	$(0.0532730)^{***}$
d_{13}	2.3644593	$(0.0793361)^{***}$
Intercept	−8.9843031	$(0.0666115)^{***}$

this model because the data exhibited over-dispersion when we applied the GLM with a Poisson distribution.

Table 2 shows the results obtained after analyzing the model. The results demonstrate that reciprocal relationships were constructed between a newcomer and an existing group member, as well as being maintained between existing group members, and that the three types of messages basically supported the reciprocal relationships. In addition, the cooperative behavior of newcomers and the three types of messages were more important for reciprocal relationships than existing group members. The results also suggest that sending messages to others may have demonstrated the cooperativeness of players in this SNG, and the construction of reciprocal relationships required more cooperation and communication than the maintenance of reciprocal relationships.

Second, we tested whether the three types of messages showed the cooperativeness of the players. We analyzed the relationships between the messaging behavior and cooperative behavior of migrants within 48 h of migration and of the existing group members within 48 h of a random time. We employed the following GLM to analyze the results:

$$C_i \sim \text{NB}(\lambda_i), \tag{2}$$
$$\begin{aligned}\ln \lambda_i = {} & \beta_1 \ln a_i H_i' \\ & + \beta_4 f_i g_i + \beta_5 (1 - f_i) g_i + \beta_6 f_i G_i + \beta_7 (1 - f_i) G_i \\ & + \beta_8 f_i b_i + \beta_9 (1 - f_i) b_i + \beta_{10t} d_{ti} + \beta_{11}.\end{aligned}$$

This model was used to explain the number of cooperative behaviors by player i (C_i) based on the interaction between a migrant flag f_i (if i migrated, then $f_i = 1$; else $f_i = 0$) and their messaging behavior (the number of direct messages by i (g_i), the number of group messages by i (G_i), and the number of forum posts by i (b_i)), and the trends in cooperative behavior on day t (d_{ti}) as dummy variables. In addition, we used the log of the product of the number of attacks by player i, a_i, and the number of help requests from their group members (H_i') to i because this value was expected to increase C_i proportionally if player i cooperated at random (see Appendix A.3), i.e. this controls i's group effect. d_{ti} was entered as covariates to control for the influence of each day. We estimated its parameters with 80, 880 relationships between players, sampled at random. $\text{NB}(x)$ shows that x followed a negative binomial distribution. We employed this model because the data exhibited over-dispersion when we applied the GLM with a Poisson distribution.

Table 3 shows the results obtained after analyzing the model. The results demonstrate that the messages sent between players basically indicated their cooperativeness. The results also suggest that the use of messaging by newcomers indicated greater cooperativeness than that by existing group members. Thus, the messaging behavior may not have been important for existing group members who had already constructed reciprocal relationships.

4   Discussion

In the present study, players constructed reciprocal relationships in a similar manner to those found in studies based on modeled environments [2,9,18,19,32].

Table 3. Results of the regression analysis based on the relationships between messaging behavior and cooperative behavior (Eq. 2). The regression coefficients of $f_i g_i$, $(1 - f_i)g_i$, $f_i G_i$, $(1 - f_i)G_i$, $f_i b_i$, and $(1 - f_i)b_i$ were positive and significant, even after controlling for the other explanatory variables. The coefficients of g_i, G_i, and b_i were positive regardless of whether $f_i = 1$ was or not, and those for $f_i = 1$ were larger than those for $f_i = 0$.

Explanatory variable	Regression coefficient	Standard error
$\ln a_i H_i'$	0.3797530	(0.0040444)***
$f_i g_i$	0.1232340	(0.0008722)***
$(1 - f_i)g_i$	0.0109966	(0.0002245)***
$f_i G_i$	0.3422500	(0.0116456)***
$(1 - f_i)G_i$	0.1180049	(0.0071251)***
$f_i b_i$	0.2045634	(0.0046320)***
$(1 - f_i)b_i$	0.0474698	(0.0015434)***
d_1	0.1919667	(0.0516471)***
d_2	0.4586802	(0.0516397)***
d_3	0.6671087	(0.0524117)***
d_4	0.6931665	(0.0516168)***
d_5	0.6556934	(0.0512680)***
d_6	0.7007019	(0.0515783)***
d_7	0.6928495	(0.0527448)***
d_8	0.7550984	(0.0534660)***
d_9	0.7368654	(0.0538759)***
d_{10}	0.6794663	(0.0548714)***
d_{11}	0.6930345	(0.0529291)***
d_{12}	0.7449962	(0.0519552)***
d_{13}	0.6131128	(0.0551394)***
d_{13}	0.6985565	(0.0824675)***
Intercept	−4.8263750	(0.0550812)***

We showed that lightweight interactions (three types of messages) were important for constructing reciprocal relationships. The messages involved low-cost signaling because they incurred no costs for the senders and recipients. Theoretical studies [22,23] have shown that low-cost signals are not guaranteed to be reliable because the senders can lie at any time using low-cost signals. However, we found that the messages sent by players demonstrated their cooperativeness (i.e., their messages were reliable signals) and their messages helped to construct and maintain their reciprocal relationships. In particular, the messages sent during initial meetings (messages from newcomers to existing group members) were more important than messages between existing group members. These results suggest that low-cost signals will be reliable in humans. The signals may be employed to increase the likelihood of cooperativeness by others in risky situations where they are not known to each other.

This evidence for low-cost signaling in humans provides insights into the mechanisms that generate and maintain large societies. Players probably use low-cost signals as a form of social grooming, which is used to construct and maintain social relationships [6]. Apes, which are closely related to humans, clean each other's fur as a form of social grooming [13]. This social grooming incurs high time costs for the groomers and provides hygiene benefits to the recipients of grooming. Therefore, their social grooming will work as a reliable signal. By contrast, social grooming by humans can be low cost such as the three types of messages used in the SNG, as well as gaze grooming [11] and one-to-many grooming (e.g., gossip) [7]. The form of social grooming practiced by apes would be too costly for humans because human groups are larger ape groups, so humans must invest time and effort in grooming others in different ways to create social relationships in large groups [6]. Therefore, the evolution of mechanisms that generate reliable signals will have facilitated the evolution of the signature social structures found in humans.

Acknowledgment. We are grateful to professor Takaya Arita at Nagoya University, assistant professor Genki Ichinose at Shizuoka University, and master's course Mitsuki Murase at Nagoya University whose comments and suggestions were very valuable throughout this study.

A Appendix

A.1 Game Information

We analyzed cooperative behavior in the SNG, "Girl Friend BETA." Table 4 presents the game information. In this SNG, players create individual decks of cards that they collect and then use their decks to perform tasks in the SNG. A powerful deck, constructed from powerful cards, provides an advantage for game play in various situations. The players' primary motivation in the SNG is to obtain powerful cards. Players can obtain powerful cards as top-ranking rewards (see details later) or by casting lots called "Gacha."

Players can communicate at any time using three types of simple text messaging. The first type was a message from one player to another (direct messaging). The second type was a message from a player to their group members (group messaging). The third involved posting on the forum for their group (forum posting). These messages had no negative effects on either the senders or receivers, but they also had few or no positive effects[2]. We limited the data to intragroup communication and cooperation.

---

[2] Players can acquire a few points for a lottery, which provided a card when the players sent messages to each other at the beginning of each day. However, the players had to pay 200 points to enter the lottery and the effect of the card was small, i.e., the points did not increase the players' abilities.

Table 4. Game information

Developer and publisher	CyberAgent Inc.
Service Name	Girl Friend BETA
URL	http://vcard.ameba.jp
Event type	Raid battle
Event time period	3/25/2013 16:00 to 4/8/2013 14:00
Analysis time period	3/25/2013 0:00 to 4/7/2013 23:59

A.2   Game Rules

Our analysis target was a raid event (Fig. 2), in which players attack large enemies[3] and acquire "event points." Players competed in the rankings based on their event points, because they received better awards as their rankings increased.

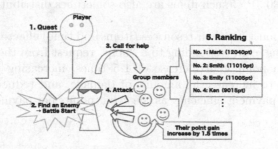

Fig. 2. Overview of raid event. A player conducts "quests" to find enemies (1). The player begins a battle upon finding an enemy and then attacks the enemy to obtain points (2). Enemies with very high hit points are strong; thus, they can call for help from other group members whom they have helped to win the battle (3). Players who helped had their point gain increased by 1.5 times (4). Players compete in rankings based on their points (5).

Players conduct quests[4] to find enemies during an event. Players begin battles when they find an enemy and then attack the enemy to obtain points. However, enemies with very high hit points are strong, making it difficult for players to win these battles unaided. Thus, they can call for help from other group members, to win the battle. Players who helped had their point gain increased by 1.5 times. Therefore, players help their fellow group members to acquire more points.

---

[3] The enemy only has hit points as an attribute, meaning that players cannot be attacked by enemies. A player must attack an enemy to acquire event points at the expense of attack points.

[4] This is one of the basic actions in SNGs. A player may encounter an enemy on performing certain action.

Players' point gains are proportional to the amount of damage caused during attacks, i.e., more powerful decks earn more event points. A player immediately acquires points upon attacking an enemy, even if the enemy is not defeated. However, a player cannot battle another enemy while already battling another enemy, and that enemies' hit points increase with each battle; therefore, players must attack enemies repeatedly in the latter half of an event. Thus, a player who finds an enemy or helps a fellow group member must defeat the enemy before taking a next action, or wait until that the enemy leaves[5].

Players increase the amount of damage caused during their attacks by launching "combo attacks," alternate attacks by two or more players in which the players need to launch attacks within ten minutes after other players[6]. The longer a chain of combo attacks, the more acquisition points are acquired. Battling enemies together with fellow group members increases the effectiveness of acquisition points.

Players must use a quarter of their attack points to attack; thus, they can attack four times when their point totals are full. There are two methods for replenishing these points: wait for the points to replenish over time or use an item that costs 100 JPY (such items are also sometimes distributed in the game as rewards).

Thus, players must use their resources (items and time) effectively to progress to a higher ranking, e.g., responding to a "help" request from their group members to acquire a point gain increase of 1.5 times, increasing the number of "combo attacks" to increase the amount of damage, and reducing the disable time. We defined payment efficiency as the event points per payment, as in game theory.

A.3   The Test Scenario

It was impossible to track every cooperative behavior, because players can exhibit various behaviors in the SNG. Hence, we focused on one easily tracked cooperative behavior, and we regarded its frequency as players' cooperativeness.

We focused on the following scenario based on these rules to define players' cooperativeness. (a) An enemy is attacked by a player and fellow group members. (b) The enemy's hit points are very few. In this scenario, players who defeat the enemy will acquire only a few event points, because their attack power is higher than the enemy's hit points. Thus, their behavior is not efficient for acquiring event points. By contrast, if the players' attack power is lower than the enemy's hit points, their behavior is efficient for acquiring event points. Furthermore, they cannot battle another enemy, if battle with one enemy is ongoing, and therefore must wait until they defeat the enemy to exhibit efficient behavior.

---

[5] The length of the disable time is set between one and two hours. It is too long to complete the rankings for middle- and higher-rank players, because other players progress in the rankings during their disabled time.

[6] If a player sequentially attacks an enemy then the attack is not count for the "combo attacks." In addition, if players do not attack during ten minutes then their chain of combo attacks are reset to 0.

Table 5. Payoff matrix for the test scenario consisting of two players and an enemy with very few hit points. The player who attacks the enemy receives S, and the other player receives T. If neither player attacks the enemy, then each receives P. Attack by both players is impossible, because either player can defeat the enemy

	Attack	Wait
Attack	-, -	S, T
Wait	T, S	P, P

In simple terms, consider that two players battled an enemy in this scenario, where their relationship is represented in Table 5. The relationship between the variables is $T > S > P$ in this payoff matrix. Attack is not efficient, when S is less than T. However, if they do not attack the enemy, they waste time by waiting for someone else to attack, i.e., P is lowest. It is not possible to cooperate both players in this scenario, because an attack on the enemy by either player immediately defeats the enemy. The values of this payoff matrix depend on each players situation, e.g., the differences between the two attack powers[7]. In the scenario, both try to avoid the worst situation (i.e., they get P), but they also do not want to pay the cost to avoid the worst situation (i.e., they do not want to get S). This social dilemma is similar to the one in the "Leader game" (Table 1). In that game, Pareto efficiency is achieved when one cooperates, and the other does not. Then, the cooperator receives S, and the noncooperator T. That is, players receive a high payoff by sharing S and T on repeated plays of the game, a process known as ST reciprocity [30]. We recognized this cooperative behavior, which provided the payoff T from one to the other, as a cooperative behavior in this scenario.

Cooperative behavior is an inefficient attack, as shown in Table 5; thus we define a_{ij} as the attack efficiency indicator: $a_{ij} = e_{ij}/M(e_i)$, where e_{ij} are the event points in player i's jth attack and $M(e_i)$ is the median of $e_i = \{e_{i1}, \cdots, e_{iN}\}$ (N is the frequency of player i's attacks). We considered cooperative behavior to be in the range of $a \leq 0.40$. Accordingly, we define c_i as the proportion of cooperative behavior ($a_i \leq 0.40$) for player i. We regarded a cooperator as a player where $c \geq 0.10$.

References

1. André, J.B.: The evolution of reciprocity: social types or social incentives? Am. Nat. **175**(2), 197–210 (2010)
2. Axelrod, R.: The Evolution of Cooperation: Revised Edition. Basic Books (2006)
3. Bainbridge, W.S.: The scientific research potential of virtual worlds. Science **317**(5837), 472–476 (2007)

---

[7] In addition, it does not mean that the relationship between the payoffs is constant. If a player is about to go to sleep, then S is larger than T, because the attack points replenish the next morning.

4. Bshary, R., Grutter, A.: Image scoring and cooperation in a cleaner fish mutualism. Nature **441**(7096), 975–978 (2006)
5. Castronova, E.: On the research value of large games: natural experiments in norrath and camelot. Games Culture **1**(2), 163–186 (2006)
6. Dunbar, R.: On the origin of the human mind. In: Carruthers, P., Chamberlain, A. (eds.) Evolution and the Human Mind, pp. 238–253. Cambridge University Press (2000)
7. Dunbar, R.: Gossip in evolutionary perspective. Rev. Gen. Psychol. **8**(2), 100–110 (2004)
8. Grujić, J., Fosco, C., Araujo, L., Cuesta, J., Sánchez, A.: Social experiments in the mesoscale: humans playing a spatial prisoner's dilemma. PLoS ONE **5**(11), e13749 (2010)
9. Grujić, J., Röhl, T., Semmann, D., Milinski, M., Traulsen, A.: Consistent strategy updating in spatial and non-spatial behavioral experiments does not promote cooperation in social networks. PLoS ONE **7**(11), e47718 (2012)
10. Hamilton, W.D.: The evolution of altruistic behavior. Am. Nat. **97**(896), 354–356 (1963)
11. Kobayashi, H., Kohshima, S.: Unique morphology of the human eye. Nature **387**(6635), 767–768 (1997)
12. Lindgren, K.: Evolutionary phenomena in simple dynamics. Artif. Life **II**, 295–312 (1991)
13. Nakamura, M.: 'Gatherings' of social grooming among wild chimpanzees: implications for evolution of sociality. J. Hum. Evol. **44**(1), 59–71 (2003)
14. Nowak, M.A.: Five rules for the evolution of cooperation. Science **314**(5805), 1560–1563 (2006)
15. Nowak, M.A., Sigmund, K.: A strategy of win-stay, lose-shift that outperforms tit-for-tat in the prisoner's dilemma game. Nature **364**(6432), 56–58 (1993)
16. Nowak, M.A., Sigmund, K.: Evolution of indirect reciprocity. Nature **437**(7063), 1291–1298 (2005)
17. Packer, C.: Reciprocal altruism in Papio anubis. Nature **265**(5593), 441–443 (1977)
18. Peysakhovich, A., Rand, D.G.: Habits of virtue: creating norms of cooperation and defection in the laboratory. Manage. Sci. **62**(3), 631–647 (2015)
19. Rand, D.G., Arbesman, S., Christakis, N.: Dynamic social networks promote cooperation in experiments with humans. Proc. Natl. Acad. Sci. **108**(48), 19193–19198 (2011)
20. Rand, D.G., Nowak, M.A.: Human cooperation. Trends Cogn. Sci. **17**(8), 413–425 (2013)
21. Riolo, R.L., Cohen, M.D., Axelrod, R.: Evolution of cooperation without reciprocity. Nature **414**(6862), 441–443 (2001)
22. Smith, J.M.: Must reliable signals always be costly? Anim. Behav. **47**(5), 1115–1120 (1994)
23. Smith, J.M., Harper, D.: Animal Signals. Oxfold University Press, Oxford (2003)
24. Szell, M., Sinatra, R., Petri, G., Thurner, S., Latora, V.: Understanding mobility in a social petri dish. Sci. Rep. **2**, 457 (2012)
25. Szell, M., Thurner, S.: Measuring social dynamics in a massive multiplayer online game. Soc. Netw. **32**(4), 313–329 (2010)
26. Szell, M., Thurner, S.: How women organize social networks different from men. Sci. Rep. **3**, 1214 (2013)
27. Takano, M., Wada, K., Fukuda, I.: Environmentally driven migration in a social network game. Sci. Rep. **5**, 12481 (2015)

28. Takano, M., Wada, K., Fukuda, I.: How do newcomers blend into a group? Study on a social network game. In: 3rd International Workshop on Data Oriented Constructive Mining and Multi-agent Simulation (DOCMAS) & 7th International Workshop on Emergent Intelligence on Networked Agents (WEIN) (workshop at WI-IAT 2015) (2015)
29. Takano, M., Wada, K., Fukuda, I.: Reciprocal altruism-based cooperation in a social network game. New Gen. Comput. **34**(3), 257–271 (2016)
30. Tanimoto, J., Sagara, H.: Relationship between dilemma occurrence and the existence of a weakly dominant strategy in a two-player symmetric game. Biosystems **90**(1), 105–114 (2007)
31. Trivers, R.L.: The evolution of reciprocal altruism. Q. Rev. Biol. **46**, 35–37 (1971)
32. Wang, J., Suri, S., Watts, D.J.: Cooperation and assortativity with dynamic partner updating. Proc. Natl. Acad. Sci. U.S.A. **109**(36), 14363–14368 (2012)
33. Wilkinson, G.: Food sharing in vampire bats. Sci. Am. **262**(2), 76–82 (1990)

Continuous Recipe Selection Model
Based on Cooking History

Shuhei Yamamoto[1]([envelope]), Noriko Kando[2], and Tetsuji Satoh[1]

[1] Faculty of Library, Information and Media Science,
University of Tsukuba, Ibaraki, Japan
{yamahei,satoh}@ce.slis.tsukuba.ac.jp
[2] Information and Society Research Division,
National Institute of Informatics, Tokyo, Japan
kando@nii.ac.jp

Abstract. Thousands of different recipes are posted on recipe sites by consumers who often refer to them when they cook. Such users occasionally select new recipes. In this paper, we propose for users a recipe selection model composed of both preference and challenging viewpoints to appropriately predict recipes that users are more likely to cook next in continuous cooking behaviors. The occurrence probability of the challenging behaviors of each user is estimated from past cooking sequences, and recipe scores are calculated by incorporating preference and challenging viewpoints. Our experimental evaluations using actual cooking histories demonstrate the high prediction performance of our method. We clarified the estimation efficiency of users who tackle challenging recipes.

Keywords: Recipe recommendation · Repertoire expansion · Challenging behavior · Sequence mining

1 Introduction

Recently, many recipes can be found on such Internet user-posting sites as COOKPAD,[1] which has almost two million recipes. Each recipe mainly consists of ingredients and cooking steps and techniques/hints. In these popular and convenient sites, users can search for recipes by category, ingredients, and preparation times.

Based on familiarization to such sites, users want to easily find recipes that fit their specific contexts such as preference of ingredients, health concerns, and the ingredients in their refrigerators. To satisfy such requirements, several studies focus on recipe recommendations [5,16,18,19]. Although these researches demonstrated high efficiency in experimental evaluations with actual users/cooks, they do not consider the user skills. We also believe that recommending appropriate recipes that match users is important because not every user can cook every recipe well.

---

[1] http://www.cookpad.com.

© Springer International Publishing AG 2016
E. Spiro and Y.-Y. Ahn (Eds.): SocInfo 2016, Part I, LNCS 10046, pp. 138–151, 2016.
DOI: 10.1007/978-3-319-47880-7_9

To enhance user cooking skills, we have to recommend recipes that not only reflect preferences but that also challenging. In our previous research [10], we proposed challenging recipe recommendation methods to expand cooking repertoires with users by estimating the high versatility recipes included ingredients that are frequently used in many recipes and users have not experienced. On the other hand, in the daily cooking process, users do not probably select recipes for repertoire expansion. To satisfy their desire for eating, users select their preference recipes from their repertoire. To adjust the nutrient requirements for their family, users might cook different recipes.

In this paper, we propose a recipe selection model to recommend recipes by considering such continuous cooking processes of user and predict the recipes that users will cook next. Our hypothesis is that users have two recipe selection viewpoints: preference and challenging. They continuously switch both viewpoints with different probability and select recipes based on each viewpoint model. In preference recipe selection, we calculate recipe scores based on TF-IDF using a user's preference ingredient model, combined with dissimilarity against recent cooking history. In the challenging recipe selection, we estimate recipe difficulty from ingredient popularity.

The remainder of our paper is organized as follows. In Sect. 2, we discuss related works, and in Sect. 3, we explain our recipe selection model and describe the score calculation method based on both preference and challenging viewpoints. In Sect. 4, the experimental evaluations for estimating the scores of the actually cooked recipes are described, including average precision. In Sect. 5, we discuss our model's effectiveness and conclude the paper by briefly describing future works in Sect. 6.

2 Related Works

2.1 Recipe Recommendations

Many studies have addressed recipe recommendations, especially personalized suggestion methods. Ueda *et al.* [16] proposed a personalized method based on food preferences and derived user preferences based on browsing activities. Their purpose was consecutive recipe recommendations and introduced weight parameters to each recipe to avoid repeatedly recommending identical recipes. To train the ingredients of the preference model of users, Yang *et al.* [19] developed a preference extraction system, which suggested dishes by images, where users click on the preference ingredients from the images. Harvey *et al.* [4] analyzed factors that influence people's food choices and clarified reasons for liking or disliking a recipe include particular ingredients or combinations, health-conscious, and the preparation time. As reasons for positive ratings of peoples, they reported the type of dish and the novelty of the recipe. Kadowaki *et al.* [5] recommended recipes by recommending foods based on the evidence of user situations. To extract recommending evidence, they analyzed tweets related to food. Yajima and Kobayashi [18] proposed a recommendation method for easy-cooking recipe by identifying such recipes by both content and user conditions. Their recommendation also considered the user contexts such current seasonal, preference seasoning, and preference ingredients. Karikome and Fujii [6] proposed a nourishment-balancing method based on the ingredients of recipes. Their recommendations are suitable for those suffering from food restrictions or allergies. Ge *et al.* also [2] proposed the food recommender system by not only user's preferences but also user's health on a mobile platform. Seki and Ono [13] identified practical recipes that are easy to understand, written concisely with sufficient description, and offer detailed tips and pointers. They analyzed popular recipes in on-line recipe communities and clarified the content characteristics, e.g., heating levels and cooking times.

These methods were developed to recommend appropriate recipes or dishes. In contrast, we propose a recipe selection model composed of both preference and challenging viewpoints.

2.2   Recipe Structure Analysis

Another research branch analyzes structures. Su *et al.* [14] extracted the relationship between cuisines and ingredients for recipe recommendations. Wang *et al.* [17] and Yu *et al.* [20] extracted a recipe graph as a workflow for cooking procedures. By using sub-graph similarity, they achieved high accuracy with their recipe recommendations. Freyne and Berkovsky [1] made a bipartite graph between recipes and ingredients to infer both user food preferences and to create special recipes based on dietary considerations. Their experiments achieved high coverage and reasonable accuracy. Teng *et al.* [15] constructed complement and substitution networks using co-occurrence ingredients in a recipe and captured user's preference for healthier variants of a recipe. They accurately predicted recipe ratings with ingredient networks. Kuo *et al.* [7] proposed an intelligent menu-planning method that recommends sets of recipes that contain user-specified ingredients as queries. They proposed a graph-based algorithm for representing the co-occurrence relationships between recipes and ingredients. Hamada *et al.* [3] built a structure of text material for cooking shows. By building an original dictionary composed of ingredients, cuisines, and cookware, they analyzed the cooking processes in recipes and made flow diagrams. Rokicki *et al.* [11] clarified differences in nutritional values between recipes posted by different user groups such as ages and genders. Especially, in gender, Rokicki *et al.* [12] focused on ingredients and preparation instructions in each recipe, clarified the gender differences, e.g., men are more innovative, women use spices more subtly, and showed to improve food recommendation by using these features. Kusmierczyk *et al.* [8] analyzed a large online food community website and found that food innovation factor depends on the season of the year and the week. They clarified the temporal dynamics in online food innovation.

As mentioned above, although studies that analyze cooking processes by focusing on ingredients and cuisines are widely known, no studies have addressed recipe selection that provides challenging recipes to expand the repertoires of cooks. In this paper, we incorporate both a recipe selection model of preferences

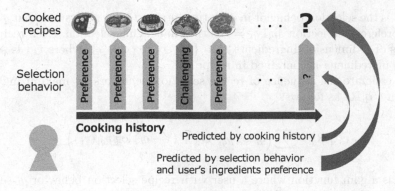

Fig. 1. Recipe selection model

and challenges and appropriately build it for the continuous recipe selection of users.

3 Continuous Recipe Selection Model

3.1 Overview

In this section, we explain our overview of a continuous recipe selection model based on the cooking histories of users who generally select recipes based on their cooking and eating tastes. We believe that users have two viewpoints in recipe selection: their personal preferences and challenging recipes that they haven't tried before. Preference recipe selection is a model that chooses a recipe that includes a user's preferred ingredients and satisfies specific tastes. The challenging recipe selection model chooses a recipe that includes ingredients whose preparation is difficult and expands cooking repertoires. Our hypothesis is that users sequentially switch between such recipe selection behaviors by probability and each time pick a recipe based on their current selection behavior. If the occurrence probability of challenging recipe selection is predicted by their cooking history, we can appropriately predict a recipe that they might prepare next time (Fig. 1).

This section of our paper consists of the following parts. Section 3.2 estimates recipe selection behaviors based on each user's cooking history and calculates the occurrence parameters of challenging recipe selection. Section 3.3 describes our recipe score calculation method that is composed of both preference and challenging viewpoints.

3.2 Recipe Selection Behavior Estimation

Here, our goal is to estimate a sequence of recipe selection behaviors $\mathbf{q} = \{q_1, q_2, \cdots, q_T\}$ from a sequence of recipes cooked by user $\mathbf{R} = \{\mathbf{r_1}, \mathbf{r_2}, \cdots, \mathbf{r_T}\}$, where T denotes the number of times a particular dish has been cooked by each user. q_t is the selection behavior in each time-stamp t, and we assume that $q_t = 0$ is the preference selection and $q_t = 1$ is the challenging selection. Each recipe $\mathbf{r_t}$ consists of I dimension ingredients $\mathbf{r_t} = \{r_{t,1}, r_{t,2}, \cdots, r_{t,I}\}$, where $r_{t,i}$ is set to 1 when ingredient i is contained in recipe $\mathbf{r_t}$.

We estimate the sequence of recipe selection behaviors by maximizing cost function $c(\mathbf{q}|\mathbf{R})$ as follows:

$$c(\mathbf{q}|\mathbf{R}) = \sum_{t=2}^{T} \Big(q_t \log p_{q_t}(\mathbf{r_t}) + (1 - q_t) \log p_{q_t}(\mathbf{r_t}) \Big). \tag{1}$$

$p_{q_t}(\mathbf{r_t})$ is a gain function where a user with recipe selection behavior q_t selects recipe $\mathbf{r_t}$ and is defined as follows:

$$\log p_{q_t}(\mathbf{r_t}) = \sum_{i=1}^{I} r_{t,i} \log p_{q_t}(t,i), \tag{2}$$

$$p_0(t,i) = \frac{n_{t,i} + 1}{t + 2} \quad, \quad p_1(t,i) = \frac{m_t + 1}{t + 2}, \tag{3}$$

where $p_0(t,i)$ denotes the gain function of preference selection ($q_t = 0$) and $n_{t,i}(= \sum_{k=1}^{t-1} r_{k,i})$ is the number of times ingredient i was cooked up to time-stamp t. $p_1(t,i)$ denotes gain function of challenging selection ($q_t = 1$) and $m_t(= \frac{1}{I_t} \sum_{k=1}^{t-1} n_{k,i})$ is the average number of cooking times for all the ingredients up to time-stamp t. I_t is the number of experience ingredients up to time-stamp t. Therefore, all ingredients have a constant probability value for each user.

Equation (1) estimates the preference selection when recipe $\mathbf{r_t}$ contains many ingredients with which users have much cooking experience. In contrast, when a recipe contains many ingredients with which users have little or no experience, it designates it as a challenging selection because p_1 has higher probability than p_0. Each gain value, which is updated based on accumulated cooking experiences, can estimate recipe selection behavior that reflects cooking experience up to time-stamp t.

By these procedures, we obtain sequence of recipe selection behavior $\mathbf{q}$ and then estimate the occurrence probability of the challenging behaviors in each user. Our assumption is that users basically select from among their preferred recipes and occasionally select new, challenging recipes to expand the range of what they know how to cook. A probability model to achieve such an assumption is suggested by cumulative exponential distribution function $\beta = 1 - e^{-\lambda x}$. In other words, users probably make challenging selections with probability β in time-interval x from the last challenging selection occurrence to the present, and this procedure is controlled by parameter λ of each user. λ is estimated by each user's sequence of recipe selection behavior. Based on the expectation value of the exponential distribution, when a time-interval sequence of the occurrence of a challenging selection is assumed to be $\Delta = \{\Delta_1, \Delta_2, \cdots, \Delta_D\}$, λ is calculated as follows: $\lambda = D / \sum_{d=1}^{D} \Delta_d$. According to λ value's magnitude, the challenging recipe selection behavior is simplified.

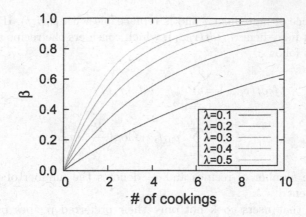

Fig. 2. Occurrence probability distributions of challenging selection behavior β in each parameter λ

Calculate λ from example sequence of recipe selection behavior as follows: $\mathbf{q} = \{0, 0, 0, 1, 0, 0, 0, 1, 0, 0, 0, 0, 1\}$. The time-interval sequence of the occurrence of challenging selection ($q_t = 1$) is $\mathbf{\Delta} = \{3, 3, 4\}$, and λ is calculated as follows: $\lambda = 3/(3+3+4) = 0.3$. Figure 2 shows the occurrence probability distribution of challenging selection behavior β in each parameter λ. The vertical and horizontal axes are probability β and number of time-intervals x. For example, a user with parameter $\lambda = 3$ makes a challenging selection behavior at approximately 0.8 probability when time-interval x is 6.

3.3   Recipe Score Calculation

In this section, we calculate each recipe's score to predict recipe $\mathbf{r_{t+1}}$ that users will select with high probability for upcoming cooking opportunity $t + 1$. When the recipe scores obtained by both preference and challenging viewpoints are $preference(\mathbf{r_{t+1}})$ and $challenge(\mathbf{r_{t+1}})$, respectively, $score(\mathbf{r_{t+1}})$, which integrates them, is defined as follows:

$$score(\mathbf{r_{t+1}}) = preference(\mathbf{r_{t+1}})^{(1-\beta)} \cdot challenge(\mathbf{r_{t+1}})^{\beta}, \tag{4}$$

where β denotes the occurrence probability of the challenging selection explained in the previous section. The calculation procedures of $preference$ and $challenge$ are separately described below.

Preference Score: We calculate the preference scores for each user for each recipe as follows:

$$preference(\mathbf{r_{t+1}}) = tfidf(\mathbf{r_{t+1}})^{\alpha} \cdot div(\mathbf{r_{t+1}})^{(1-\alpha)}, \tag{5}$$

$tfidf(\mathbf{r_{t+1}})$ denotes the TF-IDF score, which is often used in information retrieval systems to calculate term weights [9]. In our recipe selection model, TF is the importance of ingredient i and is replaceable with $p_0(i, t)$. IDF evaluates the rarity of an ingredient. $tfidf(\mathbf{r_{t+1}})$, which considers the recipe a document, is calculated as follows:

$$tfidf(\mathbf{r_{t+1}}) = \sum_{i=1}^{I} tf_i \cdot idf_i \cdot r_{t+1,i} \tag{6}$$

$$= \sum_{i=1}^{I} p_0(t, i) \cdot \log\left(\frac{R}{n_i}\right) \cdot r_{t+1,i}, \tag{7}$$

where R is the number of recipes and n_i denotes the number of recipes that contain ingredient i.

We believe that users cook not only their preferred recipes but also new recipes more than previously cooked recipes because Harvey *et al.* [4] reported the novelty of the recipe as one of the reasons for user's recipe choices. $div(\mathbf{r_{t+1}})$ in Eq. (5), which denotes the diversification scores to recommend different recipes compared with the recipes contained in the recent cooking history, is defined as follows:

$$div(\mathbf{r_{t+1}}) = \sum_{k=1}^{t} \frac{1 - sim(\mathbf{r_{t+1}}, \mathbf{r_k})}{\log_2(t - k + 2)}, \tag{8}$$

where $sim(\mathbf{r_{t+1}}, \mathbf{r_k})$ gives by cosine similarity using ingredient IDF value $(= \log \frac{R}{n_i})$ and defined as follows:

$$sim(\mathbf{r_{t+1}}, \mathbf{r_k}) = \frac{\sum_{i \in I_{t+1}} \sum_{j \in I_k} idf_i \cdot idf_j}{\sqrt{\sum_{i \in I_{t+1}} idf_i^2} \cdot \sqrt{\sum_{i \in I_k} idf_i^2}}, \tag{9}$$

where I_t denotes the ingredients set with recipe $\mathbf{r_t}$. When feature ingredients with high IDF values are not contained in another recipe, the $sim(\mathbf{r_{t+1}}, \mathbf{r_k})$ becomes low and $\log_2(t - k + 2)$ is a decaying function that emphasizes the recent cooking history.

In Eq. (5), we introduce parameter α to control both the preference and diversification measures. α varies within 0.0 and 1.0 by users. A user with optimal $\alpha \approx 1.0$ selects his preferred recipes without considering recent cooking history. A user with optimal $\alpha \approx 0.0$ more often selects different recipes compared with his recent cooking history.

Challenging Score: We assume that challenging recipes are difficult and contain not only unfamiliar ingredients but also low versatility items. Here, we can replace low versatility ingredients with high IDF ingredients. However, we suggest that users hesitate to select recipes that consist of high IDF ingredients only because they require both cooking and eating behaviors after recipe selection. We calculate the challenging scores as follows:

$$challenge(\mathbf{r_{t+1}}) = \max_{i \in I_{t+1}^-} idf_i - \min_{i \in I_{t+1}^-} idf_i, \tag{10}$$

where I_{t+1}^- denotes the ingredients set with recipe $\mathbf{r_{t+1}}$ and with non-experience up to time-stamp t $(n_{t,i} = 0)$ for a user. This formula concurrently evaluates the recipes including ingredients with both high and low IDFs.

4   Experimental Evaluations

4.1   COOKPAD Dataset

To evaluate the effectiveness of our method, we used the COOKPAD dataset,[2] which is produced by the COOKPAD Inc. and the National Institute of Informatics in Japan. In COOKPAD, user cooking histories are posted as cooking reports called Tsukurepos. Each Tsukurepo has a posted user id, a date, and a target recipe id. In this paper, we assume the Tsukurepo sequences of users are their cooking histories because these reports include all of the information to build our recipe selection model.

Figure 3 shows the frequency distribution of users who have made Tsukurepo posts. The vertical and horizontal axes respectively show the number of users and Tsukurepos by users by common logarithm $\log_{10}$. The number of users who posted Tsukurepos is approximately 10^5. Since the objective of our research is to model users with continuous cooking experiences, we randomly extracted 100 users whose number of Tsukurepos was within 500 to $1,000^3$.

For experimental evaluations, we extracted reproducible recipes that were posted at least once as a Tsukurepo by other users. The number of different kinds of recipes with at least one Tsukurepo was 805,018. From these recipes, we calculated the IDF values that are used in recipe score calculations. Figure 4 shows the frequency distribution of the Tsukurepo recipes. The vertical and horizontal axes respectively show the number of recipes and Tsukurepos by a common logarithm. The number of recipes posted once by Tsukurepo exceeds 10^5. Over 10,000 recipes were posted as Tsukurepos.

4.2   Ingredients Coherence Procedure in Japanese Characters

Ingredients are identified by various characters in Cookpad because the ingredients in each recipe are written in Japanese by users. For example, the word onion in English can be written as "たまねぎ", "玉ねぎ", or "タマネギ". Therefore, the same ingredients can probably be managed simply by their names. In this paper, we standardize the ingredients using MeCab,[4] which is the Japanese morphological analyzer, by changing all the characters to Katakana.

---

[2] http://www.nii.ac.jp/dsc/idr/cookpad/cookad.html.

[3] The number of users whose the number of Tsukurepos was within 500 to 1,000, was 1,234.

[4] Another part-of-speech and morphological analyze is, http://mecab.sourceforge.net/.

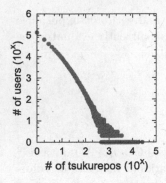

Fig. 3. Frequency distribution of users who posted Tsukurepos

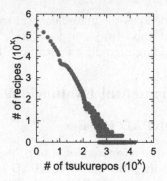

Fig. 4. Frequency distribution of recipes posted as Tsukurepos

4.3 Experiemental Procedure

To evaluate the effectiveness of our recipe selection model, we divided the Tsukurepo sequence of each user into two sets: former part 90 % and latter part 10 %. Both used parameter estimation of each user and evaluation.

Parameter Estimation: Our model needs to estimate several parameters in each user. A λ of the controlling challenging occurrence is calculated by estimating recipe selection behavior sequence $\mathbf{q}$. An α for considering the recent cooking history is optimized in each user by maximizing the recipe's score of training data in 100 randomly extracted recipes. Because the similarity against the cooking history's recipes is part of the calculation score, we update model p_0 by the recipes after the evaluation.

Evaluation Metric: In the evaluation, we prepared 100 recipes that were randomly extracted from 805,018 Tsukurepo recipes in addition to actual cooking recipes in next time-stamp $t + 1$. We calculated the scores for these recipes and ranked the scores in descending order. When many actual cooking recipes are ranked at the top, we believe that our recipe selection model's performance is appropriate. To quantitatively evaluate these processes, we calculate the mean average precision up to the top K (MAP@K) in all of the evaluation users. To evaluate the effectiveness that consists of both preference and challenging viewpoints, we create the rankings using preference score *preference* and challenging score *challenging* and compare each MAP@K.

4.4 Experimental Results

Parameter Distributions: We estimated the sequence of recipe selection behavior $\mathbf{q}$ for the cooking history of each user and calculated λ to control the

Fig. 5. Two examples of recipe selection behavior sequences **q** and estimated λ values

probability of challenging selection occurrence β. From these results, we show the sequences with the highest and lowest λ value in Fig. 5. The vertical and horizontal axes are respectively the q_t values and the number of times the recipe was cooked. When that number is low, the user frequently estimated the challenging recipe selection behavior because model p_0 cannot adequately be built. When sufficient cooking history is accumulated, the estimation of the challenging recipe selection behavior was different based on users because we adequately built model p_0, and the difference appears as λ values.

The frequency distributions of both the α and λ parameters are shown in Figs. 6 and 7. The α value was tuned in 0.1 steps and the λ value was rounded to one decimal place. The distribution of parameter α was within 0.3 and 0.6, and the value with the most users was 0.4. This result suggests that most users frequently selected unknown/new recipes by considering their own recent cooking history. The distribution of parameter λ fell within 0.1 and 0.5 and the values exceeding 80 % of the users were within 0.2 and 0.3.

MAP Score by Varying the Top K: Figure 8 shows the prediction performance result as MAP@K scores based on the parameters that were estimated for each user. The vertical and horizontal axes are the mean average precision up to the top K (MAP@K) and the top K in ranking. The top K steps are 1, 10, 20, 30, 40, 50, 60, 70, 80, 90, and 100. Hybrid scores, which are composed of both preference and challenging viewpoints, achieved the highest scores compared with only the preference and challenging scores in all the top K steps. In the preference and challenging scores, although the MAP values of both were not different at the top 1, preference showed higher MAP values than challenging from the top 10 to 100.

5   Discussion

From Fig. 8, the preference scores showed higher recipe prediction performance than the challenging scores. We show examples of recipe rank sequences esti-

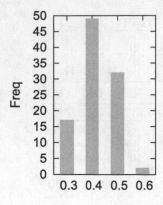

Fig. 6. Frequency distribution of α

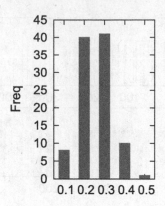

Fig. 7. Frequency distribution of λ

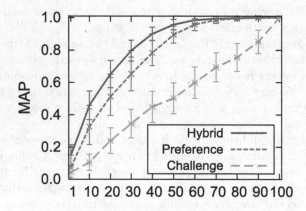

Fig. 8. MAP values by varying top K

mated by each method in Fig. 9. The vertical and horizontal axes are commonly the estimation ranks of the actually cooked recipes and the number of times they were cooked. The hybrid, which is composed of both selection models, achieved maximum estimation performance among the three methods. Although the challenging selection model made too many incorrect estimations, it successfully estimated the recipes at the top rank several times. Therefore, we confirmed that users occasionally selected the challenging recipes.

Figure 6 suggests that users selected a recipe by considering their recent cooking history because most users are included in $\alpha < 0.6$. Figure 10 shows the estimation rank sequence with an example user using optimal constant parameter $\alpha = 0.5$ and another using dynamically optimized α in each time-stamp, combined with the α value optimized in each time-stamp. The optimal α of this user is 0.5, estimated from the training data, and she considers her preference and recent cooking history with the same ratio. On the other hand, we observed that the dynamic α values in each time-stamp are optimized at both ends,

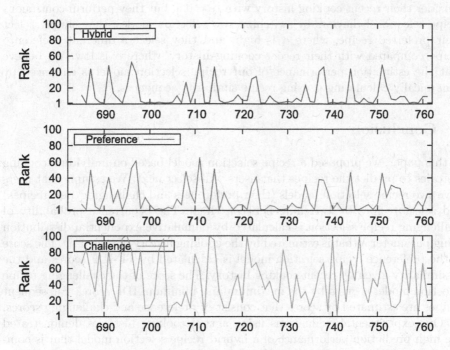

Fig. 9. Example of recipe rank sequences estimated by each method

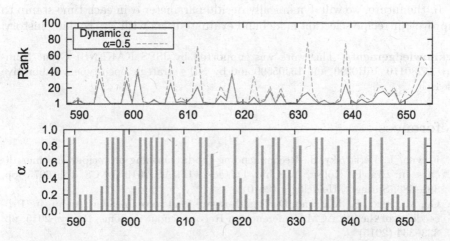

Fig. 10. Example pf recipe rank sequences estimated by dynamic α

such as $\alpha \geq 0.9$ and $\alpha \leq 0.2$. These results suggest that users don't usually consider their recent cooking history with $\alpha = 0.5$, but they perform continuous recipe selection behaviors by switching two selection-modes; they simply select their preferred recipe, where α is high, and they select a different preference recipe compared with their recent cooking history, where α is low. We believe that the estimation performance of our recipe selection model is enhanced by dynamically calculating α using recipe similarity sequences.

6 Conclusion

In this paper, we proposed a recipe selection model based on individual cooking histories to predict the recipes that users will select next. We assume that users have two recipe selection models (their preferences and challenging/new recipes) and select recipes by switching between them. The occurrence probability of challenging recipe selection is calculated by cumulative exponential distribution using parameter λ that is estimated by the cooking history of each user. The score of the preference recipe selection model is calculated by TF-IDF scores and the dissimilarity given by recent cooking history. The score of the challenging recipe selection model is given by the maximum and minimum IDF scores. Subsequent recipes are estimated by scores that consist of preference and challenging scores.

Our experimental evaluations using actual cooking histories demonstrated the high prediction performance of a hybrid recipe selection model that is composed of preference and challenging viewpoints compared with only preference and challenging scores. In the preference viewpoint, we obtained suggestions that users make continuous recipe selections by switching between two selection-modes. Users simply choose their preferred recipes and the dissimilar preference recipes based on their recent cooking histories.

In the future, we will dynamically decide parameter α in each time-stamp to improve our recipe selection model and evaluate users with less cooking history.

Acknowledgements. This work was supported by JSPS KAKENHI Grant Numbers 25280110, 16H0290, and 15J05599 and by NII's strategic open-type collaborative research.

References

1. Freyne, J., Berkovsky, S.: Recommending food: reasoning on recipes and ingredients. In: Bra, P., Kobsa, A., Chin, D. (eds.) UMAP 2010. LNCS, vol. 6075, pp. 381–386. Springer, Heidelberg (2010)
2. Ge, M., Ricci, F., Massimo, D.: Health-aware food recommender system. In: Proceedings of the 9th ACM Conference on Recommender Systems, RecSys 2015, pp. 333–334 (2015)
3. Hamada, R., Ide, I., Sakai, S., Tanaka, H.: Structural analysis of cooking preparation steps in Japanese. In: Proceedings of the IRAL 2000, New York, NY, USA, pp. 157–164 (2000)

4. Harvey, M., Ludwig, B., Elsweiler, D.: Learning user tastes: a first step to generating healthy meal plans. In: First International Workshop on Recommendation Technologies for Lifestyle Change (LIFESTYLE 2012), pp. 18–23 (2012)
5. Kadowaki, T., Yamakata, Y., Tanaka, K.: Situation-based food recommendation for yielding good results. In: 2015 IEEE International Conference on Multimedia Expo Workshops, pp. 1–6, June 2015
6. Karikome, S., Fujii, A.: A system for supporting dietary habits: planning menus and visualizing nutritional intake balance. In: Proceedings of the ICUIMC 2010, pp. 56:1–56:6 (2010)
7. Kuo, F.F., Li, C.T., Shan, M.K., Lee, S.Y.: Intelligent menu planning: recommending set of recipes by ingredients. In: Proceedings of the ACM Multimedia 2012 Workshop on Multimedia for Cooking and Eating Activities, pp. 1–6. ACM, New York (2012)
8. Kusmierczyk, T., Trattner, C., Nørvåg, K.: Temporal patterns in online food innovation. In: Proceedings of the 24th International Conference on World Wide Web, WWW 2015 Companion, pp. 1345–1350 (2015)
9. Manning, C.D., Raghavan, P., Schutze, H.: Scoring, term weighting, and the vector space model. In: Introduction to Information Retrieval, p. 100. Cambridge University Press (2008)
10. Nakaoka, Y., Satoh, T.: The proposal of the cooking repertory expansion method based on a user's cooking experience. Technical report of IEICE (2014) (in Japanese with English abstract)
11. Rokicki, M., Herder, E., Demidova, E.: What's on my plate: towards recommending recipe variations for diabetes patients. In: Proceedings of 2016 LBRS (2015)
12. Rokicki, M., Herder, E., Kuśmierczyk, T., Trattner, C.: Plate and prejudice: gender differences in online cooking. In: Proceedings of the 2016 Conference on User Modeling Adaptation and Personalization, UMAP 2016, pp. 207–215 (2016)
13. Seki, Y., Ono, K.: Discriminating practical recipes based on content characteristics in popular social recipes. In: Proceedings of the 2014 ACM International Joint Conference on Pervasive and Ubiquitous Computing, pp. 487–496 (2014)
14. Su, H., Shan, M., Lin, T., Chang, J., Li, C.: Automatic recipe cuisine classification by ingredients. In: Proceedings of the Ubicomp 2014, pp. 565–570 (2014)
15. Teng, C.Y., Lin, Y.R., Adamic, L.A.: Recipe recommendation using ingredient networks. In: Proceedings of the 4th Annual ACM Web Science Conference, WebSci 2012, pp. 298–307 (2012)
16. Ueda, M., Takahata, M., Nakajima, S.: User's food preference extraction for personalized cooking recipe recommendation. In: Proceedings of the SPIM 2011, pp. 98–105 (2011)
17. Wang, L., Li, Q., Li, N., Dong, G., Yang, Y.: Substructure similarity measurement in Chinese recipes. In: Proceedings of the WWW 2008, pp. 979–988 (2008)
18. Yajima, A., Kobayashi, I.: "Easy" cooking recipe recommendation considering user's conditions. In: Proceedings of the WI-IAT 2009, pp. 13–16 (2009)
19. Yang, L., Cui, Y., Zhang, F., Pollak, J.P., Belongie, S., Estrin, D.: PlateClick: bootstrapping food preferences through an adaptive visual interface. In: Proceedings of the 24th ACM International on Conference on Information and Knowledge Management, CIKM 2015, pp. 183–192 (2015)
20. Yu, L., Li, Q., Xie, H., Cai, Y.: Exploring folksonomy and cooking procedures to boost cooking recipe recommendation. In: Du, X., Fan, W., Wang, J., Peng, Z., Sharaf, M.A. (eds.) APWeb 2011. LNCS, vol. 6612, pp. 119–130. Springer, Heidelberg (2011)

Politics, News, and Events

Examining Community Policing on Twitter: Precinct Use and Community Response

Nina Cesare[⊠], Emma S. Spiro, Hedwig Lee, and Tyler McCormick

University of Washington, Seattle, WA 98195, USA
ninac2@uw.edu

Abstract. A number of high-profile incidents have highlighted tensions between citizens and police, bringing issues of police-citizen trust and community policing to the forefront of the public's attention. Efforts to mediate this tension emphasize the importance of promoting inter-action and developing social relationships between citizens and police. This strategy – a critical component of community policing – may be employed in a variety of settings, including social media. While the use of social media as a community policing tool has gained attention from precincts and law enforcement oversight bodies, the ways in which police are expected to use social media to meet these goals remains an open question. This study seeks to explore how police are currently using social media as a community policing tool. It focuses on Twitter – a functionally flexible social media space – and considers whether and how law enforcement agencies are co-negotiating norms of engagement within this space, as well as how the public responds to the behavior of police accounts.

Keywords: Police · Community policing · Social media · Twitter

1 Introduction

Current approaches to managing the relationship between citizens and police emphasize the importance of promoting police-community interaction and accommodating pathways of communication that place citizens and police on a more level playing field [8]. By facilitating communication regarding appropriate police practices and general community well-being, police have a greater chance of being viewed as legitimate and promoting citizen cooperation in law enforcement activity [4]. This communication may be particularly important for alleviating feelings of distrust toward the police among minority citizens [2,4,11].

Effective communication between police and citizens is the central component of a strategy known as community policing. Community policing emphasizes the importance of fostering interpersonal relationships between citizens and law enforcement, as well as training officers to take a holistic rather than incident-based approach toward evaluating community well-being [3]. The premise of this strategy is that officers should pay attention not only to instances of crime within

© Springer International Publishing AG 2016
E. Spiro and Y.-Y. Ahn (Eds.): SocInfo 2016, Part I, LNCS 10046, pp. 155–167, 2016.
DOI: 10.1007/978-3-319-47880-7_10

an area but also to overall community health – including citizens' satisfaction with their community and feelings of safety. Officers who effectively integrate themselves into the fabric of a community through community policing may also help level power dynamics between officers and citizens, thus making enforcement activity appear more just and appropriate [4].

While the interpersonal interaction central to community policing may occur through offline, face-to-face contact, it may also occur on social media sites. Social media sites provide easy-to-access common forums through which citizens can engage with law enforcement agencies by gathering information on current events or providing feedback on police activity in real time. In light of this, a recommendation from the President's Task Force on 21st Century Policing states: "Law enforcement agencies should adopt model policies and best practices for technology-based community engagement that increases community trust and access" [9].

It is unclear, however, what these intended strategies are and what best practices should be adopted. Given this deficit, this study seeks to examine community policing on social media – specifically, on Twitter – by examining the behavior and community engagement activity of law enforcement agencies, as well as reactions to these accounts from the public. Focusing on the activity of law enforcement in two cities - Seattle and New York - this study finds that a common interpretation of community policing on Twitter may still be in flux.

2   Community Policing on Twitter

There is growing interest in the adoption of social media spaces as platforms for community policing [9], however there are few common standards for how law enforcement agencies are expected to achieve this goal. This is particularly true of Twitter, which is a highly flexible and sparse social media space that many view as both a social space and a "global town square" or "microphone for the masses" designed for information collection and broadcast [1,6,7]. Strategies for using Twitter vary among users, and this may render it difficult to for precincts to establish standard Twitter usage practices to promote community policing.

As of yet, little published research has explored how law enforcement agencies use Twitter. Heverin and Zach provide the most comprehensive work on this topic by analyzing what information police choose to share on Twitter and how citizens respond to and share this information [5]. These authors find that law enforcement agencies generally tweet about events, traffic, safety awareness, and crime prevention, and that they sometimes engage with news media or other law enforcement agencies directly. They also note that citizens who mention law enforcement agencies often do so through direct retweets of police account activity. While this study provides valuable insight into the content of the conversation space occupied by law enforcement agencies on Twitter, further research is needed to better understand how law enforcement agencies utilize and co-negotiate the use of Twitter as a community policing tool.

Understanding how Twitter is used as a community policing tool is a multifaceted question that requires consideration of activity from both law enforcement accounts and the public. In this project we consider how law enforcement agencies behave on Twitter, as well as how the public reacts and response to their behavior. Through these findings we seek to uncover whether police and/or citizens appear to use Twitter as a community policing tool, and whether common standards for how to use this space are beginning to emerge. Overall, we seek to shed light upon what community policing currently means within this space to help guide future research on this topic.

3   Data

This project uses two different samples of data from Twitter: (1) data that documents online activity from selected police accounts and (2) public tweets that contain mentions of these accounts posted over a 100 day period in 2015.

We focus on law enforcement agencies in Seattle and New York City (NYC) as case studies. NYC was selected due to the large number and diverse nature of accounts associated with the New York Police Department (NYPD). Seattle was selected due to the fact that the Seattle Police Department is known to interact frequently with citizens via social media and endorses social media engagement as a form of building community involvement and trust [10]. All precincts and oversight bodies in these two locations were enumerated by members of the research team and are expected to represent a census of the law enforcement bodies within each city. Note that these accounts include a variety of entities – including commissioners, fire departments and divisions of the NYPD's Housing Bureau's Police Service Areas. The diversity of cases included in this analysis will add dimensionality to our understanding of how users engage with the police on Twitter.

Between the two locations, a total of 135 active police accounts were identified and included in this analysis (see Appendix Table 3 for a complete list of the accounts included). Police account information – including user profile metadata, user timeline data, and network information – was collected using Twitter's public Application Programming Interface (API). Public mentions of these accounts were collected via Gnip's Historical Powertrack Twitter API.[1] These data include all public mentions, but exclude mentions that were subsequently deleted by the user.

4   Findings

We examine Twitter behavior of both police and citizens interacting with police through the lens of community policing. Findings focus in part on the behavior of the police accounts, including connections between accounts and the extent to which accounts appear to engage with the public through the use of Twitter conventions. They also address how the public responds to content posted by these accounts.

---

[1] We would like to acknowledge the assistance of the University of Washington eScience Center for providing access to and assistance using the Gnip Historical Powertrack API.

4.1   Police Account Activity and Social Interaction

We begin by examining how police behave on Twitter. For this analysis we focus specifically on posting behaviors that may be used as a means of engaging with the public. Measures of community engagement considered include: the proportion of tweets that are directed at other accounts, the proportion of tweets that are retweets from other accounts, the proportion of tweets that link to outside material, the proportion of tweets that contain multimedia (such as photos of officers within a precinct, participating in community activity), and the average number of hashtags used per tweet. Each of these behaviors indicate that police accounts are making strategic use of platform conventions. Some – such as using directed messages – help to capture direct interaction between police accounts and other police accounts and/or citizens. Distributions of the proportion of police posts that contain each of these features are depicted in Fig. 1.

For the NYPD, we see a fairly normal, sometimes negatively skewed distribution in activity across accounts for selected community engagement measures. Most Seattle accounts, however, participate in almost none of the engagement activities examined. Indeed, the contrast between accounts from these two locations is striking. Closer examination reveals that Seattle accounts are often used as 'beat' accounts that keep citizens up-to-date on criminal incidents, as seen in the example tweets in Fig. 2. Such posts are often formulaic and auto-generated. If we break down posting activity statistics by these designations, as seen in

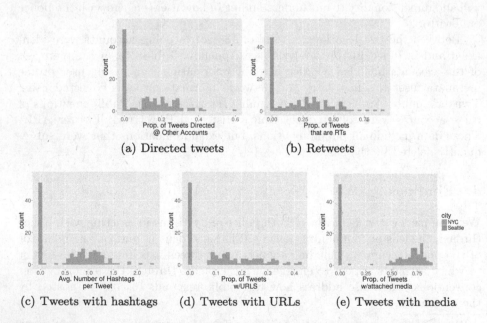

(a) Directed tweets   (b) Retweets

(c) Tweets with hashtags   (d) Tweets with URLs   (e) Tweets with media

Fig. 1. Proportion of police account tweets containing textual and content features by location - Seattle in blue and NYC in orange. (Color figure online)

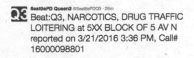
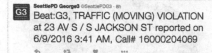

Fig. 2. Illustration of Seattle 'beat' accounts

Table 1. Interaction-based posting statistics by account designation

Account	@Mentions	RTs	URLs	Media	Hashtags
Seattle PD	0.596	0.123	0.365	0.106	0.100
Seattle beat accounts	0.000	0.015	0.000	0.000	0.000
NYPD	0.199	0.367	0.197	0.703	0.033

Table 1, we can identify these patterns clearly. The Seattle Police Department general account is highly engaged, in some respects even more so than the NYPD accounts, while the Seattle 'beat' accounts show minimal to no use of many Twitter conventions.

Next, we consider the connections between these police accounts. We may expect that as law enforcement agencies develop strategies for Twitter use, they do so by watching one another's behavior and/or co-negotiating norms of use within this space. Figure 3 displays the network of following relationships among the accounts collected. NYPD accounts - including oversight bodies and local precincts - are highly interconnected, indicating they follow one another on the platform - perhaps keeping up to date on what others are posting and how they utilize platform conventions. Most hyper-local Seattle accounts, on the other hand, are only connected to the Seattle Police Department's primary account (@SeattlePD). This account appears to act as a broker or bridge between the Seattle Police Department and NYPD accounts.

We see a similar pattern emerge when we consider connectedness in the form of a shared audience. To operationalize this, we consider account A and B connected if the have followers in common. A visualization of this network is displayed in Fig. 4. Again, SeattlePD acts as a broker between disparate clusters of Seattle PD and NYPD accounts. Despite the fact that community policing via social media is a proposed tool for national change, there seems to be little city-to-city communication regarding how this tool is intended to be used.

Overall, we see highly disparate patterns of Twitter use for law enforcement agencies in Seattle and New York City. Among NYPD accounts there is somewhat strong consistency in Twitter usage and, given following connections between NYPD accounts, a possible co-negotiation of norms within this space. Casual observation of NYPD accounts indicates that most are similar visually as well. Many feature a single figurehead displayed in the profile photo and a description that lists the commanding officer, mission and/or region of oversight, and a link to the NYPD Social Media Customer Use Policy. While the

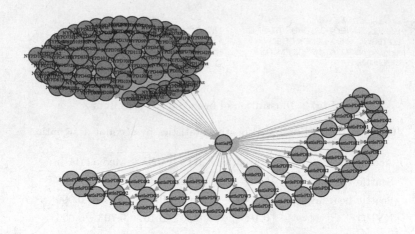

Fig. 3. Following relationships among SEA and NYC police accounts

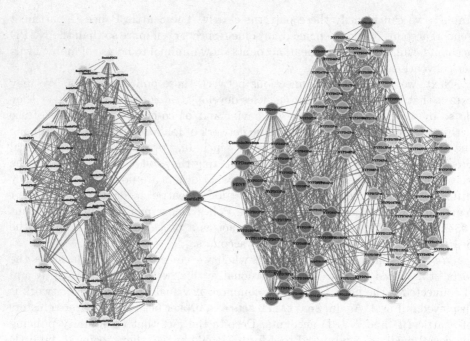

Fig. 4. Shared audience relationships among SEA and NYC police accounts

Seattle PD's primary account (@SeattlePD) appears to focus on community engagement by utilizing Twitter conventions and connecting to both NYPD and Seattle accounts, many Seattle accounts are not connected to one another and do not seem to consider community engagement at all.

4.2   Reactions from the Public

In addition to examining how law enforcement accounts use Twitter as a community policing tool, we also consider how the public reacts to this behavior. For instance, does including multimedia in Tweets encourage users to share or respond to this content? Do citizens use Twitter as a space to connect with police and raise topics of discussion regarding police conduct and community well-being?

We first consider whether the proportion of URLs and multimedia included in tweets and the average number of hashtags per tweet is positively or negatively associated the volume of citizen reactions the account receives. Our outcomes of interest are the average number of retweets and favorites that an account's tweets receive (as measured by timeline content gathered via the Twitter API). Figure 5 displays predicted values from linear regression models that explore these outcomes. Models used control for the following activity measures: the proportion of tweets with URLs in the account's timeline, the proportion of tweets containing media in the account's timeline, the average number of hashtags per tweet in the account's timeline, the account's total friends and followers, and the average number of tweets issued per month. Note that for these models we exclude Seattle 'beat' accounts, as there seems to be little interaction between citizens and these accounts. In addition to this, four outliers were removed that displayed follower counts and/or average retweet values that were two to three times magnitude of other accounts. The dependent variables were logged to correct for over-dispersion.

These results indicate that including links to outside sources and/or media in tweets is a catalyst for the spread of information. The more media and URLs an account includes in their tweets, the higher their average retweet count. Overall, it seems that engagement strategies lend themselves better to information spread (retweets) than gaining popularity (favorites).

Another critical component of the effectiveness of Twitter as a community policing tool involves the willingness of citizens to use Twitter as a platform for discussing controversial and/or important matters. Given this, we examine the overall sentiment of public posts mentioning police accounts, as well as the

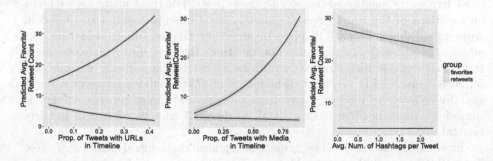

Fig. 5. Predicted average favorites/retweets given police account activity

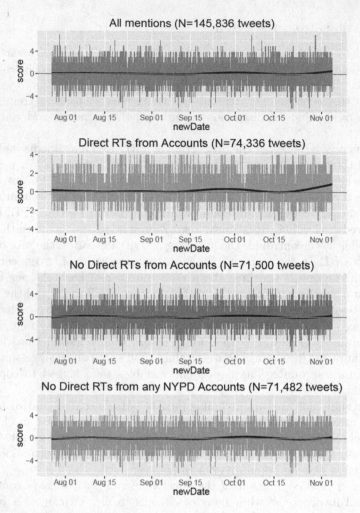

Fig. 6. Public sentiment in tweets mentioning police accounts over time

most frequently mentioned terms with these posts. For these analyses, we focus on mentions of the top 10 most frequently mentioned accounts - displayed in Appendix Table 4. Figure 6 displays positive/negative sentiment over time for: all mentions of high-activity accounts, mentions that are direct retweets, mentions that exclude direct retweets, and mentions that specifically exclude any retweets from NYPD accounts. While content that excludes retweets is occasionally collectively more negative than positive and direct retweet content is sometimes more positive than neutral, the overall content of tweets is fairly neutral.

Examining word frequencies helps contextualize this neutral commentary (see Table 2). Prior to analyzing word frequencies, text was lowered and stemmed and stop words, punctuations and URLs were removed. This analysis indicates that

Table 2. Frequently occurring terms in public tweets mentioning police accounts

Corpus	Most frequently occurring terms
All high-activity accounts	"call", "commissbratton", "day", "fdni", "fdny", "fire", "holder", "member", "neverforget", "nyc", "nypd", "nypdnew", "nypdnews", "offic", "polic", "randolph", "seattlepd", "thank", "today", "wanted", "will"
Direct Retweets	"343", "", "800577tip", "call", "commissbratton", "day", "end", "fdni", "fdny", "member", "neverforget", "nyc", "nypd", "nypdnews", "offic", "polic", "today", "tour", "year"
No Direct Retweets	"billdeblasio", "blake", "commissbratton", "cop", "fdni", "get", "jame", "nyc", "nypd", "nypdnew", "offic", "polic", "rememb", "seattlepd", "thank", "time", "today", "will", "wwe"
No NYPD Retweets	"billdeblasio", "blake", "commissbratton", "cop", "fdni", "get", "jame", "nyc", "nypd", "nypdnew", "offic", "polic", "seattlepd", "thank", "time", "today", "wwe"

the majority of terms frequently used appear commonplace or even complementary. We see some evidence of public critique through mentions of "Blake," which refers to James Blake – a professional tennis player who accused the NYPD of use of excessive force in September of 2015 – within the corpus that excludes retweeted content. However, on the whole citizens do not seem to view Twitter as a space for public debate where controversial topics may be raised and critical police-citizen communication may occur.

5   Discussion

Social media spaces are cited by many as contexts in which police and citizens can interact with one another and establish opens lines of communication. Improving police-citizen communication – a core strategy of community policing – is cited as a possible solution to alleviating tensions between police and citizens [9]. Social media sites provide up-to-the-minute communication spaces through which citizens and police may share thoughts, form relationships and engage in interaction. One such platform is Twitter, which offers flexible communication tools featuring microblog posts generated in real time. Promoting communication between citizens and police within this space might be an important component of making community policing online a successful reform strategy.

However, while some express enthusiasm regarding the use of Twitter as a community engagement space, it seems that this goal has yet to be reached. For one, we note highly inconsistent interpretations of effective Twitter use between cities. While the behavior of NYPD accounts - including oversight bodies and local precincts - is fairly consistent, relies on Twitter conventions that may promote engagement, and may be actively co-negotiated through connections between accounts, Seattle PD accounts are relatively disparate and many hyper-local accounts do not appear to engage with citizens at all. A shared expression of community policing on Twitter has yet to come into focus.

Citizens do not seem to view Twitter as a space for public discussion, either. Most tweets mentioning high-activity accounts included in this study were neutral in tone and contained relatively common terms related to policing and community. There is some evidence of critical feedback – specifically, some mentions of tennis star James Blake – but this was only seen after removing retweeted content. The lack of critical conversation within this space may be a consequence of inconsistent police-citizen engagement strategies on Twitter. The broad spectrum of police engagement strategies on Twitter may make citizens feel unclear about what is or is not appropriate to say within this space. Overall, while Twitter may help level the playing field and open dialogue between these groups, it may be a long time until this goal is achieved and/or Twitter interaction may supplement face-to-face interaction as a community policing strategy.

6   Conclusion and Future Work

Twitter may provide a space where police and citizens can interact, form relationships and discuss matters of community importance. However, without clear directives regarding the appropriate use of Twitter as a community policing tool law enforcement agencies - particularly agencies within different cities - seem to be developing very different ways of using this space. There seems to be little co-negotiation between cities regarding best practices, and we note little use of Twitter as a space for critical discussion of citizen satisfaction and community well-being.

Given that Gnip's Historical Powertrack Twitter API provides access to a large volume of longitudinal data, future work may consider if and how police behavior has evolved over time. For instance, we may ask: since citizens respond to the inclusion of multimedia in police tweets, do police accounts include more multimedia over time? Additionally, do police accounts change their profile content over time to help them appear less institutional and more personal – perhaps by making the commanding officer the 'face' of the account, as we currently see in many NYPD precinct accounts? Having this longitudinal data may help us view the nuance of normative co-negotiation that occurs within this space.

Future work may also compare the composition of law enforcement agencies' Twitter audience with that of their in-person constituents. In order to develop effective community policing strategies on Twitter, agencies must be sure they are accessing a diverse and representative group of citizens within this space. Otherwise, community policing through Twitter will do little to improve police-citizen trust, promote police legitimacy, and ensure community safety. Preliminary analyses indicate that police account audiences on Twitter may not be racially/ethnically diverse and/or representative of communities that police intend to protect. We plan to build upon our existing data to explore this possibility in greater detail.

Acknowledgements. This material is based upon work supported by, or in part by, the U. S. Army Research Laboratory and the U. S. Army Research Office under contract/grant numbers W911NF-12-1-0379 and W911NF-15-1-0270. Support also provided by NSF SES-1559778 to McCormick.

Appendix

Table 3. Cities and accounts collected

Seattle	SeattlePDC1, SeattlePDJ3, SeattlePDC3, SeattlePD1, SeattlePDW3, SeattlePDU2, SeattlePDK1, SeattlePDF1, SeattlePDM1, SeattlePDB3, SeattlePDD3, SeattlePDS3, SeattlePDW1, SeattlePDO1, SeattlePDB2, SeattlePDW2, SeattlePDG2, SeattlePDG1, SeattlePDE3, SeattlePDN3, SeattlePDL1, SeattlePDQ3, SeattlePDU1, SeattlePDJ2, SeattlePDF3, SeattlePDM3, SeattlePDE1, SeattlePDN2, SeattlePDD2, SeattlePDM2, SeattlePDL3, SeattlePDK3, SeattlePDQ1, SeattlePDF2, SeattlePDR3, SeattlePDO3, SeattlePD, SeattlePDE2, SeattlePDC2, SeattlePDO2, SeattlePDU3, SeattlePDS1, SeattlePDL2, SeattlePDQ2, SeattlePDN1, SeattlePDK2, SeattlePDR1, SeattlePDJ1, SeattlePDS2, SeattlePDR2, SeattlePDD1, SeattlePDG3
New York City	NYPD68Pct, NYPD90Pct, NYPDPSA2, NYPD66Pct, NYPDPSA6, NYPD123Pct, NYPD102Pct, NYPD6Pct, NYPD47Pct, NYPDPaws, NYPD108Pct, NYPD19Pct, NYPD101Pct, NYPD10Pct, NYPD50Pct, NYPDnews, NYPD81Pct, CommissBratton, NYPD105Pct, NYPD110Pct, NYPD100Pct, NYPD13Pct, NYPD72Pct, NYPD84Pct, NYPD49Pct, NYPD94Pct, NYPD71Pct, NYPD67Pct, NYPDMTS, NYPD88Pct, NYPD69Pct, NYPD61Pct, NYPD33Pct, NYPD46Pct, NYPD109Pct, NYPD75Pct, NYPD112Pct, NYPD7Pct, NYPD120Pct, nypdrecruit, NYPD32Pct, NYPD24Pct, NYPD122Pct, NYPD26Pct, NYPD5Pct, NYPD23Pct, NYPD106Pct, NYPD73Pct, NYPD78Pct, NYPD107Pct, NYPD115Pct, NYPD48Pct, NYPD103Pct, NYPD9Pct, NYPD121Pct, NYPD40Pct, NYPD17pct, NYPD60Pct, NYPDDetectives, NYPD83Pct, NYPD79Pct, NYPD52Pct, NYPD114Pct, NYPD20Pct, NYPD45Pct, NYPD77Pct, NYPD113Pct, NYPD104Pct, NYPD25Pct, NYPD30Pct, NYPD62Pct, NYPD42Pct, NYPD76Pct, NYPD111Pct, NYPD41Pct, NYPD63Pct, NYPD70Pct, NYPD34Pct, NYPD44Pct, NYPD1Pct, NYPD43Pct, FDNY, NYPD28Pct

Table 4. Top 10 most frequently mentioned accounts

Account	Description	Mentions in public corpus
NYPDNews	The official Twitter of the New York City Police Dept.	36167
FDNY	The official New York City Fire Department feed	28912
CommissBratton	Commissioner of the New York City Police Department	12897
SeattlePD	Seattle Police news/events	8664
NYPD19Pct	Deputy Inspector James M. Grant, Commanding Officer. The official Twitter of the 19th Precinct #UpperEastSide #UES	3770
NYPD108Pct	Captain John Travaglia, Commanding Officer. The official Twitter of the 108th Precinct	2592
NYPD78Pct	Captain Frank DiGiaComo, Commanding Officer. The official Twitter of the 78th Precinct	2546
NYPD1Pct	Captain Mark Iocco, Commanding Officer. The official Twitter of the 1st Precinct	1855
NYPDDetectives	NYPD Chief of Detectives	1791
NYPD104Pct	Captain Mark Wachter, Commanding Officer. The official Twitter of the 104th Precinct	1658

References

1. Bruns, A., Highfield, T., Lind, R.A.: Blogs, twitter, and breaking news: the produsage of citizen journalism. In: Produsing Theory in a Digital World: The intersection of Audiences and Production in Contemporary Theory, vol. 80, pp. 15–32 (2012)
2. Doherty, C., Tyson, A., Weisel, R.: Few say police forces nationally do well in treating races equally. Pew Research Center: U.S. Politics and Policy (2014)
3. Community Policing Consortium. Understanding community policing: a framework for action (1994)
4. Goldsmith, A.: Police reform and the problem of trust. Theor. Criminol. **9**(4), 443–470 (2005)
5. Heverin, T., Zach, L.: Twitter for city police department information sharing. Proc. Am. Soc. Inf. Sci. Technol. **47**(1), 1–7 (2010)
6. Leetaru, K.H.: Who's doing the talking on twitter? (2015)
7. Murthy, D.: Twitter: microphone for the masses? Media Cult. Soc. **33**(5), 779 (2011)

8. US Dept of Justice: Office of Community Oriented Policing Services (COPS). Community policing defined (2009)
9. President's Task Force on 21st Century Policing. Final report on the president's task force on 21st century policing (2015)
10. O'Toole, K.: Chief o'toole's announces new social media policy. Seattle police blotter (2015)
11. Stoutland, S.E.: The multiple dimensions of trust in resident/police relations in Boston. J. Res. Crime Delinquency **38**(3), 226–256 (2001)

The Dynamics of Group Risk Perception in the US After Paris Attacks

Wen-Ting Chung[1], Kai Wei[2], Yu-Ru Lin[3(✉)], and Xidao Wen[3]

[1] Department of Psychology in Education, University of Pittsburgh, Pittsburgh, USA
[2] School of Social Work, University of Pittsburgh, Pittsburgh, USA
[3] School of Information Sciences, University of Pittsburgh, Pittsburgh, USA
yurulin@pitt.edu

Abstract. This paper examines how the public perceived immigrant groups as potential risk, and how such risk perception changed after the attacks that took place in Paris on November 13, 2015. The study utilizes the Twitter conversations associated with different political leanings in the U.S., and mixed methods approach that integrated both quantitative and qualitative analyses. Risk perception profiles of Muslim, Islam, Latino, and immigrant were quantitatively constructed, based on how these groups/issues were morally judged as risk. Discourse analysis on how risk narratives constructed before and after the event was conducted. The study reveals that the groups/issues differed by how they were perceived as a risk or at risk across political leanings, and how the risk perception was related to in- and out-group biases. The study has important implication on how different communities conceptualize, perceive, and respond to danger, especially in the context of terrorism.

Keywords: Risk perception · Terrorist attacks · Risk analysis · Immigrants · Group identity · In- and out-group bias · Social media · Mixed methods

1 Introduction

In recent years, terrorist attacks, particularly plotted and carried out by the self-declared Islamic groups such as Al-Qaeda and Islamic State (ISIS), have complicated the policies and politics of immigrant issues globally. In an immigrant society like the U.S., its immigration policy is sensitive to disruptive events that signal potential threat of any particular group of immigrants to its national security. The recent rising "Islamic terrorism," in which terrorists proclaim their identity of being Islam believers and justify their motives and actions by Islam [1], has complicated people's attitudes toward Muslims who practice Islam.

This research aims to disentangle *risk perception* – that is, how people perceive and judge a potential harm [2], and how such risk perception changes through a major terrorist event. Characterizing risk perception is important because the collective perception drives the public's felt need of reducing the perceived danger, which leads to demanding the government's actions as policy

© Springer International Publishing AG 2016
E. Spiro and Y.-Y. Ahn (Eds.): SocInfo 2016, Part I, LNCS 10046, pp. 168–184, 2016.
DOI: 10.1007/978-3-319-47880-7_11

makers need to respond to the public's perceived risk [2]. Nevertheless, quantitatively measuring perceived risk is a challenge. Risk is not a neutral, objective, fixed concept, but a psychological perception and sociocultural construct, created and shaped through social processes [2,3]. Understanding the nature of risk perception, what factors would affect people's perception toward a specific risk target, and how these perceptions differ is fundamental to facilitate policy communication and formation.

Risk perception is particularly sensitive in the context of terrorist attacks. The terrorist incident in France occurred on a scale comparable to 9/11 attacks on New York City, altering the public perception of threat toward terrorism. In the U.S., perceived threats toward immigrant groups have been well documented [4,5], and political ideology has been studied as a determinant role in differentiating societal members' views and opinions on immigration issues [6]. After Paris attacks, in responding to this significant terrorist incident, U.S. politicians, especially the candidates for 2016 presidential election, spoke to the public. While the perceived threat toward terrorism rose high in general, their remarks conveyed distinct views of how terrorism, Muslims, and Islam should be concerned as a risk issue.

In this social process of co-constructing particular group or issue as risk, politicians are not the only persons who contributed. Differing from 9/11 a decade ago, during the Paris attacks, social media has now enabled people around the globe to participate in and contribute to, the disaster response. In the hours and days after the attacks, people in Paris and worldwide used social media, including Twitter, Facebook, Instagram and many other platforms, to serve a variety of immediate needs and supports for how we can make sense of, understand, survive and recover from such surprising, disruptive events. While these collective processes have always operated on populations impacted by terrorism, social media now makes them observable. Moreover, these expressive, communicative and conversational artifacts left on social media allow us to look deeply into how people respond to terrorism through how narrative and discourse of perceived risk is shaped and co-constructed in a social process over time.

In this study, we focus on examining *group risk perception*, referred as *how a group of people was morally judged and perceived as risk*. We develop a novel framework, utilizing both qualitative and quantitative methods, to analyze group risk perception by leveraging social and cultural psychology theories, including *moral dyad theory* [7], *moral foundation theory* [8], and *social identity theory* [9]. Our analysis is based on data collected and extracted from Twitter conversations covering the Paris attacks occurred on November 13, 2015. We are interested in the variety of patterns of ascribing and perceiving a cause of terrorism as risk, how the issues of terrorists are understood to be entangled with other social concerns, and the role of external disruptive events in shaping collective group risk perception. Our study is guided by the following **research questions:**

1. How did the public's interest of discussing a particular group of immigrants and immigrants in general change, before and after Paris attacks?

2. How were particular groups of immigrants and immigrants in general perceived as risk? Did the risk perception change before and after Paris attacks?
3. Did users with different political leanings perceive particular groups of immigrants and immigrants in general as risk differently? Did their risk perception change before and after Paris attacks?

This study uses a mixed methods approach that integrates both quantitative and qualitative analyses. Our study shows that Paris attacks boosted the public's conversations regarding Muslims and Islam. Responding to Paris attacks, people who held distinct political ideologies contributed to the construction of risk discourses distinctively. A key ideological difference revealed in our study is whether Muslims were perceived *as risk* or *as risk victims* through a development of "Islamophobia," and whether such risk concern was purveyed toward other immigrant communities.

This present work has several **key contributions:** First, it is the first empirical research on *risk perception* and *group risk perception* in the context of terrorism. Second, we propose a novel framework to investigate the construct of risk perception – which is grounded on social and cultural psychology theories, extracted by Lexicon method, and validated and expanded by an in-depth qualitative analysis on collective risk discourses through social media as platforms of social processes. Third, our findings, particularly the distinctions of risk perception between political leanings offer valuable insights for policy makers regarding what psychological mechanisms drive the public's opinions and have implications on the public's acts toward terrorism and response to relevant policies.

2   Related Work

2.1   Social Media and Risk Studies

Social media has been utilized to understand collective sense-making process during crises. In an earlier study, Cheong and Lee [10] proposed a micro-blogging-based approach to study civilian response to a terrorist attack. In the study, they followed the 2009 Jakarta and Mumbai terrorist attacks and demonstrated the utility of Twitter in terms of analyzing potential response to terrorist attacks. Following the 2013 Woolwich attack, Awan [11] examined 500 tweets from 100 users to investigate islamophobia (a sense of fear or hatred towards Muslim) on Twitter after the attack and created the typology describing group attributes for those who perceived Muslims as threats. Williams and Burnap [12] took a case study approach and studied the escalation, duration, diffusion and de-escalation of hateful speech on Twitter following the Woolwich attack. Following 2015 Paris attacks, researchers investigated attitudes towards Muslims by using crowd-sourcing to classify tweets into defending, neutral, and attacking categories, in which they found that a considerable number of tweets blaming Muslims were from western countries, such as the U.S. [13]. Researchers started to investigate the relationship between users' news sharing behaviors on Twitter and its potential relationship with their positions on issues such as Islam and immigration [14]. Most recently, researchers also started to explore ways that could potentially achieve automatic detection of cyber-hate speech from pre-defined hateful words related to race, disability and sexual orientation [15].

These previous studies have provided important insights regarding quantifying islamophobia [13], on-line hateful speech [12,15], and the influence of new exposure on attitudes [14], with the majority analyzing post-event response to Muslim group. However, few studies have examined the dynamics of perceptions towards specific groups (e.g., Latino immigrants) with respect to terrorist attacks. While there has been an arguable connection between terrorist attacks and an elevated risk perception [16], few research has provided evidence on how risk perception is manifested in different immigrant groups.

2.2 Risk Perception

The analysis of risk perception addresses how people perceive and judge a potential harm [2]. The perceived risk can be natural disaster such as tsunami, man-made hazards such as nuclear waste, or of mixed causes of nature and human such as the spreading of epidemic disease. We humans ourselves–individuals or groups can be concerned as risk, too. Kemshall [17] argued that human societies have developed into a risk culture, in which a philosophy of "better safe than sorry" has dominated how we conceptualize, perceive, and respond to danger.

Our conceptualization of *group risk perception* is grounded in social and cultural psychology studies – in particular, *moral dyad theory* [7], *moral foundation theory* [8], and *social identity theory* [9]. Moral dyad theory suggests that while evaluating a harm, humans spontaneously enact a cognitive template of dyadic morality in which we look for *an intentional agent who does harm and a suffering moral victim* to make sense and ascribe causes of harm. What criteria do people rely upon when judging immorality? Moral foundation theory suggests that morality across cultures varies but shares at least five foundations of "intuitive ethics" that our moral judgements are based on: Harm/Care, Fairness/Cheating, Loyalty/Betrayal, Authority/Subversion, and Sanctity/Degradation. The five foundations reflect the five virtues–the capacity to feel others' pain, altruism that concerns others' rights and autonomy, patriotism and self-sacrifice for the group, leadership and followership, and nobleness–that if being violated, we humans perceive immorality. It is argued that the violation of all of these moral virtues is a perceived harm while the violation of Harm/Care foundation would be perceived the most serious harm [7]. In cases that the judged harmful target is a person, however, group identity matters [3].

Social identity theory [9] suggests that people categorize social groups as ingroup or outgroup, depending on how they identify themselves as a belonged group member. People tend to favor their identified ingorup over outgroup, referred as in-group favoritism or in-group bias [18], risk perception is hence captured by how distinct groups of people within a society identify the *Self* and the *Other*, as the other is often more likely to be perceived as a dangerous threat and potential harm [3]. It is notable that the group risk perception, hence, may not relate to real harm or harm fact; yet the selection of risk is a social process reflecting a collective psychological need, e.g., to maintain particular social solidarity or a means for achieving specific political agendas [3].

Based on these theories, we developed an analysis framework through analyzing group risk perception using Twitter conversations before and after Paris attacks.

3   Method

3.1   Data Collection

To examine the group risk perception and its co-construction in social processes, we collected Twitter data over a period covering the Paris attacks. We are interested in how users change their expression reflecting the group risk perception. To mitigate the selection bias, we adopted a quasi-experiment method called *computational focus group* [19,20]. Based on the idea, we created our user panels containing users whose prior behavior showed them to be interested in information relevant to the study, and the relevant information we utilized is their political preference. The political preference was identified based on their *exclusive* interest in the party candidates for the U.S. 2016 presidential election. Specifically, users in the Democrat panel only followed candidates in the Democrat Party but not a single candidate in the Republican Party; likewise, users in the Republican panel only followed Republican candidates. Then, for every user in our panels, we collected his/her full historic tweets through the Twitter REST API. In total, we obtained 30, 804 unique users (Dem: 5, 426; Rep: 25, 378). Tweets from the Democrat panel were assigned with a "liberal" leaning, and tweets from the Republican panel were with a "conservative" leaning.

The Paris attacks occurred at 21:20:00 Paris Time (15:20:00 Eastern Standard Time) on November 13, 2015. We extracted data from our user panel historic tweets in the two weeks (one week prior and one week following the attacks), and organized data into *before* and *after* time intervals. Since we are interested in users' original expression, we removed retweets and duplicates. We identified the tweets related to specific groups ("Muslim", "Latino") or issues ("Islam", "Immigrant") using a set of selected keywords (search terms) – "immigra*," "latino(s)," "muslim(s)," and "islam*". Tweets containing any of these terms were assigned to the corresponding (non-exclusive) groups. For example, the query containing "immigra*" represents the tweets related to immigrants in general. For comparison, we created a *baseline* group containing tweets that did not match any of the search terms. Measures associated with the baseline group was used as a base rate for measures associated with all other specific groups/issues. This study included a total of 5, 164, 914 tweets. Table 1 summarizes the data for the five groups.

3.2   Defining and Extracting Data for Risk Perception

To quantitatively capture group risk perception from the tweets, we employed the moral foundation lexicon [21] based on moral foundation theories [8].

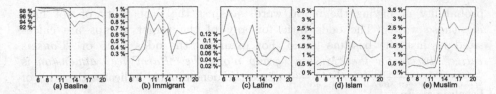

Fig. 1. The ratio of daily tweets before and after Paris attacks. The x-axes shows the dates from November 6th to 20th. The red/blue lines represent conservative/liberal tweets. The black vertical line represents the time that the attacks happened. (Color figure online)

Table 1. Number of tweets before and after the Paris attacks

	Before attacks		After attacks		
	Liberal	Conservative	Liberal	Conservative	Total
Baseline	539,808	1,859,802	541,848	2,031,708	4,973,166
Immigra*	2,917	13,160	2,732	13,873	32,682
Latino(s)	673	1,441	401	708	3,223
Muslim(s)	1,547	12,530	8,519	61,546	84,142
Islam*	1,096	9,090	7,027	54,488	71,701
Total	546,041	1,896,023	560,527	2,162,323	5,164,914

The lexicon was curated based on the psycho-linguistic lexicon LIWC (Linguistic Inquiry and Word Count) [22,23] and created specifically for the expression involving moral perception. The dictionary includes a total of 318 words that are categorized into 10 types of moral foundations: Harm vice/virtue, Fairness vice/virtue, Ingroup vice/virtue, Authority vice/virtue, and Purity vice/virtue. We used these moral words as a proxy for capturing tweet expressions indicating a group/issue was morally judged or perceived as a risk issue.

For quantitative analysis, we use *odds ratio* (OR) to measure (1) *change* in discussion about groups/issues after the attacks: the extent to which the panel users tend to mention a specific group/issue, and (2) *moral association* with groups/issues: the extent to which the panel users tend to morally associated a specific group/issue as a risk concern. For measuring the change, let $O_{s,t,d}$ be the *odds* of tweets from ideological leaning $s \in \{Liberal, Conservative\}$ that mention a particular target group/issue $t \in \{Muslim, \ Latino, Islam, Immigrant, Baseline\}$ in two time interval $d \in \{Before, After\}$, i.e. the probability of having the group terms against the probability of not having the group/issue/baseline terms. The odds ratio for mentioning any of the target groups after the attacks is calculated against the mentioning prior to the attacks, i.e., $\frac{O_{s,t,d=After}}{O_{s,t,d=Before}}$.

Similarly, for the measuring moral association in a particular time interval, let $O_{s,t,c}$ be the *odds* of tweets from ideological leaning s mentioning a particular target group t that contain any moral words in the 10 moral types, i.e. the probability of having the moral words against the probability of not having those words. The odds ratio for any of the target groups are calculated against the baseline group. For example, the odds ratio of *"Conservative" tweets for the "Muslim" group along the "HarmVice" dimension* is $\frac{O_{s=Conservative,t=Muslim,c=HarmVice}}{O_{s=Conservative,t=Baseline,c=HarmVice}}$, where the denominator indicates the odds of baseline group.

For qualitative analysis on the discourse of tweet contents, we randomly sampled 200 tweets from each moral category and for each group and ideological leaning. If the total tweets were less than 200, we sampled all the tweets in the category. The total sampled tweets for qualitative in-depth analysis were 17,913.

4   Analyses and Results

4.1   Changes in Discussing Groups After the Attacks

As an overall trend, Fig. 1 shows the ratio of daily tweets mentioning the group/issue terms before and after Paris attacks, indicating there were sudden increases in the ratio of Islam- and Muslim-related tweets.

To examine whether the interest of discussing a particular group/issue changed before and after Paris attacks, we computed odds ratios, using before the attacks as an unexposed group and after the attacks as an exposed group for baseline tweets, immigrant, Latino, Islam, and Muslim-related tweets. Figure 2 shows the changes of mentioning groups in terms of odds ratio.

After Paris attacks, there was a significant decrease of tweets related to Latinos (Liberal: $OR = 0.58$, $p < 0.001$; Conservative: $OR = 0.43$, $p < 0.001$) and immigrants (Liberal: $OR = 0.91$, $p < 0.001$; Conservative: $OR = 0.92$, $p < 0.001$), a pattern consistent with baseline tweets (Liberal: $OR = 0.34$, $p < 0.001$; Conservative: $OR = 0.30$, $p < 0.001$). However, a significant increase was observed among tweets related to Muslims (Liberal: $OR = 5.43$, $p < 0.001$; Conservative: $OR = 4.40$, $p < 0.001$) and Islams (Liberal: $OR = 6.31$, $p < 0.001$; Conservative: $OR = 5.37$, $p < 0.001$).

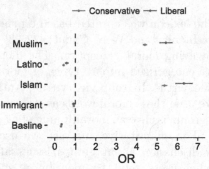

Fig. 2. The odds ratios of group/issue and baseline tweets before and after the attacks.

4.2   Risk Perception Profiles

For each group/issue, we created a *risk perception profile* to capture how a specific group/issue was perceived as risk through being morally judged by the 10 distinct moral dimensions. These risk profiles are visually summarized using radar charts (see Fig. 3), where each spoke shows the percentage of tweets mentioning the specific group/issue that is associated with a corresponding moral dimension. Table 2 shows the detailed statistics of these profiles.

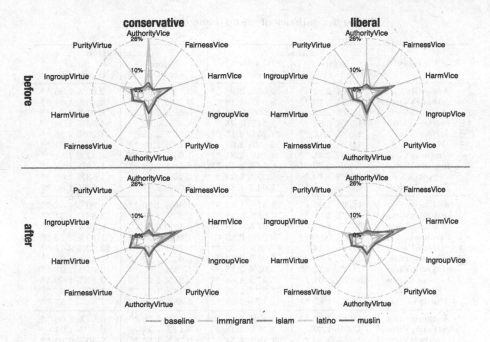

Fig. 3. Risk perception profiles. The five top-ranked dimensions are Harm vice/virtue, Authority vice/virtue, and Ingroup virtue. Group/Issue profiles differ by the significance of these dimensions and the associated political leanings, and changed after the attacks. The first leading dimension for Muslim, Islam, Latino, and immigrant were HarmVice, HarmVice, IngroupVirtue, and Authority vice/virtue, respectively.

The results show that among all of the 10 moral dimensions, the following five stand out: two are vices, HarmVice and AuthorityVice; three were virtues, HarmVirtue, AuthorityVirtue, and IngroupVirtue. Except for Latino, these are the five top-ranked dimensions for the profiles. To examine whether those tweets mentioning Muslim, Islam, Latino, and Immigrant (referring as group/issue tweets hereafter), compared with baseline tweets, are more likely to use moral words, we calculated odds ratios that indicate the odds of a group/issue tweet involving moral judgment words in type X against a non-X tweet, compared to the odds of a baseline tweet involving moral words in type X against a non-X tweet. The results confirm this pattern with a few exceptions (especially for Latino group) that both conservatives and liberals are more likely to engage moral judgement expression when discussing the four groups/issues, compared when discussing the topics irrelevant to the four groups/issues.

While the aforementioned five dimensions are prominent across profiles, the profiles differ by the order of the significance of these dimensions. Muslim and Islam are judged by HarmVice the most; Latino by IngroupVirtue the most; and Immigrant by Authority, either vice or virtue, the most. The following sessions discuss the profiles in details.

Table 2. Statistics of group/issue risk profiles.

Category	Liberal Before % tweets OR	Liberal After % tweets OR	Conservative Before % tweets OR	Conservative After % tweets OR
Muslim				
AuthorityVice	2.46% 2.82 **	2.07% 2.26 **	3.70% 2.65 **	2.87% 2.35 **
AuthorityVirtue	5.95% 1.76 **	3.93% 1.16 *	6.26% 1.70 **	4.43% 1.21 **
FairnessVice	1.16% 2.31 **	2.50% 4.55 **	0.68% 1.35 **	0.91% 2.30 **
FairnessVirtue	1.29% 0.92 **	1.44% 1.28 *	1.08% 0.82 *	1.33% 1.30 **
HarmVice	8.60% 2.16 **	11.25% 2.15 **	9.37% 2.63 **	11.26% 2.35 **
HarmVirtue	4.65% 1.79 -	5.56% 1.68 **	5.18% 1.89 **	7.36% 1.91 **
IngroupVice	1.29% 2.45 **	1.74% 1.82 **	1.62% 2.27 **	1.92% 1.51 **
IngroupVirtue	7.50% 2.90 **	6.78% 2.19 **	6.59% 2.39 **	5.73% 1.86 **
PurityVice	1.62% 1.70 *	1.33% 1.24 *	2.20% 2.25 **	1.63% 1.62 **
PurityVirtue	1.81% 2.16 **	1.64% 1.84 **	2.03% 2.41 **	1.67% 1.80 **
Islam				
AuthorityVice	1.28% 1.45 -	1.31% 1.42 **	1.86% 1.30 **	1.90% 1.54 **
AuthorityVirtue	7.12% 2.13 **	4.03% 1.18 **	6.50% 1.77 **	4.88% 1.34 **
FairnessVice	0.73% 1.44 -	1.27% 2.28 **	0.41% 0.81 -	0.54% 1.37 **
FairnessVirtue	1.37% 0.98 -	0.94% 0.83 -	1.60% 1.24 *	1.15% 1.12 **
HarmVice	12.04% 3.14 **	17.65% 3.63 **	9.70% 2.74 **	14.62% 3.17 **
HarmVirtue	4.20% 1.60 **	5.08% 1.53 **	6.06% 2.23 **	7.34% 1.91 **
IngroupVice	1.64% 3.12 **	2.05% 2.15 **	1.65% 2.32 **	2.61% 2.07 **
IngroupVirtue	5.47% 2.07 **	5.27% 1.68 **	5.91% 2.13 **	4.57% 1.46 **
PurityVice	1.37% 1.44 *	1.37% 1.28 *	1.84% 1.87 **	1.32% 1.31 **
PurityVirtue	1.55% 1.85 *	1.12% 1.25 *	1.88% 2.23 **	1.44% 1.55 **
Lation				
AuthorityVice	2.00% 3.97 **	3.42% 2.18 *	10.13% 7.76 **	6.64% 5.66 **
AuthorityVirtue	4.99% 1.39 -	4.75% 1.48 -	5.14% 1.38 -	3.81% 1.04 -
FairnessVice	2.49% 2.36 *	1.19% 4.54 **	0.76% 1.52 *	2.40% 6.19 **
FairnessVirtue	2.49% 2.50 **	3.42% 2.23 *	1.60% 1.23 **	1.13% 1.10 -
HarmVice	5.74% 1.49 *	6.09% 1.03 -	4.02% 1.07 *	6.78% 1.35 -
HarmVirtue	5.74% 0.84 -	2.23% 1.74 *	2.64% 0.94 -	4.66% 1.18 -
IngroupVice	1.25% 0.84 -	0.45% 1.30 -	0.56% 0.77 -	1.55% 1.22 -
IngroupVirtue	8.73% 4.56 **	11.29% 2.88 **	12.21% 4.72 **	7.20% 2.37 **
PurityVice	0.00% 0.31 -	0.30% 0 *	.97% 0.98 -	0.71% 0.70 -
PurityVirtue	0.75% 0.17 -	0.15% 0.83 -	0.56% 0.65 -	0.56% 0.60 -
Immigrant				
AuthorityVice	13.92% 18.12 **	8.09% 9.43 **	25.62% 23.72 **	13.77% 12.70 **
AuthorityVirtue	9.80% 3.02 **	9.19% 2.86 **	14.46% 4.30 **	11.48% 3.39 **
FairnessVice	0.55% 1.08 -	1.39% 2.50 **	0.47% 0.93 -	0.90% 2.29 **
FairnessVirtue	2.33% 1.68 **	1.39% 1.23 -	1.90% 1.46 **	1.33% 1.30 **
HarmVice	5.38% 1.30 **	3.59% 0.63 **	4.38% 1.17 **	5.33% 1.04 -
HarmVirtue	4.32% 1.65 **	5.01% 1.51 **	5.17% 1.88 **	6.55% 1.69 **
IngroupVice	0.82% 1.55 **	1.35% 1.41 *	1.09% 1.53 **	1.44% 1.13 -
IngroupVirtue	5.18% 1.95 **	7.39% 2.41 **	4.79% 1.71 **	5.38% 1.74 **
PurityVice	0.79% 0.82 -	1.06% 0.99 -	0.79% 0.80 *	0.88% 0.87 -
PurityVirtue	0.62% 0.73 -	0.48% 0.53 *	0.52% 0.61 **	0.52% 0.55 **

** indicates $p < .01$; * indicates $p < .05$

4.3 Group Risk Perception in Relation to Leanings and the Attacks

To examine the risk perception toward each group/issue in depth, we conducted qualitative discourse analysis on tweet contents. We adopted inductive approach to identify themes. The following discussions are based on both quantitative and qualitative analyses. We organize our discussions by group/issue.

Muslim. *HarmVice as the First Leading Dimension.* For both conservative and liberal tweets, HarmVice was the first leading dimension for Muslim risk profile, either before or after Paris attacks. The top used HarmVice words for both political leanings, in the order of frequency, were "kill* (kill, killed, killing)," "war," and "fight." A prominent concern was whether Muslims were perceived as either being *a risk* (potential moral agents who impose harm to others), or *at risk* (potential moral victims who are harmed by terrorists or other groups).

While Muslims were argued as a risk, the tweets involved discussions that Muslims kill, have war, and fight against others. The moral judgment toward Muslims was inclined to ascribe blame on them and suggested eliminating and excluding them all. In contrast, while Muslims were argued to be at risk, the tweets involved the discussions that other groups or agents who have intent to kill, have war, or fight against Muslims, e.g., Muslims being killed by Islamic State (ISIS) and the suffered victims too; the moral judgment toward Muslims was hence inclined to offer them care, empathizing Muslims in general.

Difference Between Political Leanings. The two views, Muslims being either a risk or at risk, were found in both conservative and liberal tweets. However, among conservative tweets, more discussions addressed Muslims, or particularly, *all* Muslims, as a group risk who to kill Americans or destroy the world, rather than at risk; e.g., "*You are Muslim. Today not a terrorist. Tomorrow asked to kill innocent people called infidels. Quran 5:32 translated #ParisAttacks #tcot.*" Moreover, conservative tweets were more likely to ascribe blames to those who were not Muslim but rejected the idea of declaring war at Muslims after the attacks and further considered those who refused to recognize a need of declaring war to Muslims as non-ingroup members.

Liberal tweets appeared more likely to decline the idea of treating all Muslims as a risk; instead, liberal tweets focused more on the Islamic religion, ISIS, or certain Muslim values and customs conflicting with other groups such as children, women, and LGBT. Compared with conservative tweets, liberal tweets were more inclined to disseminate the information indicating that terrorists or other groups do do harm to Muslims as a hate crime, and recognized Muslim immigrants as members of the U.S., who contributed to the nation as well. For example, "*There are 1.6 billion Muslims in the world, #Daech numbers @ best 30,000. They kill more Muslims than Westerners. They don't represent Islam,*" and "*Are you living in a bubble Muslims are serving in the us military Muslim countries are fighting Isis what are you doing @dbtcollector.*"

Change Before and After Paris Attacks. Before Paris attacks, HarmVice, the first leading dimension, was followed by IngroupVirtue, AurthorityVirtue, HarmVirtue, and AuthorityVice, which applied to both conservative and liberal tweets. After Paris attacks, HarmVice was still the first leading dimension. In both conservative and liberal tweets, there were increasing concerns in general of either other groups killing Muslims or being killed by Muslims. Among all the dimensions, only the tweets in three dimensions, HarmVirtue, FairnessVice, and FairnessVirtue, were more likely to appear after Paris attacks compared with baselines tweets. For conservatives, HarmVirtue, originally the fourth, became the second leading dimension, followed by IngroupVirtue, AuthorityVirtue, and

AuthorityVice. For liberals, IngroupVirtue remained the second dimension, followed by HarmVirtue and AuthorityVirtue, while FairnessVice replaced AuthorityVice as the fifth dimension. To sum up, HarmVirtue became more prominent among conservatives, and Fairness, both vice and virtue, among liberals.

Muslim Co-mentioned with Islam. "Islam" was frequently co-mentioned when the tweets mentioning "Muslim." There existed distinct patterns between liberals and conservatives in terms of how these tweets were related to the moral dimensions.

For liberals, before Paris attacks, the rate of co-mentioning of Muslim and Islam was only 0.02 %; after the attacks, the rate increased to 10.09 %. For conservatives, the rates before Paris attacks had been 9.3 %; after the attacks, the rates increased 2.2 % to 11.5 %. These statistics indicated that before the attacks, conservatives had related Islam to Muslims when talking about Muslims and the probability of doing so slightly increased after the attacks. Instead, liberals seldom related Muslim to Islam before attacks, yet the probability of doing so increased to the level of what conservatives did after the attacks. For both conservatives and liberals, after the attacks, these co-mentioning tweets were mostly likely to be related to the HarmVice dimension (liberal: 13.99 %; conservative: 14.12 %), while before the attacks, the co-mentioning tweets from conservatives were most likely to be related to AuthorityVirtue dimension (15.03 %).

Our qualitative analysis on tweet contents shows that in the co-mentioning discussions, there appears a spectrum of how the tweeters differentiated Muslim group, Islam religion, ISIS, and terrorism/terrorist from one another – on one end, Muslim, Islam, and ISIS were discussed as if they equated to one another and were the terror itself; on the other, it was argued that Muslims and Islam did not equate to ISIS, and ISIS was not real Muslim but the terrorists claiming a Islam religious root.

Among conservatives, a majority of the co-mentioning tweets did not show explicit differentiation between Islam and Muslim and in some cases, equated the risk of ISIS to the risk of all Muslims and to the risk to Islam religion; a majority of co-mentioning tweets among liberals expressed or advocated their views of distinction, and discussed the issue of "Islamphobia," an extreme fear of the whole Muslim group. Conservatives and liberals had some fights regarding the phobia issue, e.g. "*@ArcticFox2016 @BarracudaMama you are an infidel (civilized), Muslims are reaised to kill you! Your fear of Islam is rational, NOT phobia.*" Moreover, after the attacks, there appeared a distinction between Muslims/Islamist and Radical Muslims/Islamic extremist when discussing what was the source of risk, which was not found before the attacks. Conservatives and liberals shared a more common view regarding such distinction, and expressed that "extremists" or "radicals," and a perceived high risk of them.

Islam. While Islam was mentioned, it mostly referred to Islam religion or certain Islam groups (e.g., Islam Group, or Islam State). Islam profile is similar to the profile of Muslim, in which HarmVice is the first leading dimension; however,

AuthorityVice, which is the fifth dimension for Muslim, is much less likely to appear. Also, the significance of each dimension differs between conservative and liberal tweets and is influenced by the attacks too.

Before Paris attacks, for both conservatives and liberals, the second leading dimension was AuthorityVirtue (In Muslims profile, it was IngroupVirtue). For conservatives, HarmVirtue came the third and IngroupVirtue the forth; for liberals, the vice verse. In AuthorityVirtue, the top used moral words were "leader(s)," which concerned the acts and ascribed responsibilities to the leaders of primarily two sides – ISIS and anti-ISIS countries including the U.S. and Australia. The tweets concerned that the ingroup leadership of the U.S. and other anti-terrorism allies were not tough enough to defeat the ISIS leaders who schemed detrimental harms successfully; for example, *"How can the so-called leaders of the free world FALL for these psychopathic murderers? It's CLEAR they are soldiers of Islam! @DesignerDeb3;"* and *"Islamic radicals know who their enemies are. It is some of our leaders who have forgotten who is under attack.* https://t.co/7Q76vQlOMW."

Latino. *IngroupVirtue as the Leading Dimension.* IngroupVirtue was the first leading dimension in Latino profile, for both political leanings, either before or after Paris Attacks. The tweet contents had to do with how the Latino as an immigrant group/community fought against being perceived as threatening immoral community and negotiated to be recognized as an Ingroup member of the U.S. Most of the discussions were related to the Republican presidential candidate, Donald Trump. Among conservatives, there were 31 % of the IngroupVirtue tweets mentioning Trump; among liberals, there were 21 %. There was a common call for the Latino to "unite," not voting for "racist." There was no discussion among liberal tweeters supporting Trump, while conservative tweets had diverging attitudes, both in supporting and criticizing the candidate's statements about the Latino issues, e.g., *"people say trump is a threat to latino communities but those communities will be fine they just have to go be fine in another country;"* and *"@esd2000 The violence that #OperationWetback carried continued on Latino community for decades, #Trump's embrace of it is OFFENSIVE! #GOP."*

Difference Between Political Leanings. Before Paris Attacks, for liberal tweets, IngroupVirtue was a prominent dimension while other dimensions were about a half or less than a half of IngroupVirtue tweets. For conservatives, the second leading dimension, AuthorityVice, was prominent dimension as IngroupVirtue. In AuthorityVice tweets among conservatives, the most used two groups of words were "illegal* (illegal(s) and illegally)" and "protest." The words "illegal*" was mostly used in referring to undocumented Latinos as "illegal immigrants," who were judged as immoral and threatening agents with unjustifiable status, and to differentiate Latinos who had authorized immigrant statuses from those who did not; for example, *"Arrest the Latino Kids' Parents and deport them if they are illegal immigrants; and a large majority of Latinos are law abiding citizens who dislike ILLEGAL immigration as much as anyone.@AJDelgado13 @BradThor."*

Change Before and After Paris Attacks. After Paris attacks, in both of the liberal and conservative tweets, IngroupVirtue remained the first leading dimension but the percentages and odds decreased significantly (the before-after difference of odds ratio for conservatives and liberals were −2.35 and −1.68, respectively). For conservatives, tweets in HarmVice dimension increased and became as significant as AuthorityVice dimension, followed by HarmVirtue and AuthorityVirtue.

The increasing HarmVice tweets among conservatives started relating Latino to terrorists/terrorism, which is not found before Paris attacks in both conservative and liberal and only found after the attacks among conservative tweets, e.g., *"Everytime a Mexican, or other illegal Latino kills an American in the #USA, we should look at it as an act of Catholic terrorism."* Such connection entails a rising perceived risk toward Latino group after the Paris attacks among conservatives. Also, while addressing Syrian refugee issue, Latino was being commented, in some cases together with blacks as perceived trouble groups for the U.S.; for example, *"Democrat Big Plan: First they tried with blacks and Latinos to stir a race war, now they invited Muslim terrorists to create chaos #tcot."*

For liberals, the profile changed little. It is notable that FairnessVice tweets increased to about twice to be mentioned, which related the discrimination of Latino to a larger phenomenon in the society against particular groups including Muslims immigrants or refugees within the U.S. society; e.g., *"@allinwithchris The GOP Racism and Bigotry toward Blacks Latinos Gays women and now WAR TORN Refugees is DESTROYING this COUNTRY and World."* The increased moral judgment among liberals relevant to Latino was not toward Latino but conservatives who considered Latino as a threat.

Immigrant. *Authority Vice/Virtue as the Leading Dimensions.* Authority, both vice and virtue, were the most significant leading dimensions in Immigrant profile, for both conservative and liberal tweets. Particularly for conservative, about one out of every four immigrant-relevant tweets were AuthorityVice tweets (25.62%, while for liberals, the percentage was 13.92%). The odds ratios in AuthorityVice dimension were the highest ones to appear across all profiles (before Paris attacks: 23.72 and 18.12 for conservative and liberal, respectively; after Paris attacks: 12.70 and 9.43, respectively.)

Difference Between Political Leanings, Before and After Paris Attacks. Before the attacks, the contents were mostly immigrant issues in general or Latino relevant since the primary immigrant issue in current time has been of Latino in the U.S. For conservatives, 40.0% of the tweets used Authority (both vice and virtue, relevant to legal concerns); for liberals, 23.7%; however, after the attacks, the rates decreased to 25.3% and 17.3%, respectively.

Among conservatives, after Paris attack, FairnessVice was the only dimension in which the tweets increased both of the percentages and odds (odd ratio increased from 0.93 to 2.29). 15% of the tweets after the attacks were relevant to Syrian refugees, while none of tweets before the attacks relevant to refugees. Before the attacks, FairnessVice tweets primarily focused on arguing that anti-illegal immigrants were not a bias, discrimination, or a bigotry; e.g., *"@20142*

@ConnieHair There is nothing racist or bigoted about opposing illegal immigration and open borders." After the attacks, a rising perceived risk was that terrorists would utilize the U.S.'s refuge policy sending camouflages as refugees into the U.S. to enact attacks; however, there were also counter arguments against such view among conservative tweeters; e.g., "*@spongefile Plenty of anti immigration people do have a bias against Muslims and other immigrants but not me. My grandparents fled Hitler;*" and "*Syrians can immigrate to the U.S. the same way others do. No special visas for Syrians only. That's discrimination.*"

Among liberals, Fairness tweets increased too but were in Virtue dimension not in Vice; the likelihood to appear increased (odd ratio increased from 1.10 to 2.50). Before the attacks, the discussions in FairnessVirtue covered justice and human rights for immigrants who have been in the U.S., concerning that all immigrants in the U.S., documented or undocumented, deserved equal and fair treatments; no specific groups of people were stressed. After the attacks, concerns of Muslims and Syrian refugees appeared; a call for equal rights to these specific groups was advocated. For example, "*@adirado29 Muslims aren't guns. ISIS attacks do not equal Islam. Immigrants aren't evil because they are Muslims.*" To sum up, in the context of terrorism, confounding with Refugee issue due to Syrian Civil War, Muslims were not perceived as risk by liberals as much as by conservatives. The perceived risk of immigrants divided its focus from immigrants as illegal entities that would impose harm to the U.S. to Muslim newcomers who could impose death threat to Americans, of which the perceived risk was far beyond an legal issue.

5   Discussion and Conclusion

Our study shows that in the context of terrorism, how people who hold distinct political ideologies contribute differently to the construction of risk discourses. Prior research on moral foundations of human societies [21] has indicated that people holding distinct political ideologies engage in moral judgment differently. For liberals, Harm/Care–moral concern of people in pain as immoral, and empathy as moral–is the one that is primarily being relied upon to make moral judgment and justification, and Fairness/Cheating a suppurating one. Conservatives extensively use all of the five foundations, while which dimensions are primarily based upon depends on issues and contexts. Our research offered empirical evidence of that in responding to the terrorist attacks, conservatives addressed most the perceived harm and its relation to immorality; liberals focused on perceived social bias and argued the discrimination as immoral more than conservatives did. Conservatives put more emphasis on their perception of Muslims as risk, while liberals on the concern of Muslims at risk.

In our study, we found that as the perceived risk toward Muslims raised higher in general, particularly among conservatives, the identification of ingroup and outgroup members could shift. First, while a moral judgement view is found not shared with another ingroup member and the perceived risk of an outgroup is high, the ingroup member holding distinct view could be judged as a betrayer to the ingroup and labeled as a member of the risk outgroup regardless how the ingroup member perceive his/her own identity and regardless of whether the ingroup member does do harm or not. This phenomenon is observed among both liberal and conservative users, but more common among conservatives. Such observation is consistent with prior studies in public health indicating that the selection of a risk target may be more a way in which people utilize to maintain a sense of social solidarity that could decrease the fear of the perceived risk [3] rather than to reflect a real issue or fact regarding the potential danger.

Harm and Care can be two sides of one, mediated by perceived group boundary and in-group favoritism. The stronger the categorical line drawn between the self and others, the more likely that a person analyzes a risk situation by perceiving themselves as victim, shows less empathy to the perceived outgroup concerned as potential harm, and overlooks the diversities and individual differences among outgroup members.

The topic of immigration itself is not only controversial, but also polarized. Perception towards immigrants has been strongly shaped by political ideologies [24]. During the post-1965 era, conservative ideology has been linked to the higher likelihood of perceiving immigrants as threat [25] and the tendency of blaming illegal immigration [26] and Muslims [27]. Our study further reveals that the difference in risk perception across political leanings is associated with moral intuition, and may shed light on the recent Islamophobia phenomenon.

This paper is the first empirical study on risk perception. It offers insights for policy makers to understand what psychological factors dominate the public's views on terrorism. For future work, we plan to examine our analysis framework of risk perception in the context of other terrorist attack incidents, which is to test the applicability of our conceptualization of perceived risk and meanwhile, deepen and expand the understanding of the interaction of the social and cultural theories we draw upon. One potential direction is to compare and contrast how the perceived ingroup and outgroup shifts, in relation to the terrorist attacks that have happened inside and outside of the U.S.

Acknowledgement. This work is part of the research supported from NSF #1423697, #1634944 and the CRDF at the University of Pittsburgh. Any opinions, findings, and conclusions or recommendations expressed in this material do not necessarily reflect the views of the funding sources.

References

1. Sheridan, L.P.: Islamophobia pre-and post-September 11th, 2001. J. Interpers. Violence **21**(3), 317–336 (2006)
2. Jenkin, C.M.: Risk perception and terrorism: applying the psychometric paradigm. Homel. Secur. Aff. **2**(2) (2006)
3. Lupton, D.: Risk as moral danger: the social and political functions of risk discourse in public health. Sociol. Health Illness **23**, 425–435 (2001)
4. Chavez, L.: The Latino Threat: Constructing Immigrants, Citizens, and the Nation. Stanford University Press, Stanford (2013)

5. Stephan, W.G., Ybarra, O., Bachman, G.: Prejudice toward immigrants1. J. Appl. Soc. Psychol. **29**(11), 2221–2237 (1999)
6. Tichenor, D.J.: Dividing Lines: The Politics of Immigration Control in America. Princeton University Press, Princeton (2009)
7. Gray, K., Young, L., Waytz, A.: Mind perception is the essence of morality. Psychol. Inq. **23**(2), 101–124 (2012)
8. Graham, J., Haidt, J., Koleva, S., Motyl, M., Iyer, R., Wojcik, S.P., Ditto, P.H.: Moral foundations theory: the pragmatic validity of moral pluralism. In: Devine, P., Plant, A. (eds.) Advances in Experimental Social Psychology, vol. 47, pp. 55–130 (2012)
9. Tajfel, H.: Social identity and intergroup relations. Cambridge University Press, Cambridge (2010)
10. Cheong, M., Lee, V.C.S.: A microblogging-based approach to terrorism informatics: exploration and chronicling civilian sentiment and response to terrorism events via Twitter. Inf. Syst. Front. **13**(1), 45–59 (2011)
11. Awan, I.: Islamophobia and Twitter: a typology of online hate against Muslims on social media. Policy Internet **6**(2), 133–150 (2014)
12. Williams, M.L., Burnap, P.: Cyberhate on social media in the aftermath of woolwich: a case study in computational criminology and big data. Br. J. Criminol. **56**(2), 211–238 (2016)
13. Magdy, W., Darwish, K., Abokhodair, N., Quantifying public response towards Islam on Twitter after Paris attacks. arXiv preprint arXiv:1512.04570 (2015)
14. Puschmann, C., Ausserhofer, J., Maan, N., Hametner, M.: Information laundering, counter-publics: the news sources of islamophobic groups on Twitter. In: Tenth International AAAI Conference on Web and Social Media (2016)
15. Burnap, P., Williams, M.L.: Us and them: identifying cyber hate on Twitter across multiple protected characteristics. EPJ Data Sci. **5**(1), 1 (2016)
16. Huddy, L., Feldman, S., Capelos, T., Provost, C.: The consequences of terrorism: disentangling the effects of personal and national threat. Polit. Psychol. **23**(3), 485–509 (2002)
17. Kemshall, H.: Risk, social policy and welfare. McGraw-Hill Education, Buckingham (2001)
18. Dasgupta, N.: Implicit ingroup favoritism, outgroup favoritism, and their behavioral manifestations. Soc. Justice Res. **17**(2), 143–169 (2004)
19. Lin, Y.-R., Margolin, D., Keegan, B., Lazer, D., Voices of victory: a computational focus group framework for tracking opinion shift in real time. In: Proceedings of the 22nd international conference on World Wide Web, pp. 737–748. International World Wide Web Conferences Steering Committee (2013)
20. Lin, Y.-R., Keegan, B., Margolin, D., Lazer, D.: Rising tides or rising stars?: Dynamics of shared attention on Twitter during media events. PLoS ONE **9**(5), e94093 (2014)
21. Graham, J., Haidt, J., Nosek, B.A.: Liberals, conservatives rely on different sets of moral foundations. J. Pers. Soc. Psychol. **96**(5), 1029 (2009)
22. Pennebaker, J.W., Francis, M.E., Booth, R.J.: Linguistic Inquiry, Word Count: LIWC 2001, vol. 71. Lawrence Erlbaum Associates, Mahwah (2001)
23. Tausczik, Y.R., Pennebaker, J.W.: The psychological meaning of words: LIWC and computerized text analysis methods. J. Lang. Soc. Psychol. **29**(1), 24–54 (2010)
24. Massey, D.S.: International migration at the dawn of the twenty-first century: the role of the state. Popul. Dev. Rev. **25**(2), 303–322 (1999)

25. Massey, S.D., Pren, K.A.: Unintended consequences of US immigration policy: explaining the post-1965 surge from Latin America. Popul. Dev. Rev. **38**(1), 1–29 (2012)
26. Hayworth, J.D., Eule, J.: Whatever It Takes: Illegal Immigration, Border Security, and the War on Terror. Regnery Publishing, Washington, D.C. (2005)
27. Echebarria-Echabe, A., Guede, E.F.: A new measure of Anti-Arab prejudice: reliability and validity evidence. J. Appl. Soc. Psychol. **37**(5), 1077–1091 (2007)

Determining the Veracity of Rumours on Twitter

Georgios Giasemidis[1]([✉]), Colin Singleton[1], Ioannis Agrafiotis[2],
Jason R.C. Nurse[2], Alan Pilgrim[3], Chris Willis[3], and D.V. Greetham[4]

[1] CountingLab Ltd., Reading, UK
{georgios,colin}@countinglab.co.uk
[2] Department of Computer Science, University of Oxford, Oxford, UK
{ioannis.agrafiotis,jason.nurse}@cs.ox.ac.uk
[3] BAE Systems Applied Intelligence, Chelmsford, UK
{alan.pilgrim,chris.willis}@baesystems.com
[4] Department of Mathematics and Statistics,
University of Reading, Reading, UK
d.v.greetham@reading.ac.uk

Abstract. While social networks can provide an ideal platform for up-to-date information from individuals across the world, it has also proved to be a place where rumours fester and accidental or deliberate misinformation often emerges. In this article, we aim to support the task of making sense from social media data, and specifically, seek to build an autonomous message-classifier that filters relevant and trustworthy information from Twitter. For our work, we collected about 100 million public tweets, including users' past tweets, from which we identified 72 rumours (41 true, 31 false). We considered over 80 trustworthiness measures including the authors' profile and past behaviour, the social network connections (graphs), and the content of tweets themselves. We ran modern machine-learning classifiers over those measures to produce trustworthiness scores at various time windows from the outbreak of the rumour. Such time-windows were key as they allowed useful insight into the progression of the rumours. From our findings, we identified that our model was significantly more accurate than similar studies in the literature. We also identified critical attributes of the data that give rise to the trustworthiness scores assigned. Finally we developed a software demonstration that provides a visual user interface to allow the user to examine the analysis.

1 Introduction and Related Work

Nowadays, the social media play an essential role in our everyday lives. The majority of people use social networks as their main source of information [22,24]. However, sources of information might be trusted, unreliable, private, invalidated or ambiguous. Rumours, for example, might be true or false and started accidentally or perhaps maliciously. In situations of crisis, identifying rumours at an early stage in social media is crucial for decision making. The example of

© Springer International Publishing AG 2016
E. Spiro and Y.-Y. Ahn (Eds.): SocInfo 2016, Part I, LNCS 10046, pp. 185–205, 2016.
DOI: 10.1007/978-3-319-47880-7_12

London Riots of 2011 is characteristic. After the events unfolded, The Guardian provided an informative graphic of the initiation and progress of a number of rumours [27]. This analysis showed that categorising social media information into rumours, and analysing the content and their source, may shed light into the veracity of these rumours. Moreover, it could support emergency services in obtaining a comprehensive picture in times of crisis and make better-informed decisions.

Recently there has been progress from experts from different fields of academia in exploring the characteristics of the data source and content that will enable us to determine the veracity of information in an autonomous and efficient manner [13,25,31]. Key concepts include information quality, which can be defined as an assessment or measure of how fit an information object is for use, and information trustworthiness, which is the likelihood that a piece of information will preserve a user's trust, or belief, in it [20]. These concepts may overlap and indeed, increasing one (e.g., quality) may lead to an increase in the other (e.g., trustworthiness). Other relevant factors include Accuracy (Free-of-error), Reliability, Objectivity (Bias), Believability (Likelihood, Plausibility of arguments), Popularity, Competence and Provenance [8,11,15,16,30].

To date, there has been a significant number of articles, in both academia and industry, that have been published on the topic of information quality and trustworthiness online, particularly in the case of Twitter. Castillo et al. [3,4] focus on developing automatic methods for assessing the credibility of posts on Twitter. They utilise a machine learning approach to the problem and for their analysis use a vast range of features grouped according to whether they are user-based, topic-based or propagation-based. Nurse et al. [18,19] have also aimed towards developing a wider framework to support the assessment of the trustworthiness of information. This framework builds on trust and quality metrics such as those already reviewed, and outlines a policy-based approach to measurement. The key aspect of this approach is that it allows organisations and users to set policies to mediate content and secondly, to weight the importance of individual trust factors (e.g., expressing that for a particular context, location is more important than corroboration). The result is a tailored trustworthiness score for information suited to the individual's unique requirements.

Having established a view on the characteristics of information quality and trustworthiness researchers focused on designing systems for rumour detection. Kwon et al. [13] examined how rumours spread in social media and which characteristics may provide evidence in identifying rumours. The authors focused on three aspects of diffusion of a rumour, namely the temporal, the structural and the linguistic and identified key differences in the spread of rumours and non-rumours. Their results suggest that they were able to identify rumours with up to 92 % of accuracy. In [25] the authors provide an approach to identify the source of a false rumour and assess the likelihood that a specific information is true or not. They construct a directed graph where vertices denote users and edges denote information flows. They add monitor nodes reporting on data they receive and identify false rumours based on which monitoring nodes receive specific information and which do not.

Another approach is presented in [31], where the authors introduce a new definition for rumours and provide a novel methodology on how to collect and annotate tweets associated with an event. In contrast to other approaches which depend on predefining a set of rumours and then associating tweets to these, this methodology involves reading the replies to tweets and categorising them to stories or threads. It is a tool intended to facilitate the process of developing a machine learning approach to automatically identify rumours.

Finally, one of the most comprehensive works is presented in [29] where various models are tested aiming for detecting and verifying rumours on Twitter. The detection of different rumours about a specific event is achieved through clustering of assertive arguments regarding a fact. By applying logistic regression on several semantic and syntactic features, the authors are able to identify with 90 % accuracy the various assertions. Regarding the verification of a rumour, the models utilise features considering the diffusion of information; feature are elicited from: linguistics, user-related aspects and temporal propagation dynamics. Hidden Markov Models are then applied to predict the veracity of the rumour; these are trained on a set or rumours whose veracity has been confirmed beforehand based on evidence from trusted websites.

Our review on the literature indicates that user and content features are found to be helpful at distinguishing trustworthy content. Moreover the temporal aspect of the aforementioned features denoting the propagation dynamics in Twitter may provide useful insights into distinguishable differences between the spread of truthful and falsified rumours. Our reflection suggests that all the approaches are using manually annotated tweets or similar datasets for the training period. The annotations denote the veracity of the rumour and indicate the event the rumour describes. Syntactic and semantic features are then extracted to aid the process of identifying events and classifying rumours regarding these events. There is not an outperforming approach and most of the models require 6–9 h before accurately predicting the veracity. Understanding the literature on information trustworthiness and how concepts from linguistic analysis and machine learning are applied to capture patterns of propagating information is the first and decisive step towards a system able to identify and determine the veracity of a rumour. The lessons learnt from this review are the foundations for the requirements of the system.

This paper builds on existing literature and presents a novel system which is able to collect information from social media, classify this information into rumours and determine their veracity. We collected data from Twitter, categorised these into rumours and produced a number of features for the machine learning techniques. We train and validate our dataset and compare our model to other studies in the literature. We also aim to do better than existing work and are exploring a way to visualise our various findings in a user interface. In what follows, Sect. 2 reports on the data collection process and introduces the methodology used to analyse the data, while Sect. 3 describes the analysis and model selection process. Section 4 presents the outcome of our system when applied to the collected rumours and compares our results with the results of other systems publicly available. Section 5 concludes the paper and identifies opportunities for future work. Finally, to keep the discussion on the modelling aspects comprehensive and compact in these sections, we present further details and by-products of our research in the Appendices.

2   Approach and Methodology

We focus on messages and data from Twitter for three key reasons. First, Twitter is regarded as one of the top social networks [24]. Particularly, in emergency events, Twitter is the first choice of many for updated information, due to its continuous live feed and short length of the messages [22]. Second, the majority of messages on Twitter are publicly available. Third, Twitter's API allows us to collect the high volume of data, e.g. messages, users information, etc., required to build a rumour classifier.

In this study we define a rumour as [1]: *"An unofficial interesting story or piece of news that might be true or invented, and quickly spreads from person to person"*. A rumour consists of all tweets from the beginning of the rumour until its verification from two trusted sources. Trusted sources are considered news agencies with global reputation, e.g. the BBC, Reuters, the CNN, the Associated Press and a few others. For every rumour we collect four sets of data: (i) the tweets (e.g. text, timestamp, retweet/quote/reply information etc.), (ii) the users' information (e.g. user id, number of posts/followers/friends etc.), (iii) the users' followees (friends) and (iv) the users' most recent 400 tweets prior the start of the rumour, see Appendix A for a step-by-step guide on data collection. In total we collected about 100 million public tweets, including users' past tweets. We found 72 rumours, from which 41 are true and 31 are false. These rumours span diverse events, among which are: the Paris attacks in November 2015, the Brussels attacks in March 2016, the car bomb attack in Ankara in March 2016, earthquakes in Taiwan (February 2016) and Indonesia (March 2016), train incidents near London and rumours from sports and politics, see Appendix A.1 for a summary statistics of these rumours. An event may contribute with more than one rumour.

For modelling purposes we need the fraction of tweets in a rumour that support, deny or are neutral to the rumour. For this reason all the tweets are annotated as being either in favour $(+1)$, against (-1) or neutral to (0) the rumour. Regarding annotation, all retweets were assigned the same tag as their source tweet. Tweets in non-English languages that could not be easily translated were annotated as neutral. There are rumours for which this process can be automated and others which require manual tagging.

Linguistic characteristics of the messages play an important role in rumour classification [4,13]. Extracting a text's sentiment and other linguistic characteristics can be a challenging problem which requires a lot of effort, modelling and training data. Due to the complexity of this task we used an existing well-tested tool, the Linguistic Inquiry and Word Count (LIWC), version LIWC2015 [5], which has up-to-date dictionaries consisting of about 6,400 words reflecting different emotions, thinking styles, social concerns, and even parts of speech. All collected tweets were analysed through this software. For each tweet the following attributes were extracted: (i) positive and (ii) negative sentiment score, fraction of words which represent (iii) insight, (iv) cause, (v) tentative, (vi) certainty, (vii) swearing and (viii) online abbreviations (e.g. b4 for the word before).

Propagation Graph. An important set of features in the model is the propagation based set of features. All propagation features are extracted from the propagation graph or forest. A propagation forest consists of one or more connected propagation trees. A propagation tree consists of the source tweet and all of its retweets, see Appendix B for further details on making a Twitter propagation tree.

2.1   Classifiers

We approach the rumour identification problem as a supervised binary classification problem, i.e. training and building a model to identify whether a rumour is true or false. This is a well-studied problem in the machine-learning field and many different methods have been developed over the years [2,12].

A classifier requires a set of N observations $O = \{X_i, i = 1, \ldots, N\}$ where each observation, $X_i = (f_1, \ldots, f_M)$, is an M-dimensional vector of features, f_i. The observations are labelled into K distinct classes, $\{C_i, i = 1, \ldots, K\}$. Our classification problem has $N = 72$ observations (i.e. rumours) in $K = 2$ classes (true or false) and $M = 87$ features. The techniques that we use in this study are Logistic Regression, Support Vector Machines (SVM), with both a linear and a Radial Basis Function (RBF) kernel to investigate for both linear and non-linear relationship between the features and the classes [2,26], Random Forest, Decision Tree (the CART algorithm), Naïve Bayes and Neural Networks.

We assess all models using k-fold cross-validation. The purpose of the cross-validation is to avoid an overly optimistic estimate of performance from training and testing on the same data resulting in overfitting, as the model is trained and validated on different sets of data. In this study we use $k = 10$ folds. The cross-validation method requires a classification metric to validate a model. In the literature there are several classification metrics [4,23], the most popular ones are: (i) accuracy, (ii) F_1-score, (iii) area under ROC Curve (AUC) and (iv) Cohen's kappa. We choose to cross validate the models using the F_1-score.

Most of the statistical and machine learning methods that we use have already been developed and tested by the Python community. A popular and well-tested library is the *scikit-learn*[1] that we frequently use. For the neural networks classifier we used the *neurolab*[2] library.

2.2   Features

The rumour's features can be grouped into three broad categories, namely message-based, user-based and network-based. The message-based (or linguistic) features are calculated as follows. Each tweet has a number of attributes,

---

[1] http://scikit-learn.org/.
[2] https://pythonhosted.org/neurolab/.

which can be a binary indicator, for example whether the tweet contains a URL link, or a real number, e.g. the number of words in the tweet. Having calculated all attributes we aggregate them at the rumour level. This can be, for example, the fraction of tweets in the rumour containing a URL link (for categorical attributes) or the average number of words in tweets (for continuous attributes). Both of these aggregate variables become features of the rumour. However the aggregation method can be more complicated than a fraction or an average. In particular we would like to quantify the difference in the attributes between tweets that support the rumour and those that deny it. The idea is that users who support a rumour might have different characteristics, language or behaviour, from users that deny it. To capture this difference we use an aggregation function which can be represented as:

$$f_i = \frac{S^{(i)} + N^{(i)} + 1}{A^{(i)} + N^{(i)} + 1},\tag{1}$$

where f_i is the i-th feature and $S^{(i)}, N^{(i)}, A^{(i)}$ stand for support, neutral and against respectively. In mathematical terms let $D_j^{(i)}$ be the value of the i-th attribute of the j-th tweet, $j = 1, \ldots, R$, R being the total number of tweets in a rumour. This can be either $D_j^{(i)} \in \{0, 1\}$ for binary attributes or $D_j^{(i)} \in \mathbb{R}$ for continuous ones. Also, let $B_j \in \{-1, 0, 1\}$ be the annotation of the j-th tweet. Hence we define,

$$S^{(i)} = \frac{\sum_{j \in \{k|B_k=1\}} D_j^{(i)}}{\sum_{j \in \{k|B_k=1\}} 1}, \ N^{(i)} = \frac{\sum_{j \in \{k|B_k=0\}} D_j^{(i)}}{\sum_{j \in \{k|B_k=0\}} 1}, \ A^{(i)} = \frac{\sum_{j \in \{k|B_k=-1\}} D_j^{(i)}}{\sum_{j \in \{k|B_k=-1\}} 1}.$$

For a practical example on the application of the above formulae, see Appendix C. The aggregation formula (1) allows us to compare an attribute of the supporting tweets to an attribute of the against tweets. The neutral term in Eq. (1) reduces the extremities of the ratio where there are a lot of neutral viewpoints. The unit term is a regularisation term, ensuring the denominator is always strictly positive.

There are a few attributes for which we do not follow this aggregation rule. These are the fraction of tweets that support or deny the rumour where we simply use the fractions. Additionally all the sentiment attributes can take negative values, making the denominator of Eq. (1) zero or negative. For all sentiment attributes we aggregate by taking the difference between the sentiment of tweets that support the rumour and the sentiment of tweets that deny it, i.e. $S^{(i)} - A^{(i)}$. Additionally, some linguistic attributes were extracted using the popular Python library for natural language processing, $nltk^3$.

The user-based features are extracted in a similar manner focusing on the user attributes. For example, two user-attributes are whether a user has a verified account (binary) and the number of followers of a user (continuous). These attributes are aggregated to the rumour level using Eq. (1), counting each user

---

[3] http://www.nltk.org/.

who contributed to the rumour only once. If a user contributed with both a supporting and a refuting tweet then its attribute contributes to both the support, $S^{(i)}$, and against, $A^{(i)}$, terms.

The network-based features are estimated through the propagation graph, which is constructed using the *networkx*[4] Python library. It becomes evident that three propagation graphs are required; a graph consisting of tweets that support the rumour, a graph of tweets neutral to the rumour and a graph of tweets against the rumour. From each graph we calculate a number of attributes. These network-based attributes are aggregated to the rumour feature using Eq. (1).

The feature names should be treated with caution in the rest of the paper. For example, when we refer to the feature "users followers" we actually mean the feature related to the user's followers through expression Eq. (1). Exceptions are the fraction of tweets that deny the rumour, the fraction of tweets that support the rumour and all the sentiment-related features which are aggregated using expression $S^{(i)} - A^{(i)}$. Therefore when we say that the feature "user's followers" is important we don't refer to the actual number of users' followers. We hope this causes no further confusion to the reader.

Time-Series Features. One of the main goals and novel contributions of this study overall is to determine the veracity of a rumour as early as possible. We therefore wanted to find out how quickly we could deduce the veracity. For this reason we split every rumour into 20 time-intervals and extract all the features for the subset of tweets from the start of the rumour until the end of each time-period. In this way we develop a time-series of the features which we will use to estimate the time evolution of the veracity of the rumours.

3   Analysis

In this section we analyse and present the results from the cross-validation process. In particular we aim to address four key questions: (i) What is the best method for reduction of the feature space? (ii) What is the best-performing classification method? (iii) What are the optimal parameters for the selected classifiers? (iv) What are the features in the final models?

Reduction of the Feature Space. Our dataset consists of 72 observations and 87 features. The number of features is large compared to the number of observations. Using more features than necessary results in overfitting and decreased performance [9]. Feature-space reduction is a very active field of research and different techniques have been developed [9]. The importance of dimension reduction of the feature space is two-fold; first it is a powerful tool to avoid overfitting and secondly aims to reduce complexity and time of computational tasks [9,28]. We considered four methods of feature reduction, see Appendix D for further details. We apply each method to each classifier separately and assess it using k-fold cross validation. We found that the best method for reducing the feature

---

[4] https://networkx.github.io/.

space is a forward selection deterministic wrapper method, which we use in the rest of this study.

Selecting Classifier. To select the best performing classifiers, we perform a k-fold cross validation on each classifier for a number of features selected using a forward selection deterministic wrapper method. The results for *scikit-learn*'s default parametrisations are plotted in Fig. 1.

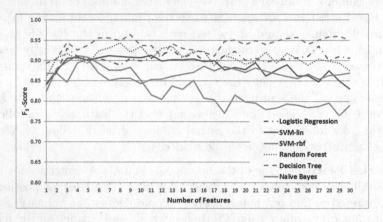

Fig. 1. F_1-score as a function of the number of features for six classification methods.

From the plot it becomes evident that, for this data set, the Decision Tree is the best performing method and the Random Forest follows. Clearly the Naïve Bayes and the SVM-rbf are under-performing. Logistic regression performs slightly better than the SVM-linear[5]. These observations are further explained and quantified in Appendix E. Therefore, we select and proceed with the three-best performing methods, i.e. Logistic Regression, Random Forest, and Decision Tree. For each of the three selected classifiers we perform further analysis to optimise its input parameters to further improve its performance. These parameters are the regularisation strength for Logistic Regression, the number of trees for Random Forest and the splitting criterion function for Decision Tree. We use those parameters that maximise the average F_1-score through cross-validation.

Best Features. Having selected the best-performing methods we now concentrate on finding the features that optimise the performance of the model. Again we focus on three classifiers; Logistic Regression, Random Forest and Decision Tree tuned to their optimal parameters (see previous section). We run 30 models; each model has a number of features ranging from 1 to 30. These features are the best-performing as selected with a forward selection deterministic wrapper described earlier. The results are plotted in Fig. 2.

---

[5] We abandoned Neural Networks at early stages as the library implementation used was very slow and the results were underperforming.

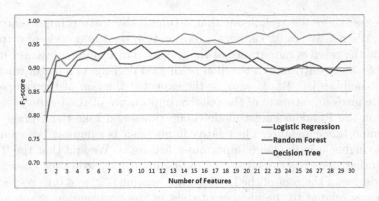

Fig. 2. F_1-score for Logistic Regression, Random Forest and Decision Tree tuned to their optimal parameters as a function of the number of input features.

Figure 2 suggests that Logistic Regression peaks at a model with seven features and then decays. Similarly, the Random Forest peaks at models with eight and ten features respectively and decays for models with more features. In contrast, the Decision Tree peaks at a model with six features but remains constant (on average) for models with more features. This overall behaviour was also observed in Fig. 1, where all classifiers, except the Decision Tree, peak their performance at models with four to eight features. Models with more than eight features decrease their accuracy on average.

For Logistic Regression these seven features are users' followers (user-based), complexity of tweets, defined as the depth of the dependency tree of the tweet (message-based), fraction of tweets denying the rumour (message-based), sentiment of users' past tweets (user-based), tenure of users in Twitter (in days) (user-based), retweets within the network (network-based) and low to high diffusion, defined as a retweet from a user with a higher number of followers than the retweeted user (network-based). However, not all of these features are statistically significant for this data set. We further run a statistical test to assess the significance of these features. We use the log-likelihood test and confidence level 0.05 [7]. We found that only three out of seven features are statistically significant. These are the fraction of tweets against the rumour, the tenure of users and users' followers.

Random Forest peaks its performance at a model with eight features, which are the number of propagation trees in the propagation graph (network-based), fraction of tweets denying the rumour (message-based), verified users (user-based), users with location information (user-based), the degree of the root of the largest connected component (network-based), number of users' likes (user-based), quotes within the network (network-based) and tweets with negation (message-based).

Random Forest is a black-box technique with no clear interpretation of the parameters of its trees. It is a non-parametric machine learning technique hence it is not straight-forward to estimate the statistical significance of its features.

Nevertheless, there are techniques that estimate the feature importance. Particularly, the relative rank, i.e. depth, of a feature used in a node of a tree can determine the relative importance of the feature. For example, features used at the top of a tree contribute to the final prediction decision of a larger fraction of the input samples [10,21]. Measuring the expected fraction of the sample they contribute gives an estimate of the relative importance of the features.

We run the Random Forest model 1,000 times and take the average of the feature importance measure. The relative importance is a number between 0 and 1 and the higher it is the more important a feature is. We find that the fraction of tweets that deny the rumour is the most important feature for classification and the degree of the root of the LCC follows, scoring 0.22 and 0.19 respectively. The features related to the number of trees in the propagation graph and the verified users seem to contribute significantly, 0.16 and 0.15 respectively, while the remaining four play a less important role, scoring less than 0.08.

Figure 2 shows that the Decision Tree algorithm peaks its performance at a model with six features. Investigating the resulting Decision Tree and its splitting rules it became evident that actually only three out of six features are used in the decision rules. These are (i) the fraction of tweets denying the rumour (message-based), (ii) users with description (user-based) and (iii) user's post frequency (user-based). A further analysis unveils that although we feed the Decision Tree with an increasing number of features the algorithm always uses only a small subset of them which never exceeds eight. This justifies why the F_1-score of the Decision Tree classifier remains constant on average as we increase the number of features in Figs. 1 and 2. Every time we add a new feature, the Decision Tree uses a small subset with similar performance to the previous models.

Examining the three models we observe that they have one feature in common, the fraction of users that deny the rumour. Logistic regression and Random Forest have a mixture of message-based, user-based and network-based features, whereas the Decision Tree uses only message-based and user-based features. We strongly believe that with the addition of more rumours and larger data sets the Decision Tree algorithm will use more features among which the network-based ones. In conclusion, a high accuracy can be achieved with a relatively small set of features for this data. This is to be expected as sets with a small number of data points are subject to overfitting when a large number of features are used.

4   Results

In the previous section, we focused on finding the best models among a variety of possibilities. Having determined the best three models, their parameters and their features, we are ready to calibrate and validate the final model. We split the data into training (60 %) and testing (40 %) sets. These two sets also preserve the overall ratio of true to false observations. The results for the test set are presented in Table 1.

We highlighted the best-performing model on the test dataset, which is the Decision Tree. It reaches a high accuracy rate, 96.6 % and a precision 1.0. This

Table 1. Classification metrics of the three models for the test set.

Test Set	Accuracy	Precision	Recall	F_1-score	AUC	kappa
Logistic Regression	0.828	0.8	0.941	0.865	0.936	0.631
Random Forest	0.897	0.938	0.882	0.909	0.971	0.789
Decision Tree	**0.966**	**1.0**	**0.941**	**0.970**	**0.971**	**0.930**

implies that there are no false-positive predictions, i.e. rumours that are false but classified as true. The recall is 0.94 implying the presence of false negatives.

Random Forest follows with accuracy close to 90%. Logistic regression achieves the lowest accuracy of the three models, 82.8%. Although the F_1-score of Random Forest model is higher than the F_1-score of the logistic Regression their precision and recall scores differ substantially. Random Forest has a higher precision but lower recall than the Logistic Regression. This suggests that the Random Forest model returns a lower number of positives but most of the positives are correctly classified. On the other hand Logistic Regression returns many positives but a higher number of incorrect predicted labels.

4.1   Comparison to Benchmark Models

As discussed previously we are principally interested in determining the veracity of rumours as quickly as possible. In order to test this we split the duration of a rumour into 20 time-intervals and extract all the features from the beginning of the rumour until the end of each time interval. This results in 20 observations per rumour. We apply the three models to each time interval and calculate the veracity of the rumours at each time-step.

We calculate the accuracy of the three models at each time step. We compare against four benchmark models. The first is a random classifier which randomly classifies a rumour either true or false at each time period. The second model is the "always true" model, which always predicts that a rumour is true. The third model, named "single attribute model 1" classifies a rumour as true if the number of tweets supporting at a given time period is greater than the number of tweets against. Otherwise it classifies it as false. The "single attribute model 2" is similar to the "single attribute model 1", but a rumour is classified true if the ratio of in-favour tweets over the against is greater than 2.22. Otherwise it is false. The number 2.22 is the ratio of total tweets in the dataset that are in favour over the total tweets that are against. This gives on average how many more supporting tweets exist in the dataset. Figure 3 shows the results.

We observe that the random classifier oscillates around 0.5 as expected. The two "single attribute" models are better than both the random classifier and the "always true" model, with the "single attribute model 2" performing slightly better than "single attribute model 1". Our machine learning models perform better than the simple models especially towards the end of the rumour. For example, in the beginning of the rumour the Random Forest and Decision Tree

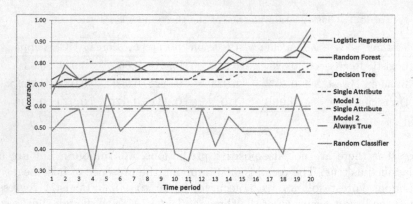

Fig. 3. The accuracy as a function of time for different models.

models have similar accuracy as the "single attribute model 2". However at the end of the rumour the Random Forest and Decision Tree have improved their accuracy by 33 % and 37 % respectively, while the simple models accuracy is improved by 14 %. The machine learning models have the ability to improve the accuracy at higher rates than any other simple model. This is a crucial result for decision making at early stages of a rumour development, before an official verification.

As a result of our analysis we also developed a visualisation tool which summarises and depicts the key observations and results. To avoid taking focus away from the modelling aspects we present further details in Appendix F.

4.2   Comparison to Literature Findings

The relevant papers to our work are summarised in Sect. 1. Here we compare our findings to the results of a few key papers [3,4,6,13,17,29]. Similar to our conclusion, some of these studies [6,17] found that the tweets which deny the rumour play a significant role in estimating the veracity of the rumour[6]. However, to the best of our knowledge, the key differences are the following. Firstly, we worked with a larger set of potential features, consisting of 87 features; particularly, none of these studies considered features related to users' past behaviour. Secondly, we aggregated the tweet, user and network attributes to rumour features using a non-trivial formula (1) which captures the difference between tweets that support and those that deny the rumour.

Thirdly, we found that we need a lower number of features than other models in the literature [3,4,13,29], varying between three and eight. Although the Logistic Regression and Random Forest models admit six and eight features respectively, about three are statistically significant or most important. It is interesting that high accuracy can be achieved with a small number of features. This can be explained by the model overfitting a relatively small number of data

---

[6] A phenomenon known as the "wisdom of the crowd".

points when more features are used. However, we expect that the number of discriminatory features might increase as the volume of data (more data points) and variety of rumours increase. The extra data also allows us to place more emphasis on early classification. This is an open question that we aim to address in the future.

Fourth, the accuracy of our classifiers varies between 82.8% (Logistic Regression) and 96.6% (Decision Tree) on "unobserved data" (validation set). To our best knowledge, the two best models, Random Forest and Decision Tree, outperform any other model in the academic literature [3,4,13,29]. We achieve a high success rate which shows the potential benefit of our model.

Last and most importantly, we considered the time-evolution of the features and hence the veracity. We built a model which is able to infer the veracity of a rumour with high accuracy at early stages, before the official confirmation from trusted sources. A time-series analysis was first attempted in [4], where the authors estimate the rumour veracity before the burst of a rumour (when the activity suddenly increases). Although this approach introduces some dynamical aspects, it lacks a full and detailed time-series analysis. Later, a proper time-series analysis of veracity was performed in [29]. The author introduces two classifiers which are specifically designed for time-series classification, the Dynamical Time Wrapping (DTW) and Hidden Markov Model (HMM) classifiers.

These two models achieve an accuracy of 71% and 75% respectively using 37 features. From these features only 17 were found to be statistically significant. The author modelled the veracity of rumours at different time-periods (as percentage of the time elapsed for each rumour). His best model does not exceed an accuracy rate 60% at a time-period half-way from the start until the trusted verification of the rumour. In contrast, we achieve a higher accuracy, at least 76%, at the same time-period, (time-period 10 in Fig. 3). This time-period, on average, corresponds to 3 h and 20 min after the outbreak of the rumour. A 76% accuracy is already reached by all of our models at one quarter of the rumour duration, which on average corresponds to 1 h and 50 min after the beginning of the rumour. However, as the time passes and more tweets and information are obtained, understandably our model accuracy increases. With more modelling time and more data, we would hope to improve early declaration still further.

5 Conclusion and Future Work

Modern lifestyle heavily relies on social media interaction and spread of information. New challenges have emerged as large volumes of information are being propagated across the internet. Assessing the trustworthiness of the news and rumours circulating in a network is the main subject of this study. The primary goal of this paper is to develop the core algorithms that can be used to automatically assign a trustworthiness measure to any communication.

We collected 72 rumours and extracted 87 features which capture three main topics and derived from our reflection on the relevant literature. The topics are the linguistic characteristics of the messages, the users' present and past behaviour and how the information propagates through the network. Furthermore, the feature space encompasses dynamic aspects for estimating rumour veracity, contributing to the literature since only one study thus far has attempted a similar approach. In addition to the modelling, we developed a graphical user interface which allows the user to investigate in details the rumour development over time.

Our overall model was significantly more accurate than similar studies due to two main reasons: (i) introduction of novel features, e.g. users' past behaviour, and (ii) the method of aggregating tweet/user attributes to rumour features. Our key findings suggest that the Decision Tree, Random Forest and Logistic Regression are the best classifiers. Additionally, the fraction of tweets that deny the rumour plays an essential role in all models. Finally, the three models require only a low number of features, varying between three and eight.

Although our paper provides the first and decisive step towards a system for determining the veracity of a rumour, there are opportunities for further research which will enhance our system. The automation of the rumour collection and tweet annotation is one area for future work. In our system the categorisation of the tweets into rumours is a manual and time-consuming task. Similarly, the annotation of the tweets require much effort and time from our side. For these reasons, we aim to build another classifier that automatically classifies the tweets into rumours and annotates them based on the content of text. This way we will be able to collect a large volume of rumours and massively scale up our dataset. Having a larger volume of data and more diverse rumours will allow us to develop more robust and accurate models.

The current models return either the probability of a rumour being true or the class itself. There is no information about the confidence levels of the results. One of the main future goals is to produce an algorithm providing uncertainty estimates of the veracity assessments. Additionally, we would like to expand our data sources and consider data from other social networks, such as the YouTube platform. Calibrating and testing our model on other sources of data will give further confidence about its validity and will extend its applicability.

Acknowledgements. This work was partly supported by UK Defence Science and Technology Labs under Centre for Defence Enterprise grant CDE42008. We thank Andrew Middleton for his helpful comments during the project. We would also like to thank Nathaniel Charlton and Matthew Edgington for their assistance in collecting and preprocessing part of the data.

Appendix

A   Data Collection Process

Our data collection process consists of four main steps:

1. Collection of tweets with a specific keyword, e.g. "#ParisAttacks" or "Brussels". The Twitter API only allows the collection of such tweets within

a ten-day window. For this reason this step must start as soon as an event happens or a rumour begins.

(a) Manual analysis of tweets and search for rumours. In this step we filter out all the irrelevant tweets. For example, if we collected tweets containing the keyword "Brussels" (due to the unfortunate Brussels attacks), we ignore tweets talking about holidays in Brussels.

(b) Collection of more tweets relevant to the story with keywords that we missed in the beginning of Step 1 (this step is optional). For example, while searching for rumours we might come across tweets talking about another rumour. We add the keyword that describes this new rumour in our tweet collection.

(c) Categorise tweets into rumours. Group all tweets referring to the same rumour.

(d) Identify all the unique users involved in a rumour. This set of users will be used in Steps 2 to 4.

2. Collect users' most recent 400 tweets, posted before the start of the rumour. This step is required because we aim to examine the users' past behaviour and sentiment, e.g. whether users' writing style or sentiment changes during the rumour, and whether these features are significant for the model. To the best of our knowledge, this set of features is considered for the first time in the academic literature in building a rumour classifier.

3. Collect users' followees (friends). This data is essential for making the propagation graph, see Sect. 2 and Appendix B.

4. Collect users' information, including user's registration date and time, description, whether account is verified or not etc.

A.1 Rumours Summary Statistics

We provide a summary statistics table of the 72 collected rumours, see Table 2. This table shows the total number, mean, median, etc., of the distributions of the number of tweets, the percentage of supporting tweets, etc., of the 72 rumours, as well as some statistics of four example rumours. We collected about a 100 million tweets, including users' past tweets. From the collected tweets, about 327.5 thousand tweets are part of rumours. These tweets contributed to the message-based features of the classification methods. The users' past tweets contributed only to the features capturing a user's past behaviour.

B   Making the Propagation Graph

Nodes in the propagation tree correspond to unique users. Edges are drawn between users who retweet messages. However the retweet relationship cannot be directly inferred from the Twitter data. Consider a scenario with three users, A, B and C. User A posts an original tweet. User B sees the tweet from user A and retweets it. Twitter API returns an edge between user A and user B. If user C sees the tweet from user B and retweets it, Twitter API returns an edge

Table 2. A summary statistics of the collected rumours. Examples 1 and 2 correspond to the rumours with the largest and second largest number of tweets respectively. Examples 3 and 4 correspond to the rumours with the second smallest and smallest number of tweets respectively.

	Tweets	% support	% against	Users	Users Tweets	Duration (h)
Total	327, 484	60.9 %	27.4 %	270, 054	95, 579, 214	N/A
Mean	4, 548	65.7 %	22.9 %	3, 751	1, 327, 489	9.02
Median	1, 660	81.5 %	2.5 %	1, 540	520, 288	3.04
Std	6, 816	34.4 %	32.2 %	5, 146	2, 005, 616	16.40
Min	23	0.3 %	0.0 %	23	9, 553	0.07
Max	46, 807	100.0 %	97.5 %	32, 529	13, 877, 121	114.22
Example 1	46, 807	76.1 %	1.3 %	32, 529	13, 877, 121	14.42
Example 2	18, 525	82.2 %	9.6 %	16, 081	5, 852, 204	1.37
Example 3	71	53.5 %	8.5 %	69	24, 303	3.84
Example 4	23	26.1 %	43.5 %	23	9, 553	3.65

between the original user A and user C, even though user A is not a friend with user C and there is no way user C could have seen the tweet from user A. To overcome this, we have collected the users followees. Therefore, in our scenario user B is connected to user C only if the retweet timestamp of user C is later than the retweet of user B and user B is in the followees list of user C.

C   A Practical Example for Using Formula (1)

Here, we elaborate on formula (1) and present a practical example. For simplicity reasons and to avoid confusion we define support, $S^{(i)}$, neutral, $N^{(i)}$, and against, $A^{(i)}$, terms in formula (1) following the example attributes given in Sect. 2.2. The generalisations are straightforward. If the attribute of the tweet is a binary indicator, for example whether a tweet contains a URL link or not, we define

$$S^{(i)} = \frac{\text{number of tweets with url that support the rumour}}{\text{total number of tweets that support the rumour}},$$

$$N^{(i)} = \frac{\text{number of tweets with url that are neutral to the rumour}}{\text{total number of tweets that are neutral to the rumour}},$$

$$A^{(i)} = \frac{\text{number of tweets with url that deny the rumour}}{\text{total number of tweets that deny the rumour}}.$$

If the attribute of the tweet is continuous, for example, the number of words in a tweet, we then define

$S^{(i)}$ = average number of words in tweets that support the rumour,

$N^{(i)}$ = average number of words in tweets that are neutral to the rumour,

$A^{(i)}$ = average number of words in tweets that deny the rumour.

These expressions are then combined through formula (1) to give the relevant feature of the rumour.

D   Feature Reduction Methods

Since our dataset consists of 72 rumours, from theoretical and experimental arguments, we expect the relevant features to be about 10. We expect models with as many as 20 features to begin to show a decrease in performance. For this reason we set the upper bound on the number of features to be 30 and aim to examine models with an increasing number of features from 1 to 30. If this bound proves to be low we will reconsider this choice. However as it becomes evident in Sect. 3, this bound is satisfactory.

In this study we use four methods which are combinations of those described so far. For filtering we use the ANOVA F-test [14].

Method 1. A combination of filter method, random wrapper and deterministic wrapper
  (a) Use ANOVA F-Statistics for filtering. Keep the 30-best scoring features.
  (b) From those 30-best we applied the classifier to 100,000 different combinations of 3 features to find the combination of 3 which maximise the F_1-score.
  (c) Add one-by-one the remaining 27 features by applying the classifier and keeping the one with the best F_1-score in each round.
Method 2. A forward selection deterministic wrapper method
  (a) Apply the classifier to all features individually and select the one which maximises the F_1-score (from all available features, no pre-filtering is required).
  (b) Scan (by applying the classifier) all remaining features to find the combination of two (one from step a.) that maximises the F_1-score.
  (c) Continue adding one-by-one the features which maximise the F_1-score until the number of features reaches 30.
Method 3. A combination of filter method and forward selection method
  (a) Use the ANOVA F-Statistics for filtering and keep the 30-best scoring features.
  (b) Apply the classifier and find the best-scoring, i.e. maximum F_1-score, from the 30-best selected from the filtering method (step a).
  (c) Continue adding one-by-one the features which maximise the classification F_1-score.
Method 4. A feature transformation method
  (a) Use a feature transformation method, the principal component analysis. Keep the 30-best components.
  (b) Start with the principal component from the 30-best selected from step a.

(c) Start adding the components one after the other.

We apply each method to each classifier separately, using *scikit-learn*'s default parameters, and assess it using k-fold cross validation. We have abandoned the Neural Network method for two reasons. First its performance was poor compared to the other methods and secondly it required long computational times which slowed down considerably the analysis of the results. We plot the F_1-score as a function of the number of features for the remaining classifiers and each feature reduction method, see Fig. 4.

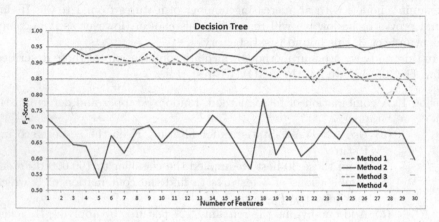

Fig. 4. F_1-score for Decision Tree versus the number of features/components selected from four methods of feature reduction. (Color figure online)

We observe that the second method (red line in Fig. 4) outperforms, in almost all cases, all the other techniques. Similar plots are produced and same conclusion is reached for the other classifiers too. Therefore we can safely conclude that the forward selection deterministic wrapper is consistently the best-performing method of feature reduction for all classifiers.

E   Further Results on Classifier Selection

In Sect. 3 we present the results from running several classifiers for thirty models, each model having an increasing number of features from one to thirty. Here we present more results that support our choice for feature selection.

In Fig. 5 we plot the average F_1-score for each method. This is a two-column plot. The first column (blue) corresponds to the average F_1-score of all 30 models. The second column (red) is the average F_1-score of the first eight models (those with number of features from 1 to 8)[7].

---

[7] We compute the average of the first eight models because this is the range where the classifiers peak their performance. As we argue in Sect. 3 all plots indicate that classifiers performance decreases when more than eight features are added.

F   Visualisation Tool

As a by-product of our modelling, we also developed a software tool which helps the user to visualise the results and gain a deeper understanding of the rumours, see Fig. 6. The tool consists of three layers. On the first layer the user selects a topic of interest (e.g. "Paris Attacks"). This directs to the second layer which displays all the relevant rumours with a basic summary (e.g. the rumour claim,

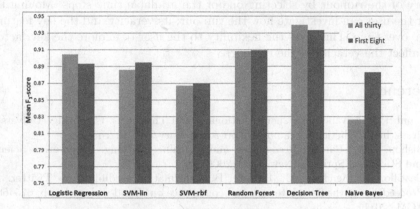

Fig. 5. Mean F_1-score of 30 (blue) and first 8 (red) models. (Color figure online)

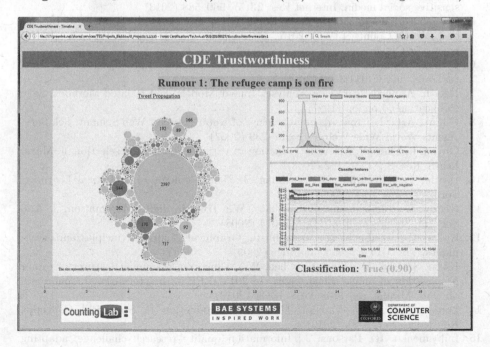

Fig. 6. The visualisation tool. Inside a rumour. (Color figure online)

timestamp of the first tweet, a word cloud, distribution of the tweets that are in favour, neutral or against the rumour and the modelled veracity). After selecting a rumour of interest, the user is navigated to the third layer, shown in Fig. 6. There, the tool shows several figures, such as the propagation forest (supporting, neutral and denying trees are coloured in green, grey and red respectively), a histogram showing the number of tweets in favour of the rumour, against the rumour, and those that are neutral, a plot of classifier's features and the rumour veracity. A time-slider is provided to allow the user to navigate through the history of the rumour by selecting one of the available time steps. Moving the slider the user can investigate how the rumour, its veracity and the key features evolve over time. This gives the flexibility to the user to explore the key factors that affect the veracity of the rumour.

References

1. Cambridge Advanced Learner's Dictionary and Thesaurus. Cambridge University Press. http://dictionary.cambridge.org/dictionary/english/rumour
2. Bishop, C.M.: Pattern Recognition and Machine Learning. Information Science and Statistics. Springer, New York (2006)
3. Castillo, C., Mendoza, M., Poblete, B.: Information credibility on Twitter. In: Proceedings of the 20th International conference on World wide web, pp. 675–684. ACM (2011)
4. Castillo, C., Mendoza, M., Poblete, B.: Predicting information credibility in time-sensitive social media. Internet Res. **23**(5), 560–588 (2013)
5. Pennebaker, J.W., Booth, R.J., Boyd, R.L., Francis, M.E.: Linguistic Inquiry and Word Count: LIWC 2015. Pennebaker Conglomerates, Austin (2015). www.LIWC.net
6. Finn, S., Metaxas, T.P., Mustafraj, E.: Investigating rumor propagation with Twitter Trails. arXiv:1411.3550 (2014)
7. Fox, J.: Applied Regression Analysis, Linear Models, and Related Methods. Sage Publications, London (1997)
8. Gil, Y., Artz, D.: Towards content trust of web resources. Web Semant. Sci. Serv. Agents World Wide Web **5**(4), 227–239 (2007)
9. Guyon, I., Elisseeff, A.: An introduction to variable and feature selection. J. Mach. Learn. Res. **3**, 1157–1182 (2003)
10. Hastie, T., Tibshirani, R., Friedman, J.: The Elements of Statistical Learning. Springer, New York (2009)
11. Kelton, K., Fleischmann, K., Wallace, W.: Trust in digital information. J. Am. Soc. Inf. Sci. Technol. **59**(3), 363–374 (2008)
12. Koller, D., Friedman, N.: Probabilistic Graphical Models: Principles and Techniques. The MIT Press, Cambridge (2009)
13. Kwon, S., Cha, M., Jung, K., Chen, W., Wang, Y.: Prominent features of rumor propagation in online social media. In 2013 IEEE 13th International Conference on Data Mining, pp. 1103–1108. IEEE (2013)
14. Lomax, G.R., Hahs-Vaughn, D.L.: An Introduction to Statistical Concepts. Routledge, New York (2012)
15. Lukyanenko, R., Parsons, J.: Information quality research challenge: adapting information quality principles to user-generated content. J. Data Inf. Qual. (JDIQ) **6**(1), 3 (2015)

16. Mai, J.: The quality and qualities of information. J. Am. Soc. Inf. Sci. Technol. **64**(4), 675–688 (2013)

17. Mendoza, M., Poblete, B., Castillo, C.: Twitter under crisis: can.we trust what we RT? In: Proceedings of the First Workshop on Social Media Analytics, pp. 71–79. ACM, New York (2010)

18. Nurse, J.R.C., Agrafiotis, I., Goldsmith, M., Creese, S., Lamberts, K.: Two sides of the coin: measuring and communicating the trustworthiness of online information. J. Trust Manag. **1**(5), 1–20 (2014). doi:10.1186/2196-064X-1-5

19. Nurse, J.R.C., Creese, S., Goldsmith, M., Rahman, S.S.: Supporting human decision-making online using information-trustworthiness metrics. In: Marinos, L., Askoxylakis, I. (eds.) HAS 2013. LNCS, vol. 8030, pp. 316–325. Springer, Heidelberg (2013). doi:10.1007/978-3-642-39345-7_33

20. Nurse, J.R.C., Rahman, S.S., Creese, S., Goldsmith, M., Lamberts, K.: Information quality and trustworthiness: a topical state-of-the-art review. In: Proceedings of the International Conference on Computer Applications and Network Security (ICCANS) (2011)

21. Pedregosa, F., Varoquaux, G., Gramfort, A., Michel, V., Thirion, B., Grisel, O., Blondel, M., Prettenhofer, P., Weiss, R., Dubourg, V., Vanderplas, J., Passos, A., Cournapeau, D., Brucher, M., Perrot, M., Duchesnay, E.: Scikit-learn: machine learning in Python. J. Mach. Learn. Res. **12**, 2825–2830 (2011)

22. Pew Research Center: The evolving role of news on Twitter and Facebook (2015). http://www.journalism.org/2015/07/14/the-evolving-role-of-news-on-twitter-and-facebook

23. Powers, D.M.W.: Evaluation: from precision, recall and F-measure to ROC, informedness, markedness and correlation. J. Mach. Learn. Technol. **2**(1), 37–63 (2011)

24. Reuters Institute for the Study of Journalism: Digital news report 2015: tracking the future of news (2015). http://www.digitalnewsreport.org/survey/2015/social-networks-and-their-role-in-news-2015/

25. Seo, E., Mohapatra, P., Abdelzaher, T.: Identifying rumors and their sources in social networks. In: SPIE Defense, Security, and Sensing, p. 83891I. International Society for Optics and Photonics (2012)

26. Smola, A.J., Scholkopf, B.: A tutorial on support vector regression. Stat. Comput. **14**, 199–222 (2004)

27. The Guardian: How riot rumours spread on Twitter (2011). http://www.theguardian.com/uk/interactive/2011/dec/07/london-riots-twitter

28. Verleysen, M., François, D.: The curse of dimensionality in data mining and time series prediction. In: Cabestany, J., Prieto, A., Sandoval, F. (eds.) IWANN 2005. LNCS, vol. 3512, pp. 758–770. Springer, Heidelberg (2005). doi:10.1007/11494669_93

29. Vosoughi, S.: Automatic detection and verification of rumors on Twitter. Ph.D. thesis, MIT (2015)

30. Wang, R.Y., Strong, D.M.: Beyond accuracy: what data quality means to data consumers. J. Manag. Inf. Syst. **12**(4), 5–33 (1996)

31. Zubiaga, A., Liakata, M., Procter, R., Bontcheva, K., Tolmie, P.: Towards detecting rumours in social media. arXiv preprint arXiv:1504.04712 (2015)

PicHunt: Social Media Image Retrieval
for Improved Law Enforcement

Sonal Goel[✉], Niharika Sachdeva, Ponnurangam Kumaraguru,
A.V. Subramanyam, and Divam Gupta

Indraprastha Institute of Information Technology, Delhi, India
{sonal1426,niharikas,pk,subramanyam,divam14038}@iiitd.ac.in

Abstract. First responders are increasingly using social media to identify and reduce crime for well-being and safety of the society. Images
shared on social media hurting religious, political, communal and other
sentiments of people, often instigate violence and create law & order situations in society. This results in the need for first responders to inspect
the spread of such images and users propagating them on social media.
In this paper, we present a comparison between different hand-crafted
features and a Convolutional Neural Network (CNN) model to retrieve
similar images, which outperforms state-of-art hand-crafted features. We
propose an Open-Source-Intelligent (OSINT) real-time image search system, robust to retrieve modified images that allows first responders to
analyze the current spread of images, sentiments floating and details of
users propagating such content. The system also aids officials to save
time of manually analyzing the content by reducing the search space on
an average by 67 %.

1 Introduction

First responders across the globe are increasingly using Online Social Networks
(OSN) for maintaining safety and law & order situations in society. Prior work
shows the role of social media to aid first responders like police, for instance;
police can use OSN to obtain actionable information like location, place, and
evidence of the crime [24]. Police have realized the effectiveness of OSN in various
activities such as investigation, crime identification, intelligence development,
and community policing [1,6,23]. However, in order to accomplish these goals
on OSN, they often needs to identify tweets and images causing safety issues
[6,35].

Researchers have explored the utility of social media platforms like Twitter
to identify text and network of people leading to law enforcement help [1,35],
similarly images on social media sites also often yield information of interest to
investigators [10]. A study shows that people don't engage equally with every
tweet, Twitter content of over 2 million tweets by thousands of verified users
over the course of a month was analyzed, and it showed that tweet with an
image present can increase user engagement by 35 % [20]. It is said that an
image is worth a thousand words. People with different backgrounds can easily

© Springer International Publishing AG 2016
E. Spiro and Y.-Y. Ahn (Eds.): SocInfo 2016, Part I, LNCS 10046, pp. 206–223, 2016.
DOI: 10.1007/978-3-319-47880-7_13

understand the main content of an image thus, increasingly becoming preferred media to reach and effect masses.

Often images shared on social media have the potential to hurt religious, political, communal, caste and other sentiments of a certain section of society. Such images intimidate people, instigate anger among them which further leads to law and order situations and security critical scenarios in the society. For example, in April 2016, a journalist tweeted a morphed image of an Indian politician touching feet of a foreign country's king, the image invited anger and backlash on all social media networks and the political party filed a complaint against the journalist for misleading public [27]. Another example, in June 2014, obscene pictures of Indian warrior-king Chhatrapati Shivaji and a late political party chief were posted on Facebook leading to riots in Maharastra, India. People went on rampage damaging public and private vehicles, pelting stones leading to severe injuries and even beating a person death [29].

With this arises the need for first responders to understand, patrol, prevent the spread of such images for maintaining law and order in the society. However, police personnel has limited exposure to technology [30] and this limitation makes it difficult to adopt findings of OSN use by police to facilitate policing needs [23]. Moreover, researchers have focussed more on textual content to aid the first responders but analyzing multimedia content like images and videos on OSN is still largely unexplored [13]. Also, the tools necessary to retrieve, filter, integrate and intelligently present relevant images and their information for better safety during security critical scenarios need to be leveraged [2,12]. These initiatives motivated us to propose a real-time image-search system which can serve the first responders by bridging the gap between research in technologies and solving real-world problems to improve security during such critical scenarios. But the modifications done on images like cropping, scaling, stitching image, wrapping around text, contrast enhancement and changing colors are one of the biggest challenges faced in image analysis. These challenges often create barriers to directly access these images and utilize information like image spread, etc., which makes image retrieval a complex and daunting process. This paper lies in the intersection of crisis informatics and social media image utility in the understudied & novel context of law enforcement. The main contributions of this work are:

1. We develop a real-time image search system, which is robust to detect similar images which are cropped, scaled, blurred, stitched with other images, wrapped around with text, brightened, or modified using other similar image processing techniques.
2. The system aims at data management for first responders, helping them reduce search space for images.
3. We analyze different techniques for retrieving similar images and experimentally show that ORB (Oriented Fast and Rotated Brief) in combination with RANSAC is the state-of-art technique in hand-crafted image features for identifying similar images.

4. We propose a supervised deep CNN model that outperforms state-of-art hand-crafted techniques to retrieve similar images.
5. We created a new human-annotated dataset of images from incidents that created law and order situations in the society.[1]

2   Problem Statement

We now formally define the problem definition and notations. Given an image I_A and set of keywords K by a user, find a set of similar images $I_B = \{I_{b1}, I_{b2}, \ldots I_{bn}\}$ using a comparison function C, from a search set of images $I_C = \{I_{c1}, I_{c2}, \ldots I_{cm}\}$ formed using a search function S on a social network S_N, where $n \leq m$:

$$I_C = S(S_N, K) \text{ and } I_B = C(I_A, I_C)$$

The search function S takes a set of keywords, and a search space S_N as input and gives a search set of images I_C. The comparison function takes the given image I_A and I_C as input and returns the required similar images set I_B.

We model the image similarity as a classification problem with four phases - database formation, feature extraction of images, feature comparison and finally similarity classification (see Fig. 1).

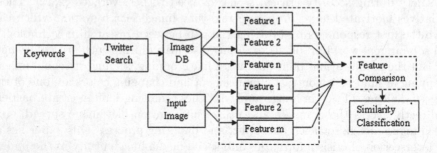

Fig. 1. Proposed methodology to find image similarity. The system takes keywords and an image from the user, using the search API of Twitter, creates a database of related images. To compare two images, their features are extracted and fed to a comparison function. Comparison score classifies image similarity.

3   Related Work

Recent studies show an increase in need of OSM as a plausible resource for first responders [3]. Police have realized the effectiveness of OSM in various activities such as investigation, identifying crime, intelligence development, and community policing [3,22,26]. Research shows the effectiveness of OSN during events

---

[1] http://precog.iiitd.edu.in/resources.html

involving law and order issues like the Boston bombings, Sichuan earthquake (2008), Haiti earthquake (2009), Oklahoma grassfires (2009), and Chile earthquake (2010) [8,9,15,28,34]. Studies have shown how OSN has been effectively used to aid police to increase community engagement, reduce crime by getting on actionable information from OSN and improve coordination between police and citizen [23–25]. But, these studies focus on the textual aspect of OSN. Research shows that the role of social media text has received attention but the role of social media images remains largely unstudied [13]. However, a report shows that if there is an image attached to a tweet it increases user engagement by 35 % [20].

There are also some studies and tools on social media image analysis in general context; like, the analysis of 581 Twitter images of the 2011 Egyptian revolution revealed that more efficacy-eliciting (crowds, protest activities, national and religious symbols) content is posted by Egyptian users than emotionally arousing (violent) content [13]. Though there have been studies on social media image search, but in most of them query is text based; for instance, a study describes how to use Wikipedia and Flickr content to improve the match of query text with database vocabulary [18]. They used a combination of Flickr and Wikipedia query model for query expansion to improve the accuracy of image retrieval systems [18]. Images are also used to visually summarise events. In [14], authors propose a technique to find most popular unique images shared on Twitter related to events like sports, law & politics, art, culture, etc. In another study, a tool was developed to show trending images of an event by extracting images from Twitter's streaming API using text as query and then detecting near duplicates using locality sensitive hashing [11]. Another hybrid approach for image retrieval combines social relevance by understanding the user's interest using social site Flickr and visual relevance by ranking the images in Google search result according to the interests of the user [2]. A study shows that image search tools such as TinEye and Google Reverse Image Search are used by journalists to find duplicates, such as other posts of the same image, and near duplicates, such as posts before or after potential Photoshop manipulations, to help find fake posts [37]. Though, some of these studies discuss image retrieval aspects but most of them take text as the query to retrieve related images, and systems like TinEye and Google Reverse Image Search gives better results for searches on the entire web than specifically for social media sites like Twitter in specific. Further, these tools do not provide knowledge management to first responders, like users propagating the visual content and sentiments floating with them. Also, most of the image-retrieval tools discussed in these studies use basic hand-crafted features for finding image similarity thus, resulting in comparatively lower accuracy.

Our work focuses on emphasizing on the needs of first responders to analyze image spread on social media, for which we deeply analyzed different techniques that can be used for image retrieval with the query as an image and developed a system to aid them to find & analyze similar images on social media.

4   Data Collection and Annotation

From October, 2015 to February, 2016, we collected data related to 5 events that created law and order issues in society. To create this dataset, we collected data from Twitter using Twitter's search API[2], filtering tweets that contain images, counting to a total of 3,725 images. Figure 2 shows images which were viral from these events and are taken as input images for evaluating different image similarity models to be studied. The details of the events are discussed below:

1. Kulkarni Ink: Black ink was sprayed on an Indian technocrat-turned-columnist Sudheendra Kulkarni by the members of a political party, ahead of the launch of former Pakistan's Foreign Minister book, 'Neither a Hawk nor a Dove: An In-sider's Account of Pakistan's Foreign Policy'. FIR was lodged against the political party workers and six workers got arrested [16]. This incidence was slammed on social media arousing political issues and the images of the man with black ink on face went viral. We collected 1,905 tweets with images using the keyword "#Kulkarni".

2. Baba Ram Rahim: An Indian self-proclaimed Godman Baba Ram Rahim posted pictures posing as Hindu God Vishnu on social media. He was accused by All India Hindu Student Federation for insulting Lord Vishnu and hurting religious sentiments of Hindus by dressing up as Lord Vishnu and lodged a complaint against him [31]. We collected 408 tweets with images using the keyword "#RamRahim".

3. Lord Hanuman Cartoon: A cartoon tweeted by an Indian politician drew much flak from other political parties of India and its affiliates for allegedly hurting religious sentiments. The cartoon purportedly showed Hindu Lord Hanuman clad in saffron robes and leader of another political party. The political party lodged a complaint alleging that the cartoon image was posted with the intent to hurt religious sentiments of Hindus [36]. We collected 664 tweets with images using keywords like "#JNU" and "Insults Hanuman".

4. Charlie Hebdo cartoon: A cartoon in the French satirical magazine Charlie Hebdo sparked outrage by publishing a cartoon attempting to satirize the Syrian refugee crisis. The cartoon imagines Alan Kurdi, the three-year-old Syrian who died in the sea in September 2015, on the way to Europe, has grown up to be a sexual abuser [7]. Many called the cartoon was racist and said it was incredibly bad. We collected 568 tweets with images using the keyword "#CharlieHebdo".

5. Shani Shingnapur protest: As many as 1,500 women, mostly homemakers and college students, planned to storm the Shani Shingnapur temple in India on Republic Day. The protesters wanted to end the age-old humiliating practice of not allowing women to enter the core shrine area [4]. The image of one of the activist, giving an interview about the protest on Republic Day to media went viral. We collected 180 tweets with images related to the keyword '#ShaniShingnapur".

---

[2] https://dev.twitter.com/rest/public/search

Fig. 2. Input images set. These images represents image which could be of interest to first responders and were viral from the event (a) Kulkarni Ink (b) Baba Ram Rahim (c) Lord Hanuman cartoon (d) Charlie Hebdo cartoon (e) Shani Shingnapur, respectively.

4.1 Data Annotation

To create the ground truth for testing the models, we obtained labels for all the 3,725 images. The annotators were given an image from each event (see Fig. 2) and from the respective event's image database they had to mark images that were similar and dissimilar to the given image. The broad definition of image similarity given to them was: images that are cropped, scaled, blurred, wrapped with text, stitched with other images, having color changes, brightness changes or contrast enhancements are considered similar. The images which had agreement from at least two annotators were marked in the final results. Table 1 shows the dataset after annotation was completed.

Table 1. Data after annotation shows the number of similar and dissimilar images for each event.

Keyword	Total images	Similar images	dissimilar images
#Kulkarni	1,905	354	1,551
#RamRahim	408	97	311
#Insults Hanuman	664	277	387
#CharlieHebdo	568	118	450
#ShaniShingnapur	180	70	110

5 Similarity Modelling

The retrieval performance of CBIR (Content Based Image Retrieval) system crucially depends on the features that represent images and their similarity measurement. In this section, we'll discuss some of the techniques we studied to compare the similarity of two images.

5.1   Hand-Crafted Features

1. Color-based feature similarity

   Color histograms are frequently used to compare images [17]. It is often done by comparing color histograms of images, which eliminates information on the spatial distribution of colors. The image descriptor is a 3D RGB color histogram with 8 bins per channel and we compare the descriptor using a similarity metric. After extracting image descriptors, we calculated Bhattacharyya distance for comparing the two image's histogram descriptors. The lesser is the distance between two color histograms, more similar the two images are.

2. Keypoint-Descriptor-based similarity

   Certain parts of an image have more information than others, particularly at edges and corners. Keypoints can be generated using this information. After finding the keypoints of an image, next step is to find their descriptors. Descriptors are fixed length vectors that describe some characteristics about the keypoints. Next, we compare each keypoint descriptors of one image to each keypoint descriptors of the other image. Since, the descriptors are vectors of numbers, we can compare them using different distance metrics. We studied techniques like DAISY [32], ORB [21] and Improved ORB [38] for computing keypoint descriptors and then comparing them.

   - The *DAISY* dense image descriptor is based on gradient orientation histograms similar to the SIFT (Scale Invariant Feature Transform) descriptor. It is formulated in a way that allows for fast dense extraction which is useful for bag-of-features image representations [33]. After extracting DAISY descriptors we calculate the distance between the descriptor vectors of two images using Brute Force matcher with KNN (K-Nearest Neighbors) as the distance metric. We use this distance as the score to measure the similarity of two images.

   - *ORB* uses improved FAST for feature detection, and these features are described using an improved Rotated BRIEF feature descriptor [38]. Since the speed of FAST and BRIEF are very fast this can be the choice for real-time systems. ORB is rotational invariant, noise invariant and uses image pyramids for scale invariance [21]. It returns binary strings to describe feature points, which is used for feature point matching using Hamming distance. We use this hamming distance as the score to evaluate the similarity of two images.

   - In order to improve the matches given by ORB we add one extra step. After comparing the keypoint descriptors given by ORB, we find homography of two images using RANSAC [38]. We extract top 30 matches having the least distance and pass these matches to RANSAC to weed out wrong matching points. This technique is called as *Improved ORB*. RANSAC returns a binary array equal to size of input matches array, where 0 represents a false match and 1 represents a true match. We define true ratio as the ratio of true matches returned by RANSAC to total matches given to it. We use this true ratio to evaluate the similarity of two images. In the given equation, t_r denotes true ratio, n is the size of matches set passed to RANSAC, A is the array returned by RANSAC and A_i represents the value of array returned by RANSAC at i^{th} index.

$$t_r = \frac{\sum_{i=0}^{n-1} A_i}{n}$$

5.2 Trained Features

In this section we discuss one important technique "deep learning" which is organised in a deep architecture and processes information through multiple stages. To that end, inspired by recent advances in neural architectures and deep learning, we choose to address CBIR using deep convolutional neural network.

For implementation of *Convolutional Neural Network (CNN)*, we took inspiration from the architecture discussed in [19]. To create the dataset for training, we used an unsupervised approach. We extracted around 25K images from popular news websites and then on each of 25K image, we applied various image processing operations like cropping, distorting, blurring, scaling, adding text, stitching image, adding noise, and changing color, to generate the corresponding similar image. Thus, we created a dataset of 50K images. To train the model we had 25K pairs of similar and dissimilar images. Similar approach of transforming images to create dataset was used by Fischer et al., 2014 [5]. As a pre-processing step we first align the images. We extract SIFT descriptors to map the matching features of both the images and align the images by applying a perspective transformation using the homography. We then club the two aligned images to form a 6 channel image which is passed to the CNN model.

The model contains 15 layers out of which 6 are trainable. First 5 of trainable layers are convolutional and one is fully connected. The output of the last fully connected layer is a 1D vector and is fed to a sigmoid activation which produces the final output between 0 and 1. Our network minimizes binary cross entropy loss objective and uses adam optimizer as the optimizer. The convolutional layers contain 64 filters 122×122 pixels, 32 filters 61×61 pixels, 16 filters 31×31 pixels, 6 filters 16×16 pixels, 1 filter 16×16 pixels. All the convolution layers are applied with a stride of 4 pixels. The fully connected layer contains 1 neuron. All the convolution layers use LeakyRelu and batch normalization except the first layer which doesn't use batch normalization and the last convolution layer directly feeds output to the fully connected layer. All the layers use a subsampling of 2×2 except the last layer, which uses a subsampling of 1×1. The initial convolution layers extract the low level features like edges and gradients. The other convolution layers learn to compare the extracted features of the two images. The similar areas of the images get propagated till top and fully connected layers calculate a similarity score using the high level features.

6 Hand-Crafted vs. Trainable Features

In this section, we evaluate different techniques discussed in the above section on the annotated test data described in Sect. 4. To evaluate the performance of the above-discussed image similarity models we calculate accuracy of their classification results, which is the ratio of true classification to the total population.

6.1 Histogram

For each of the five events, we plotted the accuracy for Bhattacharyya distance between color histogram of two images, ranging from 0.1 to 0.9. Figure 3(a) shows accuracy for different events at different histogram distance. The graph shows a lot of variation in the accuracy corresponding to the histogram distance values. We calculated average variance in accuracy at 3 distance points on the x-axis (0.2, 0.4, 0.5) as 104.24, which is high making it tough to choose a distance value as the threshold. For example, If we choose threshold distance as 0.4, the accuracy for ShaniShignapur and Lord Hanuman cartoon is close to 60 % whereas for Charlie Hebdo it is 90 %. Another drawback of this method is, it lacks spatial information, so images with very different appearances can also have similar histogram [17].

6.2 DAISY

The mean distance between DAISY descriptors of both similar and dissimilar images was in the range of 0.0 to 0.10. To choose the threshold distance values between this range, we plotted the accuracy graph for each event at different threshold value. Figure 3(b) shows the accuracy vs. threshold plot, where the threshold is the mean distance between DAISY descriptors of two images. Since the range of distance returned by DAISY descriptors is very small there was a very high overlap between distance values of similar and dissimilar images leading to high error. Due to this high error, the maximum accuracy achieved overall is less than 88 %, at a distance of 0.06 but, at this point accuracy for remaining events 4 events is less than 50 %. We also calculated the average variance in accuracy at 3 distance points on the x-axis (0.02, 0.05, 0.06) to be 196.58, which is even higher than the Histogram technique. Thus, making it hard to choose an optimal threshold value.

6.3 ORB

For all the 5 events, we plotted the accuracy for different values of mean distance between ORB descriptors. We use this plot to choose an optimal threshold distance value. Figure 4(a) represents the accuracy plot for different events at different distance between ORB descriptors. The variation in accuracy is less as compared to the Histogram and DAISY technique. Average variance in accuracy at 3 distance points on the x-axis (29, 32, 35) is 17.6. Also, the accuracy for all the events at threshold distance 29 is above 88 %. This shows that ORB is certainly a better choice than Histogram and DAISY for our data.

6.4   Improved ORB

In Improved ORB, after getting the match set of the descriptors from ORB, we pass the top thirty matches having least distance to RANSAC, which filters matches which are true. The threshold value here is the ratio of true matches returned by RANSAC to the total matches passed (thirty in our case), defined as true match ratio. Figure 4(b) shows the accuracy of Improved ORB for different true match ratio taken as the threshold. Average variance in accuracy at 3 true match ratio values on the x-axis (0.33, 0.35, 0.37) is 6.2, which is least among all the techniques discussed till now. Also, after true match ratio of 0.3, the accuracy for each event is almost constant and is above 90 % for all events. If compared with above methods Improved ORB is giving best results. Hence, Improved ORB can be termed the state-of-art technique in hand-crafted features discussed.

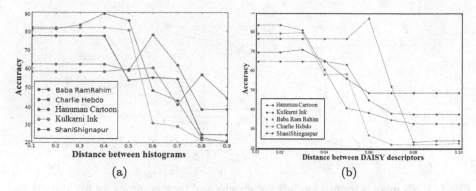

Fig. 3. (a) Accuracy for different events at different distance between histograms of two images. (b) Accuracy for different events for different distance between DAISY descriptors of two images. Both graphs shows high variance in accuracy.

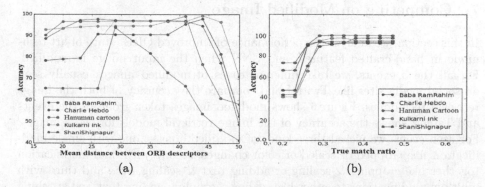

Fig. 4. (a) Accuracy for different events at different distance values between ORB descriptors, at distance value 29 on x-axis all the events have accuracy more than 88 % and (b) Accuracy for different events at different true match ratio returned by Improved ORB, at true ratio of 0.35 all events have accuracy above 90 %.

6.5   Trainable Features

We trained a deep CNN model for 25K pairs of similar and dissimilar images, and to test the model we gave input image and images from annotated set to the model for different epoch values. To choose optimal epoch for the model, we plot the accuracy for different events, for models trained on different epoch values. Figure 5 shows the accuracy for each event at different epoch value. The model trained for epoch value 35 is showing an accuracy above 97 % for all events. Also, average variance at three epoch values on the x-axis (30, 35, 40) is 0.71, which is even lower than what we achieved in Improved ORB. Since the overall accuracy for CNN model is higher and the variance is lesser than Improved ORB, we can conclude that our proposed CNN model outperforms state-of-art Improved ORB. In the next section, we'll see how these two techniques perform when the input image is modified.

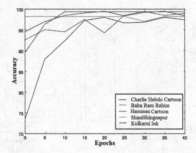

Fig. 5. CNN models trained for different epochs and their accuracy for five test events. For model trained for 35 epochs, accuracy for all events is above 97 %.

7   Competing on Modified Image

In this section, we analyze the performance of Improved ORB (state-of-art technique in hand-crafted features) and CNN when the input image is modified. For all the 5 events, we take different cases of modified images usually seen on social media sites like Twitter and compare the accuracy of both the image retrieval techniques. Figure 6 shows modified images taken from all the events and Table 2 shows the accuracy of the image retrieval models for these images. For each event, we picked three kinds of modified images: one with single modification like cropped or scaled or color changes, second with dual modification together like cropping & scaling or adding text & scaling, etc., and third with multiple modifications together like scaling, cropping, adding text and stitching image, represented as All in Table 2.

Among different modifications seen, the most common modification technique was scaling the image. Hence, we calculated the scaling factors of the modified images and found that as the scaling factor increases, the accuracy dip

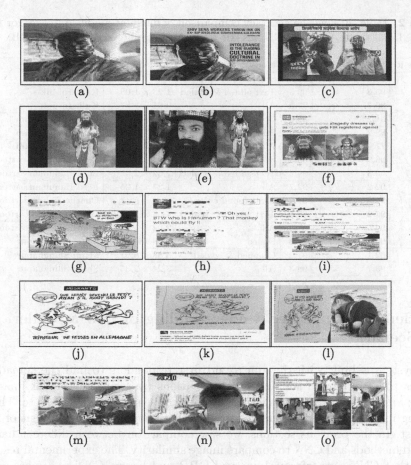

Fig. 6. Event-wise modified images. By Row: (1) Kulkarni Ink Event (2) Baba Ram Rahim (3) Lord Hanuman Cartoon (4) Charlie Hebdo Cartoon (5) Shani Shingnapur.

for Improved ORB also increases. For example, in Table 2 image 'a' is modified by a scaling factor of 7.42 × 5.23 and the accuracy of Improved ORB is 84.1% while, for the same image CNN is showing an accuracy of 99.4%. Likewise, for image 'c', 'm', 'n', 'o' also the scaling factor is high leading to low accuracy in Improved ORB but, the accuracy of CNN in all these cases outperforms Improved ORB. Another common modification affecting accuracy is cropping of an image, for instance; image 'f' has only the cropped face from the complete body of Baba Ram Rahim (original picture see in Fig. 2), and again the accuracy dip in Improved ORB is much higher than CNN. Likewise, is the case for image 'd', cropping reduced the accuracy of Improved ORB while CNN still outperforms it. Hence, we can conclude that the proposed CNN model is more robust to modifications like high scaling and high cropping than state-of-art Improved ORB.

Table 2. Table showing the accuracy comparison of different modified images for test events. In all the test cases, CNN has out-performed state-of-art Improved ORB.

Image-Id	Improved ORB	CNN	Modification	Scaling factor	Event
(a)	84.1	99.4	Scaled	7.42 × 5.23	Kulkarni ink
(b)	95.6	99.5	Text added & scaled	1.23 × 1.08	Kulkarni ink
(c)	80.7	87.3	All	9.61 × 3.87	Kulkarni ink
(d)	91.2	99.5	Cropped	–	Baba Ram Rahim
(e)	88.2	92.6	Stitched & scaled	2.31 × 0.92	Baba Ram Rahim
(f)	77.4	93.8	All	2.65 × 1.56	Baba Ram Rahim
(g)	96.9	99.3	Scaled	1.04 × 1.09	Lord Hanuman cartoon
(h)	90.3	99.5	Text added & scaled	2.32 × 2.47	Lord Hanuman cartoon
(i)	92.2	92.7	All	1.27 × 1.25	Lord Hanuman cartoon
(j)	96.8	97.2	Background color	–	Charlie Hebdo cartoon
(k)	97.2	97.6	Text added & scaled	1.04 × 1.28	Charlie Hebdo cartoon
(l)	94.5	95.3	All	2.02 × 1.54	Charlie Hebdo cartoon
(m)	90.4	95.3	Scaled	4.98 × 6.09	Shani Shingnapur
(n)	62.1	98.2	Cropped & scaled	5.88 × 5.91	Shani Shingnapur
(o)	93.7	98.3	All	3.24 × 2.35	Shani Shingnapur

8   Benefits: Improved Understanding of Law & Order Scenario

The system is built with the aim to aid first responders to find the spread of images that can create law and order situations by retrieving similar images, find & analyze the users propagating the content, sentiments floating, etc. Thus, helping in reducing their human efforts in identifying the visual content of the interest. In the previous sections, we saw comparison between different hand-crafted methods and CNN to compare image similarity. The experimental results show that CNN outperforms Improved ORB and hence, CNN is a more suitable technique to find image spread for the system. We now explain various other measures offered by the system that help in achieving goals for better law enforcement using similar image retrieval process discussed above.

Search Space Reduction: The system takes a set of keywords from the user to create a database of images and an image using which it retrieves a set similar images from the database created. Thus, reducing the search space for first responders hence, saving their time and efforts to analyse the overall dump. We were able to reduce the search space on an average by 67 % and by 65 %, using CNN and Improved ORB as image retrieval technique respectively. The maximum reduction in search space was seen for Baba Ram Rahim event followed by Charlie Hebdo cartoon. Table 3 shows the details of search space reduction for all the events using the two techniques.

Users Analysis: Users play a vital role in spreading the content on social media. Thus, it is important for security analysts to identify people who are spreading the images. The system tries to deliver this need by producing a list of users who spread the visual content and their details like, their Twitter usernames, profile pictures, description, location, and links to their Twitter profiles. In our dataset, we found maximum users for Lord Hanuman cartoon event propagating the content, for images retrieved using both CNN and Improved ORB. Table 3 shows the number of users listed when we used Improved ORB and CNN as image retrieval techniques.

Sentiment Analysis and Tweets vs. Retweets Analysis: Sentiments of the tweets play an important role in spreading the content and eliciting the reaction of people. Thus, making it an important analysis to find the percentage of content having positive, negative and neutral sentiments. The system currently analyzes the sentiment of the textual content having images using Sentiment140 API[3], widely used in the literature to study social media data. In our dataset, we found 7 % of tweets have negative sentiments and 5 % have positive sentiments and remaining have neutral sentiments which might be due to broken language, non-english, hinglish language or absence of text. Tweets vs. Retweets analysis shows the percent of retweets and original tweets. The analysis shows, most of the images in the result were spread by retweeting, average retweeted posts were 89.4 %.

Table 3. Event-wise reduction in search space and analysis on number of users after retrieving similar images using Improved ORB and CNN.

Event	Search space reduction (%)		Number of users	
	Improved ORB	CNN	Improved ORB	CNN
Kulkarni ink	51.1	54.3	770	1,144
Baba Ram Rahim	83.1	90.7	75	96
Lord Hanuman cartoon	60.1	61.6	2,210	2,657
Charlie Hebdo cartoon	74.1	75.2	2,132	2,037
Shani Shingnapur	57.9	57.9	186	186

9 Implementation and Response Time

9.1 Implementation

The proposed system takes an image and keywords as input. These keywords are then given to Twitter's Search API (Application Program Interface), part of Twitter's REST API. The API returns a set of tweets related to the keywords it takes as the query, and the system then filters and saves the tweets containing images in a database. As the images get stored in the database, the image comparison model computes the similarity score between the input image and the images in the database. We then set a threshold t, if the similarity score is above t, images are marked similar else dissimilar. After getting similar images,

---

[3] http://help.sentiment140.com/home.

the system does analysis on the retrieved set of images and finds the users propagating them, the sentiment analysis and retweets analysis. Figure 7 shows input and different output and analysis screens of the proposed system.

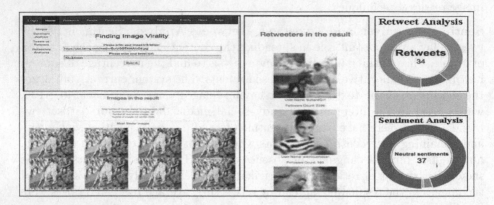

Fig. 7. Moving in anti-clockwise from top-left image the images represent (a) Input to the system is a image url and a keyword. (b) Output screen: The system retrieves similar images (c) Analysis screen: Users propagating (d) Analysis screen: Sentiment analysis on text of the resultant images (e) Analysis Screen: Retweets analysis on text of the resultant images.

9.2 Response Time

Since our aim is to build a real-time image search system, apart from the accuracy of the model we also need to select the technique that is time efficient. We compare the approximate time consumed by Improved ORB and CNN to compare one pair of image and we found that Improved ORB took 0.2 s and CNN model took 0.55 s.

10 Conclusion

In our study, we proposed a OSINT real-time image search system robust to find modified images, that aids first responders to analyze and find the current spread of images creating law & order situations in society, analyze users propagating such images and sentiments floating with them. The system performs a hybrid search by taking a keyword to create an image database and an image as input and returns similar images. The system also aims to reduce search space for first responders on average by 67 %. To find similar images we conducted an in-depth experimental analysis of different CBIR techniques. We did the experimental study on the data collected for the 5 events which created law and order situations in society and whose images were viral during the duration of the study.

We compared different hand-crafted techniques and found that Improved ORB (ORB + RANSAC) is the state-of-art technique for the hand-crafted approach. We also proposed a CNN model which outperforms the accuracy of state-of-art hand-crafted technique. Future work will focus on reducing the response time of the proposed system, improving results of textual sentiment analysis that takes broken or non-english language into consideration, including more image analytics features like, sentiments of images in result to aid maintaining safety in society during security critical scenarios.

References

1. Crump, J.: What are the police doing on Twitter? Social media, the police and the public. Policy Internet **3**(4), 1–27 (2011)
2. Cui, P., Liu, S.-W., Zhu, W.-W., Luan, H.-B., Chua, T.-S., Yang, S.-Q.: Social-sensed image search. ACM Trans. Inf. Syst. (TOIS) **32**(2), 8 (2014)
3. Denef, S., Kaptein, N., Bayerl, P.S.: ICT trends in European policing. Composite project (2011)
4. The Indian Express. Over 1500 women plan to storm Shani Shingnapur temple today, January 2016
5. Fischer, P., Dosovitskiy, A., Brox, T.: Descriptor matching with convolutional neural networks: a comparison to sift. arXiv preprint arXiv:1405.5769 (2014)
6. Gerber, M.S.: Predicting crime using Twitter and kernel density estimation. Decis. Support Syst. **61**, 115–125 (2014)
7. The Guardian: Charlie Hebdo cartoon depicting drowned child Alan Kurdi sparks racism debate, January 2016
8. Gupta, A., Kumaraguru, P.: Misinformation on Twitter during crisis events. Encyclopedia of social network analysis and mining (ESNAM) (2012)
9. Gupta, A., Lamba, H., Kumaraguru, P.: $1.00 per RT #BostonMarathon #PrayForBoston: analyzing fake content on Twitter. In: eCrime Researchers Summit (eCRS), pp. 1–12. IEEE (2013)
10. Hanson, W.: How social media is changing law enforcement, December 2011
11. Hare, J.S., Samangooei, S., Dupplaw, D.P., Lewis, P.H.: Twitter's visual pulse. In: Proceedings of the 3rd ACM Conference on International Conference on Multimedia Retrieval, pp. 297–298. ACM (2013)
12. Hoi, S.C., Pengcheng, W.: SIRE: A Social Image Retrieval Engine. In: Proceedings of the 19th ACM International Conference on Multimedia, pp. 817–818. ACM (2011)
13. Kharroub, T., Bas, O.: Social media and protests: an examination Egyptian revolution. New Media Soc. **1461444815571914**, 2015 (2011)
14. McParlane, P.J., McMinn, A.J., Jose, J.M.: Picture the scene..;: visually summarising social media events. In: Proceedings of the 23rd ACM International Conference on Conference on Information and Knowledge Management, pp. 1459–1468. ACM (2014)
15. Mendoza, M., Poblete, B., Castillo, C.: Twitter under crisis: can we trust what we RT? In: Proceedings of the First Workshop on Social Media Analytics, pp. 71–79. ACM (2010)
16. BBC News: Mumbai ink attack: India police arrest six Shiv Sena workers, October 2015

17. Pass, G., Zabih, R., Miller, J.: Comparing images using color coherence vectors. In: Proceedings of the Fourth ACM International Conference on Multimedia, pp. 65–73. ACM (1997)

18. Popescu, A., Grefenstette, G.: Social media driven image retrieval. In: Proceedings of the 1st ACM International Conference on Multimedia Retrieval, p. 33. ACM (2011)

19. Radford, A., Metz, L., Chintala, S.: Unsupervised representation learning with deep convolutional generative adversarial networks. arXiv preprint arXiv:1511.06434 (2015)

20. Rogers, S.: What fuels a tweet's engagement, March 2014

21. Rublee, E., Rabaud, V., Konolige, K., Bradski, G.: ORB: an efficient alternative to SIFT or SURF. In: 2011 International Conference on Computer Vision, pp. 2564–2571. IEEE (2011)

22. Ruddell, R., Jones, N.: Social media and policing: matching the message to the audience. Safer Communities **12**(2), 64–70 (2013)

23. Sachdeva, N., Kumaraguru, P.: Online social networks and police in India - understanding the perceptions, behavior, challenges. In: Boulus-Rødje, N., Ellingsen, G., Bratteteig, T., Aanestad, M., Bjorn, P. (eds.) ECSCW 2015: Proceedings of the 14th European Conference on Computer Supported Cooperative Work, Oslo, Norway, 19–23 September 2015, pp. 183–203. Springer, Copenhagen (2015)

24. Sachdeva, N., Kumaraguru, P.: Social networks for police and residents in India: exploring online communication for crime prevention. In: Proceedings of the 16th Annual International Conference on Digital Government Research, pp. 256–265. ACM (2015)

25. Sachdeva, N., Kumaraguru, P.: Social media - new face of collaborative policing? In: Meiselwitz, G. (ed.) SCSM 2016. LNCS, vol. 9742, pp. 221–233. Springer, Heidelberg (2016). doi:10.1007/978-3-319-39910-2_21

26. Lexis Nexis Risk Solutions: Survey of law enforcement personnel and their use of social media in investigations. Lexis Nexis (2012)

27. Standard, B.: BJP files complaint against journalist for morphed Modi photo, April 2016

28. Starbird, K., Palen, L.: Voluntweeters: self-organizing by digital volunteers in times of crisis. In: Proceedings of the SIGCHI Conference on Human Factors in Computing Systems, pp. 1071–1080. ACM (2011)

29. Sakal Times: 180 held for circulating obscene photos on internet, June 2014

30. The Economics Times: NCRB to connect police stations and crime data across country in 6 months, November 2014

31. India Today: Case against Gurmeet Ram Rahim for posing as Vishnu, will he be arrested? January 2016

32. Tola, E., Lepetit, V., Fua, P.: A fast local descriptor for dense matching. In: IEEE Conference on Computer Vision and Pattern Recognition, CVPR 2008, pp. 1–8. IEEE (2008)

33. Tola, E., Lepetit, V., Fua, P.: Daisy: an efficient dense descriptor applied to wide-baseline stereo. IEEE Trans. Pattern Anal. Mach. Intell. **32**(5), 815–830 (2010)

34. Vieweg, S., Hughes, A.L., Starbird, K., Palen, L.: Microblogging during two natural hazards events: what Twitter may contribute to situational awareness. In: Proceedings of the SIGCHI Conference on Human Factors in Computing Systems, pp. 1079–1088. ACM (2010)

35. Wang, X., Gerber, M.S., Brown, D.E.: Automatic crime prediction using events extracted from Twitter posts. In: Yang, S.J., Greenberg, A.M., Endsley, M. (eds.) SBP 2012. LNCS, vol. 7227, pp. 231–238. Springer, Heidelberg (2012). doi:10.1007/978-3-642-29047-3_28
36. Scoop Whoop: This is why the Hashtag KejriwalinsultsHanuman is trending on Twitter, February 2016
37. Wiegand, S., Middleton, S.E.: Veracity and velocity of social media content during breaking news: analysis of Paris shootings. In: Proceedings of the 25th International Conference Companion on World Wide Web, november 2015, pp. 751–756. International World Wide Web Conferences Steering Committee (2016)
38. Lei, Y., Zhixin, Y., Gong, Y.: An improved ORB algorithm of extracting and matching (2015)

TwitterNews+: A Framework for Real Time Event Detection from the Twitter Data Stream

Mahmud Hasan[✉], Mehmet A. Orgun, and Rolf Schwitter

Department of Computing, Macquarie University, Sydney, Australia
mahmud.hasan@students.mq.edu.au, {mehmet.orgun,rolf.schwitter}@mq.edu.au

Abstract. In recent years, substantial research efforts have gone into investigating different approaches to the detection of events in real time from the Twitter data stream. Most of these approaches, however, suffer from a high computational cost and are not evaluated using a publicly available corpus, thus making it difficult to properly compare them. In this paper, we propose a scalable event detection system, *TwitterNews+*, to detect and track newsworthy events in real time. *TwitterNews+* uses a novel approach to cluster event related tweets from Twitter with a significantly lower computational cost compared to the existing state-of-the-art approaches. Finally, we evaluate the effectiveness of *TwitterNews+* using a publicly available corpus and its associated ground truth data set of newsworthy events. The result of the evaluation shows a significant improvement, in terms of recall and precision, over the baselines we have used.

Keywords: Event detection · Incremental clustering · Social media · Microblog · Twitter

1 Introduction

Social networking services have gradually amassed a huge amount of users from different parts of the world. The collective information generated on these online platforms is overwhelming, in terms of the amount of content produced every moment and the diversity of the topics discussed. The real time nature of the information produced by the users has prompted researchers to analyze such content to gain insight on the current state of affairs. More specifically, the microblogging service Twitter has been a recent focus of researchers to gather information on the newsworthy events occurring in real time.

From its launch in July 2006, Twitter has seen a tremendous growth in popularity, reaching approximately 310 million active users per month, and the number of tweets sent per day exceeding 500 million[1]. The vast amount of real time information circulated through Twitter has essentially made it a host of sensors for events as they happen. However, the majority of the information propagated on Twitter is not relevant to the event detection task, and the prospect of having

---

[1] http://www.internetlivestats.com/twitter-statistics/.

© Springer International Publishing AG 2016
E. Spiro and Y.-Y. Ahn (Eds.): SocInfo 2016, Part I, LNCS 10046, pp. 224–239, 2016.
DOI: 10.1007/978-3-319-47880-7_14

a huge amount of real time data also comes with the challenge of dealing with it. Moreover, the limited context resulting from the length restriction on a tweet and the noisy nature of the data render traditional topic detection approaches ineffective [1].

Several techniques have been proposed in the recent studies to deal with the Twitter-centric event detection task, most of which can be categorized into three general approaches based on their common traits: (a) term interestingness based, (b) topic modeling based, and (c) incremental clustering based approaches (see Sect. 2). One of the major drawbacks of the approaches based on term interestingness [2–4] and topic modeling [5–7] is that they are computation intensive. Whereas, incremental clustering based approaches [8,9] are quite effective in reducing the computational cost involved in the event detection task.

Our proposed end-to-end event detection framework, *TwitterNews+*, implements a variant of the incremental clustering approach and provides a very low cost novel solution, in terms of the computational complexity, to deal with the stream of time ordered tweets for event detection and tracking in real time. The problem of event detection from the Twitter data stream in an incremental clustering context can be divided into two major stages. The first stage involves detecting a burst in the number of tweets discussing a topic/event and the second stage involves clustering the tweets that discuss the same events.

The operation of our proposed system, *TwitterNews+*, is therefore divided into two major stages. After the preprocessing of a tweet, the first stage detects whether a soft tweet burst related to a topic has occurred. The detection of a soft burst involves simply determining the novelty of the input tweet, which is achieved by storing a continuously updated but fixed number of the most recent tweets and performing a text similarity calculation on them with the input tweet. If for an input tweet a textually similar tweet can be found, then it means the input tweet is '*not unique*' and a soft tweet burst for a particular topic/event has occurred.

Different from our previously proposed system called *TwitterNews* [10], which combined a random indexing based term vector model [11] with the locality-sensitive hashing scheme proposed by Petrovic et al. [8], the operation in the first stage of *TwitterNews+* is implemented by utilizing a *term-tweets* inverted index to significantly reduce the time needed to determine the novelty of an input tweet.

A tweet decided as '*not unique*' during the first stage is sent to the second stage of the system, where similar tweets that talk about the same events are grouped together using an incremental clustering approach. Unlike *TwitterNews* [10], *TwitterNews+* utilizes a *term-eventIDs* inverted index during the second stage, to make a fast decision on assigning a tweet to an event cluster. The incorporation of the two separate and specialized inverted indices, and the manner in which they are used, allow our system to be able to scale up in a true streaming setting, while maintaining a constant space and time requirement.

The rest of this paper is organized as follows: we briefly discuss the related work in Sect. 2. Section 3 introduces the architecture of the proposed system. The various aspects of this architecture are further explained in Sects. 4 and 5. Finally, we discuss the results of our experiments and evaluation of *TwitterNews+* using a publicly available corpus in Sect. 6 and conclude in Sect. 7.

2   Related Work

For the purpose of a focused discussion on the related work, we classify various event detection methods based on the common traits they share (i.e., identifying interesting properties in tweet keywords/terms, using probabilistic topic modeling, and incremental clustering).

(a) **Term Interestingness Based Approaches.** TwitterMonitor [2] utilizes an M/M/1 queue [12] to detect emergent topics by identifying the bursty terms from the Twitter stream. A greedy search strategy is used to generate groupings for the high frequency terms that co-occur in a large number of tweets. enBlogue [3] detects emergent topics from blogs and Twitter data within a given time window, by computing statistical values for tag pairs and monitoring unusual shifts in their correlations. The strength of these shifts in tag pairs is used to rank emergent topics and the top-k ranked topics are returned by the system. Twitter Live Detection Framework (TLDF) [4] adapts the Soft Frequent Pattern Mining (SFPM) algorithm [13], to detect relevant topics within a generic macro event from the Twitter data stream. Unlike enBlogue [3] and TwitterMonitor [2], TLDF uses a dynamic temporal window size to detect events based on term co-occurrences in real time.

(b) **Topic Modeling Based Approaches.** TopicSketch [5] detects bursty events by detecting an acceleration on the whole Twitter stream, every word, and every pair of words. The system provides a low cost solution to maintain and update this information. The sketch-based topic modeling approach triggers topic inference, when an acceleration on these stream quantities is detected. As this strategy will result in data with dimensions in the order of millions, a hashing based dimension reduction scheme is utilized to address this issue. Bursty Event dEtection (BEE+) [6] is a distributed and incremental topic model that discovers bursty events by modeling the temporal information of events. The burst detection in BEE+ is similar to the approach used in TwitInfo [14]. Spatio-Temporal Multimodal TwitterLDA (STM-TwitterLDA) [7] is a topic model based framework for event detection that jointly models text, image, location, timestamp and hashtag based Twitter features to detect events. STM-TwitterLDA employs a SVM classifier to remove noisy images, a latent filter to remove general images, and uses Convolutional Neural Networks (CNN) [15] to extract visual features from images to leverage in event detection. Finally, maximum-weighted bipartite graph matching is applied to track the evolution of the detected events.

(c) **Incremental Clustering Based Approaches.** McMinn et al. [9] have utilized an inverted index for each named entity with its associated near neighbors to cluster the bursty named entities for event detection and tracking. The effectiveness of this approach, however, is dependent on the accuracy of the underlying Named Entity Recognizer [16] used by the system. Petrovic et al. [8] adapted a variant of the locality-sensitive hashing (LSH) technique to determine the novelty of a tweet by comparing it with a fixed number of previously encountered tweets. A novel tweet represents a new story, which is assigned to a newly created cluster. On the other hand, a tweet determined as 'not unique' is assigned to an existing cluster containing the nearest neighbor. Event clusters are ranked based on a combination of the entropy information and the number of unique user posts in a cluster.

Term interestingness based approaches such as TwitterMonitor [2], enBlogue [3], and TLDF [4] usually differ on the term selection methods they employ, as well as on the way in which term correlations are computed and changes in the term correlations are tracked. These approaches can often capture misleading term correlations, and measuring the term correlations can be computationally prohibitive in an online setting.

Topic modeling based approaches [5–7] suffer due to the limit imposed on the length of a tweet, and capturing good topics from the limited context is a problem yet to be addressed efficiently. Moreover, these approaches usually incur a high computational cost, and are not quite effective in handling the events that are reported in parallel [17]. Stilo and Veraldi [18] noted that LDA [19] based topic models usually can only work in an off-line manner as the temporal aspect of the events is not often considered.

Incremental clustering based approaches [8,9] are prone to fragmentation, and are usually unable to distinguish between two similar events taking place around the same time. However, despite these challenges, we believe that an incremental clustering based approach can be utilized because of its inherently low computational complexity compared to the most of the state-of-the art approaches.

3   Architecture of the TwitterNews+ System

The two main components of *TwitterNews+* are the Search Module, and the EventCluster Module (Fig. 1). The Search Module handles the operation of the first stage in our system and facilitates a fast retrieval of similar tweets from the set of most recent tweets maintained by *TwitterNews+* to provide a decision on the novelty of an input tweet. An input tweet decided as *"not unique"*, by the Search Module, assures that similar tweets have been encountered before. *TwitterNews+* uses this information to confirm the fact that either an event related tweet burst has occurred (soft burst) or the input tweet is part of an ongoing burst and needs to be tracked.

If the Search Module decides the input tweet to be *"not unique"*, then it is sent to the EventCluster Module which handles the operation of the second stage in our system. For every tweet sent to this module, a candidate event cluster to which the tweet can be assigned is searched. A tweet is assigned to an event cluster if the cosine similarity between the tweet and the centroid of the event cluster is above a certain threshold. When no such cluster is found, a new event

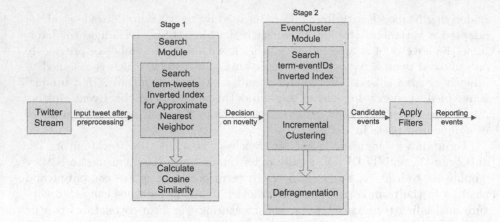

Fig. 1. *TwitterNews+* architecture

cluster is created and the tweet is assigned to the new cluster. The EventCluster Module contains a defragmentation sub-module that merges together fragmented event clusters. The defragmentation sub-module is also helpful to merge clusters that are sub-events of an event. Finally, *TwitterNews+* uses a novel Longest Common Subsequence (LCS) based scheme along with a set of different filters to retain newsworthy events from the candidate event clusters and identifies a representative tweet for each event.

4   The Search Module

To reduce the time needed to search for previously encountered tweets similar to the input tweet d, while maintaining a constant time and space requirement, we utilize a *term-tweets* inverted index (Fig. 2) maintained by *TwitterNews+* on a finite set, M, of the most recent tweets. The set M is continuously updated by replacing the oldest tweet with the latest input tweet to keep the memory requirement constant for the *term-tweets* inverted index as the number of unique terms can grow very large due to the unconstrained use of vocabulary in the streaming tweets. Each entry of the *term-tweets* inverted index contains a term and a finite set, Q, of the most recent tweets in which the term appeared. The oldest tweet is replaced with the latest tweet containing the term when the number of tweets exceeds the limit of Q. To find an approximate nearest neighbor of d, the tweet is first tokenized and part-of-speech tagged [20] as part of the preprocessing stage and an incremental $tf - idf$ based term vector is generated using the following formula:

$$tf - idf(t, d, D) = tf(t, d) \times idf(t, D) \tag{1}$$

where t is a term in the input tweet d, D is the corpus representing the tweets processed so far, $tf(t, d)$ is simply the number of times t is found in d, and

$$idf(t, D) = \log \frac{N}{|\{d \in D : t \in D\}|} \tag{2}$$

where N is the number of tweets processed so far and $|\{d \in D : t \in D\}|$ is the total number of tweets in D where the term t appears.

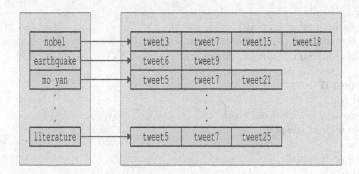

Fig. 2. *term-tweets* inverted index

Subsequently, the top-K $tf - idf$ weighted terms are selected from d, and for each of the K terms the *term-tweets* inverted index is searched to retrieve a maximum of $K \times Q$ tweets in which at least one of the K terms appeared. To elaborate this idea with an example, let us consider the input tweet *"Mo Yan wins Nobel in Literature"*, where the top three $tf - idf$ weighted terms are *"mo yan"*, *"nobel"*, and *"literature"*. Note that the term *"mo yan"* is shown in this example as a compound noun for simplicity and a similar result can be achieved when the individual parts of the compound noun are used in the index. Each of the terms is searched in the *term-tweets* inverted index (see Fig. 2) and the tweets with IDs 3, 5, 7, 15, 18, 21, and 25 are retrieved. Finally, the approximate nearest neighbor of the input tweet among the retrieved tweets is calculated using the cosine similarity measure. The cosine similarity between two tweet vectors $d1$ and $d2$ is computed using the Euclidean dot product formula, where n refers to the dimension of the vectors:

$$\cos(\theta) = \frac{\sum\limits_{i=1}^{n} d1_i \times d2_i}{\sqrt{\sum\limits_{i=1}^{n} (d1_i)^2} \times \sqrt{\sum\limits_{i=1}^{n} (d2_i)^2}} \tag{3}$$

Algorithm 1 shows the pseudocode for the Search Module and we refer to Table 1 for the empirically determined parameter settings used by *TwitterNews+*. A threshold value in the range of $[0.5 - 0.6]$ for the cosine similarity

Algorithm 1. TwitterNews+ Search Module

Require: *threshold* value

1: **for** each tweet d in twitter-stream D **do**
2:     preprocess(d)
3:     generate vector for d with incremental $tf - idf$
4:     select top-K $tf - idf$ weighted terms of d
5:     $S \leftarrow$ set of tweets that are near neighbors of d $\triangleright |S| \le K \times Q$, *retrieved from the "term-tweets" inverted index of the most recent tweets using the top-K terms*
6:     $sim_{max} \leftarrow 0$
7:     **for** each tweet d' in S **do** $\triangleright$ *parallel processing*
8:         $tempSim = $ CosineSimilarity(d, d')
9:         **if** $tempSim > sim_{max}$ **then**
10:             $sim_{max} = tempSim$
11:         **end if**
12:     **end for**
13:     **if** $sim_{max} > threshold$ **then**
14:         d is "*not unique*"
15:     **else**
16:         d is "*unique*"
17:     **end if**
18:     add new term entries and/or update *term-tweets* inverted index $\triangleright$
        d is added in each "term-tweets" inverted index entry corresponding to its top-K terms
19: **end for**

is empirically set for the Search Module to determine the novelty of the input tweet. If the cosine similarity of the approximate nearest neighbor of the input tweet is above the threshold value, then the input tweet is considered to be "*not unique*", thus confirming the occurrence of a soft burst.

The most expensive operation in Algorithm 1 (lines 7–12) is determining the approximate nearest neighbor based on the cosine similarity measure. However, using the *term-tweets* inverted index restricts the total number of tweets to compare with the input tweet within $K \times Q$. As the $K \times Q$ number of comparisons are not dependent on each other, they are ideal for parallel processing which effectively renders the computational cost to $O(1)$.

5   The EventCluster Module

The Search Module sends the tweets that are decided as "*not unique*" to the EventCluster module. Upon receiving a tweet d, the EventCluster module utilizes a *term-eventIDs* inverted index, in a manner similar to the Search module, to provide a low computational cost solution to find an event cluster in which d can be assigned. Each entry in the *term-eventIDs* inverted index contains a term and a finite set, E, of IDs of the most recent event clusters in which the term appeared. The oldest event ID is replaced with the latest event ID containing

Table 1. Parameter settings for TwitterNews+

Parameter	Value
M (Most recent tweets)	100000–200000
K (Top keywords)	10
Q (*term-tweets*)	25
E (*term-eventIDs*)	25
threshold (Serach Module)	0.5–0.6
t_{ev} (EventCluster Module)	0.6
g_{ev} (EventCluster Module)	0.05–0.07
ts_i (EventCluster Module)	15–30 min
t_{lcs} (LCS threshold)	5
Entropy	2.5
User diversity	0.0

the term when the number of stored event IDs exceeds the limit of E. For each of the top-K terms in d the *term-eventIDs* inverted index is searched to retrieve the IDs of the event clusters, not more than $K \times E$, in which at least one of the K terms appeared.

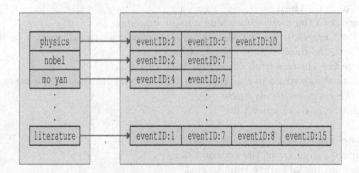

Fig. 3. *term-eventIDs* inverted index

To elaborate this idea with the same example used in the Search Module, let us consider that, the input tweet *"Mo Yan wins Nobel in Literature"* is decided as *"not unique"* and sent to the EventCluster Module. The top three $tf - idf$ weighted terms of the tweet are *"mo yan"*, *"nobel"*, and *"literature"*. Each of the terms is searched in the *term-eventIDs* inverted index (see Fig. 3) and the event clusters with IDs 1, 2, 4, 7, 8, and 15 are retrieved. Note that the total number of event clusters with which the input tweet is compared will always be within $K \times E$. Finally, the input tweet vector is compared with the centroid of each of the retrieved event clusters and assigned to the cluster with the highest cosine similarity. If the cosine similarity is below a certain threshold a new cluster is created and the tweet is added to the newly created cluster.

Algorithm 2 shows the pseudocode for the EventCluster Module, where the event threshold (t_{ev}) value for a tweet to be assigned to a cluster is empirically set to 0.6 and the defragmentation granularity (g_{ev}) value to merge fragmented events is empirically set in the range of $[0.05 - 0.07]$.

Algorithm 2. TwitterNews+ EventCluster Module

Require: tweet d decided as *"not unique"* by Algorithm 1, event threshold value t_{ev}, and defragmentaion granularity value g_{ev}

1: $C \leftarrow$ set of events containing at least one of the top-K terms in d ▷ $|C| \leq K \times E$, retrieved from the "term-eventIDs" inverted index using the top-K terms of d

2: $sim_{max} \leftarrow 0$

3: **for** each active event cluster c in C **do**   ▷ *parallel processing*

4:     $tempSim = $ findClosestCentroid(c, $d_{termVector}$)   ▷ *measures cosine similarity between the centroid of the cluster and the tweet vector*

5:     **if** $tempSim > sim_{max}$ **then**

6:         $sim_{max} = tempSim$

7:     **end if**

8:     **if** $tempSim \geq (t_{ev} + g_{ev})$ **then**

9:         $S_c \leftarrow$ assign c to the set of fragmented clusters to be merged later

10:    **end if**

11: **end for**

12: **if** $sim_{max} > t_{ev}$ **then**

13:    assign d to the closest matching cluster c_{simMax}

14:    updateCentroid(c_{simMax}, $d_{termVector}$) ▷ *cluster centroid updated by averaging with the tweet vector*

15:    merge the clusters from the set S_c with c_{simMax}

16:    update the centroid of c_{simMax} ▷ *cluster centroid updated by averaging with the event centroids in S_c*

17:    update c_{simMax} expiry time

18: **else**

19:    create a new cluster c_{new} and assign d to it

20:    assign $d_{termVector}$ as the centroid of c_{new}

21:    assign an initial expiry time to c_{new}

22: **end if**

23: add new term entries and/or update *term-eventIDs* inverted index   ▷ *ID of the event cluster containing d is added in each "term-eventIDs" inverted index entry corresponding to the top-K terms of d*

Each event cluster created by the EventCluster Module has an expiry time associated with it. When a cluster c is created, an initial expiry time ts_i for the cluster is set. Each time a new tweet is added to c, the expiry time is updated based on the average timestamp difference between the arrival of successive tweets in c. Once an event cluster has expired, it is marked as inactive to avoid similarity comparison with any subsequent tweet that arrives in the EventCluster

Module. The *term-eventIDs* inverted index is updated after a fixed interval to remove inactive events in order to maintain a fixed space requirement.

Similar to the Search Module, the most expensive operation in Algorithm 2 (lines 3–11) is finding an event cluster in which a tweet can be placed. As the *term-eventIDs* inverted index restricts the total number of event clusters to search for within $K \times E$, the time complexity of the aforementioned operation becomes $O(1)$ with parallel processing.

Any incremental algorithm, such as ours for the EventCluster Module, suffers from fragmentation when a particular event is detected multiple times as a new event, creating multiple event clusters for the same event. We have employed a defragmentation strategy to avoid cluster fragmentation as much as possible. The defragmentation strategy is also helpful to merge clusters that contain sub-events of an event resulting from the topic drifts. While searching for a cluster that is closest in similarity to the input tweet, we also keep track of the clusters in a set S_c whose cosine similarity with the input tweet is $>t_{ev} + g_{ev}$, as shown in Algorithm 2. After we assign the tweet to the closest matching cluster (given that, similarity is $>t_{ev}$), all the clusters in S_c are merged to achieve defragmentation.

From the set of candidate events formed by the EventCluster Module, newsworthy events are reported if they satisfy a few criteria as described in Sect. 6.

6   Experiment Results and Evaluation

Twitter streaming data mostly contain information irrelevant for the event detection task. A good amount of tweets contain spams, which unnecessarily slow down the processing time of an event detection system and have a detrimental effect on the precision. To improve on the precision and to reduce the number of tweets to be processed by *TwitterNews+*, a term/phrase level filter has been applied using a manually curated list of spam phrases (e.g., "click here", "free access"). Tweets containing these spam phrases are discarded. The spam phrase filter contributes to around 70 % of tweets being discarded by our system. Different from the preprocessing stage in *TwitterNews* [10], the spam phrase filter is a new feature in *TwitterNews+*.

To perform an evaluation on the result generated by *TwitterNews+*, we have conducted an experiment on the first 3 days of approximately 17 million tweets (9th to 11th of October, 2012) from the Events2012 corpus [21]. The corpus contains 120 million tweets collected from the 9th of October to the 7th of November, 2012. Once candidate event clusters were generated by *TwitterNews+*, newsworthy events were determined by applying a combination of different filters to discard the trivial events from the candidate event clusters. The first level of filters utilize the entropy [8] and the user diversity [22] information in a candidate event cluster and retains the clusters with entropy >2.5, and user diversity >0.0. The entropy threshold ensures that a minimum amount of information is contained in a cluster and a positive user diversity value ensures that the cluster contains tweets from more than one user.

Unlike *TwitterNews* [10], *TwitterNews+* incorporates a second level of filters to discard the candidate event clusters that have less than 10 tweets and do not contain a URL of a news portal from a collection of top online news entities[2]. In addition, the event clusters with tweets covering a time span of less than a minute and without a reliable news portal's URL are filtered out as well. This filter helps in removing a significant amount of trivial events while making sure that non-trivial events with a small burst of tweets are not discarded. As Lehmann et al. [23] and Yang and Leskovec [24] have shown the existence of a number of different temporal patterns of events besides the event-pattern with a bursty characteristic, the second level of filters ensure that *TwitterNews+* detects events with non-bursty temporal patterns in addition to those events with a bursty temporal pattern. This additional level of filters in *TwitterNews+* allows an improvement in terms of the precision over *TwitterNews* [10] (see Table 3).

Finally, we employ a Longest Common Subsequence (LCS) based filtering method that works on the word-level. The idea here is based on the empirical evidence found from inspecting the candidate event set. We have noticed that news propagated by the general users or the news agencies usually follow a similar sentence structure. We have applied the traditional word-level LCS algorithm on the relevant tweets of an event cluster and identified the tweet with the longest common subsequence of words. Then we use the length of the LCS to determine whether the event cluster is about a newsworthy event. If the length of the longest common subsequence of words in an event cluster c is below a certain threshold (t_{lcs}), then the tweets in c do not have an appropriate level of similarity in their sentence structure and c is not likely to be a newsworthy event.

The LCS based scheme also selects a representative tweet from an event cluster by emitting the tweet having the maximum LCS in the event. Before applying the LCS based scheme on the set of candidate events, all the tweets of each event cluster are discarded that do not contain at least one proper noun or possessive noun. Doing so reduces the total number of tweets in a cluster by discarding the tweets that do not contain any useful information.

Evaluation. Along with the corpus, McMinn et al. [21] have provided a separate file containing 506 events with their associated tweets to be used as the ground truth for evaluation. Due to the restriction imposed by Twitter, the Events2012 corpus only contains unique tweet IDs using which the tweets belonging to the corpus need to be downloaded. After downloading the tweets, we have inspected the corpus and discovered that a large number of tweets (around 30 %) belonging to the corpus were not downloaded as they are not available any more. The effect of a partially incomplete corpus, due to the unavailability of the tweets, is going to negatively impact the results produced by our system. To remedy this problem, we have decided to manually reconfirm the ground truth events provided by the authors [21]. However, there are a total of 506 ground truth events spanning from the 9th of October to the 7th of November, 2012. As this can take a substantial amount of time, we have only reconfirmed the first three

---

[2] http://www.journalism.org/media-indicators/digital-top-50-online-news-entities-2015/.

days (9th to 11th of October, 2012) of the ground truth events and manually selected a total of 41 events that belong to our selected time window. Further inspection of these 41 events were required to identify and remove the events which contained a large number of unavailable tweets. Doing so led us to a final set of 31 events to be used as the ground truth (see Table 2).

Table 2. Ground Truth for the events from the 9th to the 11th of Oct, 2012

Date	Events
09 Oct 2012	They are discussing a televised award show for the BET network.
09 Oct 2012	It is about a TV show by Keyshia Cole and her husband.
09 Oct 2012	They all like to watch Meek Millz show.
09 Oct 2012	They all discuss about fat Joe.
09 Oct 2012	Best lyricist of the year awarded to Kendrick Lamar.
09 Oct 2012	About Omarion dancing on the stage.
09 Oct 2012	2 Chainz performance.
10 Oct 2012	Penn State scandal involving imprisoned former football coach Jerry Sandusky.
10 Oct 2012	HP and Lenovo battle for top spot in the PC market of Computerworld.
10 Oct 2012	Detroit and Oakland played a postseason game.
10 Oct 2012	A court in Moscow, Russia, frees one of the three Pussy Riot members at an appeal hearing.
10 Oct 2012	Yankees win a playoff with a walkoff over the Orioles.
10 Oct 2012	BAE and EADS announce their merger talks are cancelled over political disagreements.
10 Oct 2012	Two American scientists, Robert Lefkowitz and Brian Kobilka, win the 2012 Nobel Prize in Chemistry.
10 Oct 2012	The USADA details witness-based doping claims against Lance Armstrong in its long-due report to the UCI.
10 Oct 2012	Malala Yousafzai, a 14 year old activist for women education rights is shot by Taliban gunmen in the Swat Valley.
11 Oct 2012	Chinese author Mo Yan, famous for working in the style of writing known as hallucinatory realism, wins the Nobel Prize in Literature.
11 Oct 2012	Vice presidential debate between Joe Biden and Paul Ryan.
11 Oct 2012	Jayson Werth hitting a walkoff home run for the Nationals during the playoff game against the Cardinals.
11 Oct 2012	A Cleveland bus driver punched a female passenger in the face.
11 Oct 2012	Buster Posey grand slam leads SF Giants to historic Division Series.
11 Oct 2012	Ryan Bertrand has had to pull out of the England Squad with a sore throat.

Table 2. (*continued*)

Date	Events
11 Oct 2012	A Syrian passenger plane is forced by Turkish fighter jets to land in Ankara due to the allegations of carrying weapons.
11 Oct 2012	Space shuttle Endeavour makes a final trip to a Los Angeles museum.
11 Oct 2012	Oil giant Shell is sued by Niger Delta farmers in a civil court in the Hague.
11 Oct 2012	Heavy rain in the United Kingdom causes flash flooding in the coastal village of Clovelly.
11 Oct 2012	The Marie Stopes organization is to open the first private clinic to offer abortions to women in Northern Ireland from 18 October.
11 Oct 2012	A U.S. appeals court has overturned a district court order that had banned the sale of Samsung's Galaxy Nexus in the US, delivering a winning round for Google's Android against Apple Inc.
11 Oct 2012	The topic is about a Pep rally.
11 Oct 2012	A gunman kills Qassem M. Aqlan, the Yemeni chief of security employed at the U.S. embassy in the capital, Sana'a.
11 Oct 2012	Syrian rebels claiming control of a strategic town

We have evaluated the 1523 events reported by *TwitterNews+* within the time window of three days. The baselines we have used to compare with our system are the First Story Detection (FSD) system by Petrovic et al. [8] and our previously proposed system, *TwitterNews* [10].

Table 3 summarizes the results of the evaluation, where the recall refers to the fraction of the events in the ground truth that were detected by a system, and the precision refers to the fraction of the newsworthy events out of all the events detected by a system.

Table 3. Summary of the evaluation results

Methods	Recall	Precision
FSD [8]	0.52	-
TwitterNews [10]	0.87	0.72
TwitterNews+	0.93	0.78

The FSD baseline [8] achieved a recall of 0.52 by detecting 16 events out of the 31 ground truth events. The *TwitterNews* baseline [10] achieved a recall of 0.87 (27 events out of 31), and *TwitterNews+* has detected 29 events out of 31, resulting in a recall of 0.93.

McMinn et al. [21] have noted in their later work [9] that a lot of events can be detected from the Events2012 corpus besides the set of 506 events provided as the ground truth by the authors. Hence, instead of calculating the precision with respect to the ground truth, we have used two human annotators to determine the precision of 100 randomly chosen events out of the 1523 events reported by *TwitterNews+*. The precision is calculated as a fraction of the 100 randomly chosen events that are related to realistic events. The agreement between the two annotators, measured using Cohen's kappa coefficient, was 0.77 and the precision of *TwitterNews+* reported by the annotators was 0.78 (78 out of 100 events were agreed as newsworthy events by both annotators). Table 4 shows the representative tweets of the newsworthy events, selected using the LCS based scheme, reported by *TwitterNews+*. For each day only one event's representative tweet is shown to keep Table 4 concise.

Table 4. Representative tweets of the selective newsworthy events reported by TwitterNews+

Date	Event representative tweet
09 Oct 2012	Lyricist of the year : Kendrick Lamar!
10 Oct 2012	11 teammates blow whistle on Lance Armstrong: The U.S. Anti-Doping Agency says 11 of Lance Armstrong's former te... http://bit.ly/W0x3BX
11 Oct 2012	Appeals court reverses sales ban on Samsung smartphone: WASHINGTON (Reuters) - A U.S. appeals court overturned a... http://bit.ly/TBC0SD

7   Conclusion

The approach taken in *TwitterNews+* yields a low computational cost solution to detect events from a Web scale corpora in real time, which is lacking in most of the state-of-the-art approaches. The most expensive operations in the Search Module and the EventCluster Module algorithms incur a computational complexity of $O(1)$ with parallel processing, while the rest of the parts in both algorithms also incur a constant cost in terms of time and space. The different set of filters, applied after the candidate events generation, collectively incur a computational cost of $O(n^2)$, where n refers to the number of tweets in an event cluster. However, the filters are applied as a separate process independent of the candidate event generation stages of *TwitterNews+*. To the best of our knowledge, *TwitterNews+* is one of the fastest systems to detect events from Twitter, which maintains a constant space and processing time, and achieves very good results. The different set of filters, applied to extract newsworthy events from the set of candidate events, help in detecting events with usually a non-measurable burst of a few tweets and discarding a significant amount of trivial events. This is, again, where most of the state-of-the-art approaches fail, which depend on detecting events based on burst detection. Moreover, the evaluation of *TwitterNews+*, done using a publicly available corpus, will allow different approaches to be fairly compared against our system.

As part of our future work, we intend to conduct a parameter sensitivity analysis on the various parameter settings required for the proposed system. The LCS based scheme in *TwitterNews+* selects a representative tweet from an event cluster, however, in the future we plan to provide a temporal summarization of the tweets in an event cluster.

References

1. Atefeh, F., Khreich, W.: A survey of techniques for event detection in Twitter. Comput. Intell. **31**(1), 132–164 (2015). Wiley Online Library
2. Mathioudakis, M., Koudas, N.: TwitterMonitor: trend detection over the Twitter stream. In: Proceedings of the ACM SIGMOD International Conference on Management of Data, SIGMOD 2010, NY, USA, pp. 1155–1158. ACM, New York (2010)
3. Alvanaki, F., Sebastian, M., Ramamritham, K., Weikum, G.: Enblogue: emergent topic detection in web 2.0 streams. In: Proceedings of the ACM SIGMOD International Conference on Management of Data, SIGMOD 2011, NY, USA, pp. 1271–1274. ACM, New York (2011)
4. Gaglio, S., Re, G.L., Morana, M.: A framework for real-time Twitter data analysis. Comput. Commun. **73**, 236–242 (2016). Elsevier
5. Xie, R., Zhu, F., Ma, H., Xie, W., Lin, C.: CLEar: a real-time online observatory for bursty and viral events. Proc. VLDB Endowment **7**(13), 1637–1640 (2014). VLDB Endowment
6. Li, J., Wen, J., Tai, Z., Zhang, R., Yu, W.: Bursty event detection from microblog: a distributed and incremental approach. In: Concurrency and Computation:Practice and Experience. Wiley Online Library (2015)
7. Cai, H., Yang, Y., Li, X., Huang, Z.: What are popular: exploring Twitter features for event detection, tracking and visualization. In: Proceedings of the 23rd Annual ACM Conference on Multimedia Conference, pp. 89–98. ACM (2015)
8. Petrović, S., Osborne, M., Lavrenko, V.: Streaming first story detection with application to Twitter. In: Proceedings of the Annual Conference of the North American Chapter of the Association for Computational Linguistics: Human Language Technologies, HLT 2010, ACL, Stroudsburg, PA, USA, pp. 181–189 (2010)
9. McMinn, A.J., Jose, J.M.: Real-time entity-based event detection for Twitter. In: Mothe, J., Savoy, J., Kamps, J., Pinel-Sauvagnat, K., Jones, G.J.F., SanJuan, E., Cappellato, L., Ferro, N. (eds.) CLEF 2015. LNCS, vol. 9283, pp. 65–77. Springer, Heidelberg (2015). doi:10.1007/978-3-319-24027-5_6
10. Hasan, M., Orgun, M.A., Schwitter, R.: TwitterNews: real time event detection from the Twitter data stream. PeerJ PrePrints 4, e2297v1 (2016)
11. Sahlgren, M.: An introduction to random indexing. In: Proceedings of the Methods and Applications of Semantic Indexing Workshop at the 7th International Conference on Terminology and Knowledge Engineering, TKE, vol. 5 (2005)
12. Guzman, J., Poblete, B.: On-line relevant anomaly detection in the Twitter stream: an efficient bursty keyword detection model. In: Proceedings of the ACM SIGKDD Workshop on Outlier Detection and Description, pp. 31–39. ACM (2013)

13. Petkos, G., Papadopoulos, S., Aiello, L., Skraba, R., Kompatsiaris, Y.: A soft frequent pattern mining approach for textual topic detection. In: Proceedings of the 4th International Conference on Web Intelligence, Mining and Semantics, WIMS, pp. 25: 1–25: 10. ACM (2014)

14. Marcus, A., Bernstein, M.S., Badar, O., Karger, D.R., Madden, S., Miller, R.C.: TwitInfo: aggregating and visualizing microblogs for event exploration. In: Proceedings of the SIGCHI Conference on Human Factors in Computing Systems, CHI 2011, NY, USA, pp. 227–236. ACM, New York (2011)

15. Jia, Y., Shelhamer, E., Donahue, J., Karayev, S., Long, J., Girshick, R., Guadarrama, S., Darrell, T.: Caffe: convolutional architecture for fast feature embedding. In: Proceedings of the ACM International Conference on Multimedia, pp. 675–678. ACM (2014)

16. Derczynski, L., Ritter, A., Clark, S., Bontcheva, K.: Twitter part-of-speech tagging for all: overcoming sparse and noisy data. In: Proceedings of the Recent Advances in Natural Language Processing, RANLP, pp. 198–206 (2013)

17. Aiello, L.M., Petkos, G., Martin, C., Corney, D., Papadopoulos, S., Skraba, R., Goker, A., Kompatsiaris, I., Jaimes, A.: Sensing trending topics in Twitter. IEEE Trans. Multimedia **15**(6), 1268–1282 (2013). IEEE

18. Stilo, G., Velardi, P.: Efficient temporal mining of micro-blog texts and its application to event discovery. In: Fürnkranz, J. (ed.) Data Mining and Knowledge Discovery, pp. 1–31. Springer, Heidelberg (2015)

19. Blei, D.M., Ng, A.Y., Jordan, M.I.: Latent dirichlet allocation. J. Mach. Learn. Res. **3**, 993–1022 (2003). JMLR.org

20. Owoputi, O., O'Connor, B., Dyer, C., Gimpel, K., Schneider, N., Smith, N.A.: Improved part-of-speech tagging for online conversational text with word clusters. In: Proceedings of the Annual Conference of the North American Chapter of the Association for Computational Linguistics: Human Language Technologies, HLT 2013, ACL, pp. 380–391 (2013)

21. McMinn, A.J., Moshfeghi, Y., Jose, J.M.: Building a large-scale corpus for evaluating event detection on Twitter. In: Proceedings of the 22nd ACM International Conference on Information and Knowledge Management, CIKM 2013, NY, USA, pp. 409–418. ACM, New York (2013)

22. Kumar, S., Liu, H., Mehta, S., Subramaniam, L.V.: From tweets to events: exploring a scalable solution for Twitter streams. arXiv preprint arXiv:1405.1392 (2014)

23. Lehmann, J., Gonçalves, B., Ramasco, J.J., Cattuto, C.: Dynamical classes of collective attention in Twitter. In: Proceedings of the International Conference on World Wide Web, pp. 251–260. ACM (2012)

24. Yang, J., Leskovec, J.: Patterns of temporal variation in online media. In: Proceedings of the 4th ACM International Conference on Web Search and Data Mining, pp. 177–186. ACM (2011)

Uncovering Topic Dynamics of Social Media and News: The Case of Ferguson

Lingzi Hong[1]($\boxtimes$), Weiwei Yang[2], Philip Resnik[3,4],
and Vanessa Frias-Martinez[1,4]

[1] College of Information Studies, University of Maryland, College Park, MD, USA
lzhong@umd.edu
[2] Department of Computer Science, University of Maryland, College Park, MD, USA
wwyang@umd.edu
[3] Department of Linguistics, University of Maryland, College Park, MD, USA
[4] Institute for Advanced Computer Studies, University of Maryland,
College Park, MD, USA
{resnik,vfrias}@umd.edu

Abstract. Looking at the dynamics of news content and social media content can help us understand the increasingly complex dynamics of the relationship between the media and the public surrounding noteworthy news events. Although topic models such as latent Dirichlet allocation (LDA) are valuable tools, they are a poor fit for analyses in which some documents, like news articles, tend to incorporate multiple topics, while others, like tweets, tend to be focused on just one. In this paper, we propose Single Topic LDA (ST-LDA) which jointly models news-type documents as distributions of topics and tweets as having a single topic; the model improves topic discovery in news and tweets within a unified topic space by removing noisy topics that conventional LDA tends to assign to tweets. Using ST-LDA, we focus on the unrest in Ferguson, Missouri after the fatal shooting of Michael Brown on August 9, 2014, looking in particular at the topic dynamics of tweets in and out of St. Louis area, and at differences and relationships between topic coverage in news and tweets.

1 Introduction

The cascading activation model is a widely accepted model that explores the relationship among the government, the media, and the public [4]. The model explains how the framing of information extends down from the White House to the elites, the media, and then to the public. Information moves downward along the cascade with the framing of upper layers and becomes limited to highlights to the public. The structure emphasizes heavily the direction from the media to the public, given that historically the voice from the public has been comparatively weak.

However, social networks expand sources of information for every user and enable everyone to be a potential media source. Providing a public communication platform for everyone who is accessible to the Internet, social networks

E. Spiro and Y.-Y. Ahn (Eds.): SocInfo 2016, Part I, LNCS 10046, pp. 240–256, 2016.
DOI: 10.1007/978-3-319-47880-7_15

lead to increased participation in spreading information, expressing opinion, and public activism [7,22]. During the Arab Spring, for example, Twitter promoted protest mobilization through reporting of real-time events and providing a basis for collaboration and emotional mobilization [1]. Effing et al. [3] show that political participation has been democratizing because of social media such as Facebook and Twitter, which enable more followers to engage in campaigns. So, in contrast to the traditional cascading activation model, the public may be gaining influence because of social networks.

This leads us to think about several questions. Can we observe the complex relations between the opinions of the public and the media? What does the public focus on, given highlighted topics spread by the media in an event? Are there any topics being followed by the public but not mentioned by the media? Specifically, we want to figure out what the media reports, the subjects of public attention, and the relation with and difference between these two. It is also important to observe that usually along with the evolution of an event, the topics of media and the public change over time. For example, after a gunshot accident, several relevant topics co-exist; meanwhile, the main topic may change from the description of the accident to the motivations, effect of the accident, and then to discussion about gun regulations. The changing topics form topic dynamics. By modeling the topic dynamics of media and the public, we can gain insight into the similarity, differences, and possible relationships between media topics and public topics.

In this study, we take the Ferguson unrest event of 2014 as an example, and analyze news and tweeted topics along with the unfolding of events. To make topics of news and tweets comparable, we propose a Single Topic LDA (ST-LDA) model to bring news and tweets under a unified framework, in which every tweet has only one topic, but news has a distribution of multiple topics, a novel approach that takes into account the length limitation of tweets and the greater complexity of news stories. The ST-LDA model tends to outperform LDA by removing noisy topics that conventional LDA tends to assign to tweets in a mixed collection of long and short documents. We explore our research questions by examining the shift of focus in news and tweets, specifically on the possible relation to burst, emergence, and decay of certain topics, the difference between topic dynamics of news and tweets, and whether there is strong influence of media on the public.

The contributions of this study have two main aspects:

1. We solve the technical problem of building topic models for a mixture of short and long documents. Conventional topic models such as LDA and PLSA [11] perform badly because co-occurrence patterns in short text are sparse. Our model considers the words in a tweet as a whole and assigns only one topic to a tweet, so that the main topic is more likely to be assigned. The evaluation results show that ST-LDA improves interpretability over LDA by 14%.
2. We bring the cascade activation model into a social media environment, reexamine the focus of media and the public in the Ferguson case, and bring new understanding of the influence of the media and formation of public opinions in social media.

2   Related Work

2.1   Detection of Topics in News and Social Media

The first and foremost step for comparing topics is the detection of topics in news and tweets. Tracking memes on the Internet [20] is one way to understand online information content. Memes are entities that represent units of information at the desired level of detail. The semantic units serve as clear clues for detecting dynamic change of diverse topics. However, in this approach only repeated topics can be detected.

Topic detection in tweets has been a challenge because of the short length [24]. Aggregating short tweets into a long document, such as author-based aggregation [23,25], grouping by time slices [16], and by words [12] are ways to alleviate the problem. The biterm topic model [24] directly simulates the generation of word co-occurrence patterns in a corpus, and thus leads to more coherent topics. The Word Network Topic Model [26] also uses a word co-occurrence network to solve the sparsity problem. Cataldi et al. [2] detect emerging topics on Twitter by evaluating the life cycle of Twitter terms and user authority.

To train news and tweets together, Hu et al. [13] propose a joint Bayesian model for events transcripts and tweets. It assumes that event information can impose topical influence on tweets. Gao et al. [6] create a joint topic model to extract important and complementary pieces of information across news and tweets, and generate complementary information from both.

In these models, each tweet still has a distribution of topics, despite the fact that, given Twitter's length limitation, a tweet is unlikely to have multiple topics. In contrast, we propose a model that assigns only one topic to each tweet and trains on tweets and news under a unified framework.

2.2   Topic Dynamics

Topic dynamics characterize the shift of topic proportions in a daily window. Dynamics of topics have been studied a great deal in research on the development of scientific areas [18], burst topics in publications [8], and public opinions on social media [10]. Morinaga and Yamanishi [19] employ a finite mixture model to recognize the emergency, growth, and decay of each topic in a system. Iwata et al. [15] build a sequential topic model to detect topic dynamics of document collections with multiple timescale. All these studies involve a single source of data, defining the calculation of topic dynamics in different ways. In our study, we bring tweets and news into a unified topic space so that we can compare the topic dynamics of news and tweets.

2.3   Comparison of Social Media and News

We aim to compare the topics between media and the public, a subject that has been studied both qualitatively and quantitatively. Sayre et al. [21] manually analyze thousands of videos and news media on Proposition 8 in California, and find that the post content in open social media reflects mainstream news, while posts also have influence on professional media coverage. Together they form opinion interactions between media and the public. However, the study required a large amount of human effort, and it is difficult to identify topics' weight change during the evolving process. Hua et al. [14] explore the semantic and topical relationships between news and social media to reveal topic influences among multiple datasets. However, they focus on the influence between topics based on word probability, ignoring the time element of tweets. Leskovec et al. [17] introduce a meme-tracking technique to track topic shifts in news and blogs and observe a 2.5-h lag between peaks of attention of a phrase in the news media and in blogs, suggesting possible media influence on individuals; however their characterization of memes is limited to variants of quotations rather than a broader notion of topic.

Zhao et al. [25] propose Twitter-LDA, which assigns one topic to each tweet, however its premise is that each Twitter user has a distribution of topics and the topic of each tweet is drawn from its author's topic distribution. Moreover, when they compare tweets and news media, they apply topic models separately and manually label topics for news and tweets. Topics in news and tweets with the same labels are compared. Although the meaning of topics is similar, the word distributions are actually different. This can make sense when considering all topics including arts, event, sports, etc., but the comparison is unlikely to be accurate enough for topics all focused on a single event. The Twitter-LDA approach cannot be compared to the proposed ST-LDA directly since it requires a large number of tweets per individual, which is typically not available when the tweet collection is done for events where millions of users might have just a few tweets each.

3   Methods

3.1   Dataset

We collected 13,238,863 tweets from August 10, 2014, to August 27, 2014 that contain the keyword "Ferguson" using the Twitter Streaming API. Since media may have a different influence on people who have experienced an event versus people who have not, and since perceptions of social events are usually affected by distance [5], we take geographic influence into consideration by using geo-tagged tweets as a sample of all tweets for content analysis; of the full set we collected, 110,280 (0.83 %) are geo-tagged.[1] Previous work [9] shows that temporal patterns of tweet volume do not differ significantly between geo-tagged and non-tagged tweets, nor do the proportions of more and less influential users.

It is noteworthy that the media play their role in various ways, such as news reports, TV programs, and even accounts in different social networks. We identified news stories via the links published by 108 media accounts on Twitter,

---

[1] To identify locations of tweets, we refer to the geographic boundary file of 2014 TIGER/Line, https://www.census.gov/geo/maps-data/data/tiger-line.html.

e.g. Washington Post, NBC News, ABC7News, etc., looking at all the tweets they published during the Ferguson event and identifying news reports from the links.[2] In total, we collected 1,338 news reports dated from August 11 to 27.

The same preprocessing is applied to the news and Twitter corpora, including tokenization, lemmatization, bigram detection, and removal of stop words, low frequency words, and high frequency words.[3] After preprocessing and removing empty documents, the final corpus contains 1,275 news documents and 81,553 tweets.

3.2   Single Topic lda

We introduce ST-LDA to jointly model short documents like tweets and long news documents.[4] The key intuition is that a very short document like a tweet is unlikely to be related to multiple topics; therefore it can be modeled as having all its words generated from a single topic. In contrast, long documents like news stories are likely to follow conventional LDA assumptions, so each document is modeled conventionally as having a mixture of topics. At the same time, news and tweets are likely to discuss the same events in the world, so they share the same topic-word distributions. Figure 1 shows the graphical model of ST-LDA, where superscripts N and T denote news and tweets, respectively. The corresponding generative process of ST-LDA is as follows.

1. For each topic $k \in \{1, \ldots, K\}$
   (a) Draw word distribution $\phi_k \sim \text{Dirichlet}(\beta)$
2. For each (long) news document $d \in \{1, \ldots, D^{\text{News}}\}$
   (a) Draw a topic distribution $\theta_d^{\text{News}} \sim \text{Dirichlet}(\alpha)$
   (b) For each token $t_{d,n}^{\text{News}}$ in news document d
      i. Draw a topic $z_{d,n}^{\text{News}} \sim \text{Multinomial}(\theta_d^{\text{News}})$
      ii. Draw a word $w_{d,n}^{\text{News}} \sim \text{Multinomial}(\phi_{z_{d,n}^{\text{News}}})$
3. Draw tweet background topic distribution $\theta^{\text{Tweet}} \sim \text{Dirichlet}(\alpha)$
4. For each (short) tweet document $d \in \{1, \ldots, D^{\text{Tweet}}\}$
   (a) Draw a topic $z_d^{\text{Tweet}} \sim \text{Multinomial}(\theta^{\text{Tweet}})$
   (b) For each token $t_{d,n}^{\text{Tweet}}$ in document d
      i. Draw a word $w_{d,n}^{\text{Tweet}} \sim \text{Multinomial}(\phi_{z_d^{\text{Tweet}}})$

The key to the model is the combination of conventional LDA for news and the adjusted single-topic model component for the tweets. Different from LDA, the coverage of the word plate and document plate are different, adapted according to our assumptions. First, the word plate of ST-LDA's tweet part (subscript N_d^{T}) only covers w, which means that every word in a tweet is generated from the same topic. In LDA, the corresponding plate covers both z and w, denoting

---

[2] Tweets from these media sources are filtered from our Twitter data.
[3] News tokenization is done by OpenNLP, https://opennlp.apache.org/. Tweet tokenization is done by Twokenizer, http://www.cs.cmu.edu/~ark/TweetNLP/.
[4] Code is available at https://github.com/ywwbill/YWWTools#st_lda_cmd.

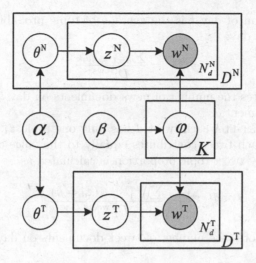

Fig. 1. Graphical model of ST-LDA

that every word has its own topic assignment and every document consists of a mixture of topics. Second, every tweet only has one topic; $\theta^{\mathbf{T}}$ is outside the document plate (subscript $D^{\mathbf{T}}$) and denotes a background topic distribution of tweets.

3.3   Posterior Inference

The Gibbs sampling equation for news documents is the same as conventional LDA. The probability of tweet d being assigned a topic k is computed as

$$\Pr\left(z_d = k \mid \boldsymbol{z_{-d}}, \boldsymbol{w}\right) \propto \left(N_k^{-d} + \alpha\right) \frac{\prod\limits_{v=1}^{V} \prod\limits_{i=0}^{N_{d,v}-1} \left(N_{k,v}^{-d} + \beta + i\right)}{\prod\limits_{i=0}^{N_{d,\cdot}-1} \left(N_{k,\cdot}^{-d} + V\beta + i\right)}, \qquad (1)$$

where N_k denotes the number of documents assigned to topic k; $N_{d,v}$ is the count of word v in document d; $N_{k,v}$ denotes the count of word v assigned to topic k. Marginal counts are denoted by $\cdot$. $^{-d}$ denotes that the count excludes document d.

3.4   Topic Dynamics

The output of ST-LDA can be used for further discovery of topic dynamics in tweets and news. Topic dynamics are characterized here as the temporal change in topics using a daily sliding window. Assuming that every news document has the same impact and contributes equally to the total media environment, the topic proportion of day t is the average of topic probabilities of all news documents on that day:

$$\bar{\theta}_{t,k}^{\text{News}} = \frac{\sum_{d=1}^{D_t^{\text{News}}} \theta_{d,k}}{D_t^{\text{News}}}, \tag{2}$$

where D_t^{News} denotes the number of news documents on day t; $\theta_{d,k}$ is topic k's proportion in document d.

In contrast, in ST-LDA each tweet d has only one topic z_d. Under the same assumption that each tweet contributes equally to the voice of the public, the aggregation of daily tweet topic proportion is calculated as

$$\bar{\theta}_{t,k}^{\text{Tweets}} = \frac{\sum_{d=1}^{D_t^{\text{Tweets}}} \mathbb{1}(z_d = k)}{D_t^{\text{Tweets}}}, \tag{3}$$

where D_t^{Tweets} denotes the number of tweet documents on day t and $\mathbb{1}(\cdot)$ is an indicator function.

Given $\bar{\theta}_t^{\text{News}}$ and $\bar{\theta}_t^{\text{Tweets}}$, where t varies from August 11 to 27, we can identify topic dynamics by the changing of daily topic proportions.

4   Quantitative Evaluation of ST-LDA

In this section, we evaluate ST-LDA and LDA quantitatively by performing topic identification task on both news and tweets.[5] We split our datasets into training (90 %) and test (10 %) sets, both for news and tweets, and evaluate the quality of topics given by ST-LDA and LDA respectively. We first align the topics given by ST-LDA and LDA using KL-divergence. The KL-divergence of topic k_1 given by LDA and topic k_2 given by ST-LDA is measured as

$$T_{k_1,k_2} = \text{KL}(\phi_{k_1}^{\mathbf{LDA}} || \phi_{k_2}^{\text{ST-LDA}}) = \sum_{v=1}^{V} \phi_{k_1,v}^{\text{LDA}} \log_2 \frac{\phi_{k_1,v}^{\text{LDA}}}{\phi_{k_2,v}^{\text{ST-LDA}}}. \tag{4}$$

Then we manually summarize each topic with a label and have two annotators annotated each tweet and news in the test set with one of the labels. To be strict, annotators are required to annotate "other" if none of the labels fit well, especially on news, because both LDA and ST-LDA give a distribution of topics on news documents. Due to the large number of tweets, we sample 5 % of test tweets for annotation. The annotation agreement rates are 71.7 % (91/127) and 79.1 % (322/407) for news and tweets, respectively. The lower agreement rate for news is due to the different opinions about the main topic, since each news document usually covers multiple topics.

We only use the data points with agreed annotations and measure the two models' accuracies. Since LDA gives a probability distribution, we consider its output as the topic with the highest probability.

---

[5] Note that ST-LDA will not outperform LDA on perplexity, since the words in a tweet are generated from the same topic. However, the sacrifice of perplexity brings improvement in topic identification.

After experimenting with different numbers of topics, we report the best results with 10 topics in Table 1. Although LDA has higher accuracy in news topic identification, the values are quite close. However, ST-LDA improves the accuracy by 14 % in assigning the main topics to test tweets, which demonstrates its efficacy.

Table 1. Topic identification accuracies

Model	News	Tweets
LDA	**0.637**	0.388
ST-LDA	0.615	**0.525**

5   Qualitative Evaluation of ST-LDA

In this section, we evaluate the quality of some of the topics assigned to news and tweets and explore their temporal evolution. Table 2 shows five matched topics of LDA and ST-LDA on mixed documents. We also set a baseline by running LDA on news, and match the topics to ST-LDA topics by KL-divergence. Common topics in LDA on news and ST-LDA on news and tweets show that ST-LDA keeps topics from news. Meanwhile the topic *Pray* only exists in the results of LDA and ST-LDA on mixed documents, which means *Pray* mainly exists in tweets. In addition, Twitter words such as *rt* and *gov* appear in the top words of topics given by ST-LDA. Therefore, ST-LDA is not biased to tweets or news and is able to discover topics from both.

Table 3 lists seven tweet examples and Table 4 shows their topic distributions. The topics are matched and numbered from 0 to 9.

The first three tweets' main topics given by LDA are the same as those given by ST-LDA. Although the main topics are consistent with the content, LDA assigns some probability to other topics. For example, though Tweet 1 contains words like *shoot*, it is not appropriate to assign this tweet to *Shooting Incident* which has top words like *street* and *Michael Brown*. Tweet 2 mainly talks about *Race and Community*, but LDA assigns the topic *Obama Talk* with probability 0.373 and small probabilities to other topics, which makes the main topic *Racism* only take 0.555.

Tweet 4 is an example in which LDA assigns the highest probability to multiple topics, namely, *Protest*, *Michael Brown*, *Shooting Incident*, *Emotion*, and *Race and Community*. In this situation, this tweet has no main topic. Tweet 5 is a case where ST-LDA assigns the right topic but LDA fails.

Tweets 6 and 7 fit in none of the ten topics, i.e. their labels are "other". Tweet 6 seems to have no clear topic. Tweet 7 talks about medical care for injuries, which is not a main topic discovered by either ST-LDA or LDA.

Next, we perform a qualitative evaluation of the topic dynamics in news and tweets provided by LDA and ST-LDA. Figures 2(a) and (b) show the changes in

Table 2. Topic examples

Model (Corpus)	Topic label	Top words
LDA (N+T)[a]	Obama Talk[b]	happen, i'm, make, thing, talk, situation, what's, what's_happen, bad, you're
	Protest	tear_gas, protester, arrest, fire, medium, rt, protestor, street, crowd, pd
	Racist	black, white, loot, protect, community, racist, stop, race, citizen, riot
	Curfew	missouri, state, obama, national_guard, call, curfew, mo, press, governor, nixon
	Pray	peace, pray, justice, stand, love, tonight, hope, stay, family, safe
ST-LDA (N+T)	Obama Talk	obama, president, law_enforcement, house, holder, make, story, post, include, community
	Protest	tear_gas, arrest, protester, fire, rt, reporter, medium, shoot, crowd, street
	Racist	black, white, make, race, america, obama, stop, happen, situation, riot
	Curfew	missouri, curfew, state, national_guard, governor, nixon, call, gov, order, make
	Pray	peace, pray, stand, justice, night, love, tonight, today, family, safe
LDA (N)	Obama Talk	obama, president, house, make, white, news, national, deal, run, defense
	Protest	st_louis, nixon, protester, shooting, county, justice, aug., investigation, state, thursday
	Racist	black, make, white, cop, time, don't, year, good, man, thing
	Curfew	protester, johnson, tear_gas, crowd, curfew, night, fire, street, missouri, shoot
	Pray	(No matching topic)

[a] N: News. T: Tweets.
[b] This topic matches *Obama Talk* in ST-LDA according to KL-divergence. However, the topic label is "question of the situation". For comparison we still name the topic *Obama Talk*.

news topic proportions from August 11 to 27 according to LDA and ST-LDA, respectively.

The topic distribution given by LDA (Fig. 2(a)) is highly skewed toward two main topics—*Shooting Incident* and *Race and Community*. Other topics take small proportions, so it is hard to identify their proportion changes. Meanwhile ST-LDA yields results that are slightly better in representing different topics. *Obama Talk* is discovered as a main topic. It keeps a relatively stable proportion

Table 3. Tweet examples

No.	Content
1	"@bkesling: "Hands up, don't shoot" after tear gas fired in #Ferguson http://t.co/9zQIh31wQg" modern day America... #PrayForFerguson
2	80 % black folks think #Ferguson raises "important issues about race that need to be discussed," only 37 % of white folks do. Very sad
3	You guys can't blame that cop in #Ferguson. Shooting your gun 6 times is literally the answer to every question in their training manual
4	#fergusongate media get it straight. U act like those who don't live in ferguson can't protest. This is for all blacks everywhere
5	But thank God for social media though. Imagine if we're dependent on the news to tell the "truth" about what's really happening in #Ferguson
6	@MikeHolmzy that's true. But I'm just talking about ferguson
7	County will not pay medical bills for toddler hurt in... http://t.co/8k8Hee5B63 via @sharethis #ferguson can you believe GA is doing this?

of 20 %, and peaks after some important events related to Obama. On August 12, Obama addressed the shooting and urged the Ferguson community to stay calm. On August 14, he gave a talk saying that there is no excuse for protests to turn into violence, which leads to the peak of topic *Obama Talk* on August 14 and 15. This demonstrates that the topics detected by ST-LDA are consistent with important events in the timeline.

There is more variance in topic dynamics of tweets according to ST-LDA than LDA, as shown in Figs. 2(c) and (d). The proportions of topics are close to each other in topic dynamics according to LDA, which makes it hard to identify the main topics for each day. In comparison, ST-LDA gives results with more variation of topics along the timeline. It is clear that after the shooting incident, *Emotion* of the public surges to a peak on August 11. After the protest event, another *Emotion* topic appears on August 14. Meanwhile, the proportion of *Pray* topic stays relatively stable from August 11 to August 24, but increases a great deal on the day when Michael Brown's funeral is held.

6 Topic Dynamics of Tweets and News

In this section, we use ST-LDA to analyze the Ferguson event. Considering the effect of distance on event perception [9], we analyze the topic dynamics for tweets in and out of St. Louis, where the Ferguson unrest took place. First, we compare the general topic dynamics to ground truth events to see the different focuses of the media and the public. Then, we compare the topic dynamics of news and tweets in and out of the St. Louis area to explore possible relations between the media and the public. For evaluation and comparison, we use the timeline of important events since Michael Brown's death. It includes information

Table 4. Tweet topic comparison. Content of the tweets can be referred in Table 3. The topics given by LDA and ST-LDA are matched.

Tweet numbers			1	2	3	4	5	6	7
LDA topic distribution	0	Obama talk	0.017	**0.373**	0.011	0.017	**0.888**	0.025	0.017
	1	Protest	**0.517**	0.009	0.011	**0.183**	0.013	0.025	0.017
	2	Racism	0.017	**0.555**	**0.233**	0.017	0.013	0.025	**0.350**
	3	Curfew	0.017	0.009	0.011	0.017	0.013	0.025	**0.350**
	4	Michael Brown	0.017	0.009	**0.567**	**0.183**	0.013	0.025	·0.017
	5	News Report	0.017	0.009	0.011	0.017	0.013	0.025	0.017
	6	Pray	0.017	0.009	0.011	0.017	0.013	0.025	0.017
	7	Shoot Incident	**0.350**	0.009	0.011	0.017	0.013	0.025	0.017
	8	Emotion	0.017	0.009	**0.122**	**0.183**	0.013	**0.775**	0.017
	9	Race and Community	0.017	0.009	0.011	**0.183**	0.013	0.025	0.183
ST-LDA topic			*1*	*2*	*4*	*8*	*5*	*1*	*6*

from different perspectives: shooting incident, looting, FBI investigation, Obama talk, protests, curfew, Michael Brown's funeral and so on.

6.1   Tweet Topics In and Out of the St. Louis Area

The tweet topic dynamics in and out of the St. Louis are shown in Fig. 3. Topic dynamics of tweets are highly related to ground truth events.

Topics *Curfew*, *News Report*, and *Michael Brown* share similar change patterns for tweets both in and out of the St. Louis area. *Curfew* increases to a peak on August 16 when Governor Nixon declared a state of emergency and imposed a curfew (see ⑤). It peaks at around 5 %, indicating that it is not the main issue of the public. The topic *Pray* shares similar dynamics with a large increase of tweets on this topic on August 25 when Michael Brown's funeral is held (see ⑨). However, more than 35 % of tweets in St. Louis are about *Pray*, while outside of St. Louis it is 20 %.

In the topic *News Report*, top words, such as *news*, *watch*, *live*, *report*, *coverage*, and *rt* indicate that the topic is mainly about the description and citation of information from news and TV. This is direct evidence of media influence on tweets. Tweets in and out of St. Louis have similar topic dynamics for *News Report*, indicating that people in and out of St. Louis paid similar attention to *News Report*.

However, differences in the topic dynamics of tweets show different perceptions of events for people in and out of the St. Louis area. More tweets in St. Louis talk about *Protest*, while out of St. Louis more tweets talk about *Racism*. From August 18, when Governor Nixon deployed the National Guard to Ferguson (see ⑥), to August 21, when the National Guard withdrew (see ⑧), protests and conflicts kept occurring. People in the Ferguson area are closer and more connected to the protests, so tweets with this topic surge to take more than 25 % of

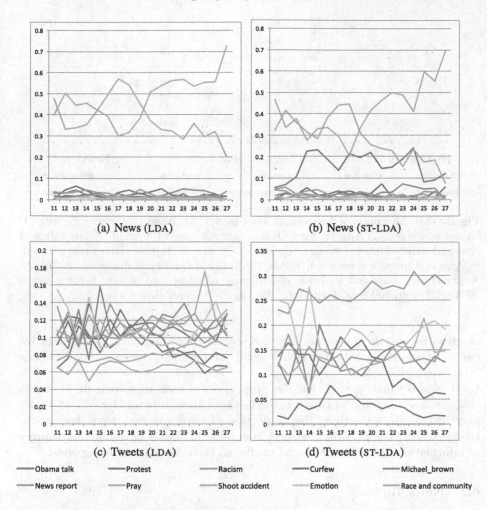

(a) News (LDA)   (b) News (ST-LDA)

(c) Tweets (LDA)   (d) Tweets (ST-LDA)

——Obama talk   ——Protest   ——Racism   ——Curfew   ——Michael_brown
——News report   ——Pray   ——Shoot accident   ——Emotion   ——Race and community

Fig. 2. Topic dynamics of news and tweets by LDA and ST-LDA

the volume. Meanwhile, the proportion of *Protest* tweets outside St. Louis is far lower. We surmise that members of the public who are not involved in the event tend to have less focus on the precise situation on the ground, and therefore are more likely to discuss the situation abstractly; thus *Racism* takes the majority most of the time.

The dynamics of the *Emotion* topic also differ geographically. Michael Brown was killed on August 9, and anger is the major topic of tweets in St. Louis; then *Emotion* tweets keep decreasing, and only take up 10 %–15 % of the volume. However, outside the St. Louis area, there is a lag effect of the *Emotion* explosion on August 14. One possible reason for this is that news takes time to spread and the public outside Ferguson needs more information to understand what happened. Another possibility is that segments of the news media emphasize

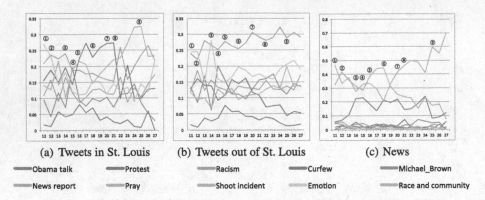

(a) Tweets in St. Louis   (b) Tweets out of St. Louis   (c) News

Obama talk   Protest   Racism   Curfew   Michael_Brown

News report   Pray   Shoot incident   Emotion   Race and community

Fig. 3. Topic Dynamics of Tweets and News by ST-LDA. Important events: ① Aug 11: Unrest continued; ② Aug 12: First Obama talk; ③ Aug 14: Second Obama talk and Nixon announced law enforcement operation; ④ Aug 15: Robbery video was released; ⑤ Aug 16: Curfew was imposed; ⑥ Aug 18: National Guard was deployed; ⑦ Aug 20: A grand jury convened to begin determining of crime and streets become quiet; ⑧ Aug 21: National Guard withdrew; ⑨ Aug 25: Michael Brown's funeral.

emotion because in the media business stories connected with strong negative emotions attract attention, which is good for business; consider the old adage "if it bleeds, it leads".

It is also worth noting that the proportion of each topic outside St. Louis changes less compared to tweets in St. Louis. Although tweets for certain topics may increase in some time, the proportion of tweets in each topic keeps relatively steady. It is possible that people outside St. Louis have different sources of information like news and social media, so their focus is more dispersed.

6.2 News Topics

As shown in Fig. 3, there are three main topic lines in news, *Obama Talk*, *Shooting Incident*, and *Race and Community*, which do not exist in tweets. According to the top words in *Shooting Incident* (*shoot, protester, michael_brown, st_louis, tear_gas*, etc.), it is quite similar to the topic *Michael Brown* (*shoot, kill, officer, unarmed, michael_brown*, etc.), which takes a certain proportion in tweets, and some proportion in news. Although news and tweets are talking about the same topic, the words they use are quite different, which leads to different topic assignments by ST-LDA. Similarly, the *Racism* topic exists mostly in tweets, while *Race and Community* mainly exists in news. These two topics are both about racism and human rights, but there is little overlap of tweets and news in the two topics. One possible reason is that the language in tweets is closer to spoken language, while news uses more formal written language; another is that media and the public frame the same event differently. According to the top words in two topics, there are more negative words in *Racism* such as *stop* and *riot*. In the topic *Race and Community*, words like *make*, *good*, and *community* are indicators of positive emotion. It seems possible that while the public tended to display negative emotion about race issues during the Ferguson unrest, the media tried to describe and lead the discussion in a constructive way.

Among the main topics, only *Obama Talk* is related to ground truth events. The proportion of topic *Obama Talk* increases from August 12, when Obama first addressed the shooting (see ②), reaching a peak on August 14 when Obama addressed the situation in Ferguson again (see ③). After that, the proportion stays steady at about 20 %. Two weeks after the shooting incident, this topic then decreases.

Of the minor topics, only *Curfew* is closely related to the occurrence of certain events. The emergence of *Curfew* in the news appears right after the day when Governor Nixon declared the curfew (see ⑤). The topic *Protest* has two peak points on August 14 and 21, which are the start and end dates of the National Guard deployment, respectively (see ③ and ⑧).

Topics *Pray* and *Emotion* take a very small part (less than 5 %). It seems reasonable that tweets are more subjective and contain more words about feelings, emotions, and praying, while news is more serious and objective, avoiding emotional leading. Despite the explosion of *Emotion* and *Pray* in tweets, there is no corresponding burst of the topic in news. This may indicate that such emotional changes in the public are not reflected in news, or that news reacts to the emotions with other topics. The relation is hard to determine with certainty and it is not clear which news topics might be in reaction to public emotions.

6.3 Influence of News on Tweets

From the comparison of topic dynamics in tweets and news, we find that tweets have more diverse topics, while news appears to have only three main themes. Topics related to Obama maintain a stable proportion in news reports, and the investigation report and discussion of race issues keep alternating dominance. On the other hand, tweet topics are more diverse and change along with the evolution of the event.

In news and tweets, both *Racism* and *Shooting Incident* are discussed quite a bit, but in different ways. In tweets, the topic *Michael Brown* takes the majority, while in news *Shooting Incident* is more dominant. Top words in *Michael Brown* mainly include *shoot, kill, officer, unarmed,* and *michael_brown,* which reflects the public paying more attention to describing the triggering incident. However, top words in *Shooting Incident* are *shoot, protester, michael_brown, st_louis,* and *tear_gas,* which all seem to reflect more of a big picture of the larger series of incidents. Meanwhile, as we have noted, discussions of race appears to be framed more positively or constructively in news than in tweets.

The above results seem to support the theory of the cascade model [4] from the perspective of the role media plays. In addition, the existence of the *News Report* topic shows that the public accepts information from the media. Meanwhile, the topic dynamics of tweets in and out of the St. Louis area show the possible influence. There is a lag effect of the *Emotion* explosion, and there is much more discussion of *Racism* among people in St. Louis, which corresponds to the major topic of *Race and Community* in news.

However, unlike the idea of a cascade leading the public to focus on what the media focuses on, topics for the two are still quite different. Under the influence of media, tweets do not simply repeat topics of news. For instance, the discussion of *Race and Community* and *Obama Talk* is rare in tweets. In addition, people on social media show more emotional change, which can be reflected by the dynamics of the *Emotion* and *Pray* topics when certain events happened. One might argue that this provides evidence of publics displaying primarily emotional rather than rational responses.

In summary, the topic dynamics of news and tweets in the Ferguson case form a picture in which:

1. The media tends to have a smaller set of topics that they emphasize consistently in coverage, in contrast to public opinions, which are more diverse and subject to change with new events;
2. Both news and tweets describe and discuss the event; however, the news tends to link events together, while tweets tend to have a quicker response to events;
3. In the context of social media, the public tends to generate information not necessarily following the news; specifically, people in St. Louis prefer to report their experiences on Twitter by quickly responding to events such as protest, while some hot topics in the news did not seem to attract much attention on Twitter;
4. Emotion tweets, such as *Emotion* and *Pray*, take a significant part; however, media reflection of these topics seems to be rare.

7   Discussion and Future Work

We introduced a new topic model, ST-LDA, that brings news and tweets under a unified topic space, so that topics of news and tweets are comparable. At the same time, it provides a solution to finding common topics from a mixed collection of long and short documents. The results show that ST-LDA is able to detect common topics in tweets and news and assign the main topic to each tweet more accurately. We plan to extend ST-LDA so that it can handle a wider range of document types.

Our analysis of dynamics showed how news and Twitter users reacted to the Ferguson case. However, it still remains to be seen whether our results generalize to other situations. Are there cases where the media and the public have different patterns of reaction? Moreover, we only used tweets and news over a limited time window and only looking at short-term influence. What about the long-term effects? Is it possible to track opinions on events like Ferguson even long after the events? We hope to address these questions in future work.

Acknowledgement. We thank anonymous reviewers for their insightful comments.

References

1. Breuer, A., Landman, T., Farquhar, D.: Social media and protest mobilization: evidence from the Tunisian revolution. Democratization **22**, 764–792 (2014)
2. Cataldi, M., Di Caro, L., Schifanella, C.: Emerging topic detection on Twitter based on temporal and social terms evaluation. In: Proceedings of Conference on Knowledge Discovery and Data Mining (2010)
3. Effing, R., van Hillegersberg, J., Huibers, T.: Social media and political participation: are Facebook, Twitter and YouTube democratizing our political systems? In: International Conference on Electronic Participation (2011)
4. Entman, R.M.: Framing: towards clarification of a fractured paradigm. J. Commun. **43**, 51–58 (1993)
5. Fujita, K., Henderson, M.D., Eng, J., Trope, Y., Liberman, N.: Spatial distance and mental construal of social events. Psycholog. Sci. **17**, 278–282 (2006)
6. Gao, W., Li, P., Darwish, K.: Joint topic modeling for event summarization across news and social media streams. In: Proceedings of the ACM International Conference on Information and Knowledge Management (2012)
7. González-Bailón, S., Borge-Holthoefer, J., Rivero, A., Moreno, Y.: The dynamics of protest recruitment through an online network. Sci. Rep. (2011)
8. He, D., Parker, D.S.: Topic dynamics: an alternative model of bursts in streams of topics. In: Proceedings of Conference on Knowledge Discovery and Data Mining (2010)
9. He, J., Hong, L., Frias-Martinez, V., Torrens, P.: Uncovering social media reaction pattern to protest events: a spatiotemporal dynamics perspective of ferguson unrest. In: International Conference on Social Informatics (2015)
10. He, Y., Lin, C., Gao, W., Wong, K.F.: Tracking sentiment and topic dynamics from social media. In: Proceedings of International Conference on Weblogs and Social Media (2012)
11. Hofmann, T.: Probabilistic latent semantic indexing. In: Proceedings of the ACM SIGIR Conference on Research and Development in Information Retrieval (1999)
12. Hong, L., Davison, B.D.: Empirical study of topic modeling in Twitter. In: Proceedings of Conference on Knowledge Discovery and Data Mining (2010)
13. Hu, Y., John, A., Wang, F., Kambhampati, S.: ET-LDA: joint topic modeling for aligning events and their Twitter feedback. In: Proceedings of the Association for the Advancement of Artificial Intelligence (2012)
14. Hua, T., Yue, N., Chen, F., Lu, C.T., Ramakrishnan, N.: Topical analysis of interactions between news and social media. In: Proceedings of the Association for the Advancement of Artificial Intelligence (2016)
15. Iwata, T., Yamada, T., Sakurai, Y., Ueda, N.: Sequential modeling of topic dynamics with multiple timescales. ACM Trans. Knowl. Discov. Data (TKDD) **5**, 19:1–19:27 (2012)
16. Lau, J.H., Collier, N., Baldwin, T.: On-line trend analysis with topic models: #Twitter trends detection topic model online. In: Proceedings of International Conference on Computational Linguistics (2012)
17. Leskovec, J., Backstrom, L., Kleinberg, J.: Meme-tracking and the dynamics of the news cycle. In: Proceedings of Conference on Knowledge Discovery and Data Mining (2009)
18. Mane, K.K., Börner, K.: Mapping topics and topic bursts in PNAS. Proc. Natl. Acad. Sci. **101**, 5287–5290 (2004)

19. Morinaga, S., Yamanishi, K.: Tracking dynamics of topic trends using a finite mixture model. In: Proceedings of Conference on Knowledge Discovery and Data Mining (2004)
20. Ratkiewicz, J., Conover, M., Meiss, M., Gonçalves, B., Patil, S., Flammini, A., Menczer, F.: Detecting and tracking the spread of astroturf memes in microblog streams. arXiv preprint arXiv:1011.3768 (2010)
21. Sayre, B., Bode, L., Shah, D., Wilcox, D., Shah, C.: Agenda setting in a digital age: tracking attention to California Proposition 8 in social media, online news and conventional news. Policy & Internet (2010)
22. Tufekci, Z., Wilson, C.: Social media and the decision to participate in political protest: observations from Tahrir Square. J. Commun. (2012)
23. Weng, J., Lim, E.P., Jiang, J., He, Q.: Twitterrank: finding topic-sensitive influential Twitterers. In: Proceedings of ACM International Conference on Web Search and Data Mining (2010)
24. Yan, X., Guo, J., Lan, Y., Cheng, X.: A biterm topic model for short texts. In: Proceedings of World Wide Web Conference (2013)
25. Zhao, W.X., Jiang, J., Weng, J., He, J., Lim, E.P., Yan, H., Li, X.: Comparing twitter and traditional media using topic models. In: Proceedings of the European Conference on Information Retrieval (2011)
26. Zuo, Y., Zhao, J., Xu, K.: Word network topic model: a simple but general solution for short and imbalanced texts. Knowl. Inf. Syst. **48**, 379–398 (2014)

Identifying Partisan Slant in News Articles and Twitter During Political Crises

Dmytro Karamshuk[1]([✉]), Tetyana Lokot[2], Oleksandr Pryymak[3], and Nishanth Sastry[1]([✉])

[1] King's College London, London, UK
{dmytro.karamshuk,nishanth}@kcl.ac.uk
[2] Dublin City University, Dublin 9, Ireland
tlokot@umd.edu
[3] Facebook, Saint Paul, USA
opr@fb.com

Abstract. In this paper, we are interested in understanding the interrelationships between mainstream and social media in forming public opinion during mass crises, specifically in regards to how events are framed in the mainstream news and on social networks and to how the language used in those frames may allow to infer political slant and partisanship. We study the lingual choices for political agenda setting in mainstream and social media by analyzing a dataset of more than 40M tweets and more than 4M news articles from the mass protests in Ukraine during 2013–2014—known as "Euromaidan"—and the post-Euromaidan conflict between Russian, pro-Russian and Ukrainian forces in eastern Ukraine and Crimea. We design a natural language processing algorithm to analyze at scale the linguistic markers which point to a particular political leaning in online media and show that political slant in news articles and Twitter posts can be inferred with a high level of accuracy. These findings allow us to better understand the dynamics of partisan opinion formation during mass crises and the interplay between mainstream and social media in such circumstances.

1 Introduction

Social media have become a crucial communication channel during mass political or civic events by shaping "a civic and democratic discourse in a vacuum of opportunities" [17]. As academic debates contest the nature of social media as an alternative public sphere [22], it is important to study the *interrelationships between mainstream media and social networks* in shaping public opinion during mass protests, especially in regards to the origin and dissemination of news frames [20]. It is also of interest to consider how *propaganda and manipulation* in the information sphere work: where partisan language and frames originate, how they spread, and whether there are certain markers that would allow to trace the distribution and paths of such content.

In this study, we analyze the role of social and mainstream media in shaping and disseminating partisan content frames during social unrest and crisis.

© Springer International Publishing AG 2016
E. Spiro and Y.-Y. Ahn (Eds.): SocInfo 2016, Part I, LNCS 10046, pp. 257–272, 2016.
DOI: 10.1007/978-3-319-47880-7_16

Specifically, we focus on the mass protests in Ukraine during 2013–2014 - known as *"Euromaidan"* - in which social media played a remarkable role, helping to raise awareness, cover, and discuss ongoing events; and the post-Euromaidan Russian occupation of Crimea and the conflict between Russian, pro-Russian and Ukrainian forces in eastern Ukraine (2014–2015), periods that were characterized by a parallel information and propaganda war occurring in mainstream media and online together with military action on the ground.

We explore the extent to which lingual choices in online discourse can illuminate the *partisan confrontation* between political factions during mass crises through the analysis of the two complementary datasets of Twitter posts and news articles. More specifically, we exploit natural language processing to single out language that points to a particular political leaning and to observe whether these markers are detectable in both mainstream media language and social media posts at scale. Our contributions can be summarized as follows:

- We exploit the word embedding approach [18,19] to identify the indicators of partisan slant in news articles and validate it over the text corpora of around 4M news articles collected during the Ukrainian conflict. Our analysis reveals a *strong use of highly polarized partisan content frames* in news articles on both sides of the conflict.
- Next, we design a machine learning approach for *detecting the markers of partisan rhetoric* in news articles with minimal efforts required for supervision. This is achieved by a "coarse-grained" labeling of the articles based on the partisan slant of the news agencies they originate from. Our approach - trained on a collection of articles from 15 representative news agencies - is able to achieve 60–77 % accuracy in distinguishing between the news articles with pro-Ukrainian, Russian pro-government and Russian independent slants during the Ukrainian conflict.
- Finally, we study the inter-relation between traditional and social media during conflicts through an analysis of individual news sharing patterns among Twitter users and find that most of the users are *exposed to a variety of news sources but with a strong partisan focus*. Using our machine learning approach, we are also able to predict the partisan leaning of Twitter users from the content of the tweets with an accuracy up of 70 %.

In summary, we demonstrate that studying the lingual choices of Twitter users and news media adds a new dimension to understanding the dynamics of information flows and partisan idea dissemination in the space between social networks and mainstream media. We also demonstrate that lingual choices-based machine learning models can be highly effective at automatically predicting the political slant of mainstream media and Twitter users, which can have serious implications for political expression in repressive and authoritarian regimes.

2   Related Literature

Computational social scientists have given substantial attention to the mainstream and social media activity around political and social change, and to the role information shared on these platforms plays in influencing political and social agendas, protest movements and events, and public opinion. Researchers have explored the role of social media and mainstream media actors in information diffusion and protest message amplification in networks through the prism of the *collective action theory* [14], as well as the role of social networks in recruitment and mobilisation during protest, revealing connections between online networks, social contagion, and collective dynamics [15].

A broad swathe of quantitative studies have focused on determining the factors that influence political leanings of social network users and metrics that allow to classify and predict this kind of political bias. Several studies have considered the *predictive power* of political news sharing habits on Twitter [3], the influence of partisan information sharing on political bias among Facebook users [5] and Twitter actors [11,16], and compared differences and biases in news story coverage, dissemination, and consumption among online mainstream and social media [9]. Others have noted the difficulty of connecting selective exposure to political news on Facebook to partisanship levels of users [4]. At the same time, researchers suggest that analysis of information consumption and distribution habits of social network users does provide data on media exposure, the relationship between various classes of media, and the diversity of media content shared on social networks [2].

Some studies have noted that reliably inferring the political orientation of Twitter users and generalizing the findings is notoriusly difficult due to it being one of the "hidden" attributes in social network data and due to differences between politically active groups of users and the general population [10]. However, other academics suggest that studying some relationships on social networks, like the co-subscriptions relationships inferred by Twitter links [1], allows for some understanding of the underlying media bias—and subsequently, political bias—of social network users. Another study showed that applying machine learning techniques to classify political leanings on Twitter based on political party messages can reveal partisanship among users [7]. A number of studies present a comparison between the predictive power of the users' social connections and their content sharing patterns for inferring political affiliation, ethnicity identification and detecting affinity for a particular business [23,24].

While the research described above uses a fairly large spectrum of methods to study, classify and find connections between social media users' behavior and their media and political preferences, most of the studies referenced employ social network analysis or related methods, focusing on relationships between actors or their behaviors within the network, such as sharing links, following other actors, etc. More recent studies have used computational methods to assess forms of political organization on social media [6], employed machine learning models to classify rumor and misinformation in crisis-related social media content [28], and used deep neural networks to identify and analyze election-related political conversations on Twitter on a longitudinal scale [26].

We propose augmenting these approaches with a focus on the *linguistic variables* present in the data, and using natural language processing and machine learning techniques to gain further insight into how political and ideological messages travel between mainstream and social media, and how these lingual choices reflect the partisan nature of mainstream media outlets and, subsequently, social media users and their content consumption and sharing habits. Such an approach would allow for a more granular understanding of how language changes allow to detect both important events and partisan leanings in mainstream and social media data.

3   Background and Datasets

The Euromaidan protests and subsequent political crisis are the outcome of a continuing trend in the post-Soviet arena. The last decade or so has seen an increase in mass protest actions in the region, with protests erupting in Russia, Belarus, Ukraine, etc. From the 2004 *Orange Revolution* in Ukraine to the *Bolotnaya rallies* of 2011–2012 in Moscow to the Euromaidan protests in Ukraine, a gradual increase in the use of digital technologies and media platforms by citizens has become evident [13,20,21]. At the same time, the region is characterized by a problematic media climate, with mainstream media often co-opted or controlled by the state or the oligarchy. The interplay and mutual influence of mainstream and social media emerge as crucial for understanding the political and civic developments in the region and thus demand more scholarly attention.

A number of researchers have already examined some of the more general aspects of Euromaidan, such as the reasons for the protest [12], who the protesters were [27], how the protest came together and evolved [21]. However,

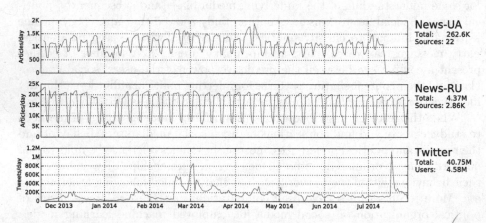

Fig. 1. Description of the Datasets. The Twitter dataset was collected via *Twitter Streaming API* (spikes in daily volume correspond to higher discussion volumes around Ukraine-related topics, February-March is the most active phase of the protest, spike in July corresponds to the MH17 airplane tragedy); the News-RU dataset was crawled from the *news.Rambler.ru* news aggregator website (periodic pattern reflects the weekly cycle); and the News-UA dataset was provided by the developers of the *Kobzi* mobile app.

few have investigated the use of civic media by Euromaidan participants beyond simply saying that social media were used in the protest as 'tools' for mobilization and information dissemination [21]. A deeper and more large-scale analysis of the relationship between mainstream media coverage of the crisis and the grassroots social media data around the political unrest, enabled by computational and big data tools and complemented by qualitative analysis, could reveal more about how the partisan agenda during the protests was formed and transformed through lingual choices and memorable memes, and who was able to exert influence on the lingual frames used by the multitudes of social media users and media outlets. Such investigation could also shed light on the reasons and mechanisms of the information war between Ukraine, Russia, and the West that gained in scale after the Euromaidan protests shifted into the crisis characterized by Russia's annexation of Crimea and the pro-Russian uprising in eastern Ukraine.

To research these questions in this paper we analyze two complementary datasets: A *social media* dataset which consists of over 40M tweets collected for the three most prominent hashtags during and after the Euromaidan protests - #euromaidan #ukraine #kyiv (and their Ukrainian and Russian equivalents) via Twitter Streaming API and a dataset of more than 4M *news articles* collected from a large Russian news aggregator (Rambler.ru) and its Ukrainian counterpart (smartphone app Kobzi). All three datasets were collected in the period between December 2013 and July 2014. The parameters of the datasets are summarized in Fig. 1.

4   Exemplar Indicators of Partisan Slant

4.1   Methodology

Our initial interest in exploring the *lingual choices* made by mainstream media sources and social media users was sparked by observing the emergence of a number of memes and buzzwords introduced during the events. In this section, we aim to measure the presence of these keywords in the rhetoric of the parties involved in the conflict and explore the extent to which this analysis can be automated.

To analyze the lingual choices of online media sources during the Ukrainian conflict we first exploit the *word embedding* methodology proposed in [18,19]. The proposed approach devises vector representations of words by analyzing the textual context in which they appear. This is achieved by training a model which for a given word, represented by a vector X_i, aims to infer the most likely $2 \times j$-surrounding words vectors which constitute the lingual context where the word was used[1], e.g., $f(X_i) = (X_{i-j}, ..., X_{i-1}, X_{i+1}, ..., X_{i+j})$. Intuitively, semantically closed words are expected to appear in similar contexts and so should produce similar outcomes when applied as the arguments of the function $f(X)$. Thus, if trained on a significantly large text corpus the algorithm is able

---

[1] We used the Skip-Gram model as it provided more interpretable results.

to assign close-by vectors to the words with similar meanings, thereby providing a powerful framework for analyzing the semantics of the word choices.

Equipped with the Word2Vec implementation[2], we build word representations for the textual corpora extracted from our datasets. Note that we mainly focus on vector spaces extracted from the news corpora (News-UA and News-RU) as Twitter's limit of 140 characters significantly constrains the applicability of this approach. In pre-processing, we remove rare words with less than 10 occurrences in each of the corpora and end up with a dictionary of 87K words trained from a content corpus of 600M words. Note that while the Ukrainian media space is generally bilingual, in our analysis we only focus on Russian-language news articles, in order to allow for a fair cross-comparison of the results, i.e., we focus on the Russian-language media sphere as it presents a sufficient diversity of news sources, with political views spread across the spectrum, ranging from pro- to anti-Kremlin and from pro- to anti-Ukraine, including both Ukrainian and Russian media outlets.

4.2 Mining Semantics of Word Choices

In Table 1 we present several examples of word associations which were mined from both our news corpora. We pick several loaded terms which were the prominent indicators of partisan rhetoric during the conflict and match these words to the most similar ones according to the trained Word2Vec dictionaries (as measured by the cosine similarity). The results illustrate that the trained model corresponds to our understanding of the semantics of chosen terms. For instance, the word *referendum* (*референдум*), frequently used in the context of the Russian-backed annexation referendum in Criméa, lies very close to its synonyms, e.g., 'plebiscite', 'voting', etc., in the devised vector space in both the News-RU and News-UA corpora, from both sides of the conflict.

On the other hand, we also notice that some of the synonyms discovered by Word2Vec reflect the propagandistic rhetoric of official Russia and Ukraine during the conflict. For instance, in the News-UA corpus (but not in the News-RU corpus), the word *referendum* (*референдум*) is associated with the words 'non-legitimate' and 'fake referendum', and reflect the official Ukrainian state's position on the plebiscite. Similarly, the word *aggressor* (*агрессор*) as used in association with Russia's actions in eastern Ukraine and Crimea is associated with the words 'cynical' and 'unprovoked' and captures the attitude of the Ukrainian government in regards to the events.

A similar partisan rhetoric is also observed in Word2Vec word representations mined from the Russian news corpus (News-RU). For instance, we observe that the loaded term *junta* (*хунта*), which was extensively used by some Russian media sources to demonize the transitional government formed in Ukraine after the Euromaidan protests, is strongly associated with the words 'neo-fascist' and 'pro-Ukrainian' in the Russian news corpus (News-RU). Similarly, the word *punisher* (*каратель*) which was used by some Russian media sources to label the

---

[2] Gensim natural language processing library https://radimrehurek.com/gensim/.

Table 1. Examples of word embeddings mined from the news datasets during the Ukrainian conflict. The table presents the word associations mined using the Word2Vec algorithm for several loaded terms (bold font) in Russian (top) and Ukrainian (bottom) news.

News-RU

каратель [punisher]		хунта [junta]	
бандера [Bandera [3]]	0.79	восстание [uprising]	0.76
безоружный [unarmed]	0.79	проукраинский [pro-Ukrainian]	0.72
расстреливать [shoot]	0.77	неофашист [neo-fascist]	0.72
терроризировать [terrorize]	0.77	неонацист [neo-Nazi]	0.71
оккупант [invader]	0.76	путчист [coupist]	0.71
фашист [fascist]	0.76	самозванец [imposter]	0.71

News-UA

референдум [referendum]		агрессор [aggressor]	
псевдо-референдум [fake referendum]	0.72	неспровоцированный [unprovoked]	0.72
плебисцит [plebiscite]	0.71	враг [enemy]	0.70
голосование [voting]	0.67	циничный [cynical]	0.66
автономный [autonomous]	0.66	развязанный [launched]	0.66
нелегитимный [non-legitimate]	0.64	извне [external]	0.65
присоединение [attachment]	0.63	оккупационный [invasive]	0.65
отсоединение [separation]	0.62	террор [terror]	0.64

Ukrainian Armed Forces and their operations in the conflict in eastern Ukraine is closely associated with 'fascism' and 'terrorize', as well as with Stepan Bandera, a controversial figure who has been acknowledged by some in the new Ukrainian government.

Such biased lingual choices are in alignment with recent findings that politically charged rhetoric and biased language were central to the discourse around the Euromaidan protests in Ukraine and the subsequent conflict in eastern Ukraine, both of which featured interference by Russian political forces [25,29].

5   Identifying Slant of News Stories

Inspired by the observations from the previous section we next question the power of the word choices to characterize the difference in partisan media at scale, i.e. are the linguistic choices of partisan media substantially different such that we can automatically differentiate them? We address this question by developing a classification machine learning model and conducting an extensive validation of the model over our news datasets.

5.1   Methodology

Using the Word2Vec word representations from the previous section, we train a supervised learning algorithm to find the best indicative words which characterize the language style of a given party. To this end, we first manually classify the

Table 2. Exemplar markers of individual news sources as identified by the words with the highest relative frequencies $\bar{\rho}_w^s$ across all words and news sources.

word	ρ_w^s	$\bar{\rho}_w^s$	News source
ъ [Ъ]	0.38	1.0	Ъ-Kommersant
m24.ru [m24.ru]	0.30	1.00	Moscow 24
известиям [izvestiam]	0.31	0.93	Izvestia
господин [gentlemen]	0.29	0.79	Ъ-Kommersant
подробнее [more]	0.36	0.69	Ъ-Kommersant
рбк [RBC]	0.27	0.68	RBC

Top-30 most popular Russian news agencies[3] as having a strong pro-government or opposition slant and complement this list by the Top-5 Russian-language sources from the Ukrainian internet segment. We achieve this by manually examining 20 or more articles per each news source for qualitative signs of slant, and by investigating public ownership records and publicly available info about the media outlets, their owners and affiliations. While granular and manual, such an approach can be replicated in other studies using media sources, as public ownership records are usually available and provide enough context for classification, while manual examination for signs of slant is based on a designated set of relevant keywords. We also remove all neutral sources from our analysis, i.e. those that have not indicated a particular partisan slant. Finally, we only consider news articles related to the Ukrainian unrest and conflict[4]. Our classification results in three categories of media outlets: UA, RU-ind, and RU-gov, exhibiting pro-Ukrainian, Russian-independent and Russian-pro-government points of view on the Ukraine unrest and conflict respectively.

Next, we cluster the Word2Vec vectors obtained from the previous section, and use the produced clusters as a feature space to describe the content of each article. In more detail, we apply k-means clustering with $N = 1000$ clusters to the word vectors of the combined News-UA and News-RU corpus. Then, for each article we calculate a 1000-items long feature vector X that represents the normalized frequencies of occurrences of words from each cluster that occur in the article, and train a function $f(X) \rightarrow \{RU-gov, Ru-ind, UA\}$ to identify the partisan slant of the article (e.g., whether coming from Russian pro-government (Ru-gov), Russian independent (RU-ind) or Ukrainian (UA) news source[5]).

5.2 Reducing News Source Bias

One crucial aspect to account for in the proposed approach is its ability to learn linguistic patterns that generalise across all news sources of a given partisan slant, rather than markers of individual news outlets. Since the training data

---

[3] as ranked by the Medialogia rating agency http://goo.gl/JNvx0Y.

[4] This is achieved through filtering the corpora by the relevant keywords, i.e., "kyiv", "ukraine", "donbass", "maidan", "crimea", "luhansk", "dnr" and "lnr". Adding a wider set of keywords had little effect on improving the recall of filtering.

[5] We refer to each of these three classes as 'party' or 'parties' in the rest of this paper.

in the above method is sourced from a selection of few news sources, without a generalisable approach, the classifier could simply learn to label the partisan slant of an article based on unique words, or *news source markers*, that are specific to a particular biased news source.

For example, it is common that the name of the news source or its correspondents are explicitly mentioned in the byline or text of the article, making it easily distinguishable among other texts. Supposing the training data contains a biased Russian pro-government news source B whose name appears in every article from B, we might learn a model that the word B is indicative of a Pro Russian-government partisan slant. Although this is useful to classify other articles from B, it does not help identify other pro-russian news sources or articles. We therefore need to adapt the method to learn labels that generalise to all news sources by ignoring news source markers.

Description of the Problem. To show that news source markers are indeed widespread among the news articles in our dataset, we measure the relative frequencies of words appearing in articles from each individual news source. More formally, we define the frequency $\rho_w^s = \frac{N_w^s}{N_s}$ of word w in news source s as a share of all articles N^s from news source s in which word w appears at least once and compare it to the sum[6] $\sum_{s \in S} \rho_w^s$ of the observed frequencies of w across all news sources $s \in S$, i.e., we define the ratio $\bar{\rho}_w^s = \frac{\rho_w^s}{\sum_{s \in S} \rho_w^s}$ to identify words which are highly unique to particular news sources.

The top news source markers are extracted as the words with the highest ratio $\bar{\rho}_w^s$ across all words and all news sources. Table 2 shows the top few markers. As expected, we observe that articles from some of the news sources - such as Ъ-Kommersant, Moscow 24, Izvestia, and RBC - contain very vivid word markers of that news source. For instance, the words "Ъ" and "m24.ru" used as abbreviations of the Ъ-Kommersant and Moscow 24 news papers appear only within the news articles originating from these two sources. More interestingly, we observe a very high relative frequency of mentioning other general words such as "господин [gentlemen]" and "подробнее [more]" in articles originating from Ъ-Kommersant, suggesting that there might be other word markers – beyond just the obvious names of newspapers – which indicate the writing style of an individual news source.

Suggested Solution. To eliminate the aforementioned news source bias in our prediction model we develop the following approach[7]. We use Random Forest

---

[6] Note that this is equivalent to using the mean $\frac{\sum_{s \in S} \rho_w^s}{|S|}$, since the sum for all words is computed over the same set of sources S.

[7] We note that a straightforward approach of removing the most prominent news source markers – as measured by the relative frequency introduced in the previous section – has proved to be inefficient for the considered classification problem. In contrast, the method we introduce in the rest of this section provides a more nuanced approach in estimating the relevance of each classification feature.

Table 3. Predicting partisan slant in news articles. The results of a supervised machine learning experiment to identify whether news articles were published by either a Ukrainian (UA), a Russian pro-government (RU-gov) or Russian independent (RU-ind) news agency.

Metric	RU-gov vs. RU-ind	RU-gov vs. UA	RU-gov vs. RU-ind vs. UA
Precision	0.66	0.78	0.57
Recall	0.65	0.77	0.60
Accuracy	0.65	0.77	0.60

classifiers known for a good performance on modeling high dimensional data and modify the underlying mechanism for constructing individual decision trees. By default, the trees of the Random Forest algorithm are constructed via a greedy search for the optimal split of the training data D which minimizes the entropy of the label classes (i.e., parties in the conflict), i.e.,

$$\underset{split}{\text{minimize}}\ H_t(D_{left}) + H_t(D_{right}) \tag{1}$$

where the entropy $H_t(D) = -\sum_{t \in T} \rho_t^D log(\rho_t^D)$ is defined on the label spaces $t \in T$ in the left D_{left} and the right D_{right} branches of the split (ρ_t^D indicates the share of instances of class t in the dataset D). In principle, an optimal split by this definition may be found around the word markers specific to individual news outlets (e.g., the one from Table 2). To penalize this unwanted behavior of the algorithm we introduce the entropy of a news source s as $H_s(D) = -\sum_{s \in S} \rho_s^D log(\rho_s^D)$ which - unlike the entropy defined on labels $H_t(D)$ - characterizes the purity of the split in terms of news sources s rather than parties t (i.e., ρ_s^D indicates the share of instances from news source s in the dataset D). Intuitively, we aim for a split that discriminates by the party t but not by the news source s and, so, we aim to find the split that minimizes the entropy $H_p(D)$ while preserving a high entropy of individual news sources within the party $H_s(D)$, i.e.:

$$\underset{split}{\text{minimize}}\ H_t(D_{left}) + H_t(D_{right}) - \alpha(H_s(D_{left}) - H_s(D_{right})) \tag{2}$$

where α is a constant controlling the effect of the proposed adjustment[8].

5.3   Cross-Source Validation

In summary, we represent each article as a 1000-long feature vector X based on relative word frequencies corresponding to $N = 1000$ clusters induced by a Word2Vec representation of the entire news corpus. We then learn a function

---

[8] The proposed approach was implemented by adapting the internal implementation of the Random Forest algorithm from the open source scikit-learn libary.

$f(X) \rightarrow \{RU - gov, Ru - ind, UA\}$ that labels the partisan slant (Russian pro-government (Ru-gov), Russian independent (RU-ind) or Ukrainian (UA) news source), whilst at the same time ensuring (using Eq. 2) that the model $f(\cdot)$ generalises beyond learning to distinguish markers specific to individual sources.

To properly validate the proposed approach we conduct a *cross-source valida-tion* as follows. From the news agencies labeled in the previous step we select all with at least 329 articles resulting in a dataset of five agencies from each party. The number of articles from each news agency is balanced by down-sampling the over-represented agencies. We further conduct a five-fold cross validation such that at each step we pick four news agencies from each class to train the classifier and use the remaining one for testing. Since we test the model over a news agency which has not been used for training we are able to control for over-fitting to the writing style and markers of specific outlets.

In Table 3 we report the average values of accuracy, precision, and recall of the proposed cross-source validation. The results in the table indicate a strong prediction performance of the algorithm (accuracy of 77 %) in classifying the content of the news articles as coming from Russian pro-government or Ukrainian news sources. This result confirms a sharp difference in the linguistic choices characteristic for the content of Russian and Ukrainian news articles as observed in Table 1 of the previous section. More interestingly, the accuracy is also high (i.e., 66 %) for the more difficult problem of distinguishing between Russian pro-government and Russian independent news sources which often shared a common view on individual episodes of the conflict (e.g., the annexation of Crimea). For the more general problem of discriminating between all three groups of news agencies, the algorithm is able to achieve an accuracy of 60 %. Note that this is significantly better than a naïve baseline of randomly guessing between the three classes (with expected accuracy of 33 %) signifying the presence of a sharp partisan slant in traditional media.

6   Understanding Partisan Slant in Twitter

Having studied the difference in linguistic choices characteristic for news agencies during the Ukrainian conflict we now switch to the analysis of the related discourse in social media. The focus of our analysis is on understanding the inter-relation between the level of exposure to different news sources among Twitter users and the linguistic choices in their posts.

6.1   Methodology

To analyze the level of exposure to various news sources among Twitter users, we rely on an established approach in the literature [3,7] and look at the news sharing patterns. To this end, we identify 248K Russian-language tweets which retweet or mention articles from one of the $Y = 22$ most popular news agencies classified in one of the three groups in our dataset (e.g., whether coming from Russian pro-government (RU-gov), Russian independent (RU-ind) or Ukrainian

(UA) news sources). Next, we measure the user focus on a partisan media and a specific news agency by computing the share of the articles he/she shared from his/her most preferred agency/party, correspondingly. We note that a similar approach has been previously used to analyze geographic bias of content access in social media [8].

More formally, for each user, we measure the number of times n_y she/he shared an article from a news source $y \in Y$ and calculate the user's *news focus* as the share of all times he/she has shared any news article from any news source in Y, i.e., $\beta = max_{y \in Y} \left(\frac{n_y}{\sum_{y \in Y} n_y} \right)$. The larger β is, the larger the fraction of the user's shares that come from a single source. Similarly, we measure the user's *party focus* β_{party} as the fraction of news articles shared from one of the three "parties": Russian pro-government (RU-gov), Russian independent (RU-ind) or Ukrainian (UA) news sources. Finally, we measure the *diversity of news sources* with which the user expressed alignment by computing the cardinality of the subset $Y_{n_y>0}$ of all sources in Y from which a user has shared at least one article, i.e., $\gamma = |Y_{n_y>0}|$.

6.2 Patterns of News Sharing

The results of the analysis are presented in Fig. 2. Firstly, we note that for the majority of users, less than half of all their shares come from a single news source (Fig. 2, left) and that the majority of users express alignment with more than 6 news sources (Fig. 2, middle). At the same time, we note that although more active users tend to focus more of their shares on a single news source, they also (occasionally) share a bigger number of news sources (as indicated by the higher diversity values for the users with more than 50 and 100 tweets in Fig. 2, middle).

However, this diversity of news sources is not seen when we look at the more coarse-grained picture at the level of 'parties' (Fig. 2, right): users' shares tend to be heavily focused on just one party – on average, more than 85 % (90 % for the very active users) of the shares of a user are for news sources in alignment with that user's main 'focus' party. In other words, although users exhibit sophisticated behaviours such as a relatively high level of diversity in sharing from multiple news sources, most of these sources have a single partisan alignment, whether Independent Russian, Pro-Russian, or Pro-Ukrainian Government. Furthermore, both the variety of news sources and the partisan focus increase as user activity levels (number of tweets made) increase.

6.3 Predicting Political Slant from Twitter Posts

Next, we attempt to draw the link between news exposure among Twitter users - as inferred by the news articles they share - and the language they use in their posts. From the results of the analysis in the previous section, we build on the fact that the vast majority of users have a strong partisan focus in the article they share and use that as a label of their political slant. To understand

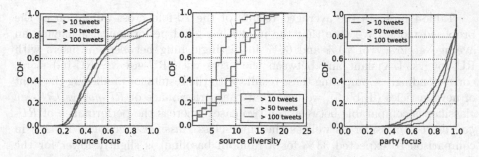

Fig. 2. Partisan news focus among Twitter users. Distributions of (left) the user focus on preferred news sources; (middle) the diversity of news sources a user has expressed alignment with; (right) the user focus on partisan media.

whether and to which extent the language choices in Twitter posts can indicate the political slant of the users we formulate a machine learning classifier to infer the latter from the former and validate it over our Twitter dataset.

To this end, we focus on the 5.9K users with at least 5 tweets, at least 3 of which contain news sources and balance the dataset by choosing an equal number of users from each party. To predict political slant of a Twitter user we formulate a supervised machine learning problem where we model language in a user's tweets with a feature vector X, using a methodology similar to the one described in the previous section, and train binary classifiers $f(X) \rightarrow \{RU-gov, RU-ind\}$ and $f(X) \rightarrow \{RU-gov, UA\}$ to identify whether a user is predominantly exposed to Russian pro-government (RU-gov), Russian independent (RU-ind) or Ukrainian (UA) news sources. Note that we remove all tweets that contain headlines of news and all retweets when constructing the language model of a user X. This ensures that we concentrate on the lingual choices in the tweets originating from the user, rather than the messages/sources that he/she (re)tweets. Also, as in the previous section, we only consider Russian-language tweets from Russian-speaking Twitter users, to ensure a fair comparison across all sides of the conflict. However, note that a large number of Ukrainian tweets are also in Russian, and our dataset contains representation from all three parties.

Table 4. Predicting partisan slant in Twitter. The results of a supervised machine learning experiment to identify partisan slant among Russian-speaking Twitter users during the Ukrainian conflict.

Metric	RU-gov vs. RU-ind	RU-gov vs. UA	RU-gov vs. RU-ind vs. UA
Precision	0.66	0.71	0.52
Recall	0.66	0.70	0.52
Accuracy	0.66	0.70	0.52

Table 4 presents the averaged results of the 10-fold cross-validation of the proposed model. We note that the model has good performance, achieving an average accuracy of 70 % and 66 % in distinguishing between the users with RU-gov vs. UA slant and between the users with RU-gov vs. RU-ind slant, correspondingly. Comparing these results with the results for predicting the slant of news articles (Table 3), we note that the performance of *RU-gov vs. RU-ind* classifier is comparable between the two cases whereas the performance of *RU-gov vs. UA* classifier as well as the three-class classifier (accuracy of 52 % in comparison to expected 33 % for a random baseline) is slightly lower for the Twitter case. This can be probably attributed to the fact that the size of the text piece in an average news article is larger than in a collection of ten 140-character-long posts collected for a median Twitter user in our training and testing sets and, so, inferring political leaning from Twitter posts seems to be a harder problem than it is for news articles.

7   Conclusions and Discussion

In this study we investigated the linguistic indicators of partisan confrontation in mainstream and social media during times of political upheaval by analyzing a dataset of more than 40M tweets and more than 4M news articles from the mass protests in Ukraine during 2013–2014—known as "Euromaidan"—and the post-Euromaidan conflict between Russian, pro-Russian and Ukrainian forces in eastern Ukraine and Crimea. We designed a natural language processing algorithm to analyze at scale the linguistic markers which point to a particular political leaning in online media and showed that partisan slant in news articles can be automatically inferred with an accuracy of 60–77 %. This difference in language in traditional news media is reflected in the word choices made by the supporters of similar partisan bent on Twitter—those who tweet news sources of a particular slant can be identified with an up to 70 % accuracy based on their lingual choices in tweets other than the retweets of particular news sources. Our results have two implications:

First, our results contribute to the debate on the role of lingual choices in traditional and social media in fostering political frames and partisan discourse during political crises, and confirm that both traditional news sources and users on social media are identifiably partisan. It would be interesting to conduct a more general study into whether the reinforcing partisan nature of the discourse and the political divisions we observe arise from a general lack of empathy and trust in conflict situations, or whether this is specific to the Ukraine conflict.

Second, the results reveal the extent to which partisan rhetoric and political leanings can be automatically inferred from lingual choices, which has implications for the use of social media as a safe platform for free speech in dangerous conflicts. Furthermore, we are able to infer all this with only a "coarse-grained" approach: The party labels (i.e., pro-Ukrainian, Russian pro-government or Russian independent) are assigned to the news articles (or social media profiles) based on the polarity of the news agencies they originate from/retweet.

The advantage of this approach is that it requires minimal manual efforts—it only requires to label the political slant of a number of news agencies (such as the 15 considered in this paper). However, the polarity and partisan rhetoric of articles may also vary between different topics and authors within a single news source, which we do not directly account for. We partially address this concern in the current paper by focusing only on the articles related to the Ukrainian events which are known for highly polarized rhetoric in both Russian and Ukrainian online media [25, 29]. Although we settled on the coarse-grained approach as a proof-of-concept model, it is clear that a more fine-grained approach could allow for greater accuracy in identifying partisan tweets and political leanings of users and news media.

Acknowledgements. This work was supported by the Space for Sharing (S4S) project (Grant No. ES/M00354X/1). We would also like to thank the developers of the Kobzi application for providing the News-UA dataset.

References

1. An, J., Cha, M., Gummadi, K.P., Crowcroft, J., Quercia, D.: Visualizing media bias through Twitter. In: Sixth International AAAI Conference on Weblogs and Social Media (2012)
2. An, J., Cha, M., Gummadi, P.K., Crowcroft, J.: Media landscape in Twitter: a world of new conventions and political diversity. In: ICWSM (2011)
3. An, J., Quercia, D., Cha, M., Gummadi, K., Crowcroft, J.: Sharing political news: the balancing act of intimacy and socialization in selective exposure. EPJ Data Sci. **3**(1), 1–21 (2014)
4. An, J., Quercia, D., Crowcroft, J.: Fragmented social media: a look into selective exposure to political news. In: Proceedings of World Wide Web Companion (2013)
5. An, J., Quercia, D., Crowcroft, J.: Partisan sharing: Facebook evidence and societal consequences. In: Proceedings of ACM COSN (2014)
6. Aragón, P., Volkovich, Y., Laniado, D., Kaltenbrunner, A.: When a movement becomes a party: computational assessment of new forms of political organization in social media. In: Tenth International AAAI Conference on Web and Social Media (2016)
7. Boutet, A., Kim, H., Yoneki, E.: What's in Twitter, I know what parties are popular and who you are supporting now! Soc. Netw. Anal. Min. **3**(4), 1379–1391 (2013)
8. Brodersen, A., Scellato, S., Wattenhofer, M.: Youtube around the world: geographic popularity of videos. In: Proceedings of WWW (2012)
9. Chakraborty, A., Ghosh, S., Ganguly, N., Gummadi, K.P.: Dissemination biases of social media channels: on the topical coverage of socially shared news. In: Tenth International AAAI Conference on Web and Social Media (2016)
10. Cohen, R., Ruths, D.: Classifying political orientation on Twitter: it's not easy! In: ICWSM (2013)
11. Conover, M.D., Gonçalves, B., Ratkiewicz, J., Flammini, A., Menczer, F.: Predicting the political alignment of Twitter users. In: 2011 IEEE Third International Conference on Privacy, Security, Risk and Trust (PASSAT) and 2011 IEEE Third Inernational Conference on Social Computing (SocialCom), pp. 192–199. IEEE (2011)

12. Diuk, N.: Euromaidan: Ukraine's self-organizing revolution. World Affairs **176**(6), 9–16 (2014)
13. Goldstein, J.: The role of digital networked technologies in the Ukrainian orange revolution. Berkman Center Research Publication (2007)
14. González-Bailón, S., Borge-Holthoefer, J., Moreno, Y.: Broadcasters and hidden influentials in online protest diffusion. Am. Behav. Sci. (2013)
15. González-Bailón, S., et al.: The dynamics of protest recruitment through an online network. Sci. Rep. **1** (2011)
16. Hegelich, S., Janetzko, D.: Are social bots on twitter political actors? Empirical evidence from a Ukrainian social botnet. In: Tenth International AAAI Conference on Web and Social Media (2016)
17. Howard, P.N.: The Digital Origins of Dictatorship and Democracy. Information Technology and Political Islam. Oxford University Press, Oxford (2010)
18. Mikolov, T., et al.: Distributed representations of words and phrases and their compositionality. In: NIPS (2013)
19. Mikolov, T., et al.: Efficient estimation of word representations in vector space. In: Proceedings of Workshop at ICLR (2013)
20. Oates, S., Lokot, T.: Twilight of the gods?: How the Internet challenged Russian television news frames in the winter protests of 2011–2012. In: International Association for Media and Communication Research Annual Conference (2013)
21. Onuch, O.: Euromaidan protests in Ukraine: social media versus social networks. Probl. Post-Communism **62**, 1–19 (2015)
22. Papacharissi, Z., Chadwick, A.: The virtual sphere 2.0: the internet, the public sphere, and beyond. In: Routledge Handbook of Internet Politics, pp. 230–245 (2009)
23. Pennacchiotti, M., Popescu, A.M.: Democrats, republicans and starbucks afficionados: user classification in twitter. In: Proceedings of the 17th ACM SIGKDD International Conference on Knowledge Discovery and Data Mining, pp. 430–438. ACM (2011)
24. Pennacchiotti, M., Popescu, A.M.: A machine learning approach to Twitter user classification. In: ICWSM 2011, vol. 1, pp. 281–288 (2011)
25. Szostek, J.: The media battles of Ukraine's Euromaidan. Digital Icons **11**, 1–19 (2014)
26. Vijayaraghavan, P., Vosoughi, S., Roy, D.: Automatic detection and categorization of election-related tweets. arXiv preprint arXiv:1605.05150 (2016)
27. Zelinska, O.: Who were the protestors and what did they want? Demokratizatsiya **23**(4), 379–400 (2015)
28. Zeng, L., Starbird, K., Spiro, E.S.: # unconfirmed: classifying rumor stance in crisis-related social media messages. In: Tenth International AAAI Conference on Web and Social Media (2016)
29. Zhukov, Y.M., Baum, M.A.: Reporting bias and information warfare. In: International Studies Association Annual Convention (2016)

Predicting Poll Trends Using Twitter and Multivariate Time-Series Classification

Tom Mirowski, Shoumik Roychoudhury, Fang Zhou, and Zoran Obradovic[✉]

Center for Data Analytics and Biomedical Informatics,
Temple University, Philadelphia, USA
{tmirowski,shoumik.rc,fang.zhou,zoran.obradovic}@temple.edu

Abstract. Social media outlets, such as Twitter, provide invaluable information for understanding the social and political climate surrounding particular issues. Millions of people who vary in age, social class, and political beliefs come together in conversation. However, this information poses challenges to making inferences from these tweets. Using the tweets from the 2016 U.S. Presidential campaign, one main research question is addressed in this work. That is, can accurate predictions be made detecting changes in a political candidate's poll score trends utilizing tweets created during their campaign? The novelty of this work is that we formulate the problem as a multivariate time-series classification problem, which fits the temporal nature of tweets, rather than as a traditional attribute-based classification. Features that represent various aspects of support for (or against) a candidate are tracked on an hour-by-hour basis. Together these form multivariate time-series. One commonly used approach to this problem is based on the majority voting scheme. This method assumes the univariate time-series from different features have equal importance. To alleviate this issue a weighted shapelet transformation model is proposed. Extensive experiments on over 12 million tweets between November 2015 and January 2016 related to the four primary candidates (Bernie Sanders, Hillary Clinton, Donald Trump and Ted Cruz) indicate that the multivariate time-series approach outperforms traditional attribute-based approaches.

1 Introduction

Traditionally public opinion has been measured via costly and time-consuming polls. The rise of social media, such as Twitter, has provided people with a new way of making their voices heard. One recent event that has generated a significant amount of buzz within the Twitterverse is the 2016 U.S. Presidential election. From November 2015 to January 2016, over 12 million tweets have been sent directly to those vying for control of the United States. On one hand, these tweets are a rich data source of information regarding each of those candidates standings within the eyes of the potential voters [5]. On the other hand, this data poses challenges to making inferences from the tweets, which is particularly important for politicians, journalist, and so on. Using these tweets many questions can be answered: *can we accurately categorize a candidate as*

E. Spiro and Y.-Y. Ahn (Eds.): SocInfo 2016, Part I, LNCS 10046, pp. 273–289, 2016.
DOI: 10.1007/978-3-319-47880-7_17

improving or deteriorating? Whether the public sentiment strength, or just the tweet volume, could aid in the prediction? Whether the public opinion from the Twitter has a clear inference related to one political party more than the other?

These questions are addressed by applying multivariate time-series classification models. Therefore, the focus of this paper is quite different from other related work. Existing works [7,12,13,15] only consider the volume of (positive or negative) tweets. In doing so the temporal aspect of these features is lost. The methods utilized in this work, however, aim to extract discriminative patterns of the fluctuations in these features over time.

Tweets express public opinion, thus we study a variety of features that are related to sentiment strength, user support and tweet volume. Each feature is examined on an hour-by-hour basis. The superiority of using a time-series classification model is that distinguishable temporal patterns can be extracted for prediction. For example, in Fig. 1, the temporal patterns of time-series when a poll score increases (Fig. 1(a), colored in blue) is quite different from the ones when a poll score decreases (Fig. 1(b), colored in red). Summing the values of a feature within a time period into a single value results in the loss of this temporal correlation, as the summing process nullifies the temporal aspect of the data and how feature values changes over time.

The goal of this work is to predict the trend of poll scores of candidates, particularly in the case of the United States presidential campaign for 2016, by examining the people's voice. One commonly used approach is based on the majority voting scheme [10]. The main idea is to conduct predictions from individual univariate time-series first, and select the majority result as the prediction for the multivariate time-series. Applied in this paper is a model using this idea, called Majority-vote Learning Shapelets Algorithm (MLS). However, the majority voting scheme assumes that the univariate time-series from different features are equally important. This may not be true in many real-world cases. Therefore, another approach is considered, called Weighted Shapelet Transformation model (WST), which is a linear classifier that learns weights of patterns extracted from multivariate time-series.

The contributions of this work include: 1. Utilizing the temporal patterns in tweets, and formulating the problem as a multivariate time-series classification. 2. Identifying multiple features to characterize public opinions and examining their individual roles for prediction. 3. Proposing a model, WST, which outperforms MLS, univariate time-series classification models, and traditional attribute-based classification models. 4. From the extensive experimental results, we found that (a) not all features are equally important; (b) Trends in Democratic candidates are easier to predict than Republicans and (c) Features based on positive sentiment tend to have higher predictive prowess than their negative counterparts.

2   Related Work

Examined in this section is work related to two relevant topics: (1) Tweets related to elections; (2) Time-series classification.

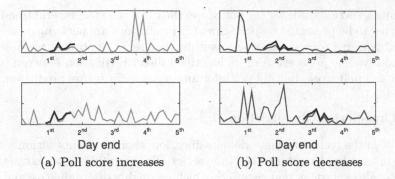

Day end

(a) Poll score increases

Day end

(b) Poll score decreases

Fig. 1. An example of time-series when (a) Poll score increases and (b) Poll score decreases. The temporal patterns (colored in black) are unique to each class. The temporal pattern in (a) has a trend of increase, and the temporal pattern in (b) has a trend of decline. (Color figure online)

2.1 Tweets Related to Elections

Support Strength. The first common theme is a focus on the support that comes from tweets [7] to determine the victor of an election campaign. For example, in [13] a linear regression model is applied to predict the value of a poll score. They consider tweet volume, overall user count, and unique user count as features within their regression model. The number of tweets about a candidate was shown to be significant in determining the outcome of election [15]. In [12], the number of unique users making posts in Twitter is considered as a feature for predicting the number of congressional seats that would be assigned to each of the political parties. U.S. politicians having proficiency in using Twitter to circulate positive information and favorable URLs about themselves is also noted [7].

Sentiment Strength. The second set of works in regards to predicting elections using Twitter focuses on the content of the tweets particularly the sentiment of them. It is noted that an improvement is observed in their regression models when incorporating sentiment from tweets [1]. Similarly, in [12] it is stated that they remove negative tweets as part of their data pre-processing step, implying they feel that negative information hurts their predictive model. Precedent established that an understanding of a candidates campaign can be achieved through analyzation tweet-content [15]. Time-series and sentiment are tied together in [9] demonstrating that the sentiment towards candidates over time is correlated to the fluctuations that can be observed in that candidates' poll score. For a detailed review of these works, please refer to [2].

Discussion. In this paper, many of the features mentioned above are used. The temporal element of these features is preserved, making use of their fluctuations over time. This work differs from other existing work in that (1) Instead of predicting an actual value for the poll score, the focus is on the general trend that is expected to be presented by poll score. (2) Predictions are performed based on detected patterns (or sub-sequences) within time-series, as tweets are observed continuously. The previous work [9] identified direct correlation between public opinion and poll score, but did not offer any method for future prediction.

2.2   Time-Series Classification

Recently, in the realm of time-series classification, short segments within a time-series are used to characterize the time-series. The main idea is to extract sub-sequence, also known as shapelets [17], which are highly discriminative and have been used for classification purposes. Most works that are related to shapelet based time-series classification are univariate time-series classification [3,4,11, 16,17]. One of the state-of-the-art univariate time-series classification models is Learning Time-series Shapelet (LTS) model [4]. In the LTS model, shapelets [17] are learned jointly with a linear classifier rather than searching over all possible time-series segments, and the shapelet tranformed data [6] is used to classify unknown univariate time-series.

Only a few works have focused on the shapelet based multivariate time-series classification. One is to concatenate univariate time-series from multi sensor (or features) together to form a longer time-series [8]. The other work is based on the majority-voting scheme [10].

3   Data

3.1   Data Collection

Tweets were gathered every twenty-four hours using Twitter's official REST API in conjunction with the httr package for R 3.2.3. More than twelve million tweets were collected from November 12th, 2015 until January 10th, 2016, for a total of 60 days.

Tweets from the account of a U.S. Presidential candidate were collected. For this work, the focus is on four U.S. Presidential candidates, Bernie Sanders and Hillary Clinton from the Democratic Party, Donald Trump and Ted Cruz from the Republican Party. While we are examining public opinion, tweets from candidates themselves were kept throughout the analysis as their thoughts and opinions have the potential to guide the conversation surrounding them.

Additionally, tweets which contain a mention (@username) to a U.S. Presidential candidate's official Twitter username were extracted. The benefit of using mentions is two-fold. The first is that other users who may be looking at tweets about that individual will see your comment, extending your reach to others in the community. The second is that it serves as an alert to the individual whom you are targeting that they are receiving a message. In the event that a tweet is extracted multiple times due to multiple mentions, duplicated tweets (identified by their unique tweet ID) have been removed before analysis, ensuring a tweet

Table 1. Tweets break-down by candidate, and the number of univariate time-series that belong to each class using the *max* labeling scheme.

Candidate	# of Tweets	# of poll increases	# of poll decreases
Bernie Sanders	2,359,938	54	36
Hillary Clinton	2,056,540	27	54
Donald Trump	7,011,224	54	27
Ted Cruz	1,234,402	81	9

only counts once per candidate. Retweets are treated as their own unique tweet, signifying agreement with the original post. This work examines over 12 million tweets within a two month period. The breakdown of how the tweets were distributed among the candidates can be found in Table 1.

3.2 Features

In this section nine features, representing various aspects of tweets with reference to a given candidate across time, are discussed in detail. Features are divided into three distinct groups: sentiment strength, user support and tweet volume.

1. Sentiment Strength
Sentiment analysis was conducted by using the sentiment analysis software "SentiStrength" [14]. Each individual tweet, regardless of length, was treated as a single document. As per SentiStrength's algorithm, each tweet is given an overall positive score [1,5] and an overall negative score [−1, −5]. Tweets were classified as positive if the overall positive score was higher than the absolute value of the overall negative score. If the tweets contained matching sentiment strength for both levels, they were classified as neutral. Otherwise, they were classified as negative. For example, if a tweet was assigned the scores [5, −3], then it was regarded as positive.

Sentiment strength is evaluated from two aspects: the average sentiment strength per tweet and the average sentiment strength per unique user.

- *Positive Average* [**PA**] refers to the average positive sentiment score among all tweets that were classified as positive within a user-specified time period, which is defined as Eq. 1. Let s_p represent the sentiment of any tweet p classified as positive, and n_h represent the total number of tweets classified as positive at time h.

$$f_h^{PA} = \frac{\sum s_p}{n_h} \tag{1}$$

- *Negative Average* [**NA**] thusly represents the average negative sentiment score among all tweets that were classified as negative within a user-specified time period.

– **Unique Positive User Average [UPUA]** represents the average sentiment among unique users who make positive postings with reference to the given candidate. First, the average sentiment of an individual user is calculated. Next, the average sentiment among all unique users is determined. Note that here only users whose tweets are all classified as positve were considered. This is represented in Eq. 2. Here $\overline{s^u}$ denotes the average sentiment score of a unique user u, and n_h^u denotes the number of unique users with positive sentiment at time h.

$$f_h^{UPUA} = \frac{\sum \overline{s^u}}{n_h^u} \qquad (2)$$

– **Unique Negative User Average [UNUA]** thusly refers to the average sentiment among unique users who make negative postings with reference to the given candidate. Similarly, only users whose tweets are all classified as negative were considered.

2. User Support

The following two features are representative of the magnitude of individuals who show support for or against a candidate.

– **Unique Positive Users [UPU]** refers to the number of unique users who made a post that was classified as positive. This accounts for the fact that a single user may have made multiple posts, so they are only accounted for once.
– **Unique Negative Users [UNU]** similarly refers to the number of users who made a post that was classified as negative.

3. Tweet Volume

– **Number of Tweets [NT]** refers to the number of tweets that a candidate issued/received within a user-specified time period. This feature was reported as a good predictor for measuring interest in a candidate [7,12,15].
– **Number of Positive Tweets [NPT]** represents the number of tweets where the overall positive sentiment was stronger than the overall negative sentiment [1].
– **Number of Negative Tweets [NNT]** thusly represents the number of tweets where the overall negative sentiment was stronger than the overall positive sentiment [1].

3.3   Univariate and Multivariate Time-Series of Poll Score Trends

A univariate time-series $T_i^q = \{f_{i,1}^q, \cdots, f_{i,L}^q\}$ is a set of time-ordered observations f_h^q starting at time $h = 1$ and ending at $h = L$, each one being the value of feature f^q recorded at time h. The label of the univariate time-series T_i^q, denoted by Y_i^q, represents which class the univariate time-series belongs to.

The univariate time-series are characterized by two user-specified values: length and granularity. Five-days is chosen as the length of time-series, as it is long enough to capture the potential changes in poll scores. Granularity represents how much information is shown by one time point. Since real-time tweets are collected, the granularity could be daily, hourly, and even by the minute. If granularity is too small, then the data will be very noisy and algorithms will overfit to detect localized temporal patterns (or sub-sequences) rather than identifying temporal patterns that give global discriminative power. If the granularity is too large, then the data will be very smooth losing potentially interesting patterns. In such a case the patterns would not be identified. In this work, 3 h is selected as granularity, because it splits the day into several time periods representative of different sections of the day, such as midday, evening or latenight. Therefore, each univariate time-series has 40 time points, with 8 points representing a single day.

A multivariate time-series $T_i = \{T_i^{PA}, T_i^{NA}, ...T_i^{NNT}\}$ is a set of univariate time-series that are related to individual features described in Sect. 3.2. Figure 2 shows the structure of our data. For each example $i \in I$, it represents a multivariate time-series $\{T_i^{PA}, T_i^{NA}, ...T_i^{NNT}\}$ with reference to a given candidate, where T_i^q represents a univariate time-series related to feature q.

The label of the multivariate time-series T_i, denoted by Y_i is same as the label of univariate time-series T_i^q. Therefore, this work predicts the label Y_i of the multivariate time-series T_i. This work only focuses on the cases when poll scores have obvious increment or decrement, that is, $Y_i \in \{+1, -1\}$. Two schemes of assigning labels are discussed in Sect. 3.4.

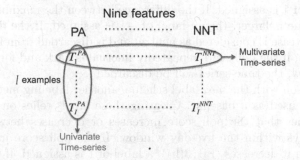

Fig. 2. Multivariate time-series data structure. T_i^{PA} represents the univariate time-series related to feature PA in the multivariate example i.

3.4   Labels of Time-Series

The research problem is a binary classification. That is, a label of 1 represents an increase in the candidate's poll score, and a label of -1 represents a decrease in the candidate's poll score. The daily average poll scores were collected from RealClearPolitics.com. These scores were used as they are an average of a wide-variety of national polls.

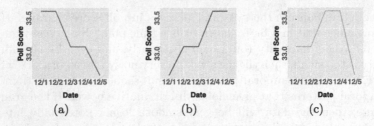

Fig. 3. Three possible poll score trends. (a) Poll scores decreased, so the label of this time period is −1. (b) Poll scores increased, so the label is 1. (c) Poll scores first increased, and then decreased. The label will be different based on the chosen scheme.

One method, which is referred to as *max*, consists of identifying when the maximum poll score occurs, and assigning an appropriate label based on it's position relative to the first and last poll score within the time period. Figure 3 illustrates three possibilities for generating a label. Figure 3(a) represents the scenario where the maximum poll score occurs on the first observed day, and then a label of −1 is assigned, representing a decrease in poll score. Figure 3(b) illustrates the case where the maximum observed poll score occurs on the last day. In this case, a label of 1 indicating a poll score increase is assigned. Figure 3(c) shows the situation where the maximum observed poll score happens on an internal date. Here, if the difference between the maximum value and the starting value is larger than the difference between the maximum value and the ending value, a label of 1 is assigned. If the difference between the maximum value and the ending value is larger, then a label of −1 is assigned. If the differences are equal, then the case is considered as neutral and is discarded from further analysis. In addition, if the poll score repeatedly fluctuates back and forth within the five-day window, the time-series will be discarded.

In comparison with the *max* label scheme, another labeling method, referred to as *count*, is used as a baseline. Creating these labels relies on counting the number of times that the poll score increases or decreases between any two consecutive days within the five-day window. If the poll score increases more frequently than it decreases (Fig. 3(b)), a label of 1 is assigned. If the poll score decreases more frequently than it increases (Fig. 3(a)), a label of −1 is assigned. If the poll score increases and decreases an equal amount, or never changes, then it is treated as a neutral case and is discarded from analysis. For example in Fig. 3(c), poll score increases once, and decreases once.

In total, after the removal of instances where a neutral label exists, we have 342 univariate time-series, that is 38 multivariate time-series. The distribution of these labeled univariate time-series instances are provided in Table 1. Assuming each time-series instance to be i.i.d, we generate three cross-validation datasets for evaluating both univariate and multivariate time-series classification performance, and report the mean performance results in the experiments.

4   Approach Overview

Two multivariate time-series classification models are presented in this Section. The first approach is based on majority-voting scheme. The second approach is a linear classifier which learns weights of patterns with respect to the class labels.

In both models, the first step is to utilize Learning Time-series Shapelet model [4] (denoted as LTS henceforth), one of the state-of-the-art univariate time-series classification models, to discover shapelets [17], which are local discriminative patterns (or sub-sequences) and are used to characterize the target class. The detailed explanation of LTS is provided in the appendix section.

4.1   Majority-Vote Learning Shapelets Model

Majority-vote Learning Shapelets (MLS) model is a majority-voting based algorithm, which effectively combines the benefits of majority-voting and the learning shapelets procedure of univariate time-series model LTS. MLS differs from LTS in that MLS aggregates the individual univariate time-series model predictions, as a univariate time-series related to a single feature often does not contain sufficient information for the prediction task.

In the framework of MLS, the LTS model is first applied to the individual univariate time-series T_t^q, and makes a prediction related to time-series T_t^q. Let t represent a multivariate time-series example in the test data, and q represent one of the features described in Sect. 3.2. The predicted value related to T_t^q is denoted as $\hat{Y}_t^q$. Then, the predicted value of the multivariate time-series example t is determined through the majority voting scheme, which selects the class that has more than half the votes. Equation 3 represents this approach mathematically.

$$\hat{Y}_t = max(\hat{Y}_t^q), \quad q \in \{PA, \cdots, NNT\} \tag{3}$$

4.2   Weighted Shapelet Transformation Model

MLS takes each univariate time-series independently and assumes that each univariate time-series have equal importance. To handle scenarios where importance is not equal, we propose one method, called Weighted Shapelet Transformation (WST), which is a linear classifier that learns weights of shapelets from multivariate time series.

In the framework of WST, the first step is to apply LTS to learn shapelets from the individual univariate time-series with respect to a particular feature. The learned shapelets can be considered as the attributes of multivariate time-series. The minimum distances between the learned shapelets and the observed time-series are calculated. These distance values are then used as the values of the attributes. The second step is to apply Logistic Regression to learn the weights of shapelets with respect to the target class.

5   Experimental Evaluation

In this section, the following questions are addressed:

Q1: How good are the predictions using multi-variate time-series? Does temporal modeling provide better predictive performance over traditional attribute-based modeling?
Q2: What are the prediction performances for individual features? Which features are good predictors?
Q3: Which labeling scheme best represents the poll trend?
Q4: Which political party is more predictable?

5.1   Experimental Setup

The experiments were conducted on 3 cross-validation datasets. Each contains 20 multivariate time-series used as training data, and 10 multivariate instances used as test data, after removing the 8 multivariate time-series that contain missing data. Internal cross-validation was conducted on the training sets in order to acquire the optimal hyper-parameters of the LTS model for shapelet extraction. These hyper-parameters were then used to train the entire training set. All training datasets were balanced in order to nullify bias in the learned model. For experiments related to individual features, the setting is same.

Evaluation Metric. Three evaluation measures were used in the conducted experiments: *sensitivity*, which refers to poll score increment detection rate; *specificity*, which refers to poll score decrement detection rate and *accuracy* which combines the sensitivity and specificity scores.

Baselines. In order to highlight the effectiveness of the proposed model, it is compared with multiple baselines. WST is compared to attribute-based models where singular values are obtained to represent the time-series in a given time period. The individual time points from univariate time-series related to a single feature are summed together within the five-day time period, and is represented as an attribute in both Logistic Regression and SVM. For features involving averages, values are adjusted using the number of positive or negative tweets and users accordingly. All attributes were then normalized. Comparisons with KNN are made to show the advantage of the shapelet-based classification over naive time-series classification methods. The proposed method is also compared against the commonly used majority vote scheme (MLS).

5.2   Performance of Multivariate Time-Series Classification

Examined here are the predictive power of two multivariate models, Majority-vote Learning Shapelets (MLS) and Weighted Shapelet Transformation (WST). Figure 4 shows that WST produces better results than MLS, with an accuracy of

70 %. This increased accuracy is due to WST learning weights for shapelets and considering univariate time-series related to different features are not equally important. MLS, which treats univariate time-series related to different features independently and equally important, has an average accuracy of 60 %. This gives evidence that not all features have similar predictive performance, which will be further discussed in Sect. 5.4.

Moreover, WST outperforms univariate time-series models with different features (see Table 2, Sect. 5.4). This leads the conclusion that making use of all features trumps focusing on individual features.

5.3   Time-Series vs. Attribute-Based Models

In this section, two multivariate time-series models, MLS and WST, are compared with two traditional attribute-based classification models, LR and SVM. Figure 4 clearly shows that on average both WST and MLS outperform both LG and SVM, and WST produces 20 % higher accuracy on average. This provides evidence that by utilizing the temporal nature of tweets, the time-series classification model produces better prediction results than traditional attribute-based models.

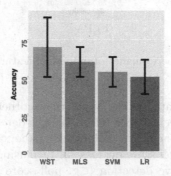

Fig. 4. Accuracy obtained from multivariate time-series models (WST, MLS) and attribute based models (SVM, LR).

5.4   Characterization of Individual Features

Next is to assess the predictive performance of each individual feature. The univariate time-series model, Learning Time-series Shapelet model (LTS) [4], was applied to each individual feature to perform prediction. For comparison, the baseline K-Nearest Neighbors (KNN) was applied, which only considers the Euclidean difference between time-series.

Features are analyzed in terms of sensitivity, specificity (Fig. 5) and accuracy (Table 2). Ideally, a method with perfect sensitivity and specificity would find

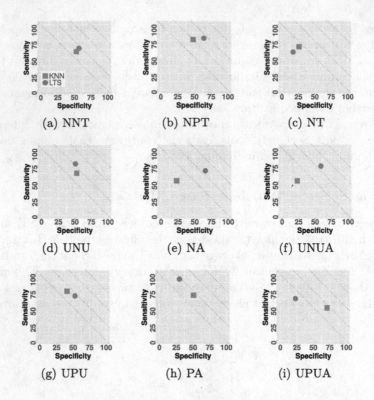

Fig. 5. The prediction performance of individual features defined in Sect. 3.2.

its corresponding icon located in the upper right corner. This region represents both a sensitivity and specificity of 100 %, perfect detection for both poll score increases and decreases. Increasing values along the x-axis indicates high true negative detection (e.g. poll score decreases). Increasing values along the y-axis indicates high true positive detection (e.g. poll score increases). Any method which finds its icon atop the main diagonal contains at least one, sensitivity or specificity, performing above random results.

LTS, indicated by a red circle, outperforms the baseline in most cases. Interesting patterns can be observed among these features. (1) In regards to the features dealing with tweet-volume, individual volume for "positive" and "negative" tweets are more accurate than the overall volume. (2) Features dealing with "negative" sentiment tend to be consistent predictors for poll increase and decrease. (3) The performance of features dealing with "positive" sentiment have more fluctuation. For example, both NPT and PA produce high accuracy, accuracies of 69.99 % and 63.33 % respectively (see Table 2), while UPUA falls just below random results. One possible justification for this is that Twitter users who post favorable information about a candidate likely have a vested interest in that candidate. Furthermore, the polls being examined in this dataset were from primary elections. It is possible that a significant portion of negative things being posted about a candidate came from users across political party lines. While their voice is heard and potentially influential, their influence on primary polls is likely to be more limited, as primary polls tend to focus on users from within the party.

5.5 Different Labeling Scheme Comparison

In this section, the labeling scheme *max* is compared with the baseline labeling scheme *count*. In Table 2, the first nine rows compare the results across all nine features, and the last two lines compare the results of all features, utilizing both labeling schemes.

For all attributes, labels generated using the *max* scheme outlined in Sect. 3.4 provide significantly more accurate predictions, with some accuracies improving by up to 30 %. The reason for this increased accuracy is the *max* scheme being more representative of the actual changes that are occuring in the poll score. For example, using the *count* labeling scheme, three small downward trends in poll scores would outweigh one very large shift upwards. The *max* scheme, by using the differences that occur between the poll scores, takes into consideration the magnitude of the overall change that occured. As per these results, all other experiments in this paper make use of the labels generated from the *max* scheme.

The bottom two rows in Table 2 show the difference when using the *max* labeling scheme versus the *count* labeling scheme in the multivariate approaches discussed in this paper. The difference between the results from two labeling schemes is large. This shows that the performance of the MLS model remains highly dependant on the performance of the univariate time-series model on each feature. This is overcome by learning and utilizing the weights of shapelets.

Table 2. Mean accuracy (± standard deviation) obtained from individual features and all features with different labeling schemes

Features	*Max*	*Count*
NumNeg	**56.6 ± 9.4**	29.1 ± 15.4
NumPos	**63.3 ± 12.4**	37.5 ± 20.4
NumTweets	**30.0 ± 8.1**	45.8 ± 15.5
UniqNegUsers	**56.6 ± 4.7**	29.1 ± 5.8
NegAvg	**56.6 ± 9.4**	16.6 ± 11.7
UniqNegUserAvg	**56.6 ± 9.4**	20.8 ± 15.5
UniqPosUsers	**60.0 ± 14.1**	33.3 ± 5.8
PosAvg	**69.9 ± 8.1**	33.3 ± 11.7
UniqPosUserAvg	**40.0 ± 8.1**	25.0 ± 10.2
All features (MLS)	**60.0 ± 10.0**	25.0 ± 12.5
All features (WST)	**70.0 ± 20.0**	45.8 ± 7.2

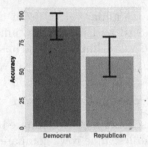

Fig. 6. Mean accuracy (± standard deviation) comparison between political parties using WST.

5.6   Characterization w.r.t. Political Party

Next we compare the prediction performance between the two primary political parties, Democrat and Republican. The model was trained and tested under two scenarios - Examining only Democrats, and examining only Republicans. The parameters of the underlying LTS model are fixed for shapelet extraction in both scenarios as the data is unbalanced within the individual parties. The accuracy for Democrat-party predictions were significantly higher.

Reflecting the election itself, there exists significant differences across political party lines. Utilizing WST the prediction accuracy is 26 % higher for Democrats. This result is not entirely shocking given the nature of the presidential race. The Democrat party tends to have a younger demographic that is more likely to take their discussion to Twitter. Furthermore, the dynamic of the election cycle was far more reactive on the Democratic side where both candidates were relatively close to one another, while the Republican race was much more one-sided in terms of polls.

While the results demonstrating the differences in predictive power between the two parties is very interesting, they should be taken with caution. One possible explanation is the difference in general patterns of poll scores between the Democratic candidates and Republican candidates. Donald Trump and Ted Cruz, both Republican, generally increased their poll score consistently across time. That is, the number of positive labels is much higher than the number of negative labels (see Table 1). Bernie Sanders and Hillary Clinton of the Democratic party experienced a much more dynamic race for the presidency in terms of poll scores. Both candidates experienced many more increases and decreases individually. Furthermore, Republican candidates on the whole had 1.8 times as many tweets as Democratic candidates as shown in Table 1, but were also less evenly distributed (Fig. 6).

6   Conclusion

In this paper, the primary research question was to address if a candidate's poll score trend could accurately be classified as increasing or decreasing using only Twitter. The temporal nature of tweets was considered in this work. Nine different features were used to characterize public opinion (both positive and negative) and were examined temporally. These features were then used in two multivariate time-series classification models, MLS and WST. Over 12 million tweets were analyzed using these models to provide the answers to this question.

From our extensive experimental results, we conclude that: (1) Our proposed approach, WST, produces higher accuracy than the MLS model. (2) Time-series based classification models outperform traditional attribute based classification models. (3) Using distance-based metric, *max*, for creating labels outperforms the simple *count* scheme, and (4) There exists a difference between the predictability across political party lines on social media. With an accuracy of 70 % when using WST, Twitter can serve as a substitute to the time-consuming polling options that are traditionally used to gather information on public opinion.

Future work includes expanding the use of time-series classification in social media. Optimal combinations of parameters will be considered, rather than the one-or-all approach currently used. Additional factors present in social media will also be considered, such as favorites, shares, whether images were included in the post, and whether URLs were shared within the tweet.

Acknowledgments. This research was supported in part by NSF BIGDATA grant 14476570 and ONR grant N00014-15-1-2729.

Appendix

Learning Time-Series Classification Model (LTS)

LTS [4] is one of the state-of-the-art univariate time-series classification models. The method discovers short time-series sub-sequences known as shapelets [17], which are local discriminative patterns (or sub-sequences) that can be used to characterize the target class, for determining the time-series class membership. In the LTS model, shapelets are learned jointly with a linear classifier rather than searching over all possible time-series segments. More specifically, the algorithm jointly learns the weights of the classifier hyper-plane as well as the generalized shapelets.

A shapelet of length W is a sub-sequence of an instance of the time-series. There can be at most $L - W + 1$ sub-sequences, and each can be represented as $\{f^q_{i,j}, ..., f^q_{i,j+W-1}\}$. K shapelets are initialized using K-Means centroid of all segments.

Equation 4 represents a linear model, where $M_{i,k}$ is the minimum distance between the i-th series in T^q and the k-th shapelet S^q_k.

$$\hat{Y}^q_i = \beta_0 + \sum_{k=1}^{K} M_{i,k}\beta_k \qquad \forall i \in \{1, ..., I\} \qquad (4)$$

The minimum distance $M_{i,k}$ is the predictor in this framework for shapelet learning and can be defined by a soft-minimum function:

$$M_{i,k} = \frac{\sum D_{i,k,j} e^{\alpha D_{i,k,j}}}{\sum e^{\alpha D_{i,k,j'}}} \tag{5}$$

where $D_{i,k,j}$ is defined as the distance between the jth segment of series i and the kth shapelet given by the formula

$$D_{i,k,j} = \frac{1}{W} \sum_{w=1}^{W} (T_{i,j+w-1}^q - S_{k,w}^q)^2 \tag{6}$$

Equation 7 shows the regularized objective function, composed of a logistic loss defined by Eq. 8 and the regularization terms.

$$argmin_{S,\beta} F(S,W) = argmin_{S,\beta} \sum_{i=1}^{I} \mathcal{L}(Y_i^q, \hat{Y}_i^q) + \lambda_\beta ||\beta||^2 \tag{7}$$

$$\mathcal{L}(Y_i^q, \hat{Y}_i^q) = -Y_i^q \, ln(\sigma(\hat{Y}_i^q)) - (1 - Y_i^q) ln(1 - \sigma(\hat{Y}_i^q)) \tag{8}$$

Equation 7 is optimized using a stochastic gradient descent algorithm. The weights β and the shapelet S^q are jointly learned to minimize the objective function. Once the model is learned, classifying an unknown instance is simply computing $\hat{Y}_t^q$ for the t-th test instance of the q-th feature and determining the class label via Eq. 9

$$\hat{Y}_t^q \leftarrow argmax_{c \in \{1,-1\}} \sigma(\hat{Y}_{t,c}^q), \tag{9}$$

where $\sigma(\cdot)$ denotes the sigmoid function.

For more details about individual gradient computation of the objective function, the reader is referred to [4].

References

1. Bermingham, A., Smeaton, A.F.: On using Twitter to monitor political sentiment and predict election results. In: Sentiment Analysis where AI meets Psychology (SAAIP), p. 2 (2011)
2. Gayo-Avello, D.: A meta-analysis of state-of-the-art electoral prediction from Twitter data. Social Science Computer Review, pp. 649–679 (2013)
3. Ghalwash, M., Radosavljevic, V., Obradovic, Z.: Utilizing temporal patterns for estimating uncertainty in interpretable early decision making. In: Proceedings of the ACM SIGKDD International Conference on Knowledge Discovery and Data Mining, pp. 402–411 (2014)
4. Grabocka, J., Schilling, N., Wistuba, M., Schmidt-Thieme, L.: Learning time-series shapelets. In: Proceedings of the 20th ACM SIGKDD International Conference on Knowledge Discovery and Data Mining, KDD 2014, pp. 392–401. ACM (2014)
5. Graham, T., Jackson, D., Broersma, M.: New platform, old habits? Candidates use of Twitter during the 2010 British and Dutch general election campaigns. New Media Soc. **18**(5), 765–783 (2016)

6. Hills, J., Lines, J., Baranauskas, E., Mapp, J., Bagnall, A.: Classification of time series by shapelet transformation. Data Min. Knowl. Disc. **28**(4), 851–881 (2014)
7. Larsson, A.O., Moe, H.: Studying political microblogging: Twitter users in the 2010 Swedish election campaign. New Media Soc. **14**, 729–747 (2012)
8. Mueen, A., Keogh, E., Young, N.: Logical-shapelets: an expressive primitive for time series classification. In: Proceedings of the 17th ACM SIGKDD International Conference on Knowledge Discovery and Data Mining, KDD 2011, pp. 1154–1162 (2011)
9. O'Connor, B., Balasubramanyan, R., Routledge, B.R., Smith, N.A.: From Tweets to polls: linking text sentiment to public opinion time series. ICWSM **11**(122–129), 1–2 (2010)
10. Patri, O.P., Sharma, A.B., Chen, H., Jiang, G., Panangadan, A.V., Prasanna, V.K.: Extracting discriminative shapelets from heterogeneous sensor data. In: 2014 IEEE International Conference on Big Data, Big Data 2014, Washington, DC, USA, 27–30 October 2014, pp. 1095–1104 (2014)
11. Roychoudhury, S., Ghalwash, M.F., Obradovic, Z.: False alarm suppression in early prediction of cardiac arrhythmia. In: 2015 IEEE 15th International Conference on Bioinformatics and Bioengineering (BIBE), pp. 1–6 (2015)
12. Sang, E.T.K., Bos, J.: Predicting the 2011 Dutch senate election results with Twitter. In: Proceedings of the Workshop on Semantic Analysis in Social Media, pp. 53–60. Association for Computational Linguistics (2012)
13. Shi, L., Agarwal, N., Agrawal, A., Garg, R., Spoelstra, J.: Predicting us primary elections with Twitter (2012)
14. Thelwall, M., Buckley, K., Paltoglou, G.: Sentiment strength detection for the social web. J. Am. Soc. Inform. Sci. Technol. **63**(1), 163–173 (2012)
15. Tumasjan, A., Sprenger, T.O., Sandner, P.G., Welpe, I.M.: Predicting elections with Twitter: what 140 characters reveal about political sentiment. ICWSM **10**, 178–185 (2010)
16. Xing, Z., Pei, J., Yu, P.S., Wang, K.: Extracting interpretable features for early classification on time series. In: SIAM International Conference on Data Mining, pp. 247–258 (2011)
17. Ye, L., Keogh, E.: Time series shapelets: a new primitive for data mining. In: Proceedings of the 15th ACM SIGKDD International Conference on Knowledge Discovery and Data Mining, KDD 2009, pp. 947–956. ACM (2009)

Inferring Population Preferences via Mixtures of Spatial Voting Models

Alison Nahm[1(⊠)], Alex Pentland[2], and Peter Krafft[2]

[1] Harvard University, Cambridge, USA
anahm@post.harvard.edu
[2] Massachusetts Institute of Technology, Cambridge, USA
{pentland,pkrafft}@mit.edu

Abstract. Understanding political phenomena requires measuring the political preferences of society. We introduce a model based on mixtures of spatial voting models that infers the underlying distribution of political preferences of voters with only voting records of the population and political positions of candidates in an election. Beyond offering a cost-effective alternative to surveys, this method projects the political preferences of voters and candidates into a shared latent preference space. This projection allows us to directly compare the preferences of the two groups, which is desirable for political science but difficult with traditional survey methods. After validating the aggregated-level inferences of this model against results of related work and on simple prediction tasks, we apply the model to better understand the phenomenon of political polarization in the Texas, New York, and Ohio electorates. Taken at face value, inferences drawn from our model indicate that the electorates in these states may be less bimodal than the distribution of candidates, but that the electorates are comparatively more extreme in their variance. We conclude with a discussion of limitations of our method and potential future directions for research.

Keywords: Probabilistic generative models · Political polarization · Demographic inference · Ideal point models · Computational social science

1 Introduction

Within a representative democracy, understanding the extent to which elected officials represent their constituencies is critical to evaluating the efficiency of the political system. Here we focus on the political system in the United States, where surveys typically evaluate the preferences of the electorate. Despite their widespread use, surveys can be costly, time consuming to execute, and often lack broad geographical coverage. Fortunately, there is an alternative source of readily available data about the preferences of the electorate—votes cast in elections.

The key challenge of using voting data to infer localized distributions of political preferences is the coarseness of the data. Consider inferring the distribution

E. Spiro and Y.-Y. Ahn (Eds.): SocInfo 2016, Part I, LNCS 10046, pp. 290–311, 2016.
DOI: 10.1007/978-3-319-47880-7_18

of political preferences of voters from votes cast in a two-candidate election. Since there are only two data points from the election (and a constraint that the two points sum to the total voting population size), inferring the distribution of preferences from these vote shares appears underdetermined.

In this paper, we introduce a model-based machine learning method to measure the political preferences of voters at a fine level of geographical granularity. To solve the underdetermination problem, we introduce a Bayesian mixture model that pools data from similar election outcomes of different geographical voting units (precincts). The method connects the distribution of preferences within each precinct to voting outcomes using a spatial voting model—a standard rational voting model from the political science literature [11,12]. Our model utilizes vote share data and a preprocessed form of campaign finance data to infer distributions of political preferences with potentially better coverage and lower cost than traditional survey methods [1,13,14,19,21].

An additional benefit of our method over surveys is that the inferred political preferences of voters are represented on the same scale as those of candidates. This is important for social science applications involving the comparison of politicians and the electorate [5]. To demonstrate the potential utility of our method and of related future work in this area, we apply our technique to understand the extent of political polarization in the Texas, New York, and Ohio electorates in comparison to that of the political candidates. While it is well-known that elected officials are highly polarized in their political positions, the political science community has not reached consensus as to whether the preferences of voters mirror this elite polarization or are comparatively moderate [1,13]. Using congressional election data for the states from the 2006, 2008, and 2010 election cycles, we find varying answers to the question depending on the polarization metric we use. We find that the distribution of the political preferences of voters is likely more extreme than that of candidates in terms of variance, while less extreme than that of candidates in terms of bimodality.

In the remainder of this paper, we begin with a discussion of related works. We then provide an overview of our novel probabilistic generative model of voting behavior. We validate this model with comparisons to results of related work and with simple prediction tasks. We then apply our model to better understand political polarization in Texas, New York, and Ohio. Finally, we conclude with a discussion of the limitations of this method and suggestions for future work.

1.1 Related Work

There has been recent work in quantitative political science that is closely related to our work. For instance, researchers recently developed a technique for estimating the preferences of the electorate and elected officials from Twitter data using a probabilistic generative network model related to the spatial voting model we use [5]. Some political scientists have used ideal point models, which are closely related to spatial voting models, to infer distributions of voter preferences from fine-grained voter data [16,22]. Unlike our work, these previous works using voting data relied on individual-level voting data, which is difficult to obtain. Other political scientists have developed meta-analysis-like methods for aggregating survey results to improve accuracy and representativeness [27], but this work still suffers from the limitation of low coverage of survey data due to collection difficulties. Thus, the methods can only consider a coarser level of geographical granularity.

Within the computer science field, our work falls closest to a growing line of research dedicated to developing novel machine learning models for computational social science. Machine learning researchers in this area have not yet addressed the exact problem we study in our work, to the best of our knowledge. However, they have been interested in similar problems and related classes of models (e.g. [2, 15, 17, 18]). More tangentially, a large body of work in computer science has been dedicated to drawing inferences from public observational data. Some researchers have suggested using social media data to better understand public opinion [26], while others have developed models based on inconsistent user behavior to infer their implicit preferences [10].

2   Model

We first discuss mathematical theories of voting behavior that inform our novel model. Then, we describe our model for inferring political preferences of voters.

2.1   Spatial Voting Models

Our model generalizes the widely used "spatial" or "Downsian" voting model, which is a standard model in political science of rational voting and turnout behavior in two-candidate majority vote elections [11,12]. The spatial voting model defines each voter and each political candidate as points in a one-dimensional policy space. The model defines the utility to a voter of a specific candidate winning as the Euclidean distance between their two points. Assuming the election involves exactly two candidates, the spatial voting model predicts that voters will select for their votes the candidates closest to them according to Euclidean distance in the single-dimensional policy space.

2.2   Mixtures of Spatial Voting Models

Our statistical model consists of a generative process for the vote shares of candidates in an election. Each precinct i with N_i total voters is associated with an election of exactly two candidates, c_{0i} and c_{1i}. In line with the spatial voting model, we assume that voters $v_{ji}, j \in \{1, \ldots, N_i\}$ and candidates $c_{ki}, k \in \{0, 1\}$, have positions in the same one-dimensional latent space, which we consider their *political preferences*. We assume the candidate positions are known to all election participants. Like other spatial voting models, we assume that each voter j of precinct i votes for the closest candidate in the one-dimensional latent policy space. In other words, voter j in precinct i votes for candidate 0 in precinct i if $|v_{ji} - c_{0i}| < |v_{ji} - c_{1i}|$.

We assume that each precinct is associated with a particular distribution over political preferences, which determines the preferences of the voters in that precinct. However, it is problematic to assume that these distributions are all distinct from and independent of each other. In this case we are limited to using only one data point per precinct to infer a distinct distribution. To solve this issue we use a mixture model to pool data across precincts. We assume that certain subsets, or *clusters*, of precincts share the same distribution of preferences. These assignment of precincts to clusters is determined dynamically during inference according to similarity in observed voting patterns. This modeling assumption seems reasonable given that it is likely neighboring precincts will have similar distributions of preferences.

The expected proportion of precincts that will be assigned to each of K clusters is given a Dirichlet prior, $\theta \sim Dirichlet(1)$ and $|\theta| = K$. The assignment of each precinct i to a particular cluster is then drawn as in a standard mixture model, $x_i \sim \theta$. The position of each voter j in each precinct i is drawn according to a component distribution associated with the cluster assignment of that precinct. The parameters of these component distributions are given weakly informative priors. The number of votes n_{ki} received by candidate k in precinct i are then given deterministically by the spatial voting model specified above. Mathematically, $n_{0i} = \sum_{j=1}^{N_i} \mathbb{1}(|v_{ji} - c_{0i}| < |v_{ji} - c_{1i}|)$, where $\mathbb{1}$ is an indicator function, and $n_{1i} = N_i - n_{0i}$.

We treat candidate positions as fixed and given since we have data on these values, but we treat voter positions, precinct assignments, and cluster distribution parameters as unknown. After marginalizing over voter positions, then conditioning on direct estimates of candidate positions and on observed vote shares per candidate, we can use Bayesian inference to arrive at likely values for the remaining unknown parameters, thus estimating the distributions of voter preferences within each precinct.

3 Data

For our empirical analysis, we use three main sources of data.

Precinct-Level Voting Results. Precincts are the finest granularity of publicly accessible aggregated vote shares. We examine congressional elections of Texas, New York, and Ohio [3]. In these cycles, Republicans won 65 % of the Texas elections, and Democrats won 80 % of the New York elections. We also analyze Ohio to test the ability of our method to generalize to more extreme voter distributions, as Ohio is commonly labeled by political scientists as a "swing state". We consider the election cycles 2006, 2008, and 2010 because they all depend on the same district geographic boundaries set by the 2000 U.S. Census. We omit the 2002 and 2004 election cycles to focus on recent elections. Future work could analyze longer periods.

Candidate CFscores. We incorporate quantitative estimates of the political preferences of candidates called campaign-finance scores (CFscores) [7]. CFscores are one-dimensional quantitative estimates of the political ideology of political candidates, with lower values indicating more liberal ideologies and higher values more conservative. CFscores are recognized as effective estimates of candidate ideology when estimates for unelected candidates are needed (in contrast to DW-NOMINATE scores which only exist for winning candidates [25]).

Geographic Precinct Boundaries. We link the above-mentioned data sets with the congressional candidates running in each precinct election. We assign any precincts to the district whose geographic center fall within the specific congressional district boundary lines using the Geospatial Data Abstraction Library (GDAL/OGR) package [23,28].

4   Inference

The goal of our inference procedure is to determine likely precinct assignments (x) and likely parameters of the K cluster distributions in the model described in Sect. 2. These estimates allow us to characterize the distribution of voter preferences of each precinct.

For more efficient inference, we first integrate out the voter positions (v) and the precinct assignments (x). All component distributions we consider allow the voter positions to be integrated out analytically. This yields the following posterior distribution:

$$P(\boldsymbol{\eta}, \boldsymbol{\theta} \mid \boldsymbol{v}) \tag{1}$$

$$\propto P(\boldsymbol{\theta})P(\boldsymbol{\eta}) \sum_{\boldsymbol{x}} P(\boldsymbol{x} \mid \boldsymbol{\theta}) \int_{\boldsymbol{y}} P(\boldsymbol{y} \mid \boldsymbol{x}, \boldsymbol{\eta}) P(\boldsymbol{v} \mid \boldsymbol{\eta}, \boldsymbol{y}) \tag{2}$$

$$\propto P(\boldsymbol{\eta}) \prod_{i=1}^{M} \left[\sum_{x_i=1}^{K} \theta_{x_i} (\Phi_{i,x_i})^{n_{0i}} (1 - \Phi_{i,x_i})^{n_{1i}} \right] \tag{3}$$

where $\boldsymbol{\eta}$ is the distribution parameters of each cluster, and Φ_{i,x_i} is the cumulative distribution of the component distribution of precinct i after integrating out $\boldsymbol{y}$, in other words $\Phi_{i,x_i} = P\left(y < \frac{c_{1i}-c_{0i}}{2} \mid \eta_{x_i}\right)$. For prior distributions over the cluster parameters, when using Normal component distributions, we use a Normal prior for μ with a mean of 0 and a variance of 100, and for σ we use an Inverse Gamma distribution with scale and shape parameters both set to 1.

We then use a Metropolis-Hasting Markov Chain Monte Carlo (MCMC) algorithm to arrive at likely values for the component distribution parameters and the mixture proportion $(\boldsymbol{\theta})^1$. To generate the results presented in this paper, we ran four independent MCMC chains from randomly generated initialized values. We then selected the set of parameter values from all generated sets that

---

[1] Code and data for analyses are available at https://github.com/anahm/inferring-population-preferences.

yield the highest posterior. We infer parameter values separately for the data of each state and election cycle combination described in Sect. 3. After inferring the parameters of the mixture distribution, we infer all precinct assignments to clusters, x, by selecting the maximum a posteriori (MAP) assignment variable for each precinct.

Our model-based method provides a better approach than analyzing vote shares because we can derive an overall distribution of preferences per precinct rather than a single point. This approach also has benefits even looking at coarser geographic granularity as well. We ultimately aggregate our inferences to district or state level for validation purposes, and inferring precinct-level distributions opens the possibility for the distribution at a less-granular level to be a complicated combination of precinct-level distributions.

5   Validation

To assess validity, we compare summary statistics of the distributions we infer with corresponding values from related works. In addition, we compare the predictive performance of our model with prediction methods based on empirical data and results of related works. In this section, we present results assuming the number of clusters, K, is 4 and underlying Normal component distributions, but we reach similar results when we vary the value of K and the component distribution type, which can be seen in Sect. B. These validation methods are meant to qualitatively and quantitatively assess the "face validity" of our proposed model.

5.1   Comparison with Related Works

We compare our results with the results of two survey-based methods for estimating district-level political preferences. This comparison would ensure that our inferred distributions are qualitatively reasonable from the perspective of the prior related work. A district is a coarser granularity geographical unit than a precinct, but precinct-level surveys are rarely implemented due to high costs. To compare, we obtain a single-point estimate of each congressional district preference in Texas, New York, and Ohio from our inferred precinct-level distributions.

To compute single-point estimates of district-level voter preferences, we use our model's assumption that each precinct has the same parameters as the inferred parameters of its assigned cluster. The district-level estimates are averages of the precinct-level inferences of precincts in the same district weighted by the population of those precincts.

Comparison with Raw Survey Results. We first compare with the Cooperative Congressional Election Study (CCES) [4]. The CCES surveys over 50,000 Americans every election cycle. Many political scientists use the CCES to understand the American public opinion. Moreover, the CCES respondents report their congressional districts, which yields more fine-grained data than most other national surveys [4]. We compare our results with the survey responses to two questions. The first question asks:

Thinking about politics these days, how would you describe your own political viewpoint? (Very Liberal, Liberal, Moderate, Conservative, Very Conservative, Not sure)

As shown in Fig. 1, we find a significant positive correlation between our district-level point estimates and the responses to this question. The correlation level of the results of our method and reported survey values is 0.3127 with a p-value less than 0.01.

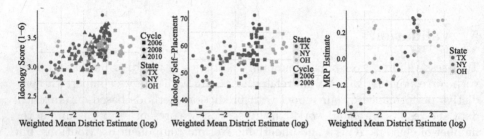

Fig. 1. (Left) Inferred district-level voter preferences compared with CCES question of self-reported ideologies on a discrete scale. (Center) Inferred preferences compared with CCES self-reported ideologies on a continuous scale from 0 to 100. (Right) District-level inferences of a decade compared with MRP estimates [27]. In all cases, the inferred district-level voter preferences are weighted mean district estimates transformed from x to $sign(x) log(|x| + 1)$.

The second CCES question asks survey respondents to score their political ideology on a continuous scale. The question is phrased:

One way that people talk about politics in the United States is in terms of left, right, and center, or liberal, conservative, and moderate. We would like to know how you view the parties and candidates using these terms. The scale below represents the ideological spectrum from very liberal (0) to very conservative (100). The most centrist American is exactly at the middle (50). Where would you place yourself?

This question was only used in the 2006 and 2008 CCES surveys [4]. As shown in Fig. 1, we find a significant positive correlation between our estimates and the responses to this CCES question. The correlation level of the log of the estimates of our method and reported survey values is 0.2535 with a p-value less than 0.01. Using a monotonic transformation is acceptable since the answers to survey questions and our inferred preferences are not necessarily on comparable scales.

Comparison with Aggregated Survey Results. In addition to CCES results, we validate our results against district-level ideological scores developed by two political scientists, Chris Tausanovitch and Christopher Warshaw. They use disaggregation and multilevel regression with post-stratification (MRP) on survey data from 2000 to 2010 to estimate mean policy preferences of congressional districts [27]. Their work is one of the recent related works understanding preferences, and they analyze election cycles in a similar time frame to this paper [27]. Further, these ideological scores might be more representative of the U.S. population because the work's inference methods account for possible sampling bias.

One caveat of the work by Tausanovitch and Warshaw is that their estimates span a decade of voter behavior rather than a single election. To ensure comparison across consistent measures, we aggregate the district-level results of our model across the three election cycles into one district-level estimate spanning 2006–2010. As shown in Fig. 1, we find a significant positive correlation of 0.4149 between the log of the results of our method and their results with a p-value less than 0.01. By contrast, the MRP compared to the two CCES responses has correlations of 0.6543 and 0.7189.

The correlations between our results and the results of the survey-based methods suggest that our method can infer similar qualitative distributions to those of prior works. Further, our method can not only recover district-level mean political preferences, but also examine more granularity, precinct-level preferences in a shorter time period.

5.2 Predictive Power

As a second method of validation, we analyze the predictive power of our model. Specifically, we derive values from a *comparison election* to predict the Democratic vote share of a separate *target election*. Here a target election is either an election occurring in a later cycle or for a different government position occurring in the same cycle.

While prediction tasks are sometimes used to argue that a model has the best predictive value compared to alternative models, that is not the goal of this section. We are aware that the predictive power could be improved by incorporating more types of data. Rather, the purpose of these prediction tasks is to demonstrate that our model achieves predictive performance comparable to reasonable alternatives. Our method has additional benefits in terms over the comparison methods, so these prediction tasks are meant to lend quantitative face validity to our model.

Methods of Prediction. We compute the expected vote share of the Democratic candidate (assigned to be candidate 0) of the target election by assuming voters follow the spatial voting model and that voter preferences are identical in the comparison and target elections. The predicted vote share of each precinct i is given by Φ_{i,x_i}, the cumulative of the inferred distribution of voter preferences at the midpoint between the CFscores of the two candidates running in the target election. We aggregate vote shares of the candidate in all precincts of the same district to facilitate comparison with less fine-grained data sources.

We compare the predictive power of values yielded by our model with three alternatives: raw vote shares of previous elections, survey data, and MRP estimates. The method using raw vote shares assumes the candidates of the same political party receive the same proportion of votes in the comparison and the target election. For instance, this naïve baseline predicts the vote share of the Democratic candidate in 2010 is equivalent to the vote share of the Democratic candidate in the previous election in 2008. We consider this the baseline prediction model.

The survey prediction method uses CCES responses to a question on political party affiliation as a proxy for votes for the Democratic candidate in the target election [4]. We approximate the percentage of Democrats as the number of reported Democrats and half the reported Independent or Other divided by the total number of responses. We assume the respondents who state Independent or Other divide equally between Democrat or Republican when faced with only those options.

We also develop a prediction method based on the MRP estimates developed by Tausanovitch and Warshaw [27]. This method, labeled in Fig. 2 as MRP Cross-Val, is a simple cross-validation leave-one-out prediction method using MRP ideological scores to predict vote share. For the Cross-Val method, we obtain predictions for each district given the remaining districts and combine error terms into one mean squared error. The previously described prediction methods predict all district preferences at once, so the methods only yield one error term, which is the mean squared error term.

Prediction Tasks. We examine two prediction tasks: next cycle and same-year. The next cycle prediction task defines target elections as congressional elections of the same district one election cycle (two years) after the comparison election cycle. In other words, we use point estimates of voter preferences from election cycle $t - 1$ to predict the results of election t.

The second prediction task, the same-year prediction task, defines target elections as elections for a different government position, a Senate seat, in the same election cycle as the comparison election. Although our method infers estimates using results of the same year as the election we are trying to predict, the target and comparison elections are for unrelated positions. We assume that voters consider their votes for different ballot items as in independent elections.

Prediction Task Results. As we can see in Fig. 2, in all but one case our model tends to do as well or better than the alternatives, which further validates our model. The MRP Cross-Val method is the most competitive alternative. However, the high performance of MRP Cross-Val is likely because the cross-validation method is optimized for prediction, whereas our model and the other alternatives are not.

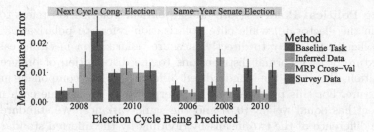

Fig. 2. Plot of the mean squared error of the actual and predicted vote share yielded by various prediction methods. Error bars are standard error of the mean.

6   Application

The results of our method distinguish themselves from the baseline methods discussed in Sect. 5 because they not only shed light on voter preferences, but also represent those preferences in the same latent policy space as known positions of political candidates. We can leverage the latter capability to assess the degree of political polarization in the electorate compared to candidates' political positions.

6.1   Background of Polarization

Previous Work. Popular media and political science communities have observed that the American political elite is becoming increasingly polarized over time, but much less work has drawn conclusions on *mass polarization* [24, 29]. Some political scientists hypothesize that the distribution of the political preferences of the U.S. electorate is unimodal and moderate compared to that of the political elite [13, 14]. Yet other evidence suggests increasing polarization in the American population [1]. The results of our method could add a valuable new perspective since we draw on data sources separate from the survey methods many of the opposing arguments used [1, 13, 14, 19, 21].

Among the related works that do not base their conclusions on survey data, Fiorina and Abrams find that most observations of mass polarization trends, such as differences in sociocultural attributes and world views, are not strong enough to make definitive claims about overall trends of mass polarization [13]. Their main critique of utilizing voting behavior as a proxy for political polarization is that the past actions and political preferences of candidates are not factored into the model. Our work addresses this by factoring in candidate CFscores in addition to voter behavior. Further, most related works only use one numeric metric, the difference of the means of subsets of the population, to measure mass polarization [21]. However, DiMaggio, et al. define mass polarization in terms of discrepancies between distributions, which can be described by more than a single number [9]. Some have analyzed polarization in a more holistic way, but their methods depend on survey data [19, 21].

Defining Political Polarization. Mass political polarization refers to polarization in the electorate, while elite polarization refers to polarization among elected officials. One way to directly measure polarization in either case is to fit a mixture of two Normal distributions to the distribution of preferences of a population, then take the standardized difference of the component means of that mixture. For this procedure to be interpretable, we assume each mixture component has equal weight (0.5) and the same variance. We standardize the absolute difference of the two means by dividing by the inferred standard deviation of each component. This unique "difference-of-means" metric is intended as a rough proxy for the probability mass missing from the center of a distribution of preferences. Related works often use a similar metric, but they compute means by aggregating survey responses or other point estimates of subsets of the population rather than fitting a mixture model to the population distribution.

We also measure political polarization in two ways previously used in the political science literature: dispersion and bimodality [9]. Dispersion represents the extent to which more varied opinions in the population increase the difficulty for a "centrist political consensus" to exist in the population [9]. DiMaggio, et al. suggest measuring dispersion with the standard deviation of the distribution of political preferences. An increase in the standard deviation signifies that voters have more extreme political preferences and less moderate preferences in the middle of the distribution. Bimodality represents the level of separate opinions of different groups that can lead to a higher chance of social conflict. Bimodality can be measured with the kurtosis of the distribution. The formal definitions of these quantities are given in the appendix in Sect. A.

6.2   Analyzing Polarization

We apply our results to the question of trends in mass polarization for Texas, New York, and Ohio. We generate state-level distributions of voter preferences similar to the method described in 5.1. The resulting inferred distributions are shown in Fig. 3

Mass Versus Elite Polarization. Using the state-level estimates, we compute the polarization metrics as described in Sect. 6.1 for the inferred voter distributions and the corresponding candidate distributions.

First, as seen in Fig. 3, the inferred distributions of voter preferences visually appear unimodal, even though our model does not make this assumption at the state level. This is supported quantitatively in Table 1, where the electorate has relatively consistently smaller difference-of-means polarization metric than the candidates. This suggests there is no mass polarization in terms of bimodality within each state. Second, we find in Table 1 that the standard deviation of the distribution of voter preferences is generally larger than that of the distribution of candidates. This consistent difference in dispersion suggests that voters often have more extreme preferences than the candidates running for office. Third, the kurtosis of the distribution of voter preferences is generally larger than that of

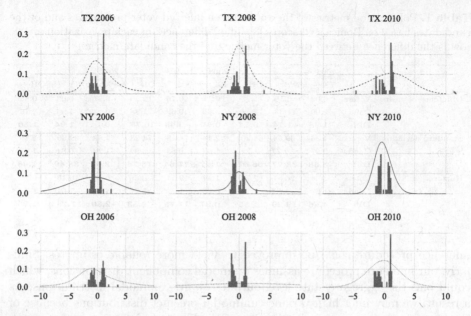

Fig. 3. Distributions of inferred voter preferences (represented by lines) and candidate preferences (histograms) based on data of the 2006, 2008 and 2010 congressional elections in Texas, New York, and Ohio.

the distribution of candidate preferences. Larger values of kurtosis indicate lower values of bimodality, so the consistent positive difference in kurtosis suggests that the voter distributions are unimodal. The results on kurtosis corroborate qualitative observation of our plots and the difference-of-means statistic.

Together these findings suggest that the distribution of political candidates may not be representative of the distribution of preferences of the electorate in the years and states we study because the extremes and the center of the voter populations are underrepresented. Further, these findings also suggest that the elites are more polarized than the electorate in terms of bimodality, but the electorate is more polarized than the elites in terms of dispersion. Inferred distributions under alternative assumed numbers of cluster components and under alternative mixture component distributions are shown in the appendix in Figs. 4 and 7. These robustness checks further support our dispersion finding but yield somewhat more mixed results on bimodality.

7   Discussion

Our methods have some limitations that could be addressed in future work. Firstly, biases inherent in our model could undermine our qualitative conclusions about polarization. Because the tails of the component distributions in our model are nearly non-identifiable, the posterior distributions in our model end up placing a large mass on preference distributions with high variances. The

Table 1. Polarization metrics of the voters given inferred voter preferences and of the candidates based on Bonica's CFscores [8]. The "difference" of each polarization metric row is the difference between the voter metric and the candidate metric.

		2006			2008			2010		
		TX	NY	OH	TX	NY	OH	TX	NY	OH
Difference-of-means	Voters	0.02	0.19	5.25	0.18	0.16	12.60	0.27	0.03	0.86
	Cand	2.00	1.33	2.11	0.36	0.62	1.88	0.13	1.96	2.16
	Diff	−1.98	−1.14	3.14	−0.17	−0.46	10.73	0.14	−1.94	−1.30
Standard deviation	Voters	4.36	39.54	68.20	2.14	78.46	74.23	3.88	39.30	2.75
	Cand	0.97	0.77	1.22	1.07	0.83	0.99	1.22	0.81	1.07
	Diff	3.38	38.77	66.97	1.07	77.64	73.24	2.66	38.49	1.68
Kurtosis	Voters	3.04	20.87	2.78	1.47	19.54	0.56	0.85	16.15	0.22
	Cand	−1.62	1.67	1.23	0.49	1.79	−1.77	3.24	−1.61	−1.73
	Diff	4.66	19.20	1.55	0.97	17.75	2.33	−2.39	17.76	1.95

inference procedure could be improved to yield more reliable estimates. Similarly, our inference procedure assumes unimodal component distributions, which could bias the aggregate state level output towards unimodal distributions. As a result, we may infer high-variance unimodal precinct distributions because of the biases in the component distributions. To address these concerns, we test the ability of our procedure to recover bimodal distributions at the state level with synthetic precinct-level data. We explicitly define bimodal precinct distributions in the synthetic data, and our inference procedure was still able to correctly recover bimodal district and state-level distributions under the assumption of unimodal component distributions. This indicates that our model is able to represent and recover bimodal preference distributions even if the model is making an incorrect assumption that individual precincts are unimodal. This result on synthetic data lends some confidence that our application results on real data are not artefactual. However, our inferences on real data still have some quirks. For example, the inferred distribution of the Ohio election in 2008 is implausibly wide. In explorations of ways to address this issue, we found that adding an informative prior on the variance helped to an extent.

Another limitation of our method is that we assume the observed vote shares of elections is tied to the underlying preference distributions. This allows the model to aggregate ideological preferences of individual voters within precincts. However, some political scientists believe voters tend to vote for candidates according to party affiliation rather than ideology [6]. While this is a large limitation to our approach, future work could extend our model by estimating the extent to which people vote along party lines as opposed to according to policy preferences. As it stands, the distributions of preferences we infer can be interpreted as the distributions of *expressed* preferences of voters, if not the distribution of actual preferences.

Finally, a more fundamental limitation is that we cannot validate the shape of the inferred precinct distributions with existing survey data, since we only have observations at the district level. Future work could create a test set through an in-depth survey of the distributions of preferences of particular precincts.

Validating the distributions we infer is critical to bringing our work to a level that would be useful to practitioners. At the moment we cannot tell to what extent the shape of the individual precinct distributions we infer is determined by the data versus biases from our model.

8   Conclusion

In this work, we present a model to address the problem of understanding public opinion in the context of voter preferences and political polarization. Our model builds on a long literature of spatial voting models. We use mixtures of spatial voting models to infer clusters of U.S. precincts that display similar voting patterns. Our model simultaneously infers preference distributions associated with those clusters utilizing data not only about voters, but also of candidates. These features allow us to analyze the distributions of voter preferences on their own and relative to distributions of candidate preferences. We infer voter preferences given precinct-level election results of three election cycles and three different states. We validate our inferences to the extent that we can using existing data. We validate against alternative measures of public opinion, as well as by comparing the predictive power of our model and alternative methods.

One extension of this work could adapt the model to account for elections with an uncontested candidate using similar ideas tried by Levendusky, et al. [20]. Our model could also be updated to include an offset accounting for the number of political parties in the election, which can address the bias of results based on a two-party system. Another direction is to explore other applications of our inference methods in the field of political science. For example, inferred precinct-level preference distributions could predict the effects of congressional redistricting, the process of assigning geographic boundary lines to congressional districts. Precincts are the building blocks for districts, so we could use precinct preference estimates of our model to predict the effects of redistricting proposals on the makeup of Congress.

Variations of our model could be applied more broadly beyond the scope of political science to understand distributions of preferences on other topics. For example, surveys are also used to understand consumer preferences on certain consumer products. A variation of our model could avoid the need for surveys or supplement surveys in this and other areas.

Acknowledgements. Special thanks to David Lazer for bringing our attention to Adam Bonica's work, to David Parkes for suggesting a reformulation of our model that enabled integrating out voter positions, and to Matt Blackwell for encouraging us to think more about validity and identifiability. This work was supported in part by the NSF GRFP under grant #1122374. Any opinions, findings, and conclusions or recommendations expressed in this material are those of the authors and do not necessarily reflect those of the sponsors.

A   Mathematical Definitions of DiMaggio's Polarization Metrics

Given the estimates of our model, we use the following analytical form of the standard deviation of a mixture model to measure political polarization in terms

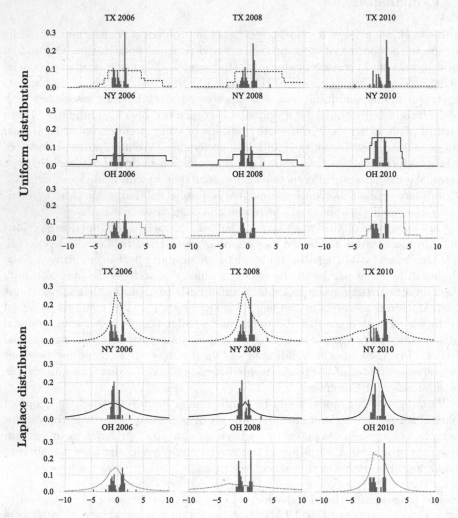

Fig. 4. Distributions of inferred voter preferences (represented by lines) under alternative assumed mixture component distributions, and candidate preferences (histograms) based on data of the 2006, 2008 and 2010 congressional elections in Texas, New York, and Ohio.

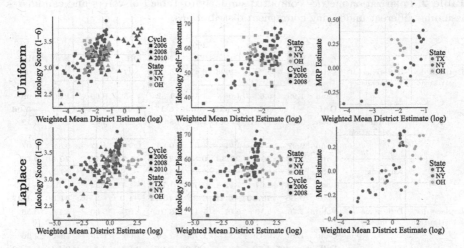

Fig. 5. In all of these cases, the inferred district-level voter preferences are based on the model varying its assumption of the underlying precinct distribution and are weighted mean district estimates transformed from x to $sign(x)\,log(|x| + 1)$. (Left) Inferences compared with CCES question of self-reported ideologies on a discrete scale. (Center) Compared with the CCES self-reported ideologies on a continuous scale from 0 to 100. (Right) District-level inferences of a full decade compared with MRP estimates [27].

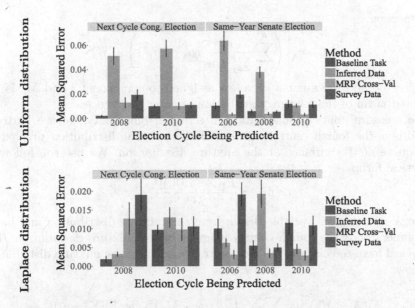

Fig. 6. Mean squared error of the actual and predicted vote share yielded by various prediction methods. Inferred data prediction method is based on our model assuming alternative underlying component distributions.

Table 2. Polarization metrics computed similarly to Table 1 of voters and candidates assuming different underlying component distributions.

			2006			2008			2010		
			TX	NY	OH	TX	NY	OH	TX	NY	OH
Uniform distribution	Difference-of-means	Voters	0.02	5.73	−0.26	0.47	6.69	3.27	0.26	3.01	0.02
		Cand	1.99	0.62	1.08	1.40	1.52	1.85	0.60	1.65	2.09
		Diff	−1.97	5.11	−0.25	−0.93	5.17	1.42	−0.34	1.37	−2.06
	Standard deviation	Voters	2.11	27.51	24.55	1.94	86.17	27.71	3.42	21.46	5.12
		Cand	0.97	0.77	1.22	1.07	0.83	0.99	1.22	0.81	1.07
		Diff	1.13	26.73	23.33	0.87	85.34	26.72	2.20	20.65	4.05
	Kurtosis	Voters	−0.77	42.17	−2.96	0.72	42.50	−2.99	0.38	39.39	−3.00
		Cand	−1.62	1.67	1.23	0.49	1.79	−1.77	3.24	−1.61	−1.73
		Diff	0.85	40.50	−4.19	0.23	40.71	−1.23	−2.86	41.00	−1.27
Laplace distribution	Difference-of-means	Voters	1.76	5.13	6.01	1.47	3.21	4.07	5.60	4.17	0.56
		Cand	1.80	0.41	1.84	1.73	1.15	2.00	1.79	1.64	2.06
		Diff	−0.04	4.72	4.17	−0.26	2.07	2.07	3.81	2.54	−1.50
	Standard deviation	Voters	19.25	61.64	14.51	10.64	86.42	26.79	193.24	53.56	1.67
		Cand	0.97	0.77	1.22	1.07	0.83	0.99	1.22	0.81	1.07
		Diff	18.28	60.87	13.29	9.57	85.60	25.80	192.03	52.74	0.60
	Kurtosis	Voters	−2.99	1.78	1.84	−2.48	1.49	2.34	−2.90	1.47	−1.42
		Cand	−1.62	1.67	1.23	0.49	1.79	−1.77	3.24	−1.61	−1.73
		Diff	−1.37	0.11	0.61	−2.97	−0.30	4.10	−6.14	3.08	0.30

of dispersion:

$$M_\sigma = \sqrt{\left(\sum_{i=1}^{K} \frac{n_i}{\sum_{j=1}^{K} n_j} (\mu_i{}^2 + \sigma_i{}^2) \right) - M_\mu} \tag{4}$$

where n_i is the total number of voters assigned to component i and M_μ is the weighted mean of the mixture distribution of voter preferences.

To measure political polarization in terms of bimodality, we use kurtosis. Kurtosis is the fourth central moment of the mixture distribution divided by the square of the variance of the mixture distribution. We use the following analytical form:

$$M_k = \frac{E[(X - M_\mu)]^4}{M_\sigma^4} \tag{5}$$

where X is a random variable drawn from the mixture distribution and hence the numerator is the fourth central moment of the mixture distribution. The analytical form to compute the z-th central moment of the mixture distribution is below.

$$E[(X - M_\mu)^z] = \sum_{i=1}^{K} \sum_{j=1}^{z} \binom{z}{j} (\mu_i - M_\mu)^{z-j} w_i \, E[(Y_i - \mu_i)^z] \tag{6}$$

where Y_i is a random variable drawn from component i of the mixture distribution, w_i is the weight of each component, and $E[(Y_i - \mu_i)^z]$ is the z-th central

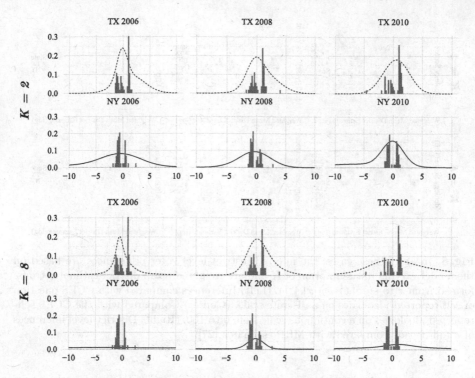

Fig. 7. Distributions of inferred voter preferences (represented by lines) under alternative assumed mixture component distributions, and candidate preferences (histograms) based on data of the 2006, 2008 and 2010 congressional elections in Texas (dotted line) and New York (solid line).

moment of the ith component distribution. In our analysis, we weight each component in the mixture distribution by the proportion of the population assigned that component.

B   Additional Results

In Sect. 5, we presented the results of our method assuming the underlying component distribution is Normal and the number of clusters (K) is 4. This section tests the robustness of these assumptions and presents our results when varying the underlying component distribution and the number of clusters.

B.1   Varying the Underlying Component Distribution

We test the inference procedure of our model not only assuming Normal component distributions, but also Uniform and Laplace component distributions. When assuming the distributions of voters follow a Laplace distribution, we use the same Normal prior defined for the mean of the Normal component for the

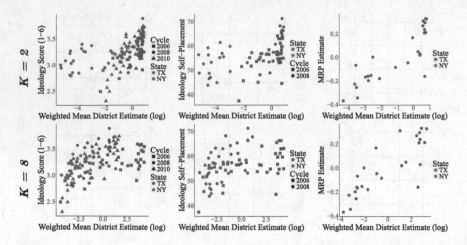

Fig. 8. In all of these cases, the inferred district-level voter preferences are based on the model varying its assumption of K and are weighted mean district estimates transformed from x to $sign(x) \, log(|x| + 1)$. (Left) Inferences compared with CÇES question of self-reported ideologies on a discrete scale. (Center) Compared with the CCES self-reported ideologies on a continuous scale from 0 to 100. (Right) District-level inferences of a full decade compared with MRP estimates [27].

location parameter and the same Inverse Gamma prior defined for the standard deviation of the Normal component for the scale parameter. When we use Uniform component distributions, we use the same Normal prior defined for the mean of the Normal component for both the minimum and the distance between the minimum and maximum parameters. The priors defined for the Normal component parameters can be found in Sect. 4. For each alternative underlying component distribution, the inferred distributions can be seen in Fig. 4, derived polarization metrics can be seen in Table 2, and prediction comparisons can be seen in Fig. 6.

Figure 5 visualizes the comparisons between the results derived of alternative component distributions and alternative data sources described in Sect. 5.1. We find significant positive correlations between our district-level point estimates and all of the alternative data sources. Assuming Laplace component distributions, our results have a correlation of 0.3216 with the responses selecting ideology given a discrete scale (left column in Fig. 5), 0.2514 with the responses selecting ideology along a continuous scale (middle column), and 0.5323 with the MRP estimates (right column). All of these correlations were significant with p-values less than 0.01. Assuming Uniform component distributions, our results have a correlation of 0.3652 with the responses selecting ideology given a discrete scale, 0.2404 with the responses selecting ideology along a continuous scale, and 0.6331 with the MRP estimates. Again, all of these correlations were significant with p-values less than 0.01.

Table 3. Polarization metrics of the voters given inferences assuming different numbers of clusters (K) and of the candidate CFscores (cand.) [8]. The "diff" of each polarization metric is the difference between the voter and candidate metric.

			2006		2008		2010	
			TX	NY	TX	NY	TX	NY
$K = 2$	Difference-of-means	Voters	0.86	2.53	0.75	2.05	1.58	1.54
		Cand	1.87	1.58	1.83	0.37	0.41	1.433
		Diff	−1.01	0.95	−1.08	1.68	1.17	0.11
	Standard deviation	Voters	2.40	7.75	2.36	188.98	2.15	4.56
		Cand	0.97	0.77	1.07	0.83	1.22	0.81
		Diff	1.42	6.98	1.28	188.15	0.94	3.74
	Kurtosis	Voters	0.49	3.13	0.27	4.07	−0.15	3.24
		Cand	−1.62	1.67	0.49	1.79	3.24	−1.61
		Diff	2.11	1.46	−0.23	2.27	−3.39	4.84
$K = 8$	Difference-of-means	Voters	9.06	9.30	3.39	6.80	1.80	1.81
		Cand	1.74	0.54	1.82	1.55	2.17	1.42
		Diff	7.32	8.76	1.56	5.25	−0.37	0.39
	Standard deviation	Voters	110.59	60.49	38.55	85.00	5.59	55.75
		Cand	0.97	0.77	1.07	0.83	1.22	0.81
		Diff	109.61	59.72	37.48	84.17	4.37	54.94
	Kurtosis	Voters	5.64	2.65	48.57	4.06	−0.03	0.50
		Cand	−1.62	1.67	0.49	1.79	3.24	−1.614
		Diff	7.26	0.98	48.07	2.26	−3.27	2.11

B.2   Varying the Number of Clusters

We also varied the number of clusters (K) used in the model. The main results section in the paper presented results assuming $K = 4$, but below we include the inferred distributions in Fig. 7, derived polarization metrics in Table 3, and prediction comparisons in Fig. 9 for $K = 2$ and $K = 8$, assuming Normal underlying precinct distributions. Due to time constraints, we were only able to generate these results given the Texas and New York congressional elections.

Figure 8 visualizes the comparisons between the results derived of alternative component distributions and alternative data sources described in Sect. 5.1. We find significant positive correlations between our district-level point estimates and all of the alternative data sources. When our model assumes 2 clusters rather than 4 clusters, the results of our model have a correlation of 0.2845 with the responses selecting ideology given a discrete scale (left column in Fig. 5), 0.2646 with the responses selecting ideology along a continuous scale (middle column), and 0.7160 with the MRP estimates (right column). All of these correlations were significant with p-values less than 0.01. When our model assumes 8 clusters, the results of our model have a correlation of 0.2925 with the responses selecting ideology given a discrete scale, 0.1867 with the responses selecting ideology along

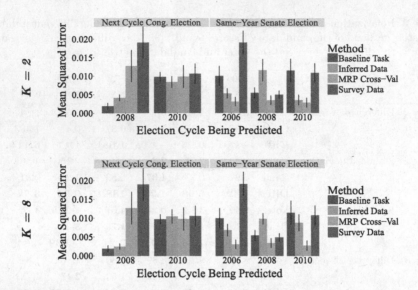

Fig. 9. Mean squared error of the actual and predicted vote share yielded by various prediction methods. Inferred data prediction method is based on our model assuming different numbers of clusters (K).

a continuous scale, and 0.6861 with the MRP estimates (right column). Again, all of these correlations were significant with p-values less than 0.01.

References

1. Abramowitz, A.I., Saunders, K.L.: Is polarization a myth? J. Polit. **70**(2), 542–555 (2008)
2. Airoldi, E.M., Blei, D.M., Fienberg, S.E., Xing, E.P.: Mixed membership stochastic blockmodels. In: Advances in Neural Information Processing Systems, pp. 33–40 (2009)
3. Ansolabehere, S., Palmer, M., Lee, A.: Precinct-level election data (2014). http://hdl.handle.net/1902.1/21919
4. Ansolabehere, S., Pettigrew, S.: Cumulative CCES Common Content (2006–2012) (2014). doi: 10.7910/DVN/26451
5. Barberá, P.: Birds of the same feather Tweet together: Bayesian ideal point estimation using Twitter data. Polit. Anal. **23**(1), 76–91 (2015)
6. Bartels, L.M.: Beyond the running tally: partisan bias in political perceptions. Polit. Behav. **24**(2), 117–150 (2002)
7. Bonica, A.: Database on ideology, money in politics, and elections: Public version 1.0 (2013). http://data.stanford.edu/dime
8. Bonica, A.: Mapping the ideological marketplace. Am. J. Polit. Sci. **58**(2), 367–386 (2014)
9. DiMaggio, P., Evans, J., Bryson, B.: Have american's social attitudes become more polarized? Am. J. Sociol. **102**, 690–755 (1996)

10. Ding, W., Ishwar, P., Saligrama, V.: Learning mixed membership mallows models from pairwise comparisons arXiv:1504.00757 (2015)
11. Downs, A.: An economic theory of political action in a democracy. J. Polit. Econ. **65**(2), 135–150 (1957)
12. Enelow, J.M., Hinich, M.J.: The Spatial Theory of Voting: an Introduction. Cambridge University Press, Cambridge (1984)
13. Fiorina, M.P., Abrams, S.J.: Political polarization in the American public. Ann. Rev. Polit. Sci. **11**, 563–588 (2008)
14. Fiorina, M.P., Abrams, S.J., Pope, J.C.: Polarization in the American public: misconceptions and misreadings. J. Polit. **70**(2), 556–560 (2008)
15. Flaxman, S.R., Wang, Y.X., Smola, A.J.: Who supported Obama in 2012? Ecological inference through distribution regression. In: Proceedings of the 21st ACM SIGKDD International Conference on Knowledge Discovery and Data Mining, pp. 289–298. ACM (2015)
16. Gerber, E.R., Lewis, J.B.: Beyond the median: voter preferences, district heterogeneity, and political representation. J. Polit. Econ. **112**(6), 1364–1383 (2004)
17. Gerrish, S., Blei, D.M.: Predicting legislative roll calls from text. In: Proceedings of the 28th International Conference on Machine Learning, pp. 489–496 (2011)
18. Krafft, P., Moore, J., Desmarais, B., Wallach, H.M.: Topic-partitioned multinetwork embeddings. In: Advances in Neural Information Processing Systems, pp. 2807–2815 (2012)
19. Lee, J.M.: Assessing mass opinion polarization in the U.S. using relative distribution method. Soc. Indic. Res. **124**(2), 571–598 (2014)
20. Levendusky, M.S., Pope, J.C., Jackman, S.D.: Measuring district-level partisanship with implications for the analysis of U.S. elections. J. Polit. **70**(3), 736–753 (2008)
21. Levendusky, M.S., Pope, J.C.: Red states vs. blue states going beyond the mean. Publ. Opin. Q. **75**(2), 227–248 (2011)
22. Lewis, J.B.: Estimating voter preference distributions from individual-level voting data. Polit. Anal. **9**(3), 275–297 (2001)
23. Lewis, J.B., DeVine, B., Pitcher, L., Martis, K.C.: Digital boundary definitions of United States congressional districts, pp. 1789–2012 (2013). http://cdmaps.polisci.ucla.edu
24. McCarty, N., Poole, K.T., Rosenthal, H.: Polarized America: The Dance of Ideology and Unequal Riches, vol. 5. MIT Press, Cambridge (2006)
25. Poole, K.T., Rosenthal, H.: A spatial model for legislative roll call analysis. Am. J. Polit. Sci. **29**(2), 357–384 (1985)
26. Ruths, D., Pfeffer, J.: Social media for large studies of behavior. Science **346**(6213), 1063–1064 (2014)
27. Tausanovitch, C., Warshaw, C.: Measuring constituent policy preferences in congress, state legislatures, and cities. J. Polit. **75**(2), 330–342 (2013)
28. United States Census Bureau: Tigerweb state-based data files: Voting districts - Census 2010 (2010). http://tigerweb.geo.census.gov
29. Zhang, Y., Friend, A., Traud, A.L., Porter, M.A., Fowler, J.H., Mucha, P.J.: Community structure in congressional cosponsorship networks. Phys. A **387**(7), 1705–1712 (2008)

Contrasting Public Opinion Dynamics and Emotional Response During Crisis

Svitlana Volkova[1,3(✉)], Ilia Chetviorkin[2], Dustin Arendt[1],
and Benjamin Van Durme[3]

[1] Pacific Northwest National Laboratory, Richland, WA, USA
svitlana.volkova@pnnl.gov, dustin.arendt@pnnl.gov
[2] Computational Mathematics and Cybernetics, Lomonosov Moscow State
University, Moscow, Russia
ilia2010@yandex.ru
[3] Center for Language and Speech Processing, Johns Hopkins University, Human
Language Technology Center of Excellence, Baltimore, MD, USA
vandurme@cs.jhu.edu

Abstract. We propose an approach for contrasting spatiotemporal
dynamics of public opinions expressed toward targeted entities, also
known as stance detection task, in Russia and Ukraine during crisis.
Our analysis relies on a novel corpus constructed from posts on the
VKontakte social network, centered on local public opinion of the ongoing Russian-Ukrainian crisis, along with newly annotated resources for
predicting expressions of fine-grained emotions including joy, sadness,
disgust, anger, surprise and fear. Akin to prior work on sentiment analysis we align traditional public opinion polls with aggregated automatic
predictions of sentiments for contrastive geo-locations. We report interesting observations on emotional response and stance variations across
geo-locations. Some of our findings contradict stereotypical misconceptions imposed by media, for example, we found posts from Ukraine that
do not support Euromaidan but support Putin, and posts from Russia that are against Putin but in favor USA. Furthermore, we are the
first to demonstrate contrastive stance variations over time across geo-
locations using storyline visualization (Storyline visualization is available
at http://www.cs.jhu.edu/~svitlana/) technique.

Keywords: Social media analytics · Spatiotemporal analysis opinion of
opinion dynamics · Emotion prediction · Storyline visualization

1 Introduction

Social media data has been extensively used for a variety of monitoring and
predictive tasks, both online activity [15] and real-world events, such as real-
time large-scale health analytics [8,9,19], multi-source disease forecasting [38,40],
stock market prediction [5], voting outcome forecasting [4,20,30], political movements and protest activity detection [3], and real-time mood changes [21]. Moreover, signals extracted from real-time social media data have been successfully

© Springer International Publishing AG 2016
E. Spiro and Y.-Y. Ahn (Eds.): SocInfo 2016, Part I, LNCS 10046, pp. 312–329, 2016.
DOI: 10.1007/978-3-319-47880-7_19

used for situational awareness e.g., to analyze online sentiments during disasters and hazard events [2,27,43,47].

In this work we study public opinion and emotions expressed through social media during the Russian-Ukrainian crisis. We formulate a problem as a stance detection task–automatically determining from posts whether the author is in favor of the given target, against the given target, or whether neither inference is likely, for example the post in favor Putin is:

— Sergey, среди моих друзей, и друзей моих друзей таких нет, зато все как один поддерживают Путина. Даже те, кто еще два года назад поддерживали всяких навальных, изменили свое мнение. Возвращение Крыма, это конечно мощный консолидирующий фактор. *My friends and friends of my friends support Putin. Even those who two years ago supported Navalniy changed their minds. The return of the Crimea is certainly a powerful consolidating factor.*

Our analysis focuses on a period of nine months–from September 2014 to March 2015. We start by collecting public data relevant to the crisis from the VKontakte (VK) social network, which is the most popular social network in Russia and Ukraine, by filtering relevant posts using topical keywords (or targets of stance). We then apply models for stance detection and develop new models for emotion classification in Russian and Ukrainian. Finally, we analyze how stance regarding targeted topics (i) evolves over time within each country by analyzing positive opinions (author is in favor) and negative opinions (author is against), and (ii) differs between contrasting populations (Russia vs. Ukraine) by measuring correlations between opinion proportions toward targeted topics. The main contributions of this work include:

– Performing large-scale contrastive targeted opinion (stance) and emotion analysis on thousands of messages posted in Russia and Ukraine during crisis on a new corpus from the VK social network.[1]
– Building a new model for fine-grained emotion prediction for low-resource languages e.g., Russian and Ukrainian.[2]
– Measuring spatiotemporal variations in targeted opinions between contrastive geo-locations e.g., Russia and Ukraine.
– Visualizing contrastive stance dynamics to qualitatively evaluate differences across geo-locations over time using storylines.
– Contrasting opinions expressed through social media with traditional polls.

Similar to recent studies on large-scale public opinion polling [30,39], the results of this work demonstrate that signals from public social media can be a faster and less expensive alternative to traditional polls [6]. Moreover, unlike the existing approaches, our study is the first to perform emotion analysis (in

---

[1] Anonymized VK corpus is available upon request at http://www.cs.jhu.edu/~svitlana/.

[2] Pre-trained models for emotion prediction and data annotated with 6 Ekman's emotions in Russian and Ukrainian can be found at http://www.cs.jhu.edu/~svitlana/.

Table 1. Sample stance targets used to filter VK posts relevant to the crisis in Russian (translated to English). We focus on named entities (Person, Organization, Location) and other event triggers.

Post-revolution period	
English	Locations: Kiev, Ukraine, Russia, USA, America, Moscow, Crimea People: Yanukovych, Putin Organizations: European Union (EU), North Atlantic Treaty Org (NATO) Events: Maidan (Independence square), Euromaidan, berkut, revolution, titushki, banderovtsy, terrorism, war, separatists
Russian	Locations: Киев, Украина, Россия, США, Америка, Москва, Крым People: Янукович, Путин Organizations: Европейский союз, Евросоюз, НАТО Events: майдан, евромайдан, революция, титушки, бандеровци, беркут, терроризм, война, сепаратисты
Conflict period	
English	Locations: Donbass, Donetsk, Debaltseve, Ilovaisk, Mariupol, Lugansk People Republic (LPR), Donetsk People Republic (DPR) Events: battle of Debaltseve, battle of Ilovaisk, rebels, junta, cyborgs, humanitarian aid, ukrops, pravy sektor, pravoseki, cease-fire, pravosek, cease-fire, negotiations, checkpoint, fire, punishers, national army, krimnash, ukrofashysty, quiet period, h**lo
Russian	Locations: Донбас, Донецк, Дебальцево, Иловайск, Мариуполь, Луганская Народная Респуб. (ЛНР), Донецкая Народная Респуб. (ДНР) Events: Дебальцевский котел, Иловайский котел, ополченцы, хунта, киборги, гуманитарная помощь, укропы, правый сектор, прекращение огня, правосек, перемирие, переговоры, блокпост, обстрел, каратели, нацгвардия, крымнаш, укрофашисты, режим тишины, х*йло

addition to traditional opinion mining) for languages other than English [12,50], and quantitatively estimate differences in targeted opinions (or stance) between contrastive populations over time using storyline visualization.

2   VKontakte Data

VK is the most popular social network in Russia and the former Soviet Union area. As of 2013, it was used by 106 million users worldwide with 9.5 million users in Moscow, 5.2 million in Saint-Petersburg and 2.75 million in Kyiv.

Sampling Real Users. Our original VK dataset of topically relevant posts – messages that include one or more keywords $k \in K$ defined in Table 1 consists of 3.3 million posts from 1 million users. The post per user ratio (PPU) is 3.3 meaning that in the original VK data many users produced only one post. However, taking into account a well-known issue of sampling biases in social

Fig. 1. Top geo-locations in Russia and Ukraine (red circle represent the number of VK posts per location). (color figure online)

media [44], we sub-sampled the original collection to eliminate bots, trolls,[3,4,5] spam, news accounts as well as hyper-active and hypo-active users as described below. We removed users without locations listed in their profiles, users with more than 500 or less than five friends, and users with more than 300 posts (hyper-active) and less than two posts (hypo-active) over nine months.

As a result, we collected "real", moderately active VK users with marked geo-locations–**49,208 unique users with 597,247 topically-relevant posts** (on average 12 posts per user).

Geo-Location Distribution. Our dataset contains 772 unique locations in Ukraine (143 locations have more than 100 posts) and 1,378 unique locations in Russia (217 locations have more than 100 posts). In Fig. 1 we report message distribution for the top 17 locations that include at least 2,500 crisis-relevant posts per location that are represented by 62 % of all posts in Russian portion of our dataset, and 77 % of all posts in Ukrainian portion.

Gender and Age Distribution. This is the first study that analyzes public posts from the VK social network. For large-scale passive polling, it is important to be aware of how gender and age distributions in the sample are different from the demographics in the population. It has been reported that VK is equally popular among men and women.[6] However, our dataset contains 64 % posts from male users, 25 % posts from female users in Russia (11 % have not reported gender), and 59 % posts from male users and 29 % posts from female users in Ukraine (12 % have not reported gender). These statistics may suggest that crisis topics are more discussed among male rather than female users in VK.

---

[3] Social Network Analysis Reveals Full Scale of Kremlin's Twitter Bot: https://globalvoices.org/2015/04/02/analyzing-kremlin-twitter-bots/.

[4] Inside Putin's Campaign Of Social Media Trolling And Faked Ukrainian Crimes: http://www.forbes.com/sites/paulroderickgregory/2014/05/11/inside-putins-campaign-of-social-media-trolling-and-faked-ukrainian-crimes/\#238cfd72629d.

[5] Ukraine conflict: Inside Russia's 'Kremlin troll army': http://www.bbc.com/news/world-europe-31962644.

[6] VK demographics: http://www.slideshare.net/andrewik1/v-kontakte-demographics.

Table 2. Age distribution in our VK data sample.

Geo-location	13–17	18–24	25–34	35–44	45–54	55–64	65 +	N/A
Ukraine	1 %	8 %	16 %	7 %	4 %	3 %	1 %	40 %
Russia	1 %	5 %	13 %	9 %	7 %	3 %	2 %	40 %

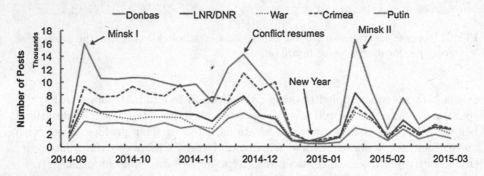

Fig. 2. Top 5 topic dynamics over time in both countries.

Table 2 reports age distribution in our VK data sample from two geo-locations – Russia and Ukraine. We observe that the sample is skewed toward younger population (18–44 y.o.) in both geo-locations. Unlike gender assignments, 40 % of posts from our sample do not have age labels available.

Topic Dynamics. Fig. 2 shows the number of posts with topical keywords produced between Sept 2014 and Mar 2015 in Russia and Ukraine. We observe that *Putin* is the least popular keyword relative to the five most discussed (trending) topics during that period–*Donbas, Crimea, DNR, LNR* and *war* (sorted by popularity). We found several spikes in keyword popularity that happened to overlap with major events relevant to the crisis – Minsk I[7] and Minsk II.[8] As expected, we captured a significant decrease in keyword popularity around the time of the New Year holiday across all keywords in both countries.

3   Approach

This section describes the approach we used for opinion classification and emotion detection, and outlines the limitations of our analysis.

3.1   Targeted Opinion Prediction

For stance prediction we used the state-of-the-art opinion classification system for Russian–POLYARNIK [18] that relies on morphological and syntactic rules,

---

[7] Minsk I: https://en.wikipedia.org/wiki/Minsk_Protocol.
[8] Minsk II: https://en.wikipedia.org/wiki/Minsk_II.

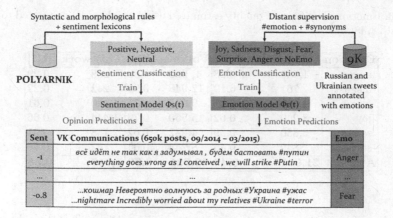

Fig. 3. Opinion and emotion classification approach: example VK posts with predicted targeted opinions (stance) and fine-grained emotions.

affect lexicons and supervised models [26]. Classification performance was measured on an external social media dataset from Twitter and achieved the F1 score of 0.62 [7,25].

We applied opinion classifiers to assign positive (in favor), negative (against) or neutral stance to every message in our dataset. Example posts in Russian with the assigned opinion scores are shown in Fig. 3. To study geo-temporal variations in public opinions between contrasting sub-populations we used several metrics as defined below.

Within within each country $c \in C$, for each keyword $k \in K$ and time period $t \in T$ we calculate **positive opinion score**:

$$s^+(k, c, t) = \frac{positive\ posts\ toward\ k\ at\ time\ t\ in\ country\ c}{total\ posts\ toward\ k\ at\ time\ t\ in\ country\ c}.$$

For each keyword $k \in K$ and time $t \in T$ we calculate **positive score ratio**:

$$\Delta s^+\left(k, \frac{UA}{RU}, t\right) = \frac{s^+(k, c = UA, t)}{s^+(k, c = RU, t)}.$$

The above metrics allow us to capture opinion drift within each country over time [29,30], as well as to contrast stance differences between two countries. More precisely, **positive opinion score** will provide the insights on how positive Ukraine is toward *Crimea* every week if $c = Ukraine$, $t = week \in \{09/2014 - 03/2015\}$, and $k = Crimea$. Similarly, **positive score ratio** will demonstrate how much more or less positive Ukraine is toward *Crimea* compared to Russia.

3.2 Emotion Classification

We constructed our emotion dataset from an independent sample of crisis-related discourse generated on Twitter. For that we bootstrapped noisy hashtag

Table 3. Emotion classification quality estimated using weighted F1 compared to other systems [28,48].

Emotion	Mohammad [EN]		Volkova [EN]		This work [RU]	
anger	1,555	0.28	4, 963	0.80	606	0.29
disgust	761	0.19	12,948	0.92	242	0.22
fear	2,816	0.51	9,097	0.77	1,120	0.61
joy	8,240	0.62	15,559	0.79	1,444	0.66
sadness	3,830	0.39	4,232	0.62	2,138	0.61
surprise	3849	0.45	8,244	0.64	167	0.56
ALL	**21,051**	**0.49**	**52,925**	**0.78**	**5,717**	**0.58**

annotations for six basic emotions put forward by Ekman[9] [11] as has been done successfully for English [28,48]. Twitter users sometimes use the hashtag *#радость (#joy)* to signal when they are happy, *#страх (#fear* when they are scared (Fig. 3). As discussed in [10,13], such annotations are extremely sparse. Therefore, we extended our emotion set with emotion synonyms collected from WordNet-Affect [46], Google Synonyms, and Roget's Thesaurus and translated them to Russian and Ukrainian.

A native speaker of Russian and Ukrainian then manually re-validated tweets annotated with emotion hashtags. In total we collected 5,717 tweets annotated with *anger* (11 %), *joy* (25 %), *fear* (20 %), *sadness* (37 %), *disgust* (4 %), and *surprise* (3 %). In addition to emotional tweets, we sampled more tweets from the 1 % Russian Twitter feed and manually validated 3,947 tweets that do not express any emotions. We applied our emotion dataset bootstrapped from Twitter to annotate VK posts similar to other predictive analytics transferred across social media e.g., from Twitter to Facebook [41].

We took a two-step approach to predict emotions in crisis-relevant posts. We first classify tweets as emotional and non-emotional, and then predict one of six Ekman's emotions in a subset of emotional posts only.

We trained our emotion classifier using an implementation of a log-linear model in scikit-learn [33]. We relied on binary word unigram features extracted from stemmed[10] posts annotated with six basic emotions as described above. In addition to *lexical features* we extracted a set of *stylistic features* including emoticons, elongated words, capitalization, and repeated punctuation and take into account negation [32]. We demonstrate model prediction quality using 10-fold cross validation on our emotion dataset and compare it to other existing datasets for English in Table 3.

---

[9] We prefer Ekman's emotion classification over others e.g., Plutchik's, because we would like to compare the performance of our predictive models to other systems.

[10] Morphological analyzer for Russian: https://pypi.python.org/pypi/pymorphy2.

3.3 Limitations

To the best of our knowledge, this is the first study that performs large-scale contrastive opinion analysis relevant to the Russian-Ukrainian crisis on the VK social network. However, our approach has several limitations. First, our opinion and emotion prediction models do not yield 100 % accuracy. Thus, some posts might be mislabeled. However, doing the analysis on such a large scale allows us to overcome noise in affect labels [30]. Second, we do not claim that our findings are representative of the whole population in Russia or Ukraine. We draw conclusions from sampled data from only one social network–VK, which by nature is more representative of younger populations in both countries.

4   Results

This section discusses spatiotemporal variations in topic popularity, target-specific opinion correlations and opinion drift over time across two countries, fine-grained emotion analysis and storyline visualization applied to our data.

4.1   Contrasting Topic Popularity Across Geo-Locations

To estimate popularity of crisis-relevant topics we measure the frequency of posts that contain one or more topics including hashtags e.g., *#putin*, *#euro-maidan*. In Fig. 4 we show the most popular topics in our VK dataset for Russia and Ukraine. We observe that *Donbas* was discussed more in Ukraine, but *war* in Russia over the same time period. There were significantly more posts about *Putin* and *USA* in Russia compared to Ukraine. On the other hand, *North Atlantic Treaty Organization (NATO)* and *European Union (EU)*, *Donetsk People's Republic (DNR)*,[11] and *Luhansk People's Republic (LNR)*[12] were discussed more in Ukraine compared to Russia.

4.2   Measuring Targeted Opinion Correlations

Overall, we found that 99 % of posts in our dataset were assigned either positive or negative polarity (in favor or against). Therefore, we correlate positive polarity scores $s^+(k, c, t)$ expressed toward targeted entities k for every week t for each country c (assuming that $s^+(k, c, t) + s^-(k, c, t) = 1$).

We calculate Pearson correlations between targeted opinion scores $s^+(k, c, t)$ in two countries. We present the results in Fig. 5, where blue represents positive correlations e.g., opinions are changing in one direction, either positive or negative, and red stands for negative correlations e.g., opinions are changing in a different direction.

Below we outline our key findings on opinion correlations between Russia and Ukraine. We found:

---

[11] Donetsk   people's   republic: https://en.wikipedia.org/wiki/Luhansk_People's_Republic.

[12] Luhansk   People's   Republic: https://en.wikipedia.org/wiki/Luhansk_People's_Republic.

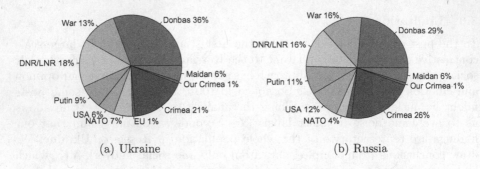

(a) Ukraine   (b) Russia

Fig. 4. Popular crisis-relevant topics in Russia and Ukraine.

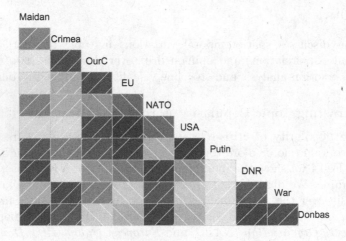

Fig. 5. Correlogram for targeted opinions (stance) between two countries (blue – positive correlations, red–negative). (Color figure online)

- Positive sentiment toward $EU\uparrow$ in Ukraine contrasts with negative toward $NATO\downarrow$, $DNR\downarrow$, and $Donbas\downarrow$ in Russia.
- Positive sentiment toward $NATO\uparrow$ in Ukraine contrasts with negative toward $USA\downarrow$ in Russia.
- Positive sentiment toward $OurCrimea\uparrow$ in Ukraine contrasts with negative sentiment toward $NATO\downarrow$, $DNR\downarrow$, and $Donbas\downarrow$ in Russia.
- Positive sentiment toward $USA\uparrow$ in Ukraine contrasts with negative toward $war\downarrow$, $DNR\downarrow$, and $Donbas\downarrow$ in Russia.

In addition, we found several major differences in stance expressed by two populations toward the same targeted entity:

- *EU* have strong negative correlations between two countries ($\rho = -0.4$).
- *USA* and *maidan* have low positive correlations ($\rho = 0.05$ and $\rho = 0.34$).
- *OurCrimea* and *Putin* have positive correlations ($\rho = 0.59$ and $\rho = 0.65$).
- *LNR, DNR, NATO, Donbas, Crimea,* and *war* have strong positive correlations (as high as $\rho = 0.94$ for the *war*).

Below we outline some examples of opinions that contradict stereotypical misconceptions imposed by media discovered in our data.

– *Negative opinions toward maidan in Ukraine*:

Яценюк говорил неправду слишком долго. Когда в феврале на Майдане лилась кровь, он с Кличко и Тягнибоком жали руку «енакиевскому лидеру» *Yatsenyuk told a lie for too long. He was with Klitschko and Tyagnibok and shook hands with "the leader of Yenakiyevo" while there has been bloodshed at the Independence square (maidan).*

– *Negative opinions toward EU in Ukraine* (in contrast, media reports Ukrainians are in favor joining EU [16,36,49]): Евросоюз кинул западенцев, а Россия - жителей Донбасса. *EU betrayed residents of Western Ukraine, and Russia–residents of Donbas.*

– *Positive opinions toward USA in Russia*:

Только санкции заставят русских оставлять деньги в России и не смотреть на запад и Америку как на единственный источник благ и цивилизаций. *Only sanctions will force Russians to leave money in Russia, and do not look to the West and America as the sole source of wealth and civilization.*

– *Negative opinions toward Putin in Russia* (in contrast, media reports the majority of Russian population are in favor Putin [22]): Есть международные договоры, которые нарушил Путин, устроив аннексию Крыма.

There are international treaties that has been violated Putin during annexation of Crimea.

4.3 Estimating Contrastive Opinion Drift over Time

To measure opinion dynamics we calculate positive score ratios $\Delta s^+(k, \frac{UA}{RU}, t)$ using location-specific targeted opinions expressed in Ukraine vs. Russia.

We present our results in Fig. 6 where blue bars show when targeted opinions are more positive in Ukraine compared to Russia, and red bars reflect when targeted opinions are more positive in Russia compared to Ukraine. We plot the number of weekly posts that contain targeted entity for both countries to reflect the amount of evidence contributing to weekly stance estimates e.g., more posts – more confident estimates. We noticed several topics including *Crimea, EU, Putin* and *USA* are more popular in Russia whereas *maidan, NATO* are more discussed in Ukraine. We observed more positive opinions were expressed toward *maidan* in Ukraine than in Russia, and more positive opinions expressed toward *Crimea, Putin* in Russia than in Ukraine.

4.4 Visualizing Contrastive Opinion Dynamics

We use storyline visualization to understand opinion dynamics at a coarse level of detail across a hand curated set of targeted entities. Storyline visualization

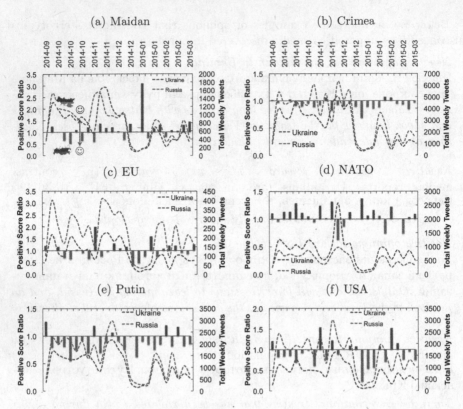

Fig. 6. Spatiotemporal variations in contrastive opinion dynamics. Blue bars show when targeted opinions are more positive in Ukraine vs. Russia. Red bars show when targeted opinions are more positive in Russia compared to Ukraine. Dashed lines represent the total number of weekly tweets with targeted entities in Russia (red) and Ukraine (blue). (Color figure online)

shows how entities interact over time [17,24,31,37,42]. This technique encodes time on the horizontal axis, and interactions on the vertical axis. When two entities are interacting, their storylines are drawn close together, otherwise they are drawn apart. Entities having the same state at the same time are assumed to be interacting with each other.

To visualize opinion dynamics, we process the original opinion time series $s+$ (k, c, t) into a format usable by the storyline visualization. For that we estimate the mean number of posts with positive and negative stance toward targeted entity per day, and normalize them by the total number of posts per day that mention targeted entity. This is done for each targeted entity, geo-location and day, by subtracting the mean and dividing by the standard deviation of that targeted entity across all days. This puts the positive and negative stance for each targeted entity on a consistent scale so that meaningful comparisons can be made across targeted entities over time.

We present some example storyline visualizations around important events relevant to the crisis – G20 meeting in Nov 2014 and Minsk II agreement in Feb 2015 in Figs. 7 and 8, respectively. We consider whether targeted opinions toward a particular topic in specific location and a day was elevated (i.e., more than one standard deviation larger than the mean) for both positive and negative opinions. This resulted in four distinct clusters corresponding to the combinations of {*normal, elevated*} × {*positive, negative*} per topic, location and day. As shown in Figs. 7 and 8 orange cluster contains stance within one standard deviation above the mean. Green cluster represents opinions that are more in favor (two or more st. dev. above). Red cluster represents opinions that are more negative. Blue cluster represents opinions that are both more positive and more negative relative to the mean. Each topic is represented with two storylines, one each for the populations from Russia and Ukraine, so that differences over time between these populations can be visually identified and understood.

More specifically, Fig. 7 demonstrates contrastive opinion drift toward *USA* in Russia and Ukraine in November 2015. Storyline visualization allows us to see how opinions toward *USA* in Russia move to a positive cluster (green) from the mean cluster (orange) on 11/7 and 11/8 (shown on the left), and opinions toward *USA* in Ukraine move to a positive cluster on 11/5. To put these findings in context, we found that trending news stories on 11/5 included "A bomb could be to blame for the Russian jet crash",[13] and on 11/7 included "Putin suspends all flights from Russia to Egypt in wake of crash".[14]

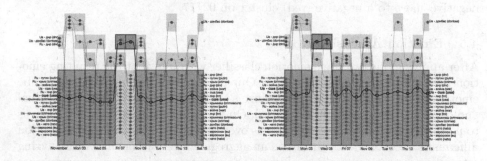

Fig. 7. Storyline visualization of opinion dynamics toward *USA* expressed in Russia (Ru) and Ukraine (Ua) during G20 summit (Nov 1–15, 2014). (Color figure online)

Figure 8 (left) highlights contrastive stance dynamics toward *Donbas* (the Eastern part of Ukraine where the conflict escalated) around G20 meeting in November 2014. We observe that on 11/16 opinions expressed in Russia toward *Donbas* moved to a negative cluster (red), on 11/17–to a positive cluster (green), and on 11/18–back to the average opinion cluster (orange). News reported on

[13] http://www.cbsnews.com/videos/cbs-news-trending-stories-for-november-5-2015/.
[14] http://theweek.com/10things/580982/10-things-need-know-today-november-7-2015.

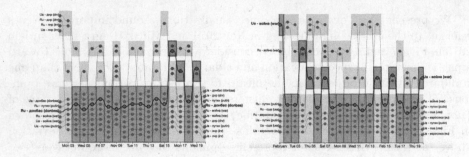

Fig. 8. Storyline visualization of opinion dynamics toward *Donbas* expressed during G20 summit between Nov 3–19, 2014 (left), and toward *war* expressed during Minsk II agreement between Feb 1–19, 2015 (right). (Color figure online)

11/17 that include targeted entities of interest e.g., *Putin, Donbas, EU, DNR, LNR* and *war* are: "E.U. to ,Toughen Sanctions on Ukrainian Separatists, but Not Russia"[15] and Poroshenko's post on Twitter "We are prepared for a scenario of total war. We don't want war, we want peace, we are fighting for EU values. Russia doesn't respect any agreement."[16]

Figure 8 (right) demonstrates contrastive stance dynamics toward *war* around Minsk II meeting on February 11, 2015. We observe that opinions toward *war* in Ukraine move to a positive (green) cluster on 02/14 and 02/16 after the agreement to cease fire. In contrast opinions toward *war* in Russia became more negative–move to a negative (red) cluster on 02/17.

4.5   Emotion Analysis

After applying our two stage emotion classifier we found that the resulting emotion predictions are extremely sparse. Our model predicts only 4 % posts to be emotional. On one hand, it further confirms that crisis-related discourse is factual and opinionated rather than emotional. On the other hand, it does not allow us to perform contrastive emotion analysis over time. Therefore, we report emotion differences between two countries on an aggregate level rather than by analyzing emotion score ratios over time. Figure 9 demonstrates that users express more emotions e.g., sadness, surprise, and anger toward *Donbas* across countries; sadness toward *USA*, fear toward *Crimea* and *Donbas* in Russia; fear toward *Putin* and *Donbas*, disgust toward *maidan* in Ukraine.

5   Discussion

Our results demonstrate that one can infer spatiotemporal variations in opinion dynamics from social media. But do these results reflect observations gathered by

---

[15] http://www.nytimes.com/2014/11/18/world/europe/eu-to-toughen-sanctions-on-ukraine-separatists-but-not-russia.html?_r=1.

[16] http://www.cnbc.com/2014/11/17/german-economy-minister-rejects-tougher-sanctions-on-russia.html#.

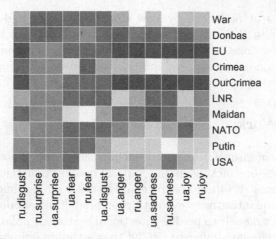

Fig. 9. Fine-grained targeted emotion analysis for Russia (ru) and Ukraine (ua). Red represents high post counts, blue represents low counts scaled over columns. (Color figure online)

traditional polling techniques? Unfortunately, there is no real time polling data on concepts related to crisis similar to the Gallup's organization "Economic Confidence" index[17] used in [30] or CDC data used in [39]. Despite that, in order to evaluate our findings we rely on recent polls from several internationally recognized organizations. Below we analyze how our findings on the most discussed and controversial topics across two countries align with public poll data.

EU, NATO, and USA. An IFES survey [16] finds that half of Ukrainians believe that their country would be better off with closer relations with Europe than with Russia, which is in line with our findings on positive sentiments in Ukraine toward *EU*, *NATO* and negative toward Putin as shown in Fig. 6. Similar polls done by the Razumkov Centre [36] and the Democratic Initiatives Foundation [49] conclude that 43.2 % of Ukrainians believe that their country will gain if it joins the EU and that 64 % of Ukrainians would vote for NATO accession in a hypothetical referendum. Regarding negative sentiments toward the USA expressed in Russia, a recent poll done by Pew Research Center reports that only 15 % of Russians have a favorable opinion of America [34].

Crimea and Putin. Posts about Crimea authored in Ukraine are less positive compared to those authored in Russia which is again in line with public polls. Russian media reports 55 % of Russian citizens strongly support and 33 % support Putin's actions on Crimea [23]. Tweets mentioning Putin in Russia are significantly more positive than those in Ukraine. Again, Russian media reports that based on recent polls 83 % of Russian citizens trust president Putin [22].

---

[17] http://www.gallup.com/poll/122840/gallup-daily-economic-indexes.aspx.

Another independent poll done by Pew Research Center finds that Russians praise Putin despite their country's economic troubles.[14] On the other hand, Gallup's interviews done in 2014 report a drastic change in approval of Russian leadership in Ukraine – the approval drops by as high as 45 %, and is now as low as 2 % [35].

6   Related Work

With the dramatic rise of text-based social media more researchers started focusing their work on tracking spatiotemporal opinion dynamics toward politicians [45,51], events [1,14], or controversial issues e.g., vaccinations [39]. Less work focused on quantitatively evaluating or predicting *opinion changes over time* e.g., [29] built models to predict directional sentiment polarity change. Our work is the most similar similar to [30,39] that estimate opinion dynamics and correlate it with real world data e.g., Gallup's polls or CDC reports. Unlike these works we not only align our findings with traditional opinion polls but also correlate opinion changes toward controversial topics between contrastive populations over time and develop a novel model for emotion detection for low resource languages.

7   Conclusions

We proposed a transparent and deterministic approach to perform large-scale contrastive opinion analysis on social media during crisis. We developed an emotion classification model for Russian and Ukrainian and qualitatively measured spatiotemporal differences in opinions and emotions between contrastive populations toward controversial topics. We developed storyline visualizations that allow to better capture contrastive opinion drift over time. We supported our findings with public opinions obtained using traditional polls.

References

1. Alves, A.L.F., de Souza Baptista, C., Firmino, A.A., de Oliveira, M.G., de Paiva, A.C.: A spatial, temporal sentiment analysis approach applied to Twitter microtexts. J. Inf. Data Manag. **6**(2), 118 (2016)
2. Ashktorab, Z., Brown, C., Nandi, M., Culotta, A., Tweedr: mining Twitter to inform disaster response. In: Proceedings of the International Conference on Information Systems for Crisis Response and Management (SCRAM) (2014)
3. Bastos, M. T., Mercea, D.: Political Twitter beyond influentials and the twitter-tariat. New Media & Society, Serial activists (2015)
4. Bermingham, A., Smeaton, A.F.: On using twitter to monitor political sentiment and predict election results. In: Joint Conference for Natural Language Processing (IJCNLP) (2011)
5. Bollen, J., Mao, H., Zeng, X.: Twitter mood predicts the stock market. J. Comput. Sci. **2**(1), 1–8 (2011)

6. Chang, L., Krosnick, J.A.: National surveys via rdd telephone interviewing versus the internet comparing sample representativeness and response quality. Pub. Opin. Quart. **73**(4), 641–678 (2009)
7. Chetviorkin, I., Loukachevitch, N.: Evaluating sentiment analysis systems in Russian. In: Proceedings of ACL (2013)
8. Chew, C., Eysenbach, G.: Pandemics in the age of Twitter: content analysis of tweets during the 2009 H1N1 outbreak. PloS one **5**(11), e14118 (2010)
9. Corley, C.D., Cook, D.J., Mikler, A.R., Singh, K.P.: Text and structural data mining of influenza mentions in web and social media. Int. J. Environ. Res. Pub. Health **7**(2), 596–615 (2010)
10. De Choudhury, M., Gamon, M., Counts, S.: Happy, nervous or surprised? Classification of human affective states in social media. In: Proceedings of ICWSM (2012)
11. Ekman, P.: An argument for basic emotions. Cogn. Emot. **6**(3–4), 169–200 (1992)
12. Fan, R., Zhao, J., Chen, Y., Ke, X.: Anger is more influential than joy: sentiment correlation in Weibo. arXiv preprint arXiv:1309.2402 (2013)
13. González-Ibáñez, R., Muresan, S., Wacholder, N.: Identifying sarcasm in Twitter: a closer look. In: Proceedings of ACL, pp. 581–586 (2011)
14. Ho, S.S., Lieberman, M., Wang, P., Samet, H.: Mining future spatiotemporal events and their sentiment from online news articles for location-aware recommendation system. In: Proceedings of the First ACM SIGSPATIAL International Workshop on Mobile Geographic Information Systems, pp. 25–32. ACM (2012)
15. Hodas, N.O., Lerman, K.: The simple rules of social contagion. Scientific reports, 4 (2014)
16. IFES: Public Opinion Survey in Ukraine, September 2015. http://www.ifes.org/surveys/september-2015-public-opinion-survey-ukraine
17. Kim, N.W., Card, S.K., Heer, J.: Tracing genealogical data with timenets. In: Proceedings of the International Conference on Advanced Visual Interfaces, pp. 241–248. ACM (2010)
18. Kuznetsova, E.S., Loukachevitch, N.V., Chetviorkin, I.I.: Testing rules for a sentiment analysis system. In: Proceedings of International Conference Dialog, pp. 71–80 (2013)
19. Lampos, V., Bie, T., Cristianini, N.: Flu detector - tracking epidemics on Twitter. In: Balcázar, J.L., Bonchi, F., Gionis, A., Sebag, M. (eds.) ECML PKDD 2010. LNCS (LNAI), vol. 6323, pp. 599–602. Springer, Heidelberg (2010). doi:10.1007/978-3-642-15939-8_42
20. Lampos, V., Preotiuc-Pietro, D., Cohn, T.: A user-centric model of voting intention from social media. In: Proceedings of ACL, pp. 993–1003 (2013)
21. Lansdall-Welfare, T., Lampos, V., Cristianini, N.: Effects of the recession on public mood in the UK. In: Proceedings of the 21st International Conference Companion on World Wide Web, pp. 1221–1226. ACM (2012)
22. Levada. Ukraine, Crimea, Sanctions (2015). http://www.levada.ru/2015/04/06/ukraina-krym-sanktsii/
23. Levada, C., Putin, V.: Trust, Evaluations, Attitudes (2015). http://www.levada.ru/2015/03/27/vladimir-putin-doverie-otsenki-otnoshenie/
24. Liu, S., Yingcai, W., Wei, E., Liu, M., Liu, Y.: Storyflow: tracking the evolution of stories. IEEE Trans. Visual Comput. Graphics **19**(12), 2436–2445 (2013)
25. Loukachevitch, N., Chetviorkin, I.: Open evaluation of sentiment-analysis systems based on the material of the Russian language. Sci. Tech. Inf. Process. **41**(6), 370–376 (2014)

26. Loukachevitch, N.V. and Chetviorkin, I.I.:Refinement of Russian sentiment lexicons using ruthes thesaurus. In: Proceedings of the 16th All-Russian Conference on Digital Libraries: Advanced Methods and Technologies, Digital Collections (2014)

27. Mandel, B., Culotta, A., Boulahanis, J., Stark, D., Lewis, B., Rodrigue, J.: A demographic analysis of online sentiment during hurricane Irene. In: Proceedings of the Second Workshop on Language in Social Media, pp. 27–36. Association for Computational Linguistics (2012)

28. Mohammad, S.M., Kiritchenko, S.: Using hashtags to capture fine emotion categories from tweets. Comput. Intell. **31**(2), 301–326 (2014)

29. Nguyen, L.T., Wu, P., Chan, W., Peng, W., Zhang, Y.: Predicting collective sentiment dynamics from time-series social media. In: Proceedings of the 1st International Workshop on Issues of Sentiment Discovery and Opinion Mining, ACM (2012)

30. O'Connor, B., Balasubramanyan, R., Routledge, B.R., Smith, N.A.: From tweets to polls: linking text sentiment to public opinion time series. In: Proceedings of ICWSM, pp. 122–129 (2010)

31. Ogawa, M., Ma, K.L.: Software evolution storylines. In: Proceedings of the 5th International Symposium on Software Visualization, pp. 35–42 (2010)

32. Pang, B., Lee, L., Vaithyanathan, S., Thumbs up?: sentiment classification using machine learning techniques. In: Proceedings of EMNLP, pp. 79–86 (2002)

33. Pedregosa, F., Varoquaux, G., Gramfort, A., Michel, V., Thirion, B., Grisel, O., Blondel, M., Prettenhofer, P., Weiss, R., Dubourg, V., Vanderplas, J., Passos, A., Cournapeau, D., Brucher, M., Perrot, M., Duchesnay, E.: Scikit-learn: machine learning in Python. J. Mach. Learn. Res. **12**, 2825–2830 (2011)

34. Poushter, J.: Key findings from our poll on the Russia-Ukraine conflict (2015). http://www.pewresearch.org/fact-tank/2015/06/10/key-findings-from-our-poll-on-the-russia-ukraine-conflict/

35. Ray, J., Esipova, N.: Ukrainian Approval of Russia's Leadership Dives Almost 90% (2015). http://www.gallup.com/poll/180110/ukrainian-approval-russia-leadership-dives-almost.aspx?g_source=Ukraine&g_medium=search& g_campaign=tiles

36. Centre Razumkov. Will Ukraine mostly gain or lose if it joins the EU? (2015). http://www.uceps.org/eng/poll.php?poll_id=675

37. Reda, K., Tantipathananandh, C., Johnson, A., Leigh, J., Berger-Wolf, T.: Visualizing the evolution of community structures in dynamic social networks. In: Computer Graphics Forum, vol. 30, pp. 1061–1070. Wiley Online Library (2011)

38. Rekatsinas, T., Ghosh, S., Mekaru, S.R., Nsoesie, E.O., Brownstein, J.S., Getoor, L., Ramakrishnan, N.: Sourceseer: forecasting rare disease outbreaks using multiple data sources. Timeline **7**, 8 (2007)

39. Salathé, M., Khandelwal, S.: Assessing vaccination sentiments with online social media: implications for infectious disease dynamics and control. PLoS Comput. Biol. **7**(10), e1002199 (2011)

40. Santillana, M., Nguyen, A.T., Dredze, M., Paul, M.J., Nsoesie, E.O., Brownstein, J.S.: Combining search, social media, and traditional data sources to improve influenza surveillance. PLoS Comput. Biol. **11**(10), e1004513 (2015)

41. Sap, M., Park, G., Eichstaedt, J.C., Kern, M.L., Stillwell, D., Kosinski, M., Ungar, L.H., Schwartz, H.A.: Developing age and gender predictive lexica over social media. In: Proceedings of EMNLP (2014)

42. Tanahashi, Y., Hsueh, C.-H., Ma, K.-L.: An efficient framework for generating storyline visualizations from streaming data. IEEE Trans. Vis. Comput. Graph. **21**(6), 730–742 (2015)

43. Terpstra, T., de Vries, A., Stronkman, R., Paradies, G.L.: Towards a realtime twitter analysis during crises for operational crisis management. In: Proceedings of ISCRAM (2012)
44. Tufekci, Z.: Big questions for social media big data: representativeness, validity and other methodological pitfalls. arXiv preprint arXiv:1403.7400 (2014)
45. Tumasjan, A., Sprenger, T.O., Sandner, P.G., Welpe, I.M.: Predicting elections with Twitter: what 140 characters reveal about political sentiment. In: Proceedings of ICWSM, pp. 178–185 (2010)
46. Valitutti, R.: Wordnet-affect: an affective extension of wordnet. In: Proceedings of LREC, pp. 1083–1086 (2004)
47. Vieweg, S., Hughes, A.L., Starbird, K., Palen, L.: Microblogging during two natural hazards events: what Twitter may contribute to situational awareness. In: Proceedings of the SIGCHI Conference on Human Factors in Computing Systems, pp. 1079–1088 (2010)
48. Volkova, S., Bachrach, Y.: On predicting sociodemographic traits and emotions from communications in social networks and their implications to online self-disclosure. Cyberpsychology Behav. Soc. Networking 18(12), 726–736 (2015)
49. Ievgen Vorobiov. Surprise! Ukraine Loves NATO (2015). http://foreignpolicy.com/2015/08/13/surprise-ukraine-loves-nato/
50. Wang, J., Zhu, T.: Classify sina weibo users into high or low happiness groups using linguistic, behavior features. arXiv preprint arXiv: 1507.01796 (2015)
51. Wang, Y., Clark, T., Agichtein, E., Staton, J.: Towards tracking political sentiment through microblog data. In: Proceedings of ACL (2014)

Social Politics: Agenda Setting and Political Communication on Social Media

Xinxin Yang[1], Bo-Chiuan Chen[1], Mrinmoy Maity[1], and Emilio Ferrara[2(✉)]

[1] Indiana University, Bloomington, IN, USA
[2] University of Southern California, Los Angeles, CA, USA
emilio.ferrara@gmail.com

Abstract. Social media play an increasingly important role in political communication. Various studies investigated how individuals adopt social media for political discussion, to share their views about politics and policy, or to mobilize and protest against social issues. Yet, little attention has been devoted to the main actors of political discussions: the politicians. In this paper, we explore the topics of discussion of U.S. President Obama and the 50 U.S. State Governors using Twitter data and agenda-setting theory as a tool to describe the patterns of daily political discussion, uncovering the main topics of attention and interest of these actors. We examine over one hundred thousand tweets produced by these politicians and identify seven macro-topics of conversation, finding that Twitter represents a particularly appealing vehicle of conversation for American opposition politicians. We highlight the main motifs of political conversation of the two parties, discovering that Republican and Democrat Governors are more or less similarly active on Twitter but exhibit different styles of communication. Finally, by reconstructing the networks of occurrences of Governors' hashtags and keywords related to political issues, we observe that Republicans and Democrats form two tight yet polarized cores, with a strongly different shared agenda on many issues of discussion.

Keywords: Social politics · Agenda setting · Social media · Political communication

1 Introduction

The widespread adoption of social media is challenging the way traditional media have been used to distribute news, and to discuss top social and political issues [23,36,41]. A large body of *Computational Social Science* research focuses on the study of individuals and their behaviors on such platforms [7,35,44]. Various seminal papers investigate social and political conversations on social platforms like Twitter [3,19,45,48] and Facebook [6,8,20]. Yet, little work has been devoted to understand how the main actors of political discussion, the politician themselves, adopt and leverage such platforms [10,29,31]. During the

© Springer International Publishing AG 2016
E. Spiro and Y.-Y. Ahn (Eds.): SocInfo 2016, Part I, LNCS 10046, pp. 330–344, 2016.
DOI: 10.1007/978-3-319-47880-7_20

2008 Presidential Election, Barack Obama used fifteen social media sites to support his campaign. His successful effort demonstrated the central role of Twitter and other social platforms as integral parts of modern political communication. Since then, online political discussion and the attention toward political candidates and political figures, and their social media presence, arose. Politicians are influential figures in the offline world, and surely can acquire a great deal of influence in the social media spheres as well. Their social media activity, in turn, can alter their success and affect their careers, especially during election time. The online campaigns preceding the 2016 Presidential Election carried out by both parties in support of various potential nominees, including Hillary Clinton, Bernie Sanders, and Donald Trump, further demonstrate the social media power to shape the political scene [51,52]. A better understanding of politicians' usage of social media channels for political conversation could therefore reveal something about the complex mechanisms of political success in the era of *social politics*.

Yet, social media are not limited to political "propaganda". The effects of social media political communication on the offline world are tangible. Examples of political campaigns that preceded mass mobilizations and civilian protests include the Arab Springs [26,32], Occupy Wall Street [13,14], and the Gezi Park protest [50]. Although it is difficult to establish a causality link, we can safely say that the "Twittersphere" can be a strong indicator of political and public opinion [49]. The open nature of Twitter[1] probably contributed to determine its *political communicative power*. The ability to communicate interesting political issues yields the opportunity to users to acquire more visibility and influence [2, 9,43], although Twitter political discussion is plagued by a number of issues related to manipulation and abuse [21,22,45].

In this paper we explore how the main actors of political discussion, the politicians, adopt Twitter to cover social and political issues. We focus on U.S. President Obama and all the 50 U.S. State Governors, and adopt the framework of *agenda-setting theory* to identify their main topics of discussion. The analysis of over one hundred thousand of their tweets reveals how Governors and the President use Twitter, what are the emerging patterns of political discussion, the top issues for each party, and finally who are the politicians who exhibit the most coherent political agenda.

2   Social Media and Politics

Twitter was born in 2006. In less than 10 years, it acquired half billion users, 310 million of which are active and produce over 500 million tweets per day as of July 2016.[2] Twitter suggests that "each tweet represents an opportunity to show one's voice and strengthen relationships with one's followers".[3] As a

---

[1] At least with respect to other platforms like Facebook where ties are mostly formed based on pre-existing offline connections [16].

[2] Twitter official data: https://about.twitter.com/company.

[3] Twitter official blog: https://blog.twitter.com/2014/what-fuels-a-tweets-engagement.

modern political toolbox, Twitter has been widely used by various Presidents, Congressmen, Governors, and other politicians all over the world. In particular in the United States, Twitter and other social media have been not only the subject of extensive research, but also the platforms used to run large-scale social experiments to study political mobilization [6]. Scholars from various disciplines have investigated the role of these platforms in modern political communication.

Generally, social media research related to politics can be categorized into two fields. The former focuses on the possibility of using social media signals to predict political elections. A large number of papers faced this challenging question, with at times promising results. For example, Gibson and McAllister's study [27] demonstrated a significant relationship between online campaigning and candidate support. Macnamara found evidence of a "significant online political engagement" in the 2008 U.S. Presidential Election [37]. Other studies covered the U.S. Presidential debate and Twitter sentiment, finding an alignment between popular opinions and votes [17,18,48]. Despite some promising work, the issue of predicting elections using social data remains debated [25].

The second area of research investigates Twitter users' behaviors, opinions and topics of political interest, at times proposing methods to identify their political alignments [11,15]. Some of these studies highlighted interesting socio-political phenomena: for example, Conover et al. [12] found that the network of political retweets exhibits a highly segregated bipartisan structure, which seems to reflect the users' political leanings, similarly to political blogs [1]. Shogan's et al. research showed that, in recent years, Republican politicians tweeted more than five times as often as Democrats, suggesting that Twitter might be particularly appealing to American opposition politicians, who use it as an instrument for voicing their dissent directly to the public [28,47]. A study conducted by Chi and Yang [10] found that Democratic congressmen tend to release information that citizens want to hear, while Republican congressmen share with the citizens their own agenda. Hemphill's work suggested that Congressmen of opposing parties use very different strategies to choose the hashtags that better reflect their framing efforts [30].

It appears that most literature either focuses on Twitter and elections, especially before and during election time, or focuses on President or Congressmen, even though "most Americans have more daily contacts with their state and local governments than with the federal government".[4]

Studies on State Governors and their social media presence are absent, and this paper aims at filling this gap. Although some research focuses on how politicians use social media before and during their election, what happens after that? Voters are excited about their party's success, and they are vocal about it. What comes after this initial excitement? We want to shed light on which Governors really follow their agenda after their election, and determine whether a framing of clear intents and goals emerges from their political channels online.

---

[4] White House: State and Local Government, 2015 https://www.whitehouse.gov/ 1600/state-and-local-government.

As of April 25, 2015, the 50 U.S. State Governors in charge collectively gathered over 3 million followers and sent out over 150,000 tweets. Though the majority of their Twitter accounts are merely political, some, such as Michigan Governor Rick Snyder's "OneToughNerd" account, show some character's personality traits, while others lend a certain intimacy, for example including family pictures like for Maryland Governor Larry Hogan, New Jersey Governor Chris Christie, Maine Governor Paul LePage and Louisiana Bobby Jingdal. Balancing personal lives and public service information makes State Governors' Twitter accounts very interesting objects to study the Governors' political stance in front of the public. This paper tries to dig into this unexplored field to analyze the State Governors' Twitter accounts by using agenda-setting theory, to understand whether the State Governors' activity on Twitter can be used to predict the popularity of parties or coalitions.

3   Agenda-Setting Theory

Twitter allows politicians to set their political agenda and reach their audience directly. Studying their behaviors brings the promising opportunity to further our understanding of *agenda setting* in digital media [46]. The agenda-setting theory is regarded as a key element to explain mass communication effects and mass media influence in long-term conditions. The primary assumptions of the theory were formulated by Maxwell McCombs and Donald Shaw in 1972 [39]. Agenda setting is one of the most widely used theories in communication studies since then [33,34,38,53,54].

Agenda setting is the filter mass media perform when selecting certain issues and portraying them frequently and prominently, which leads people to perceive those issues as more important than others. Two levels of agenda-setting theory will be used in this study. The first-level agenda setting focuses on the amount of coverage of an issue, suggesting which issues the public will be more likely to be exposed to. The second-level agenda setting, also called *framing* as suggested by McCombs, Shaw and Weaver [40], examines the influence of attribute salience, or the properties, qualities, characteristics, and relations. By making some political issues salient, agenda setting makes these specific issues more accessible than others.

The first level of agenda setting is the issue level. Though some scholars categorize top issues manually [46], we plan to use top issues listed on the White House's homepage. As of April 2015, the top seven issues listed were: economy, education, foreign policy, health care, immigration, climate change, energy and environment, and civil rights. April 2015 is also the time of our Twitter data collection. We will try to identify whether politicians give attention to these issues by analyzing how often kewords and hashtags related to these issue are mentioned on their Twitter accounts. In the second level of agenda setting, we will analyze whether Democrats and Republicans highlight different attributes of the same issue by examining the hashtags and keywords they choose when they do discuss an issue. We will also examine those hashtags and keywords relations by constructing occurrence networks to see how those hashtags and keywords are framed in the Governors' tweets.

Many researchers found different tweeting patterns among Democrats and Republicans Congressman, such as Shogan *et al.* [28,47] and Chi and Yang [10]. Our research as well aims to find whether State Governors' Twitter accounts exhibit different levels of engagement. Then, we would like to further our understanding of the general patterns of usage, applying the second level agenda-setting theory, or framing, to scrutinize the hashtags and keywords network structure. Hence, we formalize the following three research questions:

RQ 1: How frequently do Governors use Twitter to discuss their political agenda? Do party differences emerge?

RQ 2: How do Governors' Twitter accounts reflect their political agenda, and how similar political agendas are across Governors?

RQ 3: What similarities and differences emerge in hashtag usage among Governors' Twitter accounts?

4   Data Collection

We used the Governors' timelines to reference the tweets from the 50 U.S. Governors and the U.S. President Barack Obama. We collected 114,316 tweets from the Governors' timelines. We downloaded the stream of tweets for each account by querying the Twitter Public API for user timeline by using a manually-collected list of account names. This returns the entire stream of tweets for each account, avoiding sampling issues [42]. We performed the queries between January 23 and April 26, 2015, for all 51 accounts, in a systematic way and with a 100 s pause between each account. The pause was set to prevent our script from sending queries that exceed the rate limitation of the API. All data were finally stored into a JSON file and later analyzed.

We parsed each tweet to extract words and hashtags using the regular expression package *re* with Python 3. We first removed the URLs by excluding patterns starting with http, https, ftp, and mailto. Then, tweet texts were converted into lowercase for consistency. Finally, we obtained hashtags and words by another set of regular expressions. The hashtags were defined as sets of concatenated characters starting with a pound sign (#), while the words were defined as concatenated sets that start and end with alpha-numeric characters.

We identified the keywords by manually looking for the most frequent words that could be indicative of specific topics and sound meaningful to ordinary readers. To identify what could be the candidate words associated with each topic, we first manually parsed our collection of tweets and assigned the words that appeared together with the target topic as the candidate word selection for that topic.[5] For example, when we query for "health care" we will assign

---

[5] Given the massive size of the dataset, with over one hundred thousand tweets, this procedure required three annotators and countless hours of work.

each of the 17 words (we, will, fight, to, protect, the, healthcare, of, Floridians, their, right, to, be, free, from, federal, overreach) appearing in the tweet "We will fight to protect the healthcare of Floridians & their right to be free from federal overreach." as a candidate choice of keywords for health care. All the stop-words that were identified by the Python Natural Language Toolkit (NLTK) were removed. In the previous example, the set of candidate words after this further cleansing is reduced to (fight, protect, heathcare, Floridians, right, free, federal, overreach). The next step was to remove the words that are syntactically needed but not contextually meaningful. We identified the words that were a keywords of more than one topic and manually marked them to be further removed or not. Words that were shared by more than one topic were marked to be deleted if we were unable to find a potential topic for them; words that possibly related to any of the topics were marked to be kept. In the example, words to be deleted included: fight, protect, Floridians, right, federal, overreach. These words could not be attached unequivocally to any one topic. For example, the words *fight* and *protect* appeared more often attached to foreign and immigration issues, and the word *right* appeared more often related to civil right issues. Words to be kept included: *healthcare* (as well as *health care* with a space), and *freedom*, which could be assigned to health care, in particular related to the Affordable Care Act (or, ObamaCare). After we identified which words to delete or keep, we then updated the sets of each candidate keywords for each topic. We then ranked each candidate keywords by their overall frequency in our collection. The top seven candidate keywords for each category were used to identify the topic of each tweet. We assigned a tweet to a topic whenever any of the 7 keywords for a topic appeared in a tweet. The topics were not mutually exclusive: in other words, one tweet could be assigned to more than one topic when the top candidate keywords from different categories occurred in a tweet. We counted the numbers of tweets for each topic among the Governors. The agenda was finally recovered by ranking the topics by the numbers of tweets associated to them: the results are displayed in Table 1. The assessment of the quality of the agenda produced by our semi-automatic method is satisfactory: the seven topics are each clearly identified by a short list of intuitive keywords. By means of the same approach, we varied the number of keywords to include more words, finding that the results (discussed later) were substantially unaltered. Finally, the proposed method to generate the agenda was preferred over traditional topic modeling techniques that we tested, such as LDA, because of the inability of such probabilistic generative models [4] to discriminate between topics related to issues relevant to politics, and other irrelevant (for our purpose) topics that appeared in the Governors' Twitter timelines.

5   Experimental Results

5.1   Overall Tweeting Patterns

To try answer **RQ 1**, we analyze the 114,316 tweets collected from the Governor's timelines. The amount of tweets produced by each Governor ranged from

Table 1. Top words per category

Civil right	Economy	Education	Energy and Environment	Foreign	Health care	Immigration
Veterans	Economic	Education	Energy	Drug	Health	Investments
Citizens	Economy	Students	Manufacturing	Sexual	Food	Immigration
Rights	Unemployment	School	Water	Assault	Medicaid	Employment
Equal	Manufacturing	Veterans	Affordable	Campuses	Insurance	Sustainable
Marriage	Employees	Schools	Climate	Uniform	Transportation	Struggling
Defense	Transportation	Kids	Tech	Foreign	Affordable	Action
Restoration	Companies	College	Capital	Asia	Freedom	Portfolio

30 to 3,242, with a median of 2,838. These figures demonstrate that the majority of Governors is quite active on the platform. There were 46,125 tweets posted by the 19 Democrat Governors, and 68,047 by the 30 Republican ones: this suggests that, on average, each Democrat produced 2,427 tweets, and each Republican posted 2268 tweets; this difference is not particularly significant. President Obama contributed 3,242 to the Democrats, and the independent Governor of Arkansas had 144 tweets. We were able to identify 75,202 hashtags and words from the tweet texts after removing the URLs. Democrat Governors used 50,960 words while Republican governors used 41,263. The Democrats also tweeted more distinct hashtags, 6,463, while Republicans had only 4,264. A previous study conducted by Shogan *et al.* [28,47] on the House tweeting patterns suggested that Republicans tweet more, and Twitter might be particularly appealing to the American opposition politicians. Our analysis demonstrates that there are no significant differences in terms of average posting volumes between the two parties, and the larger sheer number of Republican tweets is to be attributed to the significantly greater number of Republican Governors (30 versus 19 Democrats). However some stylistic differences emerge, in that Democrat Governors seem to make a much more pervasive and diverse use of hashtags than Republicans.

5.2   Political Agenda and Keywords Usage

To answer **RQ 2**, we plan to describe each Governor's posting behavior according to the agenda we defined in Table 1. For each Governor's account, we calculated the number of times each keyword of Table 1 appeared in any of the Governor's tweets. By sorting this dictionary of keywords and relative usage in descending order, we can obtain a rank of each Governor's keyword usage. We can therefore use the ranked keyword dictionaries to perform pairwise comparisons of Governors and try capture similarities and differences in priorities regarding the categories of political discussion. Note that using rankings is preferable to using simple feature vectors of keyword counts: ranks are more amenable to direct comparisons (for example via Spearman's rank correlation) without data normalization to account for different intensity of activities and other biases.

To measure the correlation of discussion keywords between all pairs of Governors, we use Spearman's correlation applied on their ranked keyword dictionaries. Spearman's rank correlation assigns each pair $< X_i, X_j >$ a similarity score between -1 and 1, with X_i and X_j being the keyword ranks of Governors i and j respectively. Score of 1 and -1 indicate perfect positive and negative correlation, respectively, whereas a score of 0 suggests no correlation. To understand the distribution of pairwise correlation scores, we plotted Fig. 1. The range of scores spans roughly from -0.2 (indicating a slight negative correlation) to very strong positive correlation scores greater than 0.8. The skewness towards positive scores can be attributed to the fact that we have considered only seven words per category, with seven total categories, for determining the rank distributions.

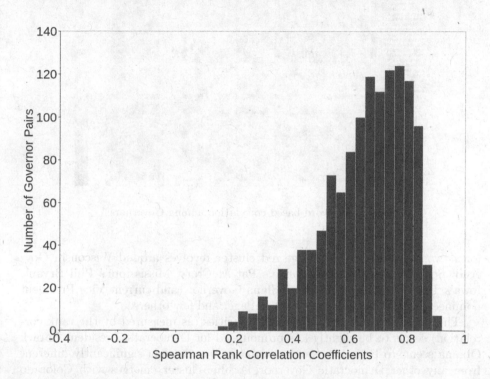

Fig. 1. Distribution of spearman rank correlation scores

Figure 2 shows the matrix of pairwise Spearman correlations among the 50 U.S. Governors plus the U.S. President Barack Obama. The visual inspection of Fig. 2 suggests the presence of a strong block structure, as groups of highly correlated accounts happen to be clearly identifiable. To further inspect this hypothesis, we generated a weighted graph of inter-Governor similarity using the matrix of Fig. 2 as adjacency matrix. The resulting graph is displayed in Fig. 3, where for visual clarity, self-loops have been removed and all edges with weights (i.e., Spearman correlation) less than 0.8 have been filtered out. Figure 3 captures the agenda similarity network among Governors. Its analysis suggests the emergence of a strong community structure, where some Republican and Democratic Governors appear to be strongly aligned on agenda priorities and

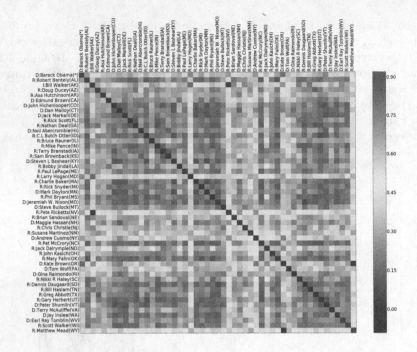

Fig. 2. Keyword-based correlation among Governors

form two tight clusters: the large red cluster revolves around Wisconsin Governor Scott Walker, North Carolina's Pat McCrory, Mississippi's Phil Bryant, Iowa's Terry Branstad, (former) Indiana Governor (and current Vice President nominee) Mike Pence, Maine's Paul LePage, and few others.

The similarity, in terms of agenda priorities (as measured by the rank correlation) seems to be slightly less pronounced for Democrats: President Barack Obama seems to be isolated and carrying out an agenda significantly different from any other Democratic Governor. A blue cluster emerges with Colorado Governor John Hickenlooper, Missouri's Jay Nixon, Kentucky's Steve Beshear, and Washington's Jay Inslee, and Vermont's Peter Shumlin and few others show some agenda similarity. All the other Governors somehow sit at the periphery of this network showing spurious alignments with some of their counterparts, and a less pronounces inter-party agenda priority sharing.

5.3   The Governor-Hashtag Graph

To address **RQ 3**, we finally explored the similarity among the governors at a hashtag level. We extracted the hashtags from each Governor's timeline and created a Governor-hashtag graph. The nodes in this bipartite graph represent the Governors and the hashtags they used. A Governor node and a hashtag node would be connected if the Governor had used the hashtag in any of his/her tweet. The weight is the number of tweets that contain that hashtag. We only extracted

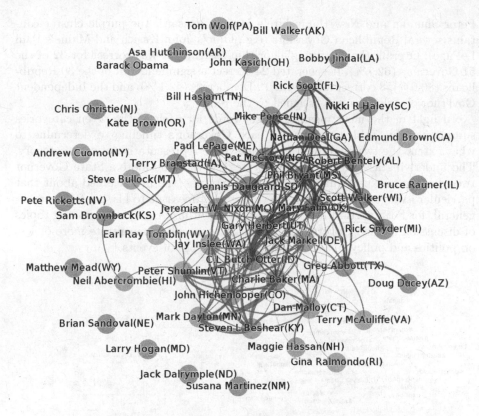

Fig. 3. Governors network through the lens of agenda setting theory

the hashtags that were used more than 10 times among all the Governors and by more than two Governors, to focus specifically on more common hashtags. We were able to identify 658 common hashtags that occurred more than 10 times and were used by more than two Governors from our collection. We also tried to recover the community structure by using the Louvain modularity maximization algorithm [5]. The result for the Governors' hashtag usage are demonstrated in Fig. 4. The graph only represents the nodes that were connected with edges with weights larger than four, for visual clarity. The large circles denoted the nodes for Governors, and the small ones were nodes for hashtags.

We were able to identify four communities using the modularity algorithm with the resolution set to 2.0. Varying the resolution limit parameter [24] provided consistent results. The four communities contained 36, 9, 3 and 3 Governors, respectively, as shown in Fig. 4. We colored the largest community in red to indicate that it's the community with the largest fraction of Republican Governors (24). The second community is colored in blue to indicate that it's the community with the largest fraction of Democrats (8). The other two communities were colored in green and purple, respectively. We believe that the green cluster should belong to the Democrats (it contains Dems like Vermont's

Peter Shumlin and New Hampshire's Maggie Hassan); the purple cluster contains several Republican Governors (e.g., Ohio's John Kasich and Maine's Paul LePage). Overall, the clustering algorithm assignment was correct for 32 of the 51 Governors (62.7 %). It generated 24 correct assignments out of the 30 Republicans (80.0 %), 8 correct among the 19 Democrats (42.1 %), and the Independent Governor of Arkansas was assigned to the reds.

In light of the most meaningful keywords for each of the seven categories summarized by Table 1, we parsed each Governor's timeline to determine to what extent the tweets of each individual were representative of each category. The underlying assumption of this strategy is that the more a State Governor tweets about any particular category, the more he/she is concerned about that particular issue, or at least wants to convey that message to his/her followers. In general, for both parties, it is quite easy to scrutinize the most recurring topics of discussion of each Governor and identify those who concentrate more or less on politics and policy related topics, or other types of events.

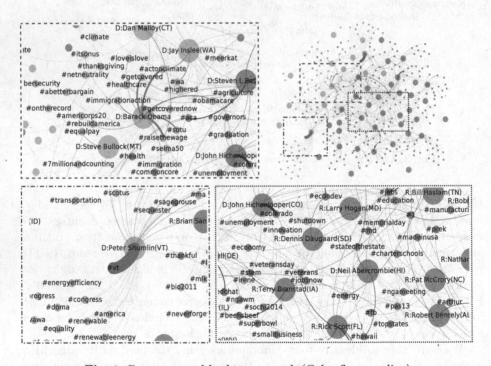

Fig. 4. Governors and hashtag network (Color figure online)

Figure 4 illustrates the most commonly occurring hashtags and issues of discussion of the two groups. Its analysis yields a good amount of insights into U.S. political discussion. One can notice the commitment of certain Governors to specific topics: for example, Vermont Governor Peter Shumlin seems pushing an agenda focused on environment, energy, and local economy issues. Other

Democrats, like Connecticut Governor Dan Malloy, Arkansas' Asa Huthinson, the U.S. President Obama, focus on issues related to climate change, equality, health care, and education.

The Republican agenda is sufficiently diverse but focuses mostly on issues related to economy (small business, innovation, "made in USA", agriculture), immigration and security (human trafficking, Texas), and civil rights (especially veterans', military, and marriage rights). A number of external events are also discussed (note that we did not remove any hashtags from the Governor-hashtag graph as long as it matched the threshold criteria explained above): some examples include reference to sport events (Nascar, Basket's March Madness, etc.), political events (2012 Elections, the GOP Convention, etc.) and tragedies (the Boston Marathon bombing, the Sandy Hook school shooting, etc.).

6. Conclusions

In this article we explored the landscape of U.S. Governors political communication on Twitter using the tool of agenda setting theory. We first collected a sizable amount of tweets (over one hundred thousand) generated by these politicians, and assessed that most of them are quite active Twitter users. Our results clarified some previous research about the usage of social media platforms by Democratic and Republican politicians, showing that Republican and Democrat Governors tend to be more or less equally active on Twitter on average, however they exhibit different styles of communication, with the Democrats significantly more inclined to use hashtags than their counterparts.

We furthered our understanding of Governors' priorities using the agenda-setting theory to identify a set of seven categories of top socio-political issues, by means of a semi-automatic annotation strategy. After inferring the priorities of each Governor, and computing the pairwise similarity among Governors, we constructed a network that reflects Governor agendas similarity. Its analysis illustrates that President Obama has a distinctive agenda-setting strategy, which has no affinity with either Democrats or Republicans.

The graph also shows that Republican Governors, such as Wisconsin Governor Walker, North Carolina's McCrory, Mississippi's Bryant, Iowa's Branstad, Indiana's Pence, Maine's Paul LePage, and few others, shared the most similar issue agenda settings. On the Democratic side, Colorado Governor Hickenlooper, Missouri's Nixon, Kentucky's Beshear, and Washington's Inslee, and Vermont's Shumlin and few others form a tight blue cluster of aligned agendas. Republican and Democratic Governors' clusters tend to be quite polarized, which confirms the intuition that the two parties share significantly different agendas (at times conflicting) and different political priorities. Similar insights emerged from the analysis of the hashtag co-occurrence networks, which allows for an easy identification of the topics of discussion of both parties.

This study displayed the high-level dynamics of adoption of Twitter by U.S. Governors based on how they set their agenda on top political issues and how they frame their conversation around it. Further studies should explore the *public agenda setting*, which means the agenda setting of the public in each State, to see if these share similar trends with their Governors' agendas. This would shed light on the effects of politicians' social media conversation on the public.

References

1. Adamic, L.A., Glance, N.: The political blogosphere and the 2004 US election: divided they blog. In: Proceedings of the 3rd International Workshop on Link Discovery, pp. 36–43. ACM (2005)
2. Bakshy, E., Hofman, J.M., Mason, W.A., Watts, D.J.: Everyone's an influencer: quantifying influence on Twitter. In: Proceedings of the Fourth ACM International Conference on Web Search and Data Mining, pp. 65–74. ACM (2011)
3. Bekafigo, M.A., McBride, A.: Who tweets about politics? Political participation of twitter users during the 2011 gubernatorial elections. Soc. Sci. Comput. Rev., 0894439313490405 (2013)
4. Blei, D.M.: Probabilistic topic models. Commun. ACM **55**(4), 77–84 (2012)
5. Blondel, V.D., Guillaume, J.L., Lambiotte, R., Lefebvre, E.: Fast unfolding of communities in large networks. J. Stat. Mech. Theory Exp. **2008**(10), P10008 (2008)
6. Bond, R.M., Fariss, C.J., Jones, J.J., Kramer, A.D., Marlow, C., Settle, J.E., Fowler, J.H.: A 61-million-person experiment in social influence and political mobilization. Nature **489**(7415), 295–298 (2012)
7. Boyd, D., Crawford, K.: Critical questions for big data: provocations for a cultural, technological, and scholarly phenomenon. Inf. Commun. Soc. **15**(5), 662–679 (2012)
8. Carlisle, J.E., Patton, R.C.: Is social media changing how we understand political engagement? An analysis of facebook and the 2008 presidential election. Polit. Res. Q. **66**(4), 883–895 (2013)
9. Cha, M., Haddadi, H., Benevenuto, F., Gummadi, P.K.: Measuring user influence in Twitter: the million follower fallacy. ICWSM **10**(10–17), 30 (2010)
10. Chi, F., Yang, N.: Twitter in congress: outreach vs transparency. Soc. Sci. **1**, 1–20 (2010)
11. Cohen, R., Ruths, D.: Classifying political orientation on Twitter: it's not easy! In: ICWSM (2013)
12. Conover, M., Ratkiewicz, J., Francisco, M.R., Gonçalves, B., Menczer, F., Flammini, A.: Political polarization on Twitter. ICWSM **133**, 89–96 (2011)
13. Conover, M.D., Davis, C., Ferrara, E., McKelvey, K., Menczer, F., Flammini, A.: The geospatial characteristics of a social movement communication network. PloS ONE **8**(3), e55957 (2013)
14. Conover, M.D., Ferrara, E., Menczer, F., Flammini, A.: The digital evolution of occupy wall street. PloS ONE **8**(5), e64679 (2013)
15. Conover, M.D., Gonçalves, B., Ratkiewicz, J., Flammini, A., Menczer, F.: Predicting the political alignment of twitter users. In: 2011 IEEE Third International Conference on Privacy, Security, Risk and Trust (PASSAT) and 2011 IEEE Third Inernational Conference on Social Computing (SocialCom), pp. 192–199. IEEE (2011)
16. De Meo, P., Ferrara, E., Fiumara, G., Provetti, A.: On facebook, most ties are weak. Commun. ACM **57**(11), 78–84 (2014)
17. Diakopoulos, N.A., Shamma, D.A.: Characterizing debate performance via aggregated Twitter sentiment. In: Proceedings of the SIGCHI Conference on Human Factors in Computing Systems, pp. 1195–1198. ACM (2010)

18. DiGrazia, J., McKelvey, K., Bollen, J., Rojas, F.: More tweets, more votes: social media as a quantitative indicator of political behavior. PloS ONE **8**(11), e79449 (2013)

19. Effing, R., Hillegersberg, J., Huibers, T.: Social media and political participation: are Facebook, Twitter and YouTube democratizing our political systems? In: Tambouris, E., Macintosh, A., Bruijn, H. (eds.) ePart 2011. LNCS, vol. 6847, pp. 25–35. Springer, Heidelberg (2011). doi:10.1007/978-3-642-23333-3_3

20. Ellison, N.B., Vitak, J., Gray, R., Lampe, C.: Cultivating social resources on social network sites: Facebook relationship maintenance behaviors and their role in social capital processes. J. Comput. Mediated Commun. **19**(4), 855–870 (2014)

21. Ferrara, E.: Manipulation and abuse on social media. ACM SIGWEB Newslett. (Spring), 4 (2015)

22. Ferrara, E., Varol, O., Davis, C., Menczer, F., Flammini, A.: The rise of social bots. Commun. ACM **59**(7), 96–104 (2016)

23. Ferrara, E., Varol, O., Menczer, F., Flammini, A.: Traveling trends: social butterflies or frequent fliers? In: Proceedings of the First ACM Conference on Online Social Networks, pp. 213–222. ACM (2013)

24. Fortunato, S., Barthelemy, M.: Resolution limit in community detection. Proc. Natl. Acad. Sci. **104**(1), 36–41 (2007)

25. Gayo-Avello, D.: I wanted to predict elections with Twitter and all i got was this lousy paper-a balanced survey on election prediction using twitter data. arXiv preprint arXiv:1204.6441 (2012)

26. Gerbaudo, P.: Tweets and the Streets: Social Media and Contemporary Activism. Pluto Press, London (2012)

27. Gibson, R.K., McAllister, I.: Does cyber-campaigning win votes? Online communication in the 2004 Australian election. J. Elections Pub. Opin. Parties **16**(3), 243–263 (2006)

28. Glassman, M., Straus, J.R., Shogan, C.J.: Social networking and constituent communication: Member use of twitter during a two-week period in the 111th congress. Congressional Research Service, Library of Congress (2009)

29. Golbeck, J., Grimes, J.M., Rogers, A.: Twitter use by the us congress. J. Am. Soc. Inform. Sci. Technol. **61**(8), 1612–1621 (2010)

30. Hemphill, L., Culotta, A., Heston, M.: Framing in social media: How the us congress uses twitter hashtags to frame political issues. Available at SSRN 2317335 (2013)

31. Hemphill, L., Otterbacher, J., Shapiro, M.: What's congress doing on Twitter? In: Proceedings of the 2013 Conference on Computer Supported Cooperative Work, pp. 877–886. ACM (2013)

32. Howard, P.N., Duffy, A., Freelon, D., Hussain, M.M., Mari, W., Maziad, M.: Opening closed regimes: what was the role of social media during the arab spring? Available at SSRN 2595096 (2011)

33. Iyengar, S.: Is Anyone Responsible? How Television Frames Political Issues. University of Chicago Press, Chicago (1994)

34. Iyengar, S., Simon, A.F.: New perspectives and evidence on political communication and campaign effects. Ann. Rev. Psychol. **51**(1), 149–169 (2000)

35. Lazer, D., Pentland, A.S., Adamic, L., Aral, S., Barabasi, A.L., Brewer, D., Christakis, N., Contractor, N., Fowler, J., Gutmann, M., et al.: Life in the network: the coming age of computational social science. Science **323**(5915), 721 (2009). (New York, NY)

36. Lerman, K., Ghosh, R.: Information contagion: an empirical study of the spread of news on Digg and Twitter social networks. In: Proceedings of the 4th International AAAI Conference on Weblogs and Social Media, pp. 90–97 (2010)

37. Macnamara, J.: The quadrivium of online public consultation: policy, culture, resources, technology. Aust. J. Polit. Sci. **45**(2), 227–244 (2010)
38. McCombs, M.: A look at agenda-setting: past, present and future. Journalism Stud. **6**(4), 543–557 (2005)
39. McCombs, M.E., Shaw, D.L.: The agenda-setting function of mass media. Pub. Opin. Q. **36**(2), 176–187 (1972)
40. McCombs, M.E., Shaw, D.L., Weaver, D.H.: Communication and Democracy: Exploring the Intellectual Frontiers in Agenda-setting Theory. Psychology Press, Mahwah (1997)
41. Metaxas, P.T., Mustafaraj, E.: Social media and the elections. Science **338**(6106), 472–473 (2012)
42. Morstatter, F., Pfeffer, J., Liu, H., Carley, K.M.: Is the sample good enough? Comparing data from twitter's streaming api with twitter's firehose. arXiv preprint arXiv:1306.5204 (2013)
43. Parmelee, J.H., Bichard, S.L.: Politics and the Twitter revolution: How tweets influence the relationship between political leaders and the public. Lexington Books (2011)
44. Pentland, A.: Social Physics: How Good Ideas Spread-the Lessons From a New Science. Penguin Press, New York (2014)
45. Ratkiewicz, J., Conover, M., Meiss, M., Goncalves, B., Flammini, A., Menczer, F.: Detecting and tracking political abuse in social media. In: Proceedings of the 5th International AAAI Conference on Weblogs and Social Media, pp. 297–304 (2011)
46. Russell Neuman, W., Guggenheim, L., Mo Jang, S., Bae, S.Y.: The dynamics of public attention: agenda-setting theory meets big data. J. Commun. **64**(2), 193–214 (2014)
47. Shogan, C.J.: Blackberries, tweets, and Youtube: technology and the future of communicating with congress. PS Polit. Sci. Polit. **43**(02), 231–233 (2010)
48. Stieglitz, S., Dang-Xuan, L.: Political communication and influence through microblogging-an empirical analysis of sentiment in Twitter messages and retweet behavior. In: 2012 45th Hawaii International Conference on System Science (HICSS), pp. 3500–3509. IEEE (2012)
49. Tumasjan, A., Sprenger, T.O., Sandner, P.G., Welpe, I.M.: Election forecasts with Twitter: how 140 characters reflect the political landscape. Soc. Sci. Comput. Rev., 0894439310386557 (2010)
50. Varol, O., Ferrara, E., Ogan, C.L., Menczer, F., Flammini, A.: Evolution of online user behavior during a social upheaval. In: Proceedings of the 2014 ACM Conference on Web Science, pp. 81–90. ACM (2014)
51. Wang, Y., Feng, Y., Zhang, X., Niemi, R., Luo, J.: Will sanders supporters jump ship for trump? Fine-grained analysis of twitter followers. arXiv preprint arXiv:1605.09473 (2016)
52. Wang, Y., Li, Y., Luo, J.: Deciphering the 2016 US presidential campaign in the twitter sphere: a comparison of the trumpists and clintonists. arXiv preprint arXiv:1603.03097 (2016)
53. Wanta, W., Ghanem, S.: Effects of agenda setting. Mass media effects research: Advances through meta-analysis, pp. 37–51 (2007)
54. Weaver, D., McCombs, M., Shaw, D.L.: Agenda-setting research: issues, attributes, and influences. In: Handbook of Political Communication Research, pp. 257–282 (2004)

Markets, Crowds, and Consumers

Preference-Aware Successive POI Recommendation with Spatial and Temporal Influence

Madhuri Debnath[✉], Praveen Kumar Tripathi, and Ramez Elmasri

University of Texas at Arlington, Arlington, TX, USA
{madhuri.debnath,praveen.tripathi}@mavs.uta.edu, elmasri@uta.edu

Abstract. There have been vast advances and rapid growth in Location based social networking (LBSN) services in recent years. Point of Interest (POI) recommendation is one of the most important applications in LBSN services. POI recommendation provides users personalized location recommendation. It helps users to explore new locations and filter uninteresting places that do not match with their interests. But traditional POI recommendation cannot suggest where a user may go the next day or next hour based on their current location or status. In this paper, we consider the task of personalized successive POI recommendation, recommending to a user the very next location where he might be interested to go next based on his current location. Multiple factors influence users to choose a POI, such as user's categorical preferences, temporal activities and location preferences, popularity of a POI as well as sequential patterns of a user. In this work, we define a unified framework that takes all these factors into consideration to build a better successive POI recommendation model. We evaluate our system with a real-world dataset collected from Foursquare. Experimental results show that our proposed framework works better than other baseline approaches.

Keywords: Successive POI recommendation · Location-based social network

1 Introduction

In recent years, location based social network (LBSN) services have gained a vast amount of attention and popularity among users. Foursquare [1], Yelp [2] and Facebook Places [3] are a few of the examples of LBSN services. LBSNs allow users to share their life experiences via mobile devices. "Check-in" is a process by which users post their arrival to a location. They also share their experiences by leaving comments or tips on that location. A *Point of Interest* (POI) location can be a "Restaurant", "Travel spot", "Park" and so on.

It was reported that there are over 30 million registered users in Foursquare. The number of check-ins posted by them by January 2013 was over 3 billion [4]. The "check-ins" contain abundant information about their daily activities as

© Springer International Publishing AG 2016
E. Spiro and Y.-Y. Ahn (Eds.): SocInfo 2016, Part I, LNCS 10046, pp. 347–360, 2016.
DOI: 10.1007/978-3-319-47880-7_21

well as their preferences among the POIs. For example, people who often visit a gym must be interested in physical exercise. Also, people who visit the same place may share similar interests. Location histories and opinions of one user can be exploited to recommend an unvisited location to another user if they share a similar interest.

The task of *POI recommendation* is to provide personalized recommendation of POI locations to mobile users. The recommended locations should match their personal interests within a geospatial range [5]. Recently, POI recommendation in LBSNs has attracted much attention in both research and industry [6,7]. However traditional POI recommendation systems consider all check-ins as a whole and generate recommendations [8–12]. They do not consider the users' sequential movement information. Therefore, they cannot suggest where a user may go in the next few hours based on their current location or status.

In this work, we consider the task of personalized successive POI recommendation. Successive POI recommendation refers to the problem of recommending users the very next location based on his current location and current time. This task recommends those locations that a user may not visit frequently or before, but he/she may like to visit at successive timestamps [13]. For example, successive POI recommendation can suggest a user location to have fun after dinner, or a location for outdoor activities in a nearby park after his work.

The essential difference between traditional recommendation system and successive POI recommendation system is that the performance of successive recommendation tasks is largely influenced by users' current visiting locations [14]. Also the shift from one location to another location depends on their categorical preferences and periodic patterns. One may go to a coffee shop to grab a cup of coffee first, then head to work or university. On a weekend people often go to shopping, then go to a restaurant for dinner or lunch.

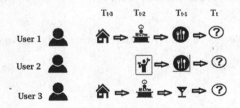

Fig. 1. Sequential check-in data of three users

Figure 1 gives examples of sequential check-in data of three users. User 1 goes to dinner from the office. User 3 goes to a bar after office. If user 1 and user 3 share similar interests, user 1 also may become interested to go to a bar after office. Thus collaborative information shared by the users can be used to recommend them the possible next locations based on their current location.

In [15], we proposed a preference-aware, location-aware and time-aware POI recommendation system. The method used *User-based Collaborative Filtering* method for POI recommendation while incorporating four other factors: (1) Categorical preferences, (2) Temporal influence, (3) Geographical preferences and (4) Popularity of POIs. In this paper, we extend this work and propose a preference-aware successive POI recommendation system with incorporating spatial and temporal influence (PLTSRS) that offers a particular user a set of POI locations based on his current location, current time and his personal interests. The contribution of this paper can be summarized as follows:

– We model personal preferences of users based on the category information of their location histories. We further analyse the temporal influence on their activities. We incorporate time dimension to model time-specific user preferences.
– We mine sequential patterns from check-in location of each user. Then, we construct personalized *Category-To-Category* transition probability matrix using first order markov chain [16].
– We analyze users' spatial behavior and incorporate spatial influence to generate spatial-aware location recommendations.
– Our recommendation model uses popularity factor of individual location by calculating time-specific popularity.
– We develope a successive POI recommendation model **PLTSRS** (**P**reference-**A**ware, **L**ocation-**A**ware and **T**ime-**A**ware **S**uccessive POI **R**ecommendation **S**ystem), which jointly considers user's personalized sequential movement information, temporal categorical preferences, location preferences and popularity of POIs. To best of our knowledge, this is the first work that uses all the factors together to build a successive POI recommendation model.
– We evaluate our proposed framework with one large scale LBSN dataset from foursquare [1].

2   Related Work

With the easy availability of users' check-in data in LBSN, many studies have been conducted for POI recommendation. In this section, we briefly introduce two lines of research related to our task: (1) Traditional POI recommendation, (2) Successive POI recommendation.

Traditional POI recommendation systems have been extensively studied in the last several years. Two popular approaches have been used to generate recommendation model: Collaborative Filtering algorithm and Non-Negative Matrix Factorization algorithm.

In [6], the *User-based CF* approach considers a combination of social influence and spatial influence. Their experiments report that geographical influence has a significant impact on the accuracy of POI recommendation, whereas the social friend link contributes little. Their results also indicate that *user-based CF* works much better than *Item-based CF*. In [10], the authors exploit spatial influence as well as temporal influence for building a recommendation model. They incorporate time factors in the basic *CF based* model by computing similarity between two users by considering check-in information at a specific time t, rather than that of all times. In [11], the authors explore user preferences with social and geographical influence for POI recommendation. They model user preferences using predefined categorical information of location data.

In [8], the authors propose a geographical probabilistic factor analysis framework for recommendation that takes various other factors into consideration, viz. user-item preferences, POI popularity and geographical preferences of individual users. In [17], the authors propose a friendship based collaborative filtering (FCF) approach for POI recommendation.

In [15], the authors propose a *User Based Collaborative Filtering* method based framework which combines 4 factors: categorical preferences, temporal influence, spatial influence and popularity of a location. They incorporate time factors by generating time specific categorical preferences. Clustering method has been used to model location preference of each user. Popularity of each location has been calculated by combining both regional factor and temporal factor.

Lately a few successive POI recommendation works have been conducted. In [7], the authors propose a probabilistic model to integrate category transition probability and POI popularity to solve the problem. But they did not incorporate spatial influence here.

In [18], the authors propose a Factorized Personalized Markov Chain (FPMC) model for next-item recommendation. In [13], the authors propose *FPMC-LR* model by extending *FPMC* model with localized region constraint to solve successive POI recommendation task. They divide the geographical space into a grid. Locations of the grid cell the user is currently visiting and its surrounding 8 grid cells are used as candidate locations. This condition is called *Localized Region Constraint*. In [19], the authors propose a personalized metric embedding method (*PRME*) to model personalized check-in sequences for next new POI recommendation.

3   Preliminaries

3.1   Data Structure

In this paper, we use one real-world LBSN dataset from *Foursquare*. This dataset has three key data structure: (1) User, (2) POI location and (3) Check-in.

(1) Each user u is represented by a unique id. Let $U = \{u_1, u_2, u_3, \ldots u_n\}$ be the set of users.
(2) Each POI location has a unique POI id and geographical position (latitude and longitude). Let $L = \{l_1, l_2, l_3, \ldots, l_m\}$ be the set of POI locations. Each location l is also associated with category information, which represents its functionality. In Foursquare, there are 8 primary categories ("Food", "Arts and Crafts" etc.). Each primary category includes other sub-categories. In this paper, we only consider the sub-category information of a location for simplicity. The word category and sub-category will be used interchangeably throughout the paper.

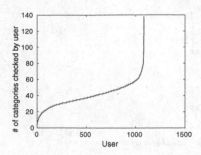 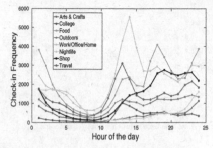

(a) Number of unique categories checked-in by users

(b) Check-in frequency at different hour of the day

Fig. 2. Data analysis: categorical preference and temporal influence

(3) *"Check-in"* is a process by which a user u announces his physical arrival or presence at a venue in location based social network. Let $Ch_{ij} = \{u_i, l_j, t\}$ be a *check-in* tuple, which represents that user u_i checked in POI l_j at time t.

3.2 Data Analysis

In this section, we present some data analysis results to see how different factors (Spatial, Temporal, Preference) influence a user to choose a location to visit.

Categorical Preference Constraints. Personal preference plays an important rule for a user to choose a POI. They prefer to visit a location only if the category of that location matches their interests. To have a better idea, we count the number of unique categories visited by users. We sort the users based on the count and plot the result (see Fig. 2a). We have a total of 252 categories. We see that the number of categories visited by most of the users is less than 60. Users generally do not visit locations of all categories, they visit a location only if they like the category. So a good POI recommendation system must recommend a location to a user that matches with his preferences.

Temporal Influence. User activities are significantly influenced by time [10]. We count the check-in frequency of 8 primary categories at different hours of the day (see Fig. 2b). Result shows that category "Shop" is more active from 3 pm to until 12 am. On the other hand, category "Nightlife" starts after 10 pm and continues until 5 am.

We have done analysis to see how frequently people visit locations. We plot the *Cumulative Distribution Function (CDF)* of the time differences between successive check-in data (see Fig. 3a). Result shows that 90 % of successive check-ins have a time difference less than 200 min.

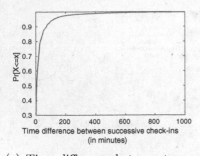
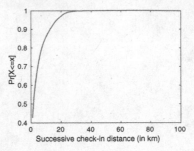

(a) Time difference between two suc-   (b) Geographical distance between two
cessive check-ins (in minutes)   successive check-ins

Fig. 3. Data analysis: spatial and temporal influence

Spatial Influence. Geographical position of a POI location plays an important
role. Figure 3b shows the *Cumulative Distribution Function (CDF)* of geographical distance between two successive check-ins. Result shows that about 90 % of
successive check-ins have a geographical distance less than 20 km.

3.3   Problem Formulation

Let U be the set of users and P be the set of locations. L_u denotes the check-in
histories of user u. Given a user $u(u \in U)$, his check-in histories L_u, his currently
visiting POI $l_{now}(l_{now} \in L)$ and corresponding visiting timestamp t_{now}, the task
is to recommend a new POI $l_{next}(l_{next} \notin L_u)$ to u to visit within time range t_{now}
to t_{next}. Here, $(t_{next} - t_{now}) \leq T_{max}$. Here, T_{max} is a user defined time interval
parameter.

3.4   User-Based Collaborative Filtering

User-based CF first finds similar users based on their interests/ratings on items
using a similarity measure. Then the recommendation score for an item is computed by the weighted combination of historical ratings on the item from similar
users [20].

Given a user $u \in U$, the recommendation score that u will check-in a POI l
that she has not visited yet is computed with the following equation,

$$R_u(l) = \frac{\sum\limits_{v \in U} w_{uv}}{|v|} \tag{1}$$

Here $v \in U$ are list of users who have visited the same location l and w_{uv} is
the similarity score between u and v.

4 PLTSRS Framework

Our proposed framework is comprised of two major steps. (1) *Offline Modeling* and (2) *Online Recommendation.*

The *Offline Modeling* step has 3 components. (1) *Learning User's Categorical Transition Probabilities,* (2) *Time-specific Personal Preference Discovery* and (3) *Calculating Time-Specific Popularity of Locations.*

In the first component, we learn each user's categorical transition probability denoted by $T_u(c_i, c_j)$. $T_u(c_i, c_j)$ is calculated using first order markov chain that indicates the probability of user u to move from a location of category c_i to location with category c_j. In the second component, we learn each user's personal categorical preference on category c denoted as $P_u(c)$. As preference depends on time, we learn time-specific categorical preference on category c at time segment t_s denoted as $P_u^{(t_s)}(c)$. In the third component, we calculate the time-specific popularity of each POI location l denoted as $\rho^{(t_s)}(l)$.

The *Online Recommendation* has two components. (1) *Spatial-Aware Candidate Selection* and (2) *Successive Location Recommendation.* The first component selects a set of candidate locations based on u's current location l_{now}. This component improves the efficiency of the approach significantly as the number of candidate locations is much smaller than the total number of locations. Given a user u, his current location l_{now} and current time t_{now}, the second component calculates the location ratings of all candidate locations based on the factors mentioned above. The *top-K* locations are recommended to user u.

4.1 Offline Modeling

Categorical Transition Probability. In this step, for each user u, we first extract the successive location pairs from his check-in sequences. A location pair (l_i, l_j) is a successive pair if the time difference between u's visit at location l_i and location l_j is less than the time interval threshold T_{max}. Then, we map the locations of the successive location pairs with corresponding category information to mine the successive category pairs (c_x, c_y). Here, c_x is the category of l_i and c_y is the category of l_j. We build a *Category-To-Category* transition probability matrix of user u denoted as T_u. The transition matrix $T_u \in [0,1]^{|C| \times |C|}$, $T_u(i,j)$ specifies the probability for a user u to move from location with category c_i to a location with category c_j. Transition probability $T_u(i,j)$ is calculated as:

$$T_u(i,j) = \frac{|\{(l_1^{(u)}, l_2^{(u)}) : l_1^{(u)} \in C_i \cap l_2^{(u)} \in C_j\}|}{|\{(l_1^{(u)}, l_2^{(u)} : l_1^{(u)} \in C_i\}|} \tag{2}$$

Figure 4a shows the transition matrices of individual users. Entries with "?" refers to missing values as there is no data to estimate probabilities. A single user generally does not visit all categories, so there may be a lot of missing values in his transition matrix. To solve this problem, we use low-rank non-negative matrix factorization [21] algorithm to factorize each transition probability matrix T_u into two low rank matrices $W_u \in \mathbb{R}^{k \times |C|}$ and $H_u \in \mathbb{R}^{|C| \times k}$, with $k \ll |C|$

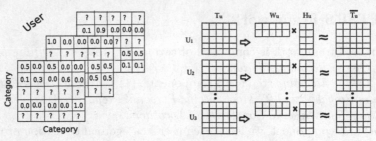

(a) Personalized Transition Matrices of users   (b) Factorized individual transition probability matrix

Fig. 4. Categorical transition probability

being the number of latent factors. After obtaining W_u and H_u, the probability matrix T_u is approximated as $\bar{T}_u$, $\bar{T}_u(i,j)$ being the approximated probability of transition from category c_i to category c_j (see Fig. 4b).

Personal Preference Discovery. In this step, we model each individual user's categorical preferences from his/her check-in history. Categorical preference of a user u denoted as $P_u(c)$ represents u's affinity to visit a location with category c. $P_u(c)$ is generated using following equation [15].

$$P_u(c) = CF(c, L_u) \times ILF(c, L) \tag{3}$$

Here $CF(c, L_u)$ is the measure of how many times user u has visited the locations with a category c. Intuitively, a user would visit more locations belonging to a category if he likes it. Here L_u is the location set visited by u. ILF handles the *Rare-Item* problem [22]. Some locations are not visited by a user very often. For example, the number of visits to a restaurant is generally more than that of a museum. If a user visits location of a category that is rarely visited by other users, it means that the user could like this category more prominently [11].

CF is calculated using Eq. (4) and ILF is calculated using Eq. (5).

$$CF(c, L_u) = \frac{|\{u.l_i : l_i \in c\}|}{|L_u|} \tag{4}$$

$$ILF(c, L) = \log \frac{|U|}{|\{u_j.l \in L : l_j.c \in u_j.C\}|} \tag{5}$$

Here, $|\{u.l_i : l_i \in c\}|$ is user u's number of visits in category c, $|L_u|$ is the total number of user's visit in all locations. $|U|$ is the number of total users in the system. $|\{u_j.l \in L : l_j.c \in u_j.C\}|$ is the number of users who visit category c among all users in U. User similarity between two users is calculated based on their categorical preferences. We use *Cosine Similarity* [23] to find the similarity between two users u and v denoted as w_{uv}.

Temporal Categorical Preference. As categorical preference may vary over time, we intend to find time-specific categorical preferences of each user. We divide the whole day into equal length of time segment (t_s). In this paper, we use time slot length $= 1$ h. So the whole day is divided into 24 time segments. Given a user u, time segment t_s, category c, temporal preference of user u on category c, denoted as $P_u^{(t_s)}(c)$ is calculated using following equation [15].

$$P_u^{(t_s)}(c) = CF^{(t_s)}(c, L_u^{(t_s)}) \times ILF(c, L^{(t_s)}) \tag{6}$$

Here $CF^{(t_s)}(c, L_u^{(t_s)})$ is the *Category Frequency* of user u and category c at time segment t_s. $L_u^{(t_s)}$ is the location set visited by u at t_s. $ILF(c, L^{(t_s)})$ is the *Inverse Location Frequency* for category c. $L^{(t_s)}$ is the list of all locations that has been visited at t_s. $CF^{(t_s)}(c, L_u^{(t_s)})$ and $ILF(c, L^{(t_s)})$ are calculated using following equations.

$$CF^{(t_s)}(c, L_u^{(t_s)}) = \frac{|\{u.l_i^{(t_s)} : l_i^{(t_s)} \in c\}|}{|L_u^{(t_s)}|} \tag{7}$$

$$ILF(c, L^{(t_s)})) = \log \frac{|U^{(t_s)}|}{|\{u_j.l \in L : l_j.c \in u_j.C\}|} \tag{8}$$

Here, $|\{u.l_i^{(t_s)} : l_i^{(t_s)} \in c\}|$ is the number of visits by user u at category c at time segment t_s. $|L_u^{(t_s)}|$ is total visits by user u at time t_s. $|U^{(t_s)}|$ is the total number of unique users in the system that has checked-in at time t_s. $|\{u_j.l \in L : l_j.c \in u_j.C\}|$ is the total number of unique users that visit at category c at time t_s.

Temporal Popularity of a Location. Popularity of a location plays a significant role to attract user. People tend to visit a more popular POI for better satisfaction. However, popularity also varies over time. For example, a bar is more popular at night, whereas people tend to visit a museum during morning or afternoon. For better recommendation, we intend to calculate popularity score of each POI on each time segment. Popularity of a POI l at time t_s is calculated using following equation [15].

$$\rho^{(ts)}(l) = \frac{1}{2} * \left\{ \frac{|U^{(t_s)}(l)|}{|U(l)|} + \frac{|Chk^{(t_s)}(l)|}{Chk(l)} \right\} \tag{9}$$

Here $|U^{(t_s)}(l)|$ is the number of users that visited l at time t_s, $|U(l)|$ is the total number of users visited l. $|Chk^{(t_s)}(l)|$ is the number of check-ins at l at time t_s and $Chk(l)$ is total number of check-ins at location l.

4.2 Online Recommendation

Spatial-Aware Candidate Selection. Geographical position of a POI plays a significant role to attract users [6,10]. People tend to visit nearby places. The

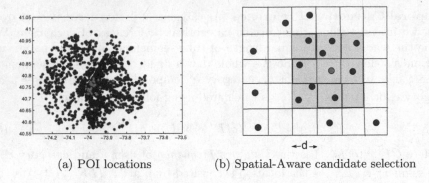

(a) POI locations   (b) Spatial-Aware candidate selection

Fig. 5. Incorporating spatial influence (Color Figure online)

propensity of a user to choose a POI decreases as the distance between the user
and the POI increases [8]. Consider the example in Fig. 5a. Black points represent
all the POI locations of NY City. Red points are the check-in distribution of a
single user. It is obvious that, this person does not move all over the city, rather
his movement data is limited to some geographical regions. We have borrowed
this example from [15]. Also Fig. 3b indicates that 90 % of successive check-ins
have distance less than 20 km.

To incorporate spatial influence, we divide the whole problem space into
square grids whose side length is d km. Locations of the grid cell the user is
currently visiting and its surrounding 8 adjacent grid cells are used as the candidate grid cells. The locations of the candidate grid cells are used as the candidate
locations for recommendation (see Fig. 5b). Distance between two points are calculated using *Haversine Formula* [24].

Successive Location Recommendation. Given a user u, his current location
l_{now} with category c_{now}, we first generate spatial-aware candidate location list
$S^{(u)}(L)$. Let the current time be t_{now}. In this section, we present the method
to rank the candidate locations. T_{max} is a user-defined time interval parameter.
Top-K locations are recommended to user u that he may want to visit within
time range t_{now} to t_{next}, where $(t_{next} - t_{now}) \leq T_{max}$.

Let, the recommended location be l_{next}. Category of l_{next} is c_{next}. In Offline
method, we have calculated time-specific categorical preference of user u at category c denoted as $P_u^{(t_s)}(c)$. Given the time interval T_{max}, we find the preference
of u for c_{next} from time range t_{now} to t_{next} defined as $P_u^{(t_{now},t_{next})}(c_{next})$.

$$P_u^{(t_{now},t_{next})}(c_{next}) = max\{P_u^{(t_s)}(c_{next})\} \tag{10}$$

where $t_s \geq t_{now}$ and $t_s \leq t_{next}$. For example, let $t_{now} = 10$ am, $T_{max} = 6$ h.
so, $t_{next} = 4$ pm. So u's preference for c_{next} from time range 10 am to 4 pm is
calculated as $max\{P_u^{(t_{10})}(c_{next}), P_u^{(t_{11})}(c_{next}), ...P_u^{(t_{16})}(c_{next})\}$

We find the popularity of location l_{next} at time range t_{now} to t_{next} denoted as $\rho^{(t_{now},t_{next})}(l_{next})$

$$\rho^{(t_{now},t_{next})}(l_{next}) = max\{\rho^{(t_s)}(l_{next})\} \tag{11}$$

where $t_s \geq t_{now}$ and $t_s \leq t_{next}$.

The rating of location l for user u, denoted as $R_u^{t_{next}}(l_{next})$ is calculated as:

$$R_u^{t_{next}}(l_{next}) = \frac{\sum\limits_{v \in U} w_{uv}}{|v|} * P_u^{(t_{now},t_{next})}(c_{next}) * \bar{T}_u(c_{now}, c_{next}) * \rho^{(t_{now},t_{next})}(l_{next}) \tag{12}$$

Here, $v \in U$ are the list of users who also visited the same location l_{next} at specified time range t_{now} to t_{next}.

5 Experiments

5.1 Dataset

We use the real-world check-in dataset from Foursquare [1]. The dataset includes 227,428 check-in data from New York City, USA. The dataset has data from 12 April 2012 to 16 February 2013 (10 months). We obtain this dataset from [25]. Each check-in Ch_{ij} contains user (u_i), location id (l_j) and time (t). Each location id l_j is associated with geographical position (lat, lon) and category c. It contains check-in data of 1,083 users and 38,383 locations. To get more effective results, we removed POIs that have lower than 10 check-ins. After preprocessing, the dataset contains 4,597 locations and 164,307 check-ins.

For experiment, we use the data of first 8 months as training set. The training data is used to learn the users' temporal categorical preferences and categorical transition probability and popularity of POIs. The rest of the data is used as a test set.

Evaluation Method. To evaluate our proposed method, we use two well-established metrics: *precision* and *recall* [26]. *Precision* and *recall* are calculated using the following equations.

$$pre@N = \frac{number\ of\ relevant\ recommendations}{N} \tag{13}$$

$$re@N = \frac{number\ of\ relevant\ recommendations}{total\ number\ of\ ground\ truths} \tag{14}$$

Here, N is the number of recommendation results. Ground truth refers to the set of locations where user has truly visited within the specified time range. So, *Pre@N* measures how many POIs in the top-N recommended POIs correspond to the ground truth POIs. *re@N* measures how many POIs in the ground truths has returned as top-N recommendation.

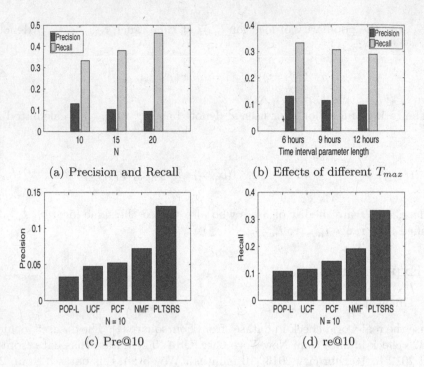

(a) Precision and Recall

(b) Effects of different T_{max}

(c) Pre@10

(d) re@10

Fig. 6. Experimental results

5.2 Experimental Results

We have used $d = 10$ km for grid cell size in all our experiments. Figure 6a shows the precision and recall value of our proposed method. We show the results for $N = 10$, 15 and 20. In this result we use $T_{max} = 6$ h.

We compare our method with the four following baseline approaches,

(1) **Popularity-Based Recommendation Method (Pop-L):** This is a spatial-aware popularity based recommendation method. Based on the current location, it first generates spatial aware candidate locations. Candidate locations are ranked based on their popularity.

(2) **Location-Based Collaborative Filtering (UCF):** This method applies Collaborative Filtering method directly over locations. This baseline utilizes the users location histories with a user-location matrix. User similarity is calculated using the location vector of users. Finally the locations are ranked using CF method. We consider the current location as a query location and generates spatial-aware candidate locations first to adapt this model for successive recommendation.

(3) **Preference-Based Collaborative Filtering (PCF):** This method is the baseline Preference-Aware approach. This method first generates users' categorical preferences from their location histories. Then it generates user-preference matrix. Similarity between two users is calculated using

their preference vector. Finally CF method is used to rank the candidate locations.

(4) **Non-negative Matrix Factorization (NMF):** This is the base-line low rank non-negative matrix facorization based recommendation method. This method first generates user-location matrix using their location histories. User-location matrix is factorized into two low rank matrices W and H.

Note that all methods use the current location as query location. We find spatial aware candidate locations first to adapt them for successive location recommendation. Figure 6c and Fig. 6d show the precision and recall values respectively. UCF works better than Pop-L. PCF approach works better than UCF as PCF can handle the data sparsity problem. NMF approach works better than PCF, but our proposed method PLTSRS outperforms all other baseline approaches.

We change the value of T_{max} to see how the value of T_{max} affects the results. Figure 6b shows the precision and recall of our algorithm for $T_{max} = 6$ h, 9 h and 12 h. $T_{max} = 6$ h gives us the best result.

6   Conclusion

In this paper, we present a novel approach for successive POI recommendation task. This approach recommends to a user a set of locations where he might be interested to visit next based on his current location and time. This method considers a combination of users' time-specific categorical preferences, categorical transition patterns, spatial influences and popularity of POIs. To the best of our knowledge, this is the first work that combines all the factors (temporal, user-preferences, categorical transition patterns, spatial and popularity) for successive POI recommendation task. Experimental results show that our method outperforms other baseline approaches. In future work, we plan to incorporate social relationships to strengthen our recommendation model.

References

1. https://foursquare.com
2. https://yelp.com
3. https://www.facebook.com/places/
4. http://statspotting.com/foursquare-statistics-20-million-users-2-billion-check-ins/
5. Zheng, Y., Zhou, X.: Computing with Spatial Trajectories. Springer Science & Business Media, New York (2011)
6. Ye, M., Yin, P., Lee, W.C., Lee, D.L.: Exploiting geographical influence for collaborative point-of-interest recommendation. In: Proceedings of the 34th International ACM SIGIR Conference on Research and Development in Information Retrieval, pp. 325–334. ACM (2011)
7. Sang, J., Mei, T., Sun, J.-T., Xu, C., Li, S.: Probabilistic sequential pois recommendation via check-in data. In: Proceedings of the 20th International Conference on Advances in Geographic Information Systems, pp. 402–405. ACM (2012)

8. Liu, B., Fu, Y., Yao, Z., Xiong, H.: Learning geographical preferences for point-of-interest recommendation. In: Proceedings of the 19th ACM SIGKDD International Conference on Knowledge Discovery and Data Mining, pp. 1043–1051. ACM (2013)

9. Horozov, T., Narasimhan, N., Vasudevan, V.: Using location for personalized POI recommendations in mobile environments. In: International Symposium on Applications and the Internet, SAINT 2006, p. 6. IEEE (2006)

10. Yuan, Q., Cong, G., Ma, Z., Sun, A., Thalmann, N.M.: Time-aware point-of-interest recommendation. In: Proceedings of the 36th International ACM SIGIR Conference on Research and Development in Information Retrieval, pp. 363–372. ACM (2013)

11. Bao, J., Zheng, Y., Mokbel, M.F.: Location-based and preference-aware recommendation using sparse geo-social networking data. In: Proceedings of the 20th International Conference on Advances in Geographic Information Systems, pp. 199–208. ACM (2012)

12. Gao, H., Tang, J., Hu, X., Liu, H.: Content-aware point of interest recommendation on location-based social networks. In: Proceedings of the 29th AAAI Conference on Artificial Intelligence (2015)

13. Cheng, C., Yang, H., Lyu, M.R., King, I.: Where you like to go next: successive point-of-interest recommendation. In: IJCAI (2013)

14. Zhang, W., Wang, J.: Location and time aware social collaborative retrieval for new successive point-of-interest recommendation. In: Proceedings of the 24th ACM International on Conference on Information and Knowledge Management, pp. 1221–1230. ACM (2015)

15. Debnath, M., Tripathi, P.K., Elmasri, R.: Preference-aware poi recommendation with temporal and spatial influence. In: FLAIRS (2016)

16. Markov, A.A.: An example of statistical investigation of the text eugene onegin concerning the connection of samples in chains. Sci. Context 19(04), 591–600 (2006)

17. Ye, M., Yin, P., Lee, W.C.: Location recommendation for location-based social networks. In: Proceedings of the 18th SIGSPATIAL International Conference on Advances in Geographic Information Systems, pp. 458–461. ACM (2010)

18. Rendle, S., Freudenthaler, C., Schmidt-Thieme, L.: Factorizing personalized Markov chains for next-basket recommendation. In: Proceedings of the 19th International Conference on World Wide Web, pp. 811–820. ACM (2010)

19. Feng, S., Li, X., Zeng, Y., Cong, G., Chee, Y.M., Yuan, Q.: Personalized ranking metric embedding for next new POI recommendation. in: Proceedings of IJCAI (2015)

20. Su, X., Khoshgoftaar, T.M.: A survey of collaborative filtering techniques. Adv. Artif. Intell. 2009, 4 (2009)

21. Lee, D.D., Seung, H.S.: Algorithms for non-negative matrix factorization

22. Jones, K.S.: A statistical interpretation of term specificity and its application in retrieval. J. Documentation 28(1), 11–21 (1972)

23. Tan, P.-N., Steinbach, M., Kumar, V.: Introduction to Data Mining. Addison-Wesley, Boston (2006)

24. Goodwin, H.: The haversine in nautical astronomy. In: Proceedings of US Naval Institute, vol. 36, pp. 735–746 (1910)

25. Yang, D., Zhang, D., Zheng, V.W., Yu, Z.: Modeling user activity preference by leveraging user spatial temporal characteristics in LBSNs. IEEE Trans. Syst. Man Cybern. Syst. 45(1), 129–142 (2015)

26. Powers, D.M.: Evaluation: from precision, recall and F-measure to ROC, informedness, markedness and correlation (2011)

Event Participation Recommendation in Event-Based Social Networks

Hao Ding[(✉)], Chenguang Yu, Guangyu Li, and Yong Liu

ECE Department, New York University, New York, USA
{hao.ding,chenguang.yu,guangyu.li,yongliu}@nyu.edu

Abstract. Event-based Social Networks (EBSN) have experienced rapid growth in recent years. Event participation recommendation is to recommend a list of users who are most likely to participate in a new event. Due to the nature of new event and severe data sparsity in EBSN, the traditional recommender systems do not work well for event participation recommendation. In this paper, we first conduct a study of Meetup users to understand the major factors impacting their event participation decisions. We then develop a sliding-window based machine-learning model that effectively combines user features from multiple channels to recommend users to new events. Through evaluation using the Meetup dataset, we demonstrate that our model can capture the short-term consistency of user preferences and outperforms the traditional popularity-based and nearest-neighbor based recommendation models. Our model is suitable for real-time recommendation on practical EBSN platforms.

Keywords: Event-based social networks · Social network analysis · Event participation recommendation · Temporal recommendation

1 Introduction

Event-based Social Networks (EBSN), such as Meetup (www.meetup.com) and Plancast (www.plancast.com), have gained momentum and experienced rapid growth in recent years. They provide online social platforms for users to create, organize and participate in social events of any kind. With Meetup, user can join different online groups, and can interact with other group members by making comments and sharing photos. Notably, users can participate in *offline* events, which are setup by event hosts for people to physically get together and have face-to-face social interactions. Prior research studies focused on event recommendation for users in EBSN. Very few studies have considered recommending users to new events, which is a more practical and important task in EBSN. The *NY Tech Meetup* group, for instance, is the largest group on Meetup website, which has about 50,000 group members. Due to the capacity limit of an event venue, the event hosts cannot always guarantee every group member can attend an event. Additionally, a large fraction of offline events require group members to purchase tickets. Consequently, predicting which users are more likely to participate in a new event can help event hosts better plan and organize events

© Springer International Publishing AG 2016
E. Spiro and Y.-Y. Ahn (Eds.): SocInfo 2016, Part I, LNCS 10046, pp. 361–375, 2016.
DOI: 10.1007/978-3-319-47880-7_22

in terms of capacity, budget, as well as time and location. Formally, the event participation recommendation task we want to tackle is: *given an upcoming new event in EBSN, predict a list of users that are more likely to participate in this event based on past event participation history.*

However, most of the existing recommendation models have limitations on this event participation recommendation task: (1) Traditional recommendation models focus on items that have been consumed or rated by other users, or users who have explicit or implicit feedbacks history. In EBSN, each event to be held is new to all group members and there is no user participation record that can be leveraged for recommendations. Needless to say, it makes no sense to recommend users to old events in a real EBSN system. (2) The history data of users' attendance at past events is dramatically sparser than the traditional online user-item rating dataset. This is mainly because a user's participation at offline events in real life takes more commitment and effort than participating in pure online activities, e.g., commenting on a picture or reading a news article. Most collaborative filtering methods cannot perform well on EBSN dataset due to its sparsity.

To tackle these limitations, we collected an extensive dataset of users, events, and groups in Meetup over twelve years, and conducted data analysis to understand the major factors impacting a user's decision to participate in an event. We then proposed a machine-learning model that effectively combined user features from multiple channels, such as social ties established through offline events, user time and location preferences, and users' activity levels, to predict user participation at new events. We further developed a sliding-window based training and testing framework to mine time-varying user interests and preferences to make real-time recommendation for emerging new events. Compared with the existing recommenders in EBSN, our model has the following merits: (1) instead of just mining users' RSVPs to events, our model exploits the explicit structure between users, events and groups in Meetup, and mines the rich meta-data of all entities to address the severe data-sparsity issue; (2) our model splits the dataset into training and testing sets along the timeline, which makes it capable of conducting *temporal* recommendations on practical EBSN platform; (3) our sliding-window based model not only improves recommendation accuracy by capturing the *short-term consistency* of user behaviors and preferences in EBSN, but also significantly reduces the data volume and time complexity of the recommendation algorithm. This makes it particularly appealing for real-time recommendations on the EBSN platform that must handle large datasets in a short time.

The rest of the paper is organized as follows: In Sect. 2 we discuss related work on recommender systems for EBSN. We demonstrate, in Sect. 3, that our analysis of the collected Meetup dataset identifies various user patterns. Building on the identified user features, we present our sliding-window based event participation recommender in Sect. 4. In Sect. 5, we evaluate our method and compare it with baseline models. Section 6 concludes the paper.

2   Related Work

Traditionally, recommendation problems are usually solved using collaborative filtering [11] or matrix factorization [3]. But in these methods only two entities – users and items are involved. Meetup consists of multiple types of entities: users, groups, and events. Each entity has rich meta-data information (such as tags, descriptions, time and locations), and is connected by complex structures. The traditional recommendation tasks and models are not well-suited for Meetup. Different types of recommendation tasks have recently been proposed. Event-to-user recommendations were studied in [2,4,7,9,12], where users were chosen as the recommendation target and events were the candidate items. Item-to-group recommendations were approached using a customized latent topic model in [6]. A general graph method was designed in [10] to make group-to-user recommendations, tag-to-group recommendations and event-to-user recommendations. To exploit the highly structured meta-data in Meetup, various graph-based methods were proposed. In [5] a "meta-path algorithm" was designed to address user-event and user-group relations. A user's preference for an event is estimated through different information paths. The graph-based recommendation method was extended in [10] to solve recommendation problems for all three entities – user, event, and group.

Significant efforts have been made to figure out the key factors affecting a user's decision-making process for attending events. In [13,14], the location factor was incorporated into a matrix factorization based latent factor model to improve recommendation accuracy. The time factor has been studied in [8] to capture a user's schedule preference, such as day of a week and hour of a day, to go to certain events. Event topics were also exploited in [1] and [8] by calculating the semantic similarity between the events attended by a user and a new event based on event descriptions. Social influences between users who attended common events and groups may be another key factor leading to event participation. In [5], offline and online social networks were combined to give a better estimate of mutual influence between users. One important factor that has been omitted by most of the previous studies is the intrinsic time-varying nature of users' interests and preferences. Our model effectively combines multiple user features to address data sparsity, and employs a sliding-window based training and testing framework to mine time-varying user interests and preferences for making real-time recommendations.

3   Meetup User Analysis

In this section, we will discuss how we conducted data analysis of event participation for Meetup users. The analysis helped us identify important features for our recommendation model in Sect. 4. A user's decision to attend one event may be influenced by many factors. In this section, we focus on four potentially important factors: *event-based social networks*, *personal time preferences*, *personal location preferences*, and *activity levels*.

Table 1. Dataset statistics.

Meetup dataset	NYC
# Groups	17,234
# Users	1,101,336
# Events	1,025,719
# RSVPs	8,338,382
# Venues	93,643
Avg. Members per group	274.13
Avg. Groups a user joins	3.54
Avg. Events per group	72.26
Avg. Participants per event	5.67
Avg. Events per active user	9.38

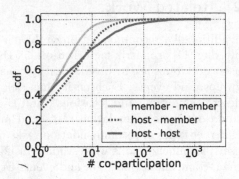

Fig. 1. CDF of co-participation of 3 types of social links.

3.1 Dataset

We used the Meetup API to crawl all groups located within 50 miles of New York City (NYC), all the events in these groups from March 2003 to February 2015, and all the related meta-data. Each group had its own group members. Group members RSVPed to each event in the group with "yes" or "no". Table 1 summarizes the salient statistics of the collected dataset.

3.2 Event-Based Offline Social Network

Different from online social networks, Meetup users have to physically engage with others to participate in offline events and gain face-to-face interaction. Social links among users are formed through repetitive event participation in real life. We say a pair of users are connected by an *offline social link* in a group if they participated in some common event(s) in that group. We can measure the strength of an offline social link between a pair of users by the number of events they co-participated in. Our statistics showed that the average number of co-participated events by a pair of connected users in a Meetup group was 1.72.

Each Meetup event has one or more event hosts who are also group members. They create events, RSVP with "yes" automatically and send out invitations to other normal group members. In our dataset, normal group members had stronger social links with event hosts than with other normal group members. Figure 1 gives the cumulative distribution of number of co-participation of three types of social links: links between two event hosts, between event host and normal member, and between two normal members. Only 7 % of the links between two normal members had more than 10 co-participations, but nearly 20 % of the links between normal member and event host had more than 10, which indicates that the social link between normal member and event host is statistically stronger than the link between two normal members. These results

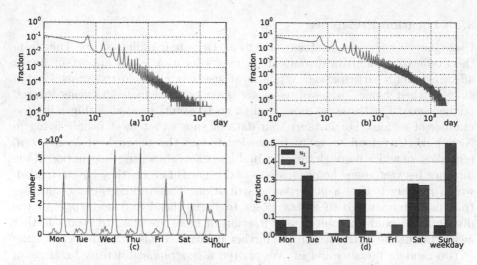

Fig. 2. (a) Distribution of time gap between two consecutive events held in a group. (b) Distribution of time gap between two consecutive events a user participated in. (c) Event time histogram over different hours of a week. (d) Participated event time distributions in one week of two members -u_1 and u_2- in the group *New York Singles Social Network*.

lead us to focus on the contribution of social connections between event hosts and normal members when we designed our recommendation model in Sect. 4.

3.3 Time Preference

When event hosts create a new event, the event time is an important factor they consider so other group members can participate in this event. Furthermore, when a normal group member replies to the invitation of a new event, he/she will also consider whether the event time complies with his personal schedule. The statistics in our dataset directly demonstrated that the event time and user's participation exhibited periodical temporal patterns on a weekly basis. Figure 2(a) gives the distribution of time gap between two consecutive events held in a group (distribution taken over data collected from all groups). The distribution curve had a peak for almost every 7 days. The distribution of a time gap between two consecutive events that a user participated in (Fig. 2(b)) has a similar weekly periodical pattern. It is clear from Fig. 2(c) that weekday events' numbers peak at around 2pm, while on the weekend, event numbers have two peaks around 11am and 8pm. To illustrate the time preference of users, we plotted the weekly time distribution of all the participated events of two randomly selected users u_1 and u_2 in the group, *The New York Singles Social Network* (in Fig. 2(d)). The figure suggests that u_1 tends to attend events on Tuesday, Thursday and Saturday, while u_2 prefers to attend weekend events.

3.4 Location Preference

Users join event-based social networks to participate in offline events. Therefore, the event venue and its distance to a user's location is an important factor affecting event participation. While most users tend to go to events located close to their homes, some of users are open to participate in events located farther away. Furthermore, users tend to go to events located within a limited number of regions. For instance, our dataset shows that most people living in New York City tend to go to events located in the borough of Manhattan, regardless of which borough they live in. Figure 3(a) shows the distribution of the distance between users' home locations and venues of events they have attended, which roughly follows a power-law distribution. This validates our conjecture that users tend to go to events closer to them (head and body part of the distribution), and a non-negligible fraction of users are open to events farther away (tail of the distribution). We further illustrated this through a case-study of two random users u_3 and u_4. We plotted a geographical density heatmap of events participated by them in Fig. 3(b). The red marker marks the location of user u_3 and the red and green heat spots are for the events u_3 attended. It is obvious that u_3 attended events only in these two regions and they are very close to u_3's home location. Different from u_3, the blue heat spots are for events u_4 attended. It shows that u_4 tends to go to events mostly in one region, even though it is far away from u_4's home location (blue marker).

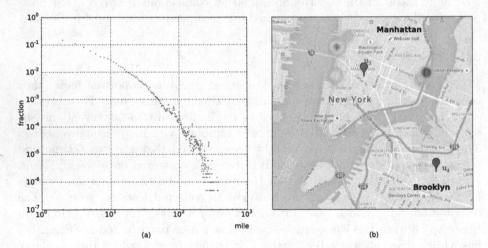

(a)   (b)

Fig. 3. (a) Distribution of distance between users and their attended events. (b) Geographical density heatmap of two users- u_3 and u_4. (Color figure online)

3.5 Activity Level

Within each group, a user's past activity level may also influence his/her participation in future events. To represent a user's activity level, we computed his/her

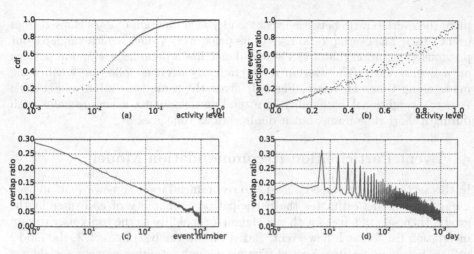

Fig. 4. (a) Cumulative distribution of user's activity level. (b) User's new event participation ratio vs. his past activity level. (c) Overlap ratio of events over the number of events in between them. (d) Overlap ratio of events over the time gap in between them.

participation ratio among all the events in one group. As illustrated in Fig. 4(a), 82 % of users have participation ratios lower than 5 %. Only the top 7 % of users have participated in over 20 % of all events in the group. These highly active users are more likely to participate in new events. In Fig. 4(b), for users at different activity levels (calculated based on events before time T_0), we computed their participation ratio for the events held after T_0. It is obvious that a user's probability of attending new events is almost directly proportional to his/her activity level in the past.

3.6 Consistency over Time

User's participation exhibits *short-term consistency* over time. Our statistics indicated that users who have attended a previous event have a higher probability of attending subsequent events. However, this probability diminishes over time. To quantify the short-term consistency of a user's participation, we first sorted all events chronologically for each group. Then we calculated the user overlap ratio of two events i and j (assuming i is before j) as the ratio of number of users that participated in both i and j over the number of participants at event i. The overlap ratio can also be interpreted as the conditional probability of a user that attended event j given that he/she attended a previous event i. Figure 4(c) plots the overlap ratio as a function of the number of events between i and j. It shows that adjacent events have a high overlap of participants (0.28), and remains above 0.2 when there are 10 events between the two events. The overlap ratio gradually declines when more events take place in between i and j. Thus, given a user has attended one event, his/her probability of attending the next event is as high as 0.28, and that of attending subsequent events will decay. Figure 4(d) plots the overlap ratio between event i and j as a function of days elapsed from event i to j. The overlap ratio rises to its peak every 7 days, which validates our previous observation that a user's attendance has a periodical temporal pattern on a weekly basis. The overlap ratio gradually declines following a vibrating mode along time. Therefore, Fig. 4(d) gives the same conclusion as Fig. 4(c): A user's conditional probability of attending subsequent events remains high during a short time frame, but it diminishes as time goes by.

4   Event Participation Recommendation Model

In this section, we developed a model to recommend users to new events in each group. Our model estimates the participation probability of each user in the group, then selects k users with the highest probability as the top-k user recommendation list for each new event. Based on our analysis in Sect. 3, the model utilizes four main features derived from users' past activities, namely, social links in event-based offline social networks, time preferences, location preferences, and activity levels. We first describe how we obtained each of the four features, then we present our sliding-window based model for real-time recommendation.

4.1   Offline Social Link Feature

Each offline social link between two users is based on how many common events they have co-participated in. Given a group g, we define the social network as $G^g = \langle U^g, A^g \rangle$, where U^g is the set of all users in this group and A^g is the set of social links. User u_i and u_j are connected if they co-participated at some events. The social link weight between u_i and u_j is calculated as:

$$c_{i,j} = \sum_{\forall k: u_i \in U_k \wedge u_j \in U_k} \frac{1}{|U_k|}, \tag{1}$$

where $|U_k|$ is the number of users participated in event e_k. The underline assumption is that the personal interaction between users at an event decreases with the number of users present at the event. When a new event is created, we can start user recommendation immediately. Since the event hosts RSVP with "yes" by default, we recommend normal group members according to their social links with these event hosts. We denote the event hosts for a new event e_j as H_j, and we calculate the *social relevance score* of user u_i to the new event e_j as:

$$s_n(u_i, e_j) = \frac{1}{|H_j|} \sum_{u_k \in H_j} c_{i,k} \tag{2}$$

4.2   Time Preference Feature

A group has its own temporal pattern of hosting events. Similarly, each user has his/her own time schedule for attending events. They both exhibit periodical temporal patterns on a weekly basis. We use kernel density estimation to model a user's time preference within a week. Given all the events one user has participated in and the timestamps of these events (in terms of the hour in one week), we treat these event times as i.i.d. samples drawn from the user's time preference probability distribution f_t. If we denote E_{u_i} as all the events user u_i has attended, our kernel density function $\hat{f}_t$ can be calculated as:

$$\hat{f}_t(t) = \frac{1}{|E_{u_i}|} \sum_{e_k \in E_{u_i}} K_h\big(t - t(e_k)\big), \qquad (3)$$

where $t(e_k)$ is the time for event e_k, and $K_h(\cdot)$ is the kernel function, which in our paper is a Gaussian kernel $K_h(t) = \frac{1}{\sqrt{2\pi}h} e^{-\frac{t^2}{2h^2}}$ with h being the standard deviation of all the event times user u_i has attended. Thus, the *time preference relevance score* of the user u_i to the new event e_j can be calculated as:

$$s_t(u_i, e_j) = \hat{f}_t\big(t(e_j)\big) \qquad (4)$$

4.3 Location Preference Feature

The statistical result in Sect. 3.4 shows that a user's location preference exhibits two significant patterns: (1) a user tends to attend events close to his/her home location, (2) the events that a user has attended are geographically clustered into a limited number of regions. These patterns lead us to propose the Gaussian kernel density estimator to model a user's location preference in a similar way to time preferences. Given a user u_i and all the events he/she has attended (E_{u_i}), we denote $l(e_k)$ as the location of event e_k, which is a 2-dimensional coordinate of latitude and longitude. The kernel density function $\hat{f}_l$ is defined as:

$$\hat{f}_l(l) = \frac{1}{|E_{u_i}|} \sum_{e_k \in E_{u_i}} K_{\boldsymbol{\Sigma}}\big(l - l(e_k)\big), \qquad (5)$$

where $K_{\boldsymbol{\Sigma}}(\cdot)$ is a bi-variate Gaussian kernel,

$$K_{\boldsymbol{\Sigma}}(l) = \frac{1}{\sqrt{|\boldsymbol{\Sigma}|(2\pi)^2}} exp\big(-\frac{1}{2} l^T \boldsymbol{\Sigma}^{-1} l\big), \qquad (6)$$

where $\boldsymbol{\Sigma}$ is a 2×2 covariance matrix with respect to the latitude and longitude variables. Consequently, we calculate the *location relevance score* of a user u_i to a new event e_j as:

$$s_l(u_i, e_j) = \hat{f}_l\big(l(e_j)\big) \qquad (7)$$

4.4 Activity Level Feature

A user's activity level in the past may also influence his/her attendance at future events. Section 3.5 gives statistical evidence that active group members tend to be more engaged in new events; they have higher new event participation ratios on average compared to inactive users. To incorporate this feature into our model, we calculate the activity level of user u_i in one group as:

$$p(u_i) = \frac{|E_{u_i}|}{total\ number\ of\ past\ events} \tag{8}$$

Thus, the *activity level relevance score* of user u_i to the event e_j is defined as:

$$s_a(u_i, e_j) = p(u_i) \tag{9}$$

4.5 Sliding-Window Based Recommendation Model

We now present a sliding-window based machine-learning model to recommend users to new events. Given a new event, we utilize all history RSVP data prior to the event, obtaining user features to make a prediction. Since user behaviors and preferences naturally vary over time, it is more accurate to focus on recent event RSVP data. One way to achieve this is to give higher weights for more recent events in feature calculation and model training. Another way is to calculate user features based on a window of events immediately prior to the current event. We adopt the latter approach and propose a sliding-window based recommendation model as illustrated in Fig. 5. All events are ordered based on time. The task of our recommendation model is to utilize user features calculated from events in a feature window to predict user participation at events in a subsequent label window. The widths of feature window and label window can be measured either in time or the number of events, and are tunable parameters. If the feature window is too small, we will not have enough samples to build stable user features; if the feature window is too large, user behaviors and preferences in remote history might just introduce noises to predicting their current preferences. We will demonstrate this trade-off in Sect. 5.

To train the sliding-window based recommendation model, we can no longer randomly partition our dataset into training and testing set. Instead, we set a time-separation line. All events before the line are used for training, and all events after it are used for testing. As illustrated in Fig. 5, for training, we have many non-overlapping label windows, each of which has an associated feature window immediately before it. Training is done using all labeled events from all training label windows. Likewise, for testing, we also have many non-overlapping testing label windows, each of which has an associated feature window before it. Our model is tested on all labeled events from all testing label windows.

Given a group g, the relevance score of a group member u_i to a new event e_j is calculated as:

$$\hat{s}(u_i, e_j) = \mathbf{w^g} \cdot \mathbf{f}(u_i, e_j), \tag{10}$$

where the weights $\mathbf{w^g} = \langle w_n^g, w_t^g, w_l^g, w_a^g \rangle$ are group-dependent parameters to be trained, and feature vector $\mathbf{f}(u_i, e_j) = \langle s_n(u_i, e_j), s_t(u_i, e_j), s_l(u_i, e_j), s_a(u_i, e_j) \rangle$ are various relevance scores of user u_i to the new event e_j calculated from (2), (4), (7), and (9) respectively. We denote $y(u_i, e_j)$ as the observed attendance value of user u_i to the event e_j. $y(u_i, e_j) = 1$ if u_i RSVP with "yes" and $y(u_i, e_j) = 0$

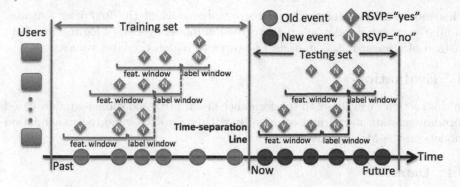

Fig. 5. Training set and testing set of sliding-window based recommendation model.

if his RSVP is "no". We observe very high sparsity in the attendance matrix Y, i.e., many users don't respond to RSVP invitations. If we only use RSVP "yes" and "no" labels, we would have had too limited data which could have easily run into model over-fitting. We should enrich the label data by imputing values for missing entries in the Y matrix. A missing entry might be due to the user missed the event invitation, or he decided not to attend the event but didn't bother to RSVP "no". Therefore, these missing labels should not be simply treated as "no" and get 0 attendance value, nor should they be treated as *missing-at-random* and get the average value of the explicit "yes" and "no" labels. If we assume that a member missed an event invitation with a probability of α_g, then it is with the probability of $1 - \alpha_g$ that he/she read the invitation, but decided not to attend. Under this assumption, the imputed attendance value for a missing label with respect to user u_i and event e_j can be calculated as:

$$y^{im}(u_i, e_j) = \alpha_g p(u_i) + (1 - \alpha_g) \times 0 = \alpha_g p(u_i), \qquad (11)$$

where $p(u_i)$ is the user's activity level defined in Sect. 4.4. The rationale is if a user is not aware of the invitation, his/her likelihood of attending is imputed as his/her event participation ratio in the past.

Given the label values, our training objective is to find the optimal weight vector $\mathbf{w^g}$ to minimize the loss function defined as:

$$L_g = \sum_{u_i, e_j} \left(\hat{s}(u_i, e_j) - y_{ij}^{o\&i} \right)^2 + \lambda_1 ||\mathbf{w^g}||^2 + \lambda_2 \alpha_g^2, \qquad (12)$$

where the last two terms are for parameter regularization, and the observed attendance scores are denoted as $y^{obs}(u_i, e_j)$. The final attendance scores after imputing are

$$y_{ij}^{o\&i} = \begin{cases} y^{obs}(u_i, e_j) & if\ RSVP\ of\ u_i\ to\ e_j\ is\ observed \\ y^{im}(u_i, e_j) & otherwise \end{cases} \qquad (13)$$

This model is essentially a linear-regression model of the four user features. Similarly, we can develop a logistic-regression model by using a logistic function, instead of a linear function, of user features to match the label values.

5   Evaluation

In this section we evaluate the performance of our proposed sliding-window based recommendation model and compare it with several baseline recommendation models on the Meetup dataset.

5.1   Dataset

In Sect. 3.1 we describe how we collected the dataset from Meetup.com. The whole dataset is split into two parts along a timeline. The first part covers all the RSVPs of events before Aug 20th, 2011, and takes 80 % of the total dataset. It is used as the training set. The second part starts from Aug 21st, 2011 to Oct 14th, 2014, which takes 20 % and is used for testing.

5.2   Evaluation Metrics

To evaluate the quality of event participation recommendation, we let a recommender system generate the top-k ($k = 5$, 10, 20, 50, 100, 200, 400, and 800) list of users who are most likely to participate in a new event, and compute the recalls for this event at each k value, which is defined as below:

$$top\text{-}k\ recall = \frac{number\ of\ users\ in\ top\text{-}k\ list\ who\ RSVP\ with\ ``yes"}{number\ of\ all\ users\ who\ RSVP\ with\ ``yes"}. \quad (14)$$

Then the final top-k recall for the model is obtained by averaging the top-k recalls of all the events in the testing set.

5.3   Baseline Methods

To validate the accuracy of the proposed model, we compared it with the following generic baseline recommendation models.

- *Most-Popular (***pop***):* All members in a group are ranked in a descending order of their popularity, which is defined as the number of events a user has attended in the training set. The top-k recommendation list is simply the first k users with the highest popularity.
- *K-Nearest-Neighbors in User Space (***userKNN***):* We can calculate the *Jaccard similarity* between users based on the events they have attended in the training set. The nearest-neighbor collaborative filtering method predicts each candidate user's likelihood to attend a new event based on his/her similarity with the event hosts. The first k users with the highest likelihood will be recommended to the event.
- *Window-based Most-Popular (***win_pop***) and User-KNN (***win_userKNN***):* Instead of using the whole training set to calculate user popularity/similarity, we use only events in a sliding feature window to generate most-popular and user-KNN recommendations.

5.4   Experimental Results

Impact of Feature and Label Window Sizes: Figure 6(a) and (b) show the performance of the proposed method over different combinations of feature window sizes and label window sizes. Since different groups have different frequencies for creating events, we used the number of events, instead of time, to measure window sizes. (a) indicates that, given a fixed label window size of one event, the recall of our model does not change significantly over different feature window sizes. Feature window size of 20 events gives the best performance. In (b), we fix the feature window size to be 20, the recall decreases as the label window size increases. We repeated our experiments for 16 combinations of four feature window sizes and four label window sizes, the combination of feature window size of 20 and label window size of 1 gives the best performance. This result confirms our conjecture that *user's preferences vary over time; it is best to use users' recent RSVP data to predict their participations at the very next event.*

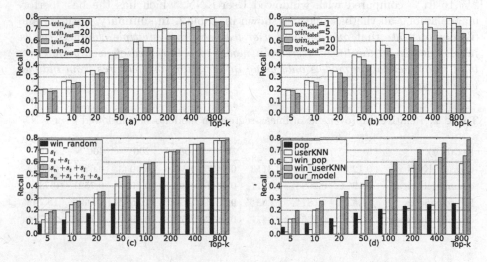

Fig. 6. (a) Top-k recall over different feature window sizes when label window size is 1. (b) Top-k recall over different label window sizes when feature window size is 20. (c) Top-k recall over different feature combinations and their comparison with random method. (d) Top-k recall over our model and different baselines.

Effectiveness of Selected Features: To validate the contribution of each feature in our model, we generated four model variations by using different combinations of the four features and compared them with a sliding-window based random recommendation approach that recommends randomly selected active users in the sliding feature window to each new event. Figure 6(c) shows that even if the model only uses the location preference feature s_l, it still out-performs the random recommendation. Every time we add one more feature into the model, the top-k recall improves, especially for k from 5 to 50. When k becomes larger, top-k recalls for all the model variations are closer to each other. This is mainly because most of the events did not have too many RSVPs in the testing set. The recalls become saturated as the recommended user list becomes longer, regardless of which recommender being used.

Performance Improvements over Baselines: Figure 6(d) shows the performance comparison of our model with other baselines. Without a sliding window, Most-Popular method gives us a higher recall than user-KNN, and their performance gap diminishes when the recommendation list expands to 800 users. Sliding-window significantly improves both performances. Their recalls with sliding-window are more than twice of their recalls without sliding window. This again confirms that *our sliding-window based recommendation model can effectively capture time-varying user behaviors and preferences to predict user participation in event-based offline social networks, such as Meetup.* Our linear-regression model gives the best performance among all the methods (even better than our logistic-regression model). The relative improvement ranges from 9 % to 51 % compared with windowed User-KNN, which has the best performance among all the baselines, as shown in Table 2. In summary, both (c) and (d) demonstrate that *our proposed user features from multiple channels can be fused together by the linear regression model to better predict user preferences than recommendation models only using information and features from individual channels.*

Table 2. Relative improvement of linear-regression model over windowed User-KNN.

Recall @k	5	10	20	50	100	200	400	800
win_userKNN	0.1297	0.2126	0.3113	0.4475	0.5376	0.6038	0.6374	0.6522
our_model(linear)	0.1964	0.2696	0.3517	0.4888	0.6045	0.6976	0.7521	0.7813
Improvements	51.43 %	26.81 %	12.98 %	9.23 %	12.44 %	15.53 %	17.99 %	19.79 %
our_model(logistic)	0.1631	0.2552	0.3477	0.4877	0.5993	0.6902	0.7436	0.7739

6   Conclusions

In this paper, we studied the event participation recommendation problem in EBSN. Through a study of Meetup users, we identified four features impacting a user's event participation decision, namely, offline social links, time/location preferences, and user activity levels. We developed a sliding-window based machine-learning model that effectively combines four user features to predict the probability for a user to attend a new event. Through evaluation using the Meetup dataset, we demonstrated that our model can capture the short-term consistency of user preferences and outperforms the traditional popularity-based and nearest-neighbor based recommendation models. Our model can be used for real-time user recommendation on practical EBSN platforms.

References

1. Ding, Y., Jiang, J.: Modeling social media content with word vectors for recommendation. In: Liu, T.-Y., Scollon, C.N., Zhu, W. (eds.) SI 2015. LNCS, vol. 9471, pp. 274–288. Springer, Heidelberg (2015). doi:10.1007/978-3-319-27433-1_19
2. Khrouf, H., Troncy, R.: Hybrid event recommendation using linked data and user diversity. In: Proceedings of the 7th ACM Conference on Recommender Systems, pp. 185–192. ACM (2013)
3. Koren, Y., Bell, R., Volinsky, C., et al.: Matrix factorization techniques for recommender systems. Computer **42**(8), 30–37 (2009)
4. Liao, G., Zhao, Y., Xie, S., Yu, P.S.: An effective latent networks fusion based model for event recommendation in offline ephemeral social networks. In: Proceedings of the 22nd ACM International Conference on Conference on Information and Knowledge Management, pp. 1655–1660. ACM (2013)
5. Liu, X., He, Q., Tian, Y., Lee, W.C., McPherson, J., Han, J.: Event-based social networks: linking the online and offline social worlds. In: Proceedings of the 18th ACM SIGKDD International Conference on Knowledge Discovery and Data Mining, pp. 1032–1040. ACM (2012)
6. Liu, X., Tian, Y., Ye, M., Lee, W.C.: Exploring personal impact for group recommendation. In: Proceedings of the 21st ACM International Conference on Information and Knowledge Management, pp. 674–683. ACM (2012)
7. de Macedo, A.Q., Marinho, L.B.: Event recommendation in event-based social networks. In: Hypertext, Social Personalization Workshop, pp. 3130–3131 (2014)
8. Macedo, A.Q., Marinho, L.B., Santos, R.L.: Context-aware event recommendation in event-based social networks. In: Proceedings of the 9th ACM Conference on Recommender Systems., pp. 123–130. ACM (2015)
9. Minkov, E., Charrow, B., Ledlie, J., Teller, S., Jaakkola, T.: Collaborative future event recommendation. In: Proceedings of the 19th ACM International Conference on Information and Knowledge Management, pp. 819–828. ACM (2010)
10. Pham, T.A.N., Li, X., Cong, G., Zhang, Z.: A general graph-based model for recommendation in event-based social networks. In: 2015 IEEE 31st International Conference on Data Engineering, pp. 567–578. IEEE (2015)
11. Schafer, J.B., Frankowski, D., Herlocker, J., Sen, S.: Collaborative filtering recommender systems. In: Brusilovsky, P., Kobsa, A., Nejdl, W. (eds.) The Adaptive Web. LNCS, vol. 4321, pp. 291–324. Springer, Heidelberg (2007). doi:10.1007/978-3-540-72079-9_9
12. Wang, Z., He, P., Shou, L., Chen, K., Wu, S., Chen, G.: Toward the new item problem: context-enhanced event recommendation in event-based social networks. In: Hanbury, A., Kazai, G., Rauber, A., Fuhr, N. (eds.) ECIR 2015. LNCS, vol. 9022, pp. 333–338. Springer, Heidelberg (2015). doi:10.1007/978-3-319-16354-3_36
13. Yin, H., Sun, Y., Cui, B., Hu, Z., Chen, L.: LCARS: a location-content-aware recommender system. In: Proceedings of the 19th ACM SIGKDD International Conference on Knowledge Discovery and Data Mining, pp. 221–229. ACM (2013)
14. Zhang, W., Wang, J., Feng, W.: Combining latent factor model with location features for event-based group recommendation. In: Proceedings of the 19th ACM SIGKDD International Conference on Knowledge Discovery and Data Mining, pp. 910–918. ACM (2013)

An Effective Approach to Finding a Context Path in Review Texts Using Pathfinder Scaling

Erin Hea-Jin Kim[(⊠)] and SuYeon Kim

Department of Library and Information Science, Yonsei University, Seoul, Korea
{erin.hj.kim, suyeon}@yonsei.ac.kr

Abstract. Customer reviews feature opinions or sentiments that a review writer has given, and these opinions or sentiments have an impact on the reader. Identifying and presenting word associations that indicate a sentiment orientation and semantics can aid in selecting the best review for providing the information customers are seeking. In this paper, we attempted to discover the context structure and the context path presenting explicit semantics in review texts. To this end, we extracted word co-occurrences and converted them to a cosine adjacency matrix. Then a co-word network applied by Pathfinder scaling was constructed. Finally, we measured the context score and presented context paths from the context structure in the review texts. In results, our approach found that a compound noun is easy to detect by network analysis. The extracted context paths remain intact, a sentiment polarity derived from review texts. The evaluative expression for a certain aspect of a product or service is clearer and more specified within the context path. Furthermore, it is not necessary to train reference words to detect the sentiment orientations.

Keywords: Content analysis · Context structure · Context path · Pathfinder network (PFNet) · Review mining

1 Introduction

With the use of the Internet and social network services increasing today, more customers are searching and reading online-review texts that other customers have written about the products and services in which they are interested. Reviews feature opinions or sentiments that a review writer has given, and these opinions or sentiments have an impact on the reader. However, sometimes a review is long and contains too much content for readers to examine in detail. This poses an obstacle when it comes to selecting the best review for providing the information customers are seeking. Review-text mining has attempted to solve the problems associated with extracting linguistic features and predicting opinions, such as sentiment polarity, i.e., positive or negative. Through review-text mining, a recommendation is made for the most useful review for readers so that customers can experience the benefits of a reduction in their search time as well as decision-making support [1–3].

Several studies have considered semantics in language for detecting sentiment orientations [4–6]. However, a limitation exists in detecting a linguistic phenomenon, such as a polarity shift in a sentence or passage [7], in the studies that exploited a bag of

© Springer International Publishing AG 2016
E. Spiro and Y.-Y. Ahn (Eds.): SocInfo 2016, Part I, LNCS 10046, pp. 376–388, 2016.
DOI: 10.1007/978-3-319-47880-7_23

words (BOW) [8–11] or manually selected reference words, such as "excellent" for positives and "poor" for negatives [5, 6]. Investigating semantic patterns may be one of the solutions to this problem. To capture semantic patterns, Choi et al. [12] suggested decomposing a sentence's structure to mine opinion expressions. Thus, identifying word associations that indicate a sentiment orientation is important for review mining. In the present paper, a context path refers to word associations that bear explicit semantics, and a context structure refers to a network consisting of context paths derived from preprocessed review texts.

The goal of the present paper is to propose an effective method for investigating the context structure of polarity texts. We extracted a context path that reflects the text polarity. In other words, we attempted to discover a context structure suiting the positive or negative review texts, and we detected a context path from the structure. The context path we detected consists of natural language that the customer wrote by himself/herself; thus, a reader can perceive the context of the reviews from the context path. Our research question is how different the context structure is among polarity corpora. To explore this question, we extracted word co-occurrences and converted them to a cosine adjacency matrix. We applied Pathfinder scaling [13] to a cosine adjacency matrix to construct a co-word network for discovering the context structure. Finally, we measured the context score and presented context paths in the review texts.

The rest of the paper is structured as follows: We briefly present related work. We then describe our methodology, and next, we report the experimental results and findings. Finally, we draw a conclusion with the highlights and implications of our study.

2 Related Work

A machine-learning approach is mostly adopted in review mining, and linguistic features are identified [2, 8–11]. To detect opinion words containing subjectivity, Ghose and Ipeirioti [1] categorized the information that a word contains into objective and subjective information. The former is provided by a company, whereas the latter is written by a customer who personally describes a product or service. They suggested review-ranking systems that analyze reviews' subjective information and predict its usefulness. In a similar fashion, Jin et al. [11] identified positive or negative expressions and product entities from product reviews. The proposed system removes non-subjective sentences and non-effective opinioned sentences.

Topic models on unlabeled data can help with finding the aspects of products or services that customers may consider [3, 14, 15]. Mei et al. [14] extracted sentiment sentences using discovered topics related to the product and observed the sentiment change over time. Jo and Oh [15] considered a different usage of sentiment words, such as adjectives, in review writing. They investigated the pair of an aspect of the product and the sentiment about the aspect using the "aspect and sentiment unification model (ASUM)". The model trains the provided sentiment seed words [6] and predicts sentiments toward a discovered aspect of the product.

Pathfinder network (PFNet) algorithm was developed for structural modeling the proximity matrices from psychological data [13]. The analysis for detecting an explicit network structure can take advantage of the PFNet network scaling [16]. McCain [17] introduced the PFNet to co-classify biotechnological patents. Chen [18] analyzed the author co-citation networks in digital library and the Pathfinder scaling helped understanding the semantic structure of the results. White [19] applied the Pathfinder scaling to reconstruct the core information scientist networks of his preceding research [20].

3   Methodology

In this section, we describe the proposed methodology, which consists of collecting data, preprocessing, generating a similarity adjacency matrix, and constructing the context structure of the distinct polarity texts. Figure 1 depicts the workflow of our proposed approach for finding the context structure of the polarity texts.

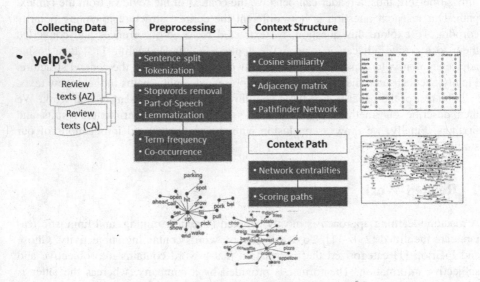

Fig. 1. Workflow for finding the context structure

3.1   Datasets

To conduct our experiment, we prepared two sets of the regional review data: Phoenix (AZ) and Los Angles (CA). We collected user-generated texts from Internet rating and review site yelp.com. A total of 369,941 reviews for both were collected, with the reviewers' five-rating score shown in Table 1. In the review texts, reviewers measured their experiences with the places they visited or the services they received, and each provided an appropriate number of stars between one and five, which represents "strongly negative" and "strongly positive", respectively. The review writers were skilled and active reviewers who had received Elite badges on their profile accounts from the *Yelp* site. We filtered the Elite reviews for our study. In Table 1, the average number of reviews per Elite reviewer of CA is much larger than that of AZ. This is because the AZ dataset is an open dataset that *Yelp*[1] provided, whereas the CA dataset was collected by targeting Elite reviews using crawling software Webzip 7.1. However, the proportion of the five ratings was similar between the two regional sets regardless of the data officially provided or randomly collected. As shown in Table 1, the total number of terms is 7,595,882 (AZ) and 21,599,141 (CA). The average length of the preprocessed terms (tokens) per review of the AZ data is shorter, 75, than that of the CA data, 80. In light of Table 1, writers who write more reviews tend to write longer reviews.

Table 1. Statistics of collected datasets

	Phoenix (AZ)	Los Angeles (CA)
No. of review texts	101,073	268,868
No. of 5-rated reviews (R5)	27,397 (27 %)	71,825 (27 %)
No. of 4-rated reviews (R4)	41,908 (41 %)	106,392 (40 %)
No. of 3-rated reviews (R3)	20,059 (20 %)	57,581 (21 %)
No. of 2-rated reviews (R2)	8,113 (8 %)	22,318 (8 %)
No. of 1-rated reviews (R1)	3,596 (4 %)	10,752 (4 %)
No. of reviewers (Elite)	5,008	2,196
Average no. of reviews	20	122
No. of business	11,448	98,565
No. of terms	7,595,882	21,599,141
Average no. of tokens	75	80
Period	2005 to 2014	2005 to 2012

3.2 Proposed Approach

The first step is data preprocessing, including sentence split, tokenization, stopword removal, part-of-speech (POS) tagging, and lemmatization using Stanford NLP[2]. We left nouns, verbs, adjectives, and adverbs at the stage of POS tagging and removed several excessive adverbs, such as "too", "really", "also", "very", "much", etc. We then extracted 5-rating and 1-rating reviews to analyze because our objective was to discover the context structure in the polarity corpora.

The second step is to count a term frequency and extract co-occurrences of term pairs to calculate the cosine similarity of a term pair. We set a sentence for the window of co-occurrence. We then generated the adjacency matrix of term pairs as an input to the Pathfinder scaling algorithm to construct the context structure of the polarity texts. Our adjacency matrix was made from pair-wise term similarity; thus, the constructed network is an undirected network. The node is a term, and the edge is a similarity weight between two nodes.

---

[1] www.yelp.com/dataset_challenge .

[2] http://nlp.stanford.edu/ .

The third step is to create the context structure. The Pathfinder scaling method is used to generate a context network structure of review texts. Pathfinder scaling finds a network by removing the edge between two nodes when the edge weight does not conform the triangle inequality [19].

The last step is to find a context path, denoted $context_{path}$, among the context structure. We put all term pairs and their similarity into Neo4 J^3, a graph database, and extracted paths whose total lengths were over two to be the possible context path. Among the extracted candidate paths, we figured out a context path. For this, we defined a context score that was a score of the context path (formula 1). We used the summation of all path weight (w) between a start node (i) and an end node (n) and the mean value of the path length from node i to node n:

$$context_{score} = \frac{1}{n-1} \sum_i^n path_{iw} \qquad (1)$$

For example, in Fig. 2, we can find four possible context paths and their context scores as follow:

- $context_{path}1$: reasonably-price-quality (0.1485)
- $context_{path}2$: reasonably-price-selection (0.1448)
- $context_{path}3$: reasonably-price-selection-wine (0.1254)
- $context_{path}4$: price-selection-wine (0.0670)

Thus, we chose $context_{path}1$ as the best context path among the paths.

4   Result

4.1   Network Analysis

Table 2 shows a comparison of the statistics between a genuine co-occurrence term network and a Pathfinder network consisting of the top 400 term nodes. We cut the sizes of the datasets by term co-occurrences, for instance, co-occurrences of 25 and over (R1 of AZ) or co-occurrences of 80 and over (R1 of CA), depending on the data amounts. Thus, the node numbers are not consistent in the four datasets. After Pathfinder scaling, the edge numbers decreased because Pathfinder retained only the edges pertaining to the triangle inequality. Consequently, the sizes of the edges are similar to those of the nodes.

We interpreted the context structure built via Pathfinder scaling using the network properties: degree, betweenness, and closeness centralities. Degree centrality refers to the degree of the number of connections to a given node. Betweenness centrality refers to the degree of the number of shortest paths that traverse a given node. Closeness centrality refers to the degree of the average of the shortest distance to all other nodes from a given node [21]. The node whose largest value of degree in the context structure

---

[3] https://neo4j.com/.

Fig. 2. Example of context score

Table 2. Numbers of nodes and edges

Rate	Dataset	No. of nodes	No. of edges	
			Co-occurrence	Pathfinder scaling
Rating 5 (R5)	AZ	401	2142	384
	CA	410	1910	390
Rating 1 (R1)	AZ	412	2316	411
	CA	397	2110	392

in the four datasets is "not" shown in Fig. 3. However, the term's topology is different between the positive and negative reviews. Degrees of "not" are much lower in the positive reviews (R5) than in the negative reviews (R1), 112 (AZ_R5), 124 (CA_R5), 170 (AZ_R1), and 187 (CA_R1), respectively. Furthermore, among the four datasets, the term "not" has the largest value of betweenness; this result shows the same pattern as that mentioned above—that the values of positive reviews are lower, 0.1872 (AZ_R5) and 0.3750 (CA_R5), but the values of negative reviews are higher, 0.8774 (AZ_R1) and 0.6547 (CA_R1). This result has an important implication. The term "not" cannot be a discriminative word for negation, but the term "not" should be considered along with co-occurred words in the context for the negative detection.

Furthermore, the 14 reference words indicating polarity in a previous study [6] do not show up in the top 400 terms determined via term co-occurrence. This practical result means that manually selected reference words do not fit perfectly when measuring the sentiment orientations, and the words customers use vary in practice.

Table 3 shows a different aspect of the term's topology in the Pathfinder network. The highest closeness centrality is 1 among the four networks; however, the five-rating reviews have many more term pairs where the value of closeness centrality for each is 1.0, which means the pair is always shown together. The values of similarity vary from 0.7574 (foie-gras) to 0.0156 (music-not). Among those, only five pairs in AZ_R5, nine pairs in CA_R5, and one pair in CA_R1 are connected by closeness, 1. Especially, the similarity of the pair of "shopping" and "center" is very low, 0.1895, which ranks 39th on similarity in the AZ_R5 dataset, but they are connected by the highest closeness in the context structure.

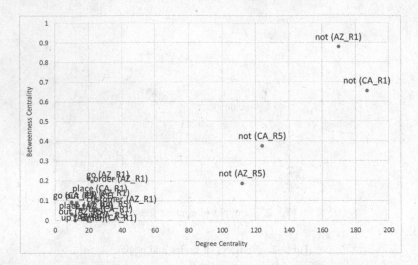

Fig. 3. Degree and betweenness centralities in Pathfinder network of four datasets

Table 3. The nodes where the closeness centrality for each is 1.0

Rate	Dataset	No. of nodes	Node label
R5	AZ	10	Shopping, center, credit, card, peanut, butter, no, longer, carne, asada
	CA	18	Credit, card, peanut, butter, no, longer, carne, asada, farmer, market, clam, chowder, foie, gras, strip, mall, hole, wall
R1	AZ	0	
	CA	2	Salsa, chip

Tables 4 and 5 list the top 30 term pairs' similarity with regard to positive reviews and negative reviews, respectively. The top 30 term pairs can be placed in three categories: food or store names, review aspects, and opinion words. "Carne-asada", "ice-cream", "foie-gras", "pork-belly", "mash-potato", etc., refer to food names. "Parking-lot", "customer-service", "question-answer", "pick-up", etc., refer to review aspects. "Highly-recommend", "melt-mouth", "pleasantly-surprise", "rip-off", "piss-off", etc., refer to opinion words (in bold).

However, it is still hard to capture the review context with only term pairs because many term pairs such as "write-review", "give-star", "avoid-cost", "save-money", and "minute-later", are not obvious to present the opinions in the review texts.

4.2   Context Path

Tables 6 and 7 present the instances of the context paths by context scores that we proposed, from positive reviews, and negative reviews, respectively. In both tables, the

Table 4. The top 30 term pairs of the positive reviews (R5)

Rank	AZ_R5		CA_R5	
	Term pair	Weight	Term pair	Weight
1	carne - asada	0.7073	foie - gras	0.7574
2	ice - cream	0.6721	carne - asada	0.7040
3	**more - truth**	0.5635	**highly - recommend**	0.5468
4	**highly - recommend**	0.5597	clam - chowder	0.4699
5	strip - mall	0.5228	no - longer	0.4592
6	no - longer	0.4593	credit - card	0.4136
7	peanut - butter	0.4551	**melt - mouth**	0.4131
8	question - answer	0.4217	farmer - market	0.4080
9	**melt - mouth**	0.4146	peanut - butter	0.3851
10	spring - training	0.3974	question - answer	0.3824
11	**pleasantly - surprise**	0.3687	pork - belly	0.3795
12	check - out	0.3283	strip - mall	0.3305
13	want - truth	0.3058	**pleasantly - surprise**	0.3213
14	credit - card	0.2986	hole - wall	0.2954
15	salsa - chip	0.2830	mash - potato	0.2876
16	mac - cheese	0.2774	mac - cheese	0.2771
17	mash - potato	0.2721	check - out	0.2758
18	goat - cheese	0.2608	shave - ice	0.2585
19	pork - belly	0.2522	**write - review**	0.2473
20	**look - forward**	0.2473	parking - lot	0.2328
21	**blow - away**	0.2457	customer - service	0.2325
22	**reasonably - price**	0.2423	pick - up	0.2263
23	**write - review**	0.2412	goat - cheese	0.2237
24	grocery - store	0.2392	**anywhere - else**	0.2196
25	pick - up	0.2349	**look - forward**	0.2178
26	parking - lot	0.2326	spring - roll	0.2154
27	taste - bud	0.2305	taste - bud	0.2120
28	**customer - service**	0.2288	**reasonably - price**	0.2113
29	year - ago	0.2287	come - back	0.2112
30	**anywhere - else**	0.2266	end - up	0.2061

sentiment orientation is perceivable through the context path. For example, the context path, "write-review-star-give", is shown in the positive reviews (R5), and a similar pattern of the context path is in the negative reviews (R1), but in this case, there is one more word with "not", i.e., "write-review-star-give-not." In fact, there is a co-occurrence of "give" and "not" in R5, whose value is 1165, and its similarity is 0.0263. However, the link between the two words is gone in the context structure affected by Pathfinder scaling. On the contrary, the pair of "give" and "not" in negative reviews remains in the context construct; its co-occurrence is 175, and its similarity is 0.0747.

Table 5. The top 30 term pairs of the negative reviews (R1)

Rank	AZ_R1		CA_R1	
	Term pair	Weight	Term pair	Weight
1	**protestor - customer**	0.8382	credit - card	0.5321
2	**customer - protest**	0.5159	ice - cream	0.4819
3	credit - card	0.4591	no - longer	0.3674
4	no - longer	0.4350	milk - tea	0.3033
5	ice - cream	0.3740	**somewhere - else**	0.2949
6	parking - lot	0.3574	pick - up	0.2830
7	salsa - chip	0.3381	parking - lot	0.2609
8	**somewhere - else**	0.3158	call - phone	0.2597
9	pick - up	0.2863	end - up	0.2550
10	spring - roll	0.2763	**never - again**	0.2483
11	customer - cop	0.2742	salsa - chip	0.2459
12	gift - card	0.2680	come - back	0.2447
13	oil - change	0.2588	year - ago	0.2282
14	**write - review**	0.2459	read - review	0.2208
15	mac - cheese	0.2411	**rip - off**	0.2198
16	end - up	0.2350	**give - star**	0.2052
17	read - review	0.2335	**avoid - cost**	0.2004
18	**minute - later**	0.2209	oil - change	0.1894
19	stop - in	0.2184	**minute - later**	0.1892
20	**never - again**	0.2154	drop - off	0.1885
21	**give - star**	0.2110	**wait - minute**	0.1754
22	fry - rice	0.2083	**piss - off**	0.1674
23	**rip - off**	0.1993	grocery - store	0.1661
24	**avoid - cost**	0.1942	come - out	0.1641
25	**not - even**	0.1913	fish - taco	0.1605
26	grocery - store	0.1874	fry - rice	0.1555
27	year - ago	0.1830	sit - down	0.1548
28	drop - off	0.1782	**waste - money**	0.1527
29	**spend - money**	0.1764	go - back	0.1504
30	rice - bean	0.1670	**save - money**	0.1477

In the context path, the aspect term of the review content, such as "parking-lot", has a more explicit positivity with the term "plenty" (Table 6). "Phone-call" and "make-reservation" have a more explicit negativity associated to "mistake" (Table 7). We found the context path containing the term "spend" in both of R5 and R1. The context path found in R5 bears positive aspect terms such as "spend-time-take-home-feel." On the contrary, the negative context path is connected to "spend-money-waste-time" in R1.

Figure 4 illustrates a part of the context structure surrounding the word association, "line-wait-minute" and analyzes the context paths inside. The three terms, "line" to

Table 6. The representative instances in the context pairs of the positive reviews (R5)

No	Dataset	Context path	Score
1	AZ_R5	highly - recommend - definitely - return	0.3133
2	AZ_R5	highly - recommend - definitely - back	0.2570
3	CA_R5	highly - recommend - definitely - back	0.2533
4	AZ_R5	highly - recommend - definitely - back - again	0.2260
5	AZ_R5	highly - recommend - definitely - back - soon	0.2188
6	CA_R5	highly - recommend - definitely - back - again	0.2186
7	AZ_R5	highly - recommend - definitely - back - again - visit	0.1898
8	CA_R5	plenty - parking – lot	0.1611
9	AZ_R5	reasonably - price - quality	0.1485
10	CA_R5	write - review - star - give	0.1429
11	CA_R5	importantly - most - ever - taste - bud	0.1384
12	CA_R5	read - review - star - give	0.1194
13	AZ_R5	want - eat - ever - taste - bud	0.0921
14	AZ_R5	go - place - love - absolutely	0.0892
15	AZ_R5	perfectly - cook - not - disappoint - never	0.0880
16	CA_R5	only - minute - wait - long - line	0.0771
17	CA_R5	perfectly - cook - not - worry	0.0759
18	AZ_R5	spend - time - take - home - feel	0.0744
19	CA_R5	only - minute – wait	0.0735
20	AZ_R5	only - minute – later	0.0723

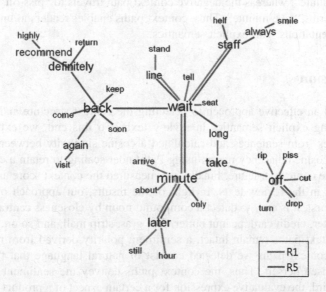

Fig. 4. The representative instances in the context pairs of the negative reviews (R1)

Table 7. The representative instances in the context pairs of the negative reviews (R1)

No	Dataset	Context path	Score
1	AZ_R1	write - review - star - give - not	0.1531
2	AZ_R1	go- never – again	0.1507
3	AZ_R1	read - review - star - give - not	0.1500
4	AZ_R1	spend - money - waste - time	0.1494
5	AZ_R1	ever - again - never - back	0.1466
6	CA_R1	stand - line - wait - minute - later	0.1420
7	CA_R1	stand - line - wait - minute - later - hour	0.1338
8	AZ_R1	order - minute - later - hour	0.1296
9	AZ_R1	order - minute - later - still	0.1217
10	AZ_R1	yelp - review - star - give - not	0.1185
11	AZ_R1	stand - line - wait - minute - later	0.1170
12	AZ_R1	stand - line - wait - minute - later - hour	0.1099
13	CA_R1	spend - money - waste - time - go - home	0.1082
14	AZ_R1	stand - line - wait - minute - later - still	0.1052
15	AZ_R1	piss - off - take - minute	0.1051
16	AZ_R1	piss - off - take - minute - wait	0.0985
17	AZ_R1	phone - call - reservation - make - mistake	0.0969
18	AZ_R1	finally - come - manager - speak	0.0933
19	CA_R1	feel - make – mistake	0.0779
20	CA_R1	help - ask - manager - speak	0.0770

"minute" appear in both of R5 and R1. However, the positive context path is associated with "only-minute", whereas the negative context path travels to "piss-off" or "rip-off" through the terms "take-minute". Thus, context paths enables readers to understand the sentiment orientations with explicit semantics.

5   Conclusion

We presented an effective approach to detecting the context structure and the context path presenting explicit semantics in review texts. To this end, we extracted word co-occurrences from sentences and calculated a cosine similarity between term pairs. We scaled a cosine adjacency matrix using Pathfinder scaling to retain a core network to discover the context structure. Finally, we measured the context score and presented context paths in the review texts. In light of the results, our approach offers several advantages: First, it is easy to detect a compound noun by closeness centrality, such as shopping center, credit card, peanut butter, foie gras, strip mall, and so on. Second, the extracted context paths remain intact, a sentiment polarity derived from review texts, because the context paths we detected consist of natural language that the customer wrote by himself/herself. Thus, the context paths convey the sentiment orientations explicitly. Third, the evaluative expression for a certain aspect of a product or service is clearer and more specified within the context path. Lastly, it is not necessary to train reference words to detect the sentiment orientations.

The proposed approach does not restrict specific domains and has many potential applications. The context path can be used for review summarization and recommendation. Converting the context path to the context vector can also be used for the classification and prediction tasks of sentiment orientations.

Acknowledgements. This work was supported by the Ministry of Education of the Republic of Korea and the National Research Foundation of Korea (NRF-2015S1A3A2046711).

References

1. Ghose, A., Ipeirotis, P.G.: Designing novel review ranking systems: predicting the usefulness and impact of reviews. In: 9th International Conference on Electronic Commerce, pp. 303–310. ACM (2007)
2. Hu, X., Downie, J.S., West, K., Ehmann, A.: Mining music reviews: promising preliminary results. In: 6th International Symposium on Music Information Retrieval (2005)
3. Titov, I., McDonald, R.: Modeling online reviews with multi-grain topic models. In: 17th International Conference on World Wide Web, pp. 111–120. ACM (2008)
4. Hatzivassiloglou, V., McKeown, K.R.: Predicting the semantic orientation of adjectives. In: 35th Annual Meeting of the Association for Computational Linguistics and 8th Conference of the European Chapter of the Association for Computational Linguistics, Association for Computational Linguistics, pp. 174–181 (1997)
5. Turney, P.D.: Thumbs up or thumbs down? Semantic orientation applied to unsupervised classification of reviews. In: 40th Annual Meeting on Association for Computational Linguistics, Association for Computational Linguistics, pp. 417–424 (2002)
6. Turney, P.D., Littman, M.L.: Measuring praise and criticism: inference of semantic orientation from association. ACM Trans. Inf. Syst. (TOIS) **21**(4), 315–346 (2003)
7. Xia, R., Xu, F., Yu, J., Qi, Y., Cambria, E.: Polarity shift detection, elimination and ensemble: a three-stage model for document-level sentiment analysis. Inf. Process. Manag. **52**(1), 36–45 (2016)
8. Pang, B., Lee, L., Vaithyanathan, S.: Thumbs up? Sentiment classification using machine learning techniques. In: ACL 2002 Conference on Empirical Methods in Natural Language Processing, Association for Computational Linguistics, vol. 10, pp. 79–86 (2002)
9. Ye, Q., Zhang, Z., Law, R.: Sentiment classification of online reviews to travel destinations by supervised machine learning approaches. Expert Syst. Appl. **36**(3), 6527–6535 (2009)
10. Dave, K., Lawrence, S., Pennock, D.M.: Mining the peanut gallery: opinion extraction and semantic classification of product reviews. In: 12th International Conference on World Wide Web, pp. 519–528. ACM (2003)
11. Jin, W., Ho, H.H., Srihari, R.K.: OpinionMiner: a novel machine learning system for web opinion mining and extraction. In: 15th ACM SIGKDD, pp. 1195–1204. ACM (2009)
12. Choi, Y., Cardie, C., Riloff, E., Patwardhan, S.: Identifying sources of opinions with conditional random fields and extraction patterns. In: Proceedings of the Conference on Human Language Technology and Empirical Methods in Natural Language Processing, Association for Computational Linguistics, pp. 355–362 (2005)
13. Schvaneveldt, R.W., Durso, F.T., Dearholt, D.W.: Network structures in proximity data. Psychol. Learn. Motiv. **24**, 249–284 (1989)

14. Mei, Q., Ling, X., Wondra, M., Su, H., Zhai, C.: Topic sentiment mixture: modeling facets and opinions in weblogs. In: 16th International Conference on World Wide Web, pp. 171–180. ACM (2007)

15. Jo, Y., Oh, A.H.: Aspect and sentiment unification model for online review analysis. In: Proceedings of the Fourth ACM International Conference on Web Search and Data Mining, pp. 815–824. ACM (2011)

16. Börner, K., Chen, C., Boyack, K.W.: Visualizing knowledge domains. Annu. Rev. Inf. Sci. Technol. **37**(1), 179–255 (2003)

17. McCain, K.W.: The structure of biotechnology R&D. Scientometrics **32**(2), 153–175 (1995)

18. Chen, C.: Visualising semantic spaces and author co-citation networks in digital libraries. Inf. Process. Manag. **35**(3), 401–420 (1999)

19. White, H.D.: Pathfinder networks and author cocitation analysis: a remapping of paradigmatic information scientists. J. Am. Soc. Inf. Sci. Technol. **54**(5), 423–434 (2003)

20. White, H.D., McCain, K.W.: Visualizing a discipline: an author co-citation analysis of information science, 1972–1995. J. Am. Soc. Inf. Sci. **49**(4), 327–355 (1998)

21. Sun, J., Tang, J.: A survey of models and algorithms for social influence analysis. In: Social Network Data Analytics, pp. 177–214 (2011)

How to Find Accessible Free Wi-Fi at Tourist Spots in Japan

Keisuke Mitomi[1]([✉]), Masaki Endo[1,2], Masaharu Hirota[3], Shohei Yokoyama[4],
Yoshiyuki Shoji[1], and Hiroshi Ishikawa[1]

[1] Graduate School of System Design, Tokyo Metropolitan University, Tokyo, Japan
mitomi-keisuke@ed.tmu.ac.jp, {y_shoji,ishikawa-hiroshi}@tmu.ac.jp
[2] Division of Core Manufacturing, Polytechnic University, Tokyo, Japan
endou@uitec.ac.jp
[3] National Institute of Technology, Oita College, Oita, Japan
m-hirota@oita-ct.ac.jp
[4] Faculty of Informatics, Shizuoka University, Shizuoka, Japan
yokoyama@inf.shizuoka.ac.jp

Abstract. We propose a method of finding spots at tourist attractions that do not have accessible Free Wi-Fi by using social media data. Although it is an important issue for the government to determine where they should install Free Wi-Fi equipment, it involves a high human cost. We focused on the difference in usage of social network services (SNSs) to find where there was a lack of Free Wi-Fi. We posed two simple hypotheses: (1) uploaded photos on Flickr, where batch-time SNS reflects the popularity of attractions from the travelers' perspective, and (2) posts on Twitter, where real-time SNS reflects the communications environment. Differences in the distributions of posts in these SNSs indicate the gap in needs and the current status of communications infrastructures. Experimental results obtained from fieldwork in the Yokohama area clarified that although our method could locate places that were popular with tourists, some of these locations did not have Free Wi-Fi equipment installed there.

1 Introduction

One of the worst things in traveling is missing Free Wi-Fi spots. The facilitation of Free Wi-Fi environments is a key problem in governmental tourism policy, especially for countries that want to attract foreign travelers. The importance of inbound tourism is currently becoming increasingly significant because visitors from foreign countries usually have a positive economic effect. A typical instance is Tokyo in Japan. Interest in the information and communications technology (ICT) environment has been escalating, which is aimed at helping foreign visitors in Japan, particularly as we head toward the Tokyo Olympics in 2020. The Japanese Government has also implemented a campaign called the "Visit Japan Campaign", which is a promotional scheme to encourage foreign tourists to visit Japan. As a result of this promotion, the number of foreign visitors to Japan annually reached over 20 million people in 2015.

© Springer International Publishing AG 2016
E. Spiro and Y.-Y. Ahn (Eds.): SocInfo 2016, Part I, LNCS 10046, pp. 389–403, 2016.
DOI: 10.1007/978-3-319-47880-7_24

Moreover, the government has been publicizing Japan's appeal overseas to increase the number of foreign visitors traveling to Japan annually even more. Thus, increasing both the numbers of new foreign visitors and repeaters is an important issue for the Japanese Government. It must effectively tackle this issue not only by promoting tourist spots and recommending personalized venues, but also by reducing dissatisfaction that foreign visitors may feel. The Japan Tourism Agency, which is a section of the Ministry of Land, Infrastructure, and Transport administered an extensive questionnaire to foreign tourists in 2011 to clarify the disadvantages of tourism in Japan. According to the results obtained from this survey, the highest percentage (36.7%) of disaffection was caused by "the lack of free public wireless LAN". This indicates a phenomenon that it is still difficult for most foreign visitors to achieve mobile Wi-Fi overseas. The second most common problem was "communication" (24.0%), and the third most common was the "acquisition of process information on public transportation to a destination" (20.0%). In this regard, we expect that if everybody could connect their devices to a free public wireless LAN, "communication" problems and "acquisition of process information on public transportation to a destination" problems could be solved by online services, such as free online machine translation systems, transportation navigation services, and Web search engines. Therefore, the dissatisfaction with "the lack of free public wireless LAN" is a crucial issue for foreigner visitors.

This concern is not only limited to Japan but also to all countries who accept tourists because tourists to a country other than their home country may have no mobile Wi-Fi. Furthermore, some Free Wi-Fi spots are often not entirely free because fees need to be paid to connect to them. This creates complicated and annoying procedures, such as having to input e-mail addresses, provide payment information, and comply with many other requirements. As this depends on Free Wi-Fi service providers whose operations are not standardized, visitors must input their information to individual providers. Therefore, they cannot easily connect to many Free Wi-Fi spots. Moreover, there are many locations that have no Free Wi-Fi spots.

The Japanese Government has weighed heavily on this issue and is now trying to increase convenient Free Wi-Fi spots to foreigners to enable easy travel as the nation approaches the Tokyo Olympics in 2020. The Ministry of Public Management, Home Affairs, Posts, and Telecommunications published an action plan that was aimed at fulfilling three requirements for the ICT environment to be: selectable, accessible, and high-quality. These activities to enable more Free Wi-Fi spots to be provided would contribute to reducing dissatisfaction from foreigner visitors to Japan. The best case could be achieved by installing huge numbers of Free Wi-Fi spots to enable people to connect to the Internet from everywhere, although this is not realistic in terms of cost. An important issue is how to select effective spots to install Free Wi-Fi. The first step would be to focus on "accessible" in the three requirements of the action plan. The two requirements (selectable and high-quality) then become crucial issues after people can connect to Free Wi-Fi spots from anywhere. Therefore, this paper proposes a method of inexpensively satisfying the lack of Free Wi-Fi spots by using social media posts.

The main aim of our research was to detect spots that were tourist attractions, were visited by large numbers of foreign visitors, but had no accessible Free Wi-Fi available. This may help the government and Free Wi-Fi providers to make decisions where they should immediately install new Free Wi-Fi spots. Additionally, it would also help foreign visitors because they could identify where they could obtain Free Wi-Fi before visiting a place, they could stay away from areas without Free Wi-Fi, or they could prepare enough prior information before arriving there. It has been pointed out that it is difficult and expensive for central and local governments to identify all the locations of Free Wi-Fi spots. Although some local governments provide free Wi-Fi spots, a few of them provide that information as open data online. Additionally, there are some private companies and organizations that have provided Free Wi-Fi spots by themselves. It is difficult for governments to track their locations by using private Free Wi-Fi providers.

Social Network Services (SNSs) posts on location information, on the other hand, have been increasing, accompanied by the rapid spread of smart phones. Twitter is one of the most well-known micro-blog services because it is easy to use on smart phones, and there are many users who immediately post their own actions and thoughts on the fly. Further, since these posts can be provided with location information (*i.e.*, geo-tags on Twitter), it is possible to know when and where a user is staying. However, since many foreign visitors do not have mobile lines, it is not possible to tweet in areas that have no Free Wi-Fi. Another SNS, Flickr, which is a photo-sharing service, is also well-known. Users can upload photos that were taken at tourist destinations since it is possible to impart position information when taking photographs (*i.e..*, exchangeable image file format (EXIF) information) with recent digital devices. This makes it possible to identify both where and when users are staying. However, they cannot immediately upload their photos from some places because there may not have any Internet connection. Nevertheless, Flickr users can batch all their trip photos and upload them after they return to their hotels or homes, which is different from Twitter. We analyzed the differences in the distributions of post areas of foreign visitors who were Twitter and Flickr users in this research to find where Free Wi-Fi spots were lacking. The different natures of these two SNSs enabled areas to be visualized that did not have any accessible Free Wi-Fi. We posed two simple hypotheses:

H1 Spots where lots of photos had been taken on Flickr must be popular places with travelers and

H2 Spots from where Twitter posts had been sent must have some Free Wi-Fi equipment installed.

These hypotheses enabled us to deduce the assumption below:

– Places with a lot of Flickr posts but few Twitter posts must have a critical lack of accessible Free Wi-Fi.

The method we propose has two advantages. First, it is inexpensive because it only needs SNS data but does not need investigators to be dispatched, and second, it can detect both Free Wi-Fi spots made available by the government and private organizations because it does not use any official information from providers.

The rest of this paper is organized as follows. Section 2 describes related research and Sect. 3 describes a method of using the differences between the number of contributions in each Twitter and Flickr region to determine low-frequency areas of Free Wi-Fi use and frequently used areas throughout Japan. In addition, we describe a method of determining the presence or absence of mobile communication means for users. Section 4 presents the results of visualization for areas of high-and low-frequency Free Wi-Fi use we describe in the discussion. Section 5 has a summary of this study.

2   Related Work

The main aim of our research was to find useful information for tourism by using social sensing techniques. This section introduces related work published in the area of social sensing tourism by using SNS.

2.1   Social Sensing

There have been many studies on the detection of real-world events and analysis of users' impressions of events [6,11,13,14] that have been obtained by analyzing the text data of tweets and temporally analyzing the number of tweets. Meladianos et al. [11] dealt with the task of sub-event detection in evolving events using posts collected from Twitter streams. Hu et al. [6] addressed this question by first operationalizing a person's Twitter engagements in real-world events such as posting, retweeting, or replying to tweets about such events.

However, there have also been studies on analyzing the spread of data by using geo-data [1,7,8,15]. Becker et al. [1] found that the resolution of social issues, such as urban planning and public transport, was important in understanding human movement trajectories. They focused on geo-temporal photo-trails from different cities from Flickr that were produced by humans when taking sequences of photos in urban areas. Kamath et al. [8] analyzed tweets with geo-data and hash tags with respect to location, time, and distance to understand memes for diffusing and propagating information all around the world.

2.2   Tourism

Some studies have focused on extracting tourist data by using information on the Web. For example, some studies have analyzed tourism images by using data on the Web [4,16,18] and the impressions of tourists [17]. Choi et al. [4] analyzed tourism images in Macau from information related to travel found on the Web. Wenger [17] extracted tourism images of Australia by analyzing the blogs of travelers. In addition, Cheng *et al.* [3] estimated the residences of users by using tweet content. Burger and Henderson [2] estimated the ages of users by using blog content and metadata. These studies estimated the attributes of users as features, such as profile and text information and the content of user tweets. Our study can contribute to further research on recommendations for travel plans and routes. Previous researchers exposed how people moved in tourism spots and analyzed this from social media sites like Flickr and Twitter. Some studies have analyzed tourist routes and plans using popularity and user preferences [5,9,10].

These previous researchers recommended travel plans and routes based on various features like the popularity of venues and user preferences. As we previously explained, it is important to lower tourist stress from Free Wi-Fi spots not working to decrease visitors' levels of dissatisfaction. Therefore, we think that by considering whether tourists can connect their devices to Free Wi-Fi spots or not to search for adequate travel plans and routes would contribute to positive recommendations. This study also found tourism areas where tourists cannot connect to Free Wi-Fi using tourism spots extracted from Flickr. Therefore, this study had limitations on various types of extraction areas. Although our approach can be used to extract such tourism areas, it cannot extract intermediate areas between two tourism spots. This was due to our hypotheses. Our method regarded areas in which many photos had been taken as tourism spots and areas with high demand as locations that had Free Wi-Fi installed. Therefore, our method was not able to determine which areas with fewer photos could not connect to Free Wi-Fi spots or which areas did not have high demand. However, our research used the analysis of tourism routes, and could profit by taking into consideration such intermediate areas with fewer photos.

3   Proposed Method

This section describes our proposed method of detecting areas where users cannot connect to accessible Free Wi-Fi by using posts by foreign travelers on social media.

Our method uses differences in the characteristics of two types of SNSs and we focused on two of these:

Real-time: Immediate posts and experience of things and photos (*e.g.*, Twitter)
Batch-time: Posts stored data later to devices (*e.g.*, Flickr)

Twitter users can only post tweets when they can connect devices to Wi-Fi or wired networks. Therefore, travelers can post tweets in areas with Free Wi-Fi outside of tourism, or when they have mobile communications. In other words, we can only obtain tweets with geo-tags posted by foreign travelers from such places. Therefore, areas where we can obtain huge numbers of tweets posted by foreign travelers are places that are attractive for them to sightsee, and areas where they can connect to accessible Free Wi-Fi. Flickr users, on the other hand, take many photographs by using digital devices regardless of networks, but whether they are able to upload photographs on-site depends on the conditions of the network. As a result, almost all users can upload photographs after returning to their hotels or home countries. However, geo-tags annotated to photographs can indicate when they were taken. Therefore, although it is difficult to obtain information (activities, destinations, or routes) on foreign travelers from Twitter, Flickr can be used to observe such information. We based our hypothesis in this study of "A place that has a lot of Flickr posts but few Twitter posts must have a critical lack of accessible Free Wi-Fi" by using these characteristics of SNSs. We extracted areas that were tourist attractions for foreign travelers, but from which they could not connect to accessible Free Wi-Fi. What our method aimed to find was accessible Free Wi-Fi, although some of this Wi-Fi was not entirely free.

There are two main reasons for areas from where foreign travelers cannot connect to Free Wi-Fi. The first is areas where there are no Wi-Fi spots. The second is areas where users can use Wi-Fi but it is not accessible. We treated them both the same as inaccessible Free Wi-Fi because both areas were unavailable to foreign travelers. Since we conducted experiments focused on foreign travelers, we could detect actual areas without accessible Free Wi-Fi. In addition, our method extracted areas with accessible Free Wi-Fi, and other locations were regarded as regions without accessible Free Wi-Fi.

3.1   Extracting Users Who Are Foreign Travelers

This subsection describes a method of extracting foreign travelers using Twitter and inFlickr. We obtained and analyzed tweets posted in Japan from Twitter using Twitter's Streaming application programming interface (API)[1]. We used the method proposed by Saeki *et al.* [12] to extract foreign travelers. Their method classifies types of Twitter users who posted in Japan as foreigners or Japanese. First, the method detects the principal language of the user and defines it as a language that meets two conditions. The first condition is that the language must be used in more than half of all the user's tweets. We used the same Language-Detection Toolkit that Saeki *et al.* used. The second is that the language must be selected by the user in his/her account settings. This means that he/she claims that they use that language. If the results obtained from this analysis for a Twitter user is a language other than a non-Japanese language used in Japan, we regard the user as a foreign traveler.

We obtained photographs with geo-tags taken in Japan from Flickr using Flickr's API[2]. We extracted foreign travelers who had taken photographs in Japan. We regarded Flickr users who had set their profiles of habitation on Flickr as Japan or geographical regions, as the users lived in Japan; otherwise, they were regarded as foreign travelers. We used the tweets and photographs that foreign travelers had created in Japan in the analysis that followed.

---

[1] https://dev.twitter.com/streaming/overview.
[2] https://www.flickr.com/services/api/.

3.2 Extracting Tourism Spots and Non-free Wi-Fi Areas

Our method envisaged trying to find places that met two conditions:

- Spots where there was no accessible Free Wi-Fi and
- Spots that many foreign visitors visited

We use the number of photographs taken at locations to extract tourism spots. Many people might take photographs of subjects such as landscapes based on their own interests. They might then upload those photographs to Flickr. As these were locations at which many photographs had been taken, these places might also be interesting places for many other people to sightsee or visit. We have defined such places as tourism spots in this paper. We specifically examined the number of photographic locations to identify tourism spots to find locations where photographs had been taken by many people. We mapped photographs that had a photographic location onto a two-dimensional grid based on the location at which a photograph had been taken to achieve this. Here, we created individual cells in a grid that was 30-m square. Consequently, all cells in the grid that was obtained included photographs taken in a range. We then counted the number of users in each cell. We regarded cells with greater numbers of users than the threshold as tourism spots. We also used the same procedure to extract areas where foreign travelers could connect to accessible Free Wi-Fi.

4 Experiment

We conducted an experiment to extract spots that seemed to have Free Wi-Fi and spots for tourist attractions on the basis of the proposed method. This section describes the procedure, the data set, and the results obtained from the experiment.

4.1 Experiment Method

We conducted an experiment to extract Free Wi-Fi spots as tourist attractions along Sakuragicho Station in Yokohama as the target area based on the proposed method. Two participants visited there, and performed fieldwork. To verify our method, we made a list of eight places where has/does not have Flickr photos and Twitter tweets. The participants went each spots, and executed three tasks below:

- Count the number of foreign visitors,
- Check the Free Wi-Fi condition by using smartphone, and
- Do a questionnaire if they are visitor or not.

There are several reasons why we chose Sakuragicho area for the fieldwork venue. The first reason is thet Yokohama is a good model to explain Japanese typical sightseeing area. Yokohama is visited by 37 million tourists a year and had more than a thousand foreign lodgers in 2015[3]. Sakuragicho Station in Yokohama has been described as a popular spot on Yokohama Visitors' Guide Site[4]. Around

---

[3] http://www.city.yokohama.lg.jp/bunka/.

[4] http://www.welcome.city.yokohama.jp/ja/tourism/courses/.

Sakuragicho Station, there is many famous venues. The Red Brick Warehouse in the vicinity of the Sakuragicho Station is a well-known tourist spot. Osanbashi Pier to the the southeast of the Red Brick Warehouse is also a tourist spot. Osanbashi Pier next to the Red Brick Warehouse has been recommended as a tourist route by Yokohama City's tourist information. Therefore, it has been assumed that tourists visit either the Red Brick Warehouse or Osanbashi Pier. Second reason is the scale of the city. Tokyo, the biggest city in Japan, is too big to validate the method proposed. In particular, at the stations in Tokyo, there are too many people to count them up. Sakuragicho Station is suitable to fieldwork, i.e., not too large and not too small. The fieldwork participants can overlook all the users of the station. The third reason is that, it is close to the authors' university; the participants are familiar with the area. Figure 1 shows a list of major tourist attraction spots in Sakuragicho. The spots marked with a star-shaped mark are spots where we actually visited.

In addition, we administered a questionnaire to confirm whether there was accessible Free Wi-Fi around Sakuragicho Station in field work. We administered the questionnaire to foreign visitors and foreign residents in Japan.

Ease of use to determine whether Free Wi-Fi was accessible was defined according to five conditions:

Completely Free: Registration was not required,
SNS Registration: Registration was not required if the visitors used other SNS information,
Registration Required: Users need to input their information on site,
Software Required: Users have to install specific software (i.e. application for smartphone, or authentication software for computer) for their devices, and
None: There are no Free Wi-Fi access points available.

We confirmed whether there were actually accessible Free Wi-Fi spots around Sakuragicho Station by using these classifications.

4.2   Dataset

This subsection describes the dataset used in the experiment. We collected more than 4.7 million data items with geo-tags from July 1, 2014 to February 28, 2015 in Japan. We detected tweets tweeted by foreign visitors by using the method proposed by Saeki et al. [12]. The number of tweets that was tweeted by foreign visitors was more than 1.9 million. The number of tweets that was tweeted by foreign visitors in the Yokohama area was more than 7,500. We collected more than 5,600 photos with geo-tags from July 1, 2014 to February 28, 2015 in Japan. We detected photos that had been posted by foreign visitors to Yokohama by using our proposed method. Foreign visitors posted 2,132 photos.

4.3   Experimental Results

This subsection provides the experimental results. Table 1 summarizes the rankings for the ratio of the number of foreign visitors who tweeted in a cell and the total number of foreign visitors. We empirically set the threshold to 0.02; thus, when the value in the cell exceeded 0.02, there was accessible Free Wi-Fi in that cell. The ratio of the number of foreign visitors who tweeted in a cell was referred to as the **Twitter Score**. The twitter score $ts(c)$ of a cell c is as below:

$$ts(c) = \frac{|Tw(c)|}{\sum_{n \in C(a)} |Tw(n)|}, \tag{1}$$

where $Tw(c)$ is all tweets in the cell c, and $C(a)$ is all cells in area a. Thus, $ts(c)$ represents the ratio of the number of tweet in the cell c to the number of all tweets in area a. Figure 2 shows the results obtained from visualizing accessible Free Wi-Fi spots. When the twitter score was high, the figure displayed cells in colors close to red.

Table 2 lists the ranking of the ratio of the number of foreign visitors taking photos in a cell and the total number of foreign visitors. We empirically set the threshold to 0.02; thus, as the values in the cells exceeded 0.02, those cells were deemed popular. The ratio of the number of foreign visitors taking photos in a cell and the total number of foreign visitors in all cells was referred to as a **Popularity Score**. The popularity score $ps(c)$ of a cell c is as below:

$$ps(c) = \frac{|Ph(c)|}{\sum_{n \subset C(a)} |Ph(n)|}, \tag{2}$$

where $Ph(c)$ is all Flickr photos posted in the cell c, and $C(a)$ represents all cells in area a. Hence $ps(c)$ denotes the ratio of the number of photos in the cell c to the number of all photos in area a. Figure 3 shows the results obtained from visualizing popular spots. When the popularity score was high, the Fig. 3 displayed cells in colors close to red. Figure 4 represents the difference of the twitter score and the popularity score for each cells in Sakuragicho area. The cell which has a lack of Free Wi-Fi displayed in dark color.

Table 3 summarizes the results obtained from the fieldwork that indicate whether there were actually accessible Free Wi-Fi spots available. In our result, we did not find any **Completely Free** Wi-Fi spots, but could find many accessible Free Wi-Fi access points.

5   Discussion

The cell containing the Red Brick Warehouse has been visualized in Fig. 2, but the cell containing Osanbashi Pier has not. Since the Red Brick Warehouse has had Free Wi-Fi installed for foreigners visitors to Japan since 2014, they can use the Free Wi-Fi when visiting the Warehouse. However, since Osanbashi Pier has not had Free Wi-Fi installed for visiting foreigners, even users who have mobile communication devices cannot post Tweets. However, there is a Starbucks Coffee shop in the vicinity of Sakuragicho and Minato Mirai Station,[5] where Free Wi-Fi

---

[5] http://www.starbucks.co.jp/.

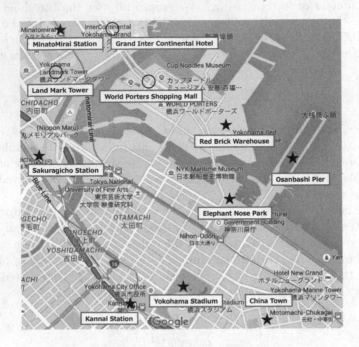

Fig. 1. Tourist attraction spots in Sakuragicho area

Table 1. The rankings for the ratio of the number of foreign visitors who tweeted in a cell and the total number of foreign visitors

Venue name	Twitter score
Land Mark Tower	0.0255
Red Brick Warehouse	0.0248
Grand InterContinental Hotel's immediate neighbor to west	0.0242
Minatomirai Station's immediate neighbor to west	0.0233
World Porters' Shopping Mall	0.0231
Minato Mirai Station	0.0228
Sakuragicho Station	0.0220
Land Mark Tower's immediate neighbor to south	0.0212
World Porters' Shopping Mall's immediate neighbor to west	0.0212
Grand InterContinental Hotel	0.0212
China Town	0.0212
Land Mark Tower's immediate neighbor to east	0.0202

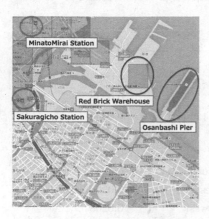

Fig. 2. Frequent tweet spots: Reflect accessible free Wi-Fi spots (Color figure online)

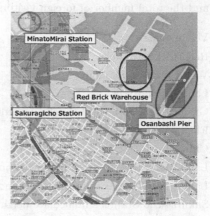

Fig. 3. Frequent Flickr spots: Reflect tourist attraction spots (Color figure online)

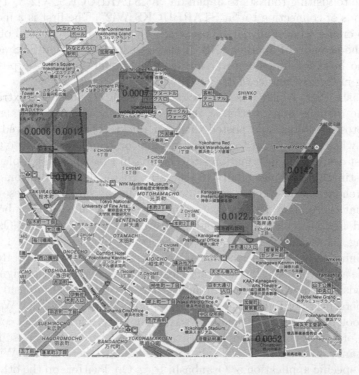

Fig. 4. Popularity score - Twitter score (Color figure online)

Table 2. The ranking of the ratio of the number of foreign visitors taking photos in a cell and the total number of foreign visitors

Venue name	Popularity score
Red Brick Warehouse	0.0244
World Porters' Shopping Mall	0.0238
Land Mark Tower	0.0238
Grand InterContinental Hotel's immediate neighbor to west	0.0238
Elephant Nose Park	0.0236
China Town	0.0222
Land Mark Tower's immediate neighbor to south	0.0218
Land Mark Tower's immediate neighbor to west	0.0208
Sakuragicho Station's immediate neighbor to west	0.0208
Land Mark Tower's immediate neighbor to west	0.0208
Osanbashi Pier	0.0206
MinatoMirai Station's immediate neighbor to north-west	0.0206

is available to visiting tourists to Japan on "at_STARBUCKS_Wi2"[6]. There are two methods of connecting to "at_STARBUCKS_Wi2". The first is a method of logging to create an account for this service. The second is a method of logging in using the account of another SNS. If the user has any plural SNS accounts, there is no need for him/her to create a new account. It may be possible to simply log in to existing accounts.

Osanbashi Pier, which was not extracted in Fig. 2, has been extracted in Fig. 3. The main reason for this is that it had been assumed that users had taken photos of interest and posted them on Flickr. They are believed to have been used when searching for tourist destinations on trips. According to a Dubai-user tourist destination information site, Osanbashi Pier was in 11th place in the tourist destination rankings of all 346 places in Yokohama. In addition, the Red Brick Warehouse in its vicinity was in 3rd place in the rankings. These findings suggest that there many foreign visitors have visited Osanbashi Pier because it is assumed that they visited these sights together. The cell including Osanbashi Pier in Fig. 2 has not been extracted when comparing Figs. 2 and 3, but the cell containing Osanbashi Pier has been extracted in Fig. 3. Since the Red Brick Warehouse and Osanbashi Pier are very close, we considered that foreign visitors who had visited the Red Brick Warehouse had also visited Osanbashi Pier. In fact, the percentage of Flickr users who took photos at the Red Brick Warehouse and also took photos at Osanbashi Pier was about 50%. Therefore, we considered that 50% of Twitter users who had visited the Red Brick Warehouse had also visited Osanbashi Pier. In fact, Table 3 indicates that there is Free Wi-Fi that requires a specific application at Osanbashi Pier. On Twitter, on the other hand,

---

[6] http://starbucks.wi2.co.jp/pc/index_jp.html.

Table 3. Fieldwork summary

Venue name	# visitors	Wi-Fi condition
Sakuragicho Station	53	Registration required
		SNS Registration
Minato Mirai Station	15	Registration required
		SNS registration
Red Brick Warehouse	36	Registration required
Osanbashi Pier	34	Software required
Elephant Nose Park	30	None
China Town	28	SNS registration
Yokohama Stadium	10	None
Kannai Station	6	None

the percentage of Twitter users who tweeted from the Red Brick Warehouse and Osanbashi Pier was 10 % or less. One of the main reasons that users who tweeted from the Red Brick Warehouse did not tweet from Osanbashi Pier was that no Free Wi-Fi was accessible to foreign visitors from Osanbashi Pier.

Further, while cells can be found at Sakuragicho and Minato Mirai Station in Fig. 2, no cells can be found in Fig. 3 Since there is accessible Free Wi-Fi such as that "at_STARBUCKS_Wi2", foreign visitors used accessible Free Wi-Fi for Twitter, as was previously explained. In fact, when we confirmed tweet content manually, we found that there were many tweets that had content that represented visitors arriving at Sakuragicho or Minato Mirai Station and stopping by Starbuck's Coffee.

The interesting discovery is that, the answers of the questionnaire indicate that the **SNS Registration** is not an obstacle to connect to the Free Wi-Fi but visitors hesitate to connect to the **Registration Required** spots. It is possibly caused by two reasons. The first reason is the features of the mobilephone; It is troublesome to input long registration information (*e.g.*, name, addresses and phone numbers) with a small screen and keyboard. The second reason is the security; Users do not want to tell their real name and private information. Telling the ID of the SNS is easier than telling their real personal information.

Conclusively, we found some knowledge from the experiment. Our method has high precision but there is still some room to improve recall. For instance, we found many tweets posted by foreign visitors around Sakuragicho Station, but we found few photos around there. This phenomenon indicates that our Hypothesis 1 is partially incorrect. Although, even if

– tourist spots which have many photos posted on batch-time SNSs are popular for foreign travelers

is correct, it is not always true that

– all the popular tourist spots have many photos posted on batch-time SNS.

One of the most typical instances are the stations as they appeared in our experimental result. In transportation hubs or facilities that host public and private services (*e.g.*, stations, hospitals and hotels), foreign travelers need Free Wi-Fi spots. They actually connect to them if they are available. Another instance inferred from the experimental result is a place where taking photos is strictly prohibited. Some of the tourist spots and attractions such as museums, religious buildings and department stores disallow taking photos inside. In such situation, the method we proposed does not work. This causes the reduction in the recall rate of finding a spot which has high priority to install the Free Wi-Fi equipment. On the other hand, the number of posts on Twitter reflects accurately the communication environment around the tourist spot. Through the fieldwork, we could find accessible Free Wi-Fi spots in areas which has many posts posted to Twitter, and we found the lack of Free Wi-Fi in spots which has no Tweets. Thus, our method can find some of the spots at tourist attractions that do not have accessible Free Wi-Fi.

6   Conclusion

Our method could be used to classify spots where many photos had been taken on Flickr to identify popular destinations for travelers and those that were not. Our method could be used to classify spots that had some Free Wi-Fi equipment for Twitter posts and those that did not. In addition, we confirmed whether there were usable accessible Free Wi-Fi spots by actually conducting tests around Sakuragicho Station and administering a questionnaire to foreign visitors.

6.1   Future Work

We conducted experiments from July 1, 2014 to February 28, 2015 in Yokohama in this study. Therefore, we concluded that the amount of data we collected was insufficient. Evaluation experiments using large-scale data will be necessary in the future.

In addition, the process in this study was not carried out by using language. Therefore, we intend to conduct experiments in a foreign country in future work.

Acknowledgments. This work was supported by JSPS KAKENHI Grant Numbers 16K00157 and 16K16158, and a Tokyo Metropolitan University Grant-in-Aid for Research on Priority Areas involving "Research on Social Big Data."

References

1. Becker, M., Singer, P., Lemmerich, F., Hotho, A., Helic, D., Strohmaier, M.: Photowalking the city: comparing hypotheses about urban photo trails on Flickr. In: Liu, T.-Y., Scollon, C.N., Zhu, W. (eds.) SocInfo 2015. LNCS, vol. 9471, pp. 227–244. Springer, Heidelberg (2015). doi:10.1007/978-3-319-27433-1_16
2. Burger, J.D., Henderson, J.C.: An exploration of observable features related to blogger age, pp. 15–20 (2006)

3. Cheng, Z., Caverlee, J., Lee, K.: You are where you tweet: a content-based approach to geo-locating Twitter users. pp. 759–768 (2010)
4. Choi, S., Lehto, X.Y., Morrison, A.M.: Destination image representation on the web: content analysis of Macau travel related websites. Tourism Manag. **28**(1), 118–129 (2007)
5. Crandall, D.J., Backstrom, L., Huttenlocher, D., Kleinberg, J.: Mapping the world's photos, pp. 761–770 (2009)
6. Yuheng, H., Farnham, S., Talamadupula, K.: Predicting user engagement on twitter with real-world events (2015)
7. Jurgens, D., Finethy, T., McCorriston, J., Yi Tian, X., Ruths, D.: Geolocation prediction in twitter using social networks: a critical analysis and review of current practice (2015)
8. Kamath, K.Y., Caverlee, J., Lee, K., Cheng, Z.: Spatio-temporal dynamics of online memes: a study of geo-tagged tweets, pp. 667–678 (2013)
9. Kisilevich, S., Mansmann, F., Keim, D.: P-DBSCAN: a density based clustering algorithm for exploration and analysis of attractive areas using collections of geo-tagged photos, pp. 38:1–38:4 (2010)
10. Lacerda, Y.A., Feitosa, R.G.F., Esmeraldo, G.Á.R.M., de Souza Baptista, C., Marinho, L.B.: Compass clustering: a new clustering method for detection of points of interest using personal collections of georeferenced and oriented photographs, pp. 281–288 (2012)
11. Meladianos, P., Nikolentzos, G., Rousseau, F., Stavrakas, Y., Vazirgiannis, M.: Degeneracy-based real-time sub-event detection in Twitter stream, pp. 248–257 (2015)
12. Saeki, K., Endo, M., Hirota, M., Kurata, Y., Yokoyama, S., Ishikawa, H.: Tourism of foreign twitter users visited the attribute-based analysis. IPSJ SIG Tech. Rep. **11**(1), 45–56 (2015)
13. Sakaki, T., Okazaki, M., Matsuo, Y.: Earthquake shakes Twitter users: real-time event detection by social sensors, pp. 851–860 (2010)
14. Shamma, D.A., Kennedy, L., Churchill, E.F.: Tweet the debates: understanding community annotation of uncollected sources, pp. 3–10 (2009)
15. Shi, Y., Serdyukov, P., Hanjalic, A., Larson, M.: Personalized landmark recommendation based on geotags from photo sharing sites. ICWSM **11**, 622–625 (2011)
16. Stepchenkova, S., Morrison, A.M.: The destination image of Russia: from the online induced perspective. Tourism Manag. **27**(5), 943–956 (2006)
17. Wenger, A.: Analysis of travel bloggers' characteristics and their communication about Austria as a tourism destination. J. Vacat. Mark. **14**(2), 169–176 (2008)
18. Xue, P., et al.: A study on the tourism destination image of Japan in the Chinese market. The International Centre for the Study of East Asian Development, Working Paper Series, vol. 9, pp. 1–16 (2013)

Privacy, Health and Wellbeing

Mobile Communication Signatures
of Unemployment

Abdullah Almaatouq[1]([⊠]), Francisco Prieto-Castrillo[1,2,3], and Alex Pentland[1]

[1] Media Lab, Massachusetts Institute of Technology, Cambridge, MA, USA
amaatouq@mit.edu
[2] The New England Complex Systems Institute, Cambridge, MA, USA
[3] Harvard T.H. Chan School of Public Health, Boston, MA 02115, USA

Abstract. The mapping of populations socio-economic well-being is highly constrained by the logistics of censuses and surveys. Consequently, spatially detailed changes across scales of days, weeks, or months, or even year to year, are difficult to assess; thus the speed of which policies can be designed and evaluated is limited. However, recent studies have shown the value of mobile phone data as an enabling methodology for demographic modeling and measurement. In this work, we investigate whether indicators extracted from mobile phone usage can reveal information about the socio-economical status of microregions such as districts (i.e., average spatial resolution <2.7 km). For this we examine anonymized mobile phone metadata combined with beneficiaries records from unemployment benefit program. We find that aggregated activity, social, and mobility patterns strongly correlate with unemployment. Furthermore, we construct a simple model to produce accurate reconstruction of district level unemployment from their mobile communication patterns alone. Our results suggest that reliable and cost-effective economical indicators could be built based on passively collected and anonymized mobile phone data. With similar data being collected every day by telecommunication services across the world, survey-based methods of measuring community socioeconomic status could potentially be augmented or replaced by such passive sensing methods in the future.

1 Introduction

As is well known, a major challenge in the development space is the lack of access to reliable and timely socio-economic data. Much of our understanding of the factors that affect the economical development of cities has been traditionally obtained through complex and costly surveys, with an update rate ranging from months to decades, which limits the scope of the studies and potentially bias the data [15]. In addition, as participation rates in unemployment surveys drop, serious questions regarding the declining accuracy and increased bias in unemployment numbers have been raised [18]. However, recent wide-spread adoption of electronic and pervasive technologies (e.g., mobile penetration rate of 100 % in most countries) and the development of Computational Social Science [19], enabled these 'bread-crumbs' of digital traces (e.g., phone records, GPS traces,

© Springer International Publishing AG 2016
E. Spiro and Y.-Y. Ahn (Eds.): SocInfo 2016, Part I, LNCS 10046, pp. 407–418, 2016.
DOI: 10.1007/978-3-319-47880-7_25

credit card transactions, webpage visits, and online social networks) to act as in situ sensors for human behavior; allowing for quantifying social actions and the study of human behavior on an unprecedented scale [5, 11, 12, 25].

Scientists have long suspected that human behavior is closely linked with socioeconomical status, as many of our daily routines are driven by activities related to maintain, to improve, or afforded by such status [6, 13, 14]. Recent studies provided empirical support and investigated these theories in a vast and rich datasets (e.g., social media [20], phone records [10, 29]) with varying scales and granularities [8, 24].

In this work, we provide empirical results that support the use of Call Detail Records (CDRs) individual communication patterns to infer district-level behavioral indicators and examine their ability to explain unemployment as a socioeconomic output. In order to achieve this, we combine a large dataset of CDRs with records from the unemployment benefit program. We quantify individual behavioral indicators from over 1.8 billion logged mobile phone activities generated by 2.8 million unique phone numbers and distributed among 148 different districts in Riyadh, the capital of Saudi Arabia. We extract aggregated mobile extracted indicators (e.g., activity patterns, social interactions, and spatial markers) and examine the relationship between the district level behaviors and unemployment rates. Then, we address whether the identified variables with strong correlation suffice to explain the observed unemployment. As results, we explore the performance of several predictive models in reconstructing unemployment at the district level. Our approach is different from prior work that has already examined the relation between regional wealth and regional phone use (i.e., city [10, 29] or municipality [24] level), as we focus on microregions composed of just a few households with unprecedentedly high quality ground truth labels. This type of work can provide critical input to social and economic research and policy as well as the allocation of resources.

In summary, we frame our contributions as follows:

- We find that CDRs indicators are consistently associated with unemployment rates and that this relationship persists even when we include detailed controls for a district's area, population, and mobile penetration rate.
- We compare several categories of indicators with respect to their performance in predicting unemployment rates at the districts.

2   Datasets

For this study, we used an anonymized mobile phone meta data known as Call Detail Records (CDRs) and combined this with records from unemployment benefit program.

2.1   The CDRs Dataset

Consists of one full month of records for the entire country, with 3 billion mobile activities to over 10,000 unique cell towers, provided by a single telecommunication service provider [1,2]. Each record contains: (i) an anonymized user identifier; (ii) the type of activity (i.e., call or data etc.); (iii) the identifier of the cell tower facilitating the service; (iv) duration; and (v) timestamp of the activity. Each cell tower is spatially mapped to its latitude and longitude and the reception area is approximated by a/the corresponding Voronoi cell. The dataset studied records the identity of the closest tower at the time of activity; thus, we can not identify the position of a user within a Voronoi cell. For privacy considerations, user identification information has been anonymized by the telecommunication operator. Unlike standard CDRs, this dataset does not include the cell tower identity of the receiver end of the activity (i.e., only the location of the caller is approximated). The operator that provided the call data records had around 48 % market share at the time of data acquisition.

2.2   The Unemployment Benefit Program Dataset

The database contains more than 4 million applications for the benefit, of which 1.4 million applications were approved, accounting for $\approx 7\%$ of the total national population. Each record contains anonymized applicant information including their home address (down to the district level). Hence, we are able to derive spatial socio-economic status of unemployed populations at the regional level (i.e., 13 Administrative areas), city level (i.e., 61 cities), and down to the district level (i.e., 1277 districts). In the present work, we focus on the 148 districts within Riyadh.

2.3   Census Information

Riyadh census data was obtained from the High Commission for Development of Arriyadh (ADA) at the Traffic Analysis Zones (TAZs) level. The administrative areas and city level census information were matched using their identifier codes. The district level information was obtained by mapping the TAZ information to the district boundaries. The average spatial resolution (i.e., square root of the land area divided by the number of land units) for the districts and TAZs in Riyadh is 2.6 km and 0.04 km, respectively.

2.4   Mapping Census Population to Districts

For the ith TAZ denoted by τ_i we have the population P_{τ_i} and demographic breakdown (i.e., gender and nationality), as well as housing data (i.e., number of houses, villas, apartments etc.) provided by the ADA. However, the finest resolution for the unemployment data is at the district level. Therefore, for each district d_i we estimate the population P_{d_i} as follows:

$$P_{d_i} = \sum_{j=0}^{|\tau|} \frac{A_{(d_i \cap \tau_j)} P_{\tau_j}}{A_{\tau_j}}$$

where $|\tau|$ is the total number of TAZ units, A_{τ_j} is the area of the jth TAZ unit and $A_{(d_i \cap \tau_j)}$ is the intersection area of d_i and τ_j.

2.5   Mapping Mobile Population to Districts

For each cell tower c_j, we know the total number of different users T_{c_j} with home location (i.e., the tower where a user spends most of the time at night; as in [26]) being the jth tower. When one makes a phone call, the network usually identifies nearby towers and connects to the closest one. The coverage area of a tower c_j thus was approximated using a Voronoi-like tessellation. The Voronoi cell associated with tower c_j is denoted by v_j. Therefore, we can compute the penetration rate σ_{d_i} for district i as follows:

$$\sigma_{d_i} = \frac{1}{P_{d_i}} \sum_{j=0}^{|v|} \frac{A_{(d_i \cap v_j)} T_{c_j}}{A_{v_j}}$$

where $|v|$ is the total number of Voronoi cells, A_{v_j} is the area of the jth Voronoi cell (associated with the jth cell tower) and $A_{(d_i \cap v_j)}$ is the intersection area of d_i and v_j.

Figure 1 shows the scaling relationships between the district population versus unemployment rate and also population versus mobile users. These results are consistent with previous studies indicating that scaling with population is indeed a pervasive property of urban organization [7,23].

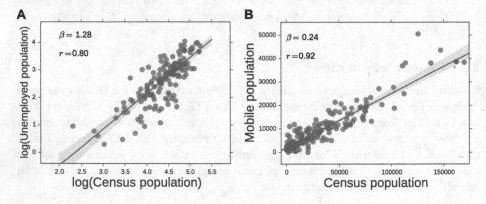

Fig. 1. Panel **(A)** shows the relationship between unemployment and population in 2012, for the 148 districts. Panel **(B)** shows the number of homes detected, for the 148 districts versus district census population. Best-fit scaling relations are shown as solid lines and the 95 % confidence intervals are shown as a shaded area.

3   Extracting Behavioral Indicators

The goal of this work is to investigate how behavioral indicators from mobile phone meta data can be extracted and then related back to the economical well-being of geographical regions (i.e., districts). To this end, we define three groups of indicators that have been widely explored in fields like economy or social sciences. Several of these indicators have been implemented in the bandicoot toolbox [21][1]. All the indicators are computed at the individual level and then aggregated and standardized (i.e., scaled to have mean zero and variance one) at the district level.

3.1 Activity Patterns

Activity patterns quantify factors related to the aggregate patterns of mobile usage for each district such as volume (average number of records per user), timing (the average percentage of night calls – nights are 7pm–7am), and duration (average duration of calls).

3.2 Social Interactions

The social interaction indicators capture the structure of the individual's contact network. We focus on the egocentric networks around the individual in order to examine the local structure and signify the types of interactions that develop within their circle [4].

Let $E_i = (\mathcal{C}_i, \mathcal{E}_i)$ be the directed egocentric-graph that represents the topological structure of the ith individual where $\mathcal{C}_i$ is the set of contacts (total number of contacts is $k = |\mathcal{C}_i|$) and $\mathcal{E}_i$ is the set of edges. A directed edge is an ordered pair (i, j) with associated call volume w_{ij} (and/or (j, i) with w_{ji} volume) representing the interaction between the ego i and a contact $j \in \mathcal{C}_i$. Note that by definition $w_{ij} + w_{ji} \in \mathbb{Z}^+$, must be satisfied, otherwise $j \notin \mathcal{C}_i$. Therefore, the volume w_{ij} is set to 0 when $(i, j) \notin \mathcal{E}_i$, alternatively $w_{ji} = 0$, if the (i, i) direction does not exist. From this we can compute several indicators for an individual within its egocentric network. We define $I(i)$ and $O(i)$ as the set of incoming interactions to (respectively, initiating from) individual i. That is,

$$I(i) = \{j \in \mathcal{C}_i | (j, i) \in \mathcal{E}_i\}, \quad \text{and} \quad O(i) = \{j \in \mathcal{C}_i | (i, j) \in \mathcal{E}_i\}.$$

Percentage of Initiated Interaction is a measure of directionality of communication. We define it as

$$\mathcal{I}(i) = |I(i)| / (|I(i)| + |O(i)|)$$

Balance of contacts is measured through the balance of interactions per contact. For an individual i, the balance of interactions is the number of outgoing interactions divided by the total number of interactions.

$$\beta(i) = \frac{1}{k} \sum_{j \in I(i) \cup O(i)} \frac{w_{ij}}{w_{ji} + w_{ij}}$$

Social Entropy captures the social diversity of communication ties within an individual's social network, we follow Eagle's et al. [10] approach by defining

---

[1] Bandicoot can be found at http://bandicoot.mit.edu/docs/.

social entropy, $D_{social}(i)$, as the normalized Shannon entropy associated with the ith individual communication behavior:

$$D_{social}(i) = - \sum_{j \in C_i} log(p_{ij})/log(k)$$

Where $p_{ij} = w_{ij}/\sum_{\ell=1}^{k} w_{i\ell}$ is the proportion of i's total call volume that involves j. High diversity scores imply that an individual splits his/her time more evenly among social ties.

3.3   Spatial Markers

The spatial markers captures mobility patterns and migration based on geospatial markers in the data. In this work, we measure the number of visited locations, which captures the frequency of return to previously visited locations over time [28] (time in our case is the entire observational period). We also compute the percentage of time the user was found at home.

3.4   Unsupervised Clustering

We use the standard form of self-organizing maps (SOMs) as an unsupervised clustering analysis tool [17, 30].

In Fig. 2A, the codebook vectors from the resulting SOMs are shown in a segments plot, where the grayscale background color of a cluster corresponds to its index (i.e., number of clusters = 9 arranged in a rectangular grid). Districts having similar characteristics based on the multivariate behavioral attributes are positioned close to each other, and the distance between them represents the degree of behavioral similarity or dissimilarity. High average spatial entropy with small percentage of time being at home, for example, is associated with districts projected in the bottom left corner of the map (i.e., cluster index one – black color). On the other hand, districts with low social entropy, percentage of initiated calls, and balance of contacts are associated with the clusters at the top column of the map. On the geographic map (see Fig. 2B), each district is assigned a color, where the meaning of the color can be interpreted from the corresponding codebook vector. We can see that at the center of the city, most districts are assigned to clusters with dark backgrounds, and the color gets lighter as we move towards the periphery of the city. As expected from the description of the corresponding codebook vectors, districts projected in the bottom of the map (dark color background) are associated with lower unemployment rates (see Fig. 2C). It is indeed the case that districts with similar behavioral attributes have similar unemployment rates (see Fig. 2D).

3.5   Statistical Correlation

As we can see in Fig. 3, all the extracted indicators exhibit at least moderate statistical correlations with unemployment. In addition, we find that the indicators relationship with unemployment persists for most indicators even when

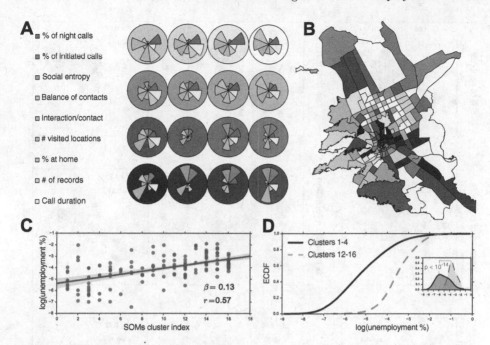

Fig. 2. Panel **(A)** shows the plot of the color-encoded codebook vectors of the 4-by-4 mapping of the districts behaviors. Panel **(B)** shows the combined view of attribute space (i.e., SOMs results) and geographic space (choropleth map). Panel **(C)** demonstrates the relationship between the clustered behaviors and unemployment rates. Finally, panel **(D)** shows that the Empirical Cumulative Distribution Function (ECDF) of the unemployment rates for contrasting groups behaviors. (Color figure online)

we include controls for a district's area, population, and mobile penetration rate (see Table 1). These results suggest that several of those indicators are sufficient to explain the observed unemployment. For instance, we find the percentage of night calls to have the highest effect size and explanatory power ($R^2 = 23\%$; model 4). This is expected, as regions with very different unemployment patterns should exhibit different temporal activities. Since working activities usually happen during the day, we would expect that districts with high levels of unemployment will tend to have higher proportion of their activities during the night.

Previous study [27] have found that the duration spent at either home or work is relatively flat distributed with peaks around time spans of 14 h at home and 3.5–8.6 h at work. Therefore, we hypothesize that the lack of having a work location for the unemployed would lead to an increase in the duration spent at home (i.e., % home), and/or reduce the tendency for revisiting locations (i.e., higher visited locations). We indeed find that the percentage of being home and number of visited locations to be associated with unemployment in our dataset.

We also find the percentage of initiated interactions to be negatively correlated with unemployment. This indicators has been shown to be predictive of

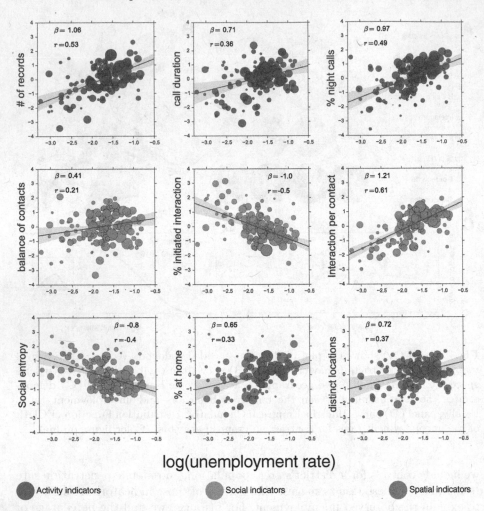

Fig. 3. Relations between the mobile extracted indicators (district average) for the 148 districts against its unemployment rate. Size of the points is proportional to the population in each district. Solid lines correspond to linear fits to the data and the shaded area represents the 95 % confidence intervals.

the Openness (i.e., the tendency to be intellectually curious, creative, and open to feelings) personality trait [16,21], which in return is predictive of success in job interviews [9].

As in [10,20], we find that districts with high unemployment rates have less diverse communication patterns than areas with low unemployment. This translates in a negative coefficient for social entropy and positive coefficient for the interaction per contact indicator. The balance of contacts factor was not found to be significant ($p > 0.1$).

Table 1. Regression table explaining the districts' unemployment rate as a function of the activity patterns, social interactions, and spatial markers, with the inclusions of controls for a district's area, population, and mobile penetration rate.

	Dependent variable: log(Unemployment Rate)								
	(1)	(2)	(3)	(4)	(5)	(6)	(7)	(8)	(9)
Population	0.15	0.16	0.12	−0.07	0.09	0.12	0.12	0.16	0.16
	(0.10)	(0.11)	(0.11)	(0.11)	(0.11)	(0.11)	(0.11)	(0.11)	(0.11)
Area	0.07	0.15*	0.12	0.28***	0.14*	0.17*	0.11	0.20**	0.19**
	(0.09)	(0.09)	(0.08)	(0.08)	(0.08)	(0.08)	(0.09)	(0.08)	(0.08)
Penetration rate	0.02	0.00	0.04	0.04	−0.06	0.02	0.05	−0.02	0.02
	(0.11)	(0.11)	(0.11)	(0.10)	(0.10)	(0.11)	(0.11)	(0.11)	(0.11)
# of records	0.31***								
	(0.09)								
Call duration		0.17**							
		(0.08)							
% initiated inter.			−0.28***						
			(0.09)						
% night calls				0.49***					
				(0.09)					
% at home					0.30***				
					(0.08)				
Social entropy						−0.18**			
						(0.09)			
Inter. per contact							0.27***		
							(0.09)		
Balance of contacts								−0.01	
								(0.08)	
visited locations									0.14*
									(0.08)
(Intercept)	−0.00	0.00	−0.00	0.00	−0.00	−0.00	−0.00	−0.00	−0.00
	(0.08)	(0.08)	(0.08)	(0.07)	(0.08)	(0.08)	(0.08)	(0.08)	(0.08)
R^2	0.15	0.10	0.14	0.23	0.16	0.10	0.13	0.07	0.09
Adj. R^2	0.13	0.08	0.12	0.21	0.13	0.07	0.11	0.05	0.07
BIC	421.97	430.53	424.03	407.25	420.75	430.65	425.16	434.73	431.86
Num. obs	147	147	147	147	147	147	147	147	147

***$p < 0.01$, **$p < 0.05$, *$p < 0.1$. Standard errors in parentheses.

3.6 Supervised Predictive Model

Here we are interested in the predictability of unemployment rates of microregions based on the mobile phone extracted indicators and independently of additional census information such as population, gender, income distribution, etc. Such additional information is often unavailable in developing nations, which by itself represents a major challenge to policy-makers and researchers. Therefore, it is of utmost importance to find novel sources of data that enables new approaches to demographic profiling.

We analyze the predictive power of the indicators using Gaussian Processes (GP) to predict unemployment based on mobile phone indicators solely. We train and test the model in K-fold-cross validation ($K = 5$) and compute the coefficient of determination R^2 as a measure of quality for each category of indicators

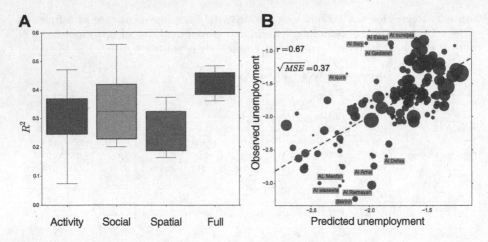

Fig. 4. Panel (**A**) shows the average performance of each indicator category in predicting unemployment. Panel (**B**) depicts the cross-validated prediction of unemployment rates versus the observed ones, $r = 0.68$. The predicted values are based on the prediction that was obtained for that district when it was in the test set. Dashed line correspond to the equality line.

(i.e., activity, social, and spatial) and also for the full indicators (involving all mobile extracted indicators presented in this work). The advantage for using Gaussian Processes (GP) to regress unemployment rates is that the model produces probabilistic (Gaussian) predictions so that one can compute empirical confidence intervals and probabilities that might be used to refit (online fitting, adaptive fitting) the prediction in some region of interest [3,22,31].

In Fig. 4A we find that the social interaction indicators to be very predictive of unemployment with an average $R^2 = 0.43$ (95 % CI: 0.37–0.48), which is more predictive than the activity pattern indicators $R^2 = 0.29$ (95 % CI: 0.15–0.39) and the spatial indicators $R^2 = 0.26$ (95 % CI:0.19–0.33). It is worth mentioning, that the composite model performed significantly better than single category models with an average $R^2 = 0.43$ (95 % CI: 0.37–0.48). Figure 4B compares the predicted and observed unemployment rate for each based on the prediction that was obtained for that district when it was in the test set.

4   Summary and Future Work

In this paper we have demonstrated that mobile phone indicators are associated with unemployment rates and that this relationship is robust to the inclusion of controls for a district's area, population, and mobile penetration rate. Following this analysis, we also investigated the predictability of unemployment rates with respect to three categories of indicators, namely, activity patterns, social interactions, and spatial markers. The results of these analyses highlighted the importance of social interaction indicators for predicting unemployment.

Note that we are not stating a causality arrow between the indicators and the unemployment rate as we do not have individual level mapping of unemployment with which to test for individual differences. In this work, our goal is to show that aggregate behavioral indicators of the members of a district represent a strong statistical signature that can be used as alternative measuring approach with a real translation in the economy.

In our future work, we intend to intersect the Call Detail Records (CDR) and unemployment data derived from the unemployment benefit program at the individual level. This will allow for the study of how the behavioral signature of a single individual can be used to predict that same individual's employment status. This could reveal the key determinants of unemployed people to find a job and allow for designing personalized intervention mechanisms.

Acknowledgments. The authors thank the Center for Complex Engineering Systems (CCES) at KACST and MIT and the Media Lab at MIT for their support.

References

1. Aleissa, F., Alnasser, R., Almaatouq, A., Jamshaid, K., Alhasoun, F., González, M.C., Alfaris, A.: Wired to connect: analyzing human communication and information sharing behavior during extreme events. In: KDD Workshop on Learning about Emergencies from Social Information (2014)
2. Alhasoun, F., Almaatouq, A., Greco, K., Campari, R., Alfaris, A., Ratti, C.: The city browser: utilizing massive call data to infer city mobility dynamics. In: SIGKDD International Workshop on Urban Computing (2014)
3. Almaatouq, A.: Complex systems and a computational social science perspective on the labor market. arXiv preprint arXiv:1606.08562 (2016)
4. Almaatouq, A., Alabdulkareem, A., Nouh, M., Shmueli, E., Alsaleh, M., Singh, V.K., Alarifi, A., Alfaris, A., Pentland, A.S.: Twitter: who gets caught? Observed trends in social micro-blogging spam. In: Proceedings of the 2014 ACM Conference on Web Science, pp. 33–41. ACM (2014)
5. Almaatouq, A., Radaelli, L., Pentland, A., Shmueli, E.: Are you your friends friend? Poor perception of friendship ties limits the ability to promote behavioral change. PloS One **11**(3), e0151588 (2016)
6. Becker, G.S.: The Economic Approach to Human Behavior. University of Chicago Press, Chicago (1976)
7. Bettencourt, L.M., Lobo, J., Helbing, D., Kühnert, C., West, G.B.: Growth, innovation, scaling, and the pace of life in cities. Proc. Natl. Acad. Sci. **104**(17), 7301–7306 (2007)
8. Blumenstock, J., Cadamuro, G., On, R.: Predicting poverty and wealth from mobile phone metadata. Science **350**(6264), 1073–1076 (2015)
9. Caldwell, D.F., Burger, J.M., et al.: Personality characteristics of job applicants and success in screening interviews. Pers. Psychol. **51**(1), 19–136 (1998)
10. Eagle, N., Macy, M., Claxton, R.: Network diversity and economic development. Science **328**(5981), 1029–1031 (2010)
11. Eagle, N., Pentland, A.: Reality mining: sensing complex social systems. Pers. Ubiquit. Comput. **10**(4), 255–268 (2006)

12. Gonzalez, M.C., Hidalgo, C.A., Barabasi, A.L.: Understanding individual human mobility patterns. Nature **453**(7196), 779–782 (2008)
13. Granovetter, M.: Economic action and social structure: the problem of embeddedness. Am. J. Sociol. **91**, 481–510 (1985)
14. Granovetter, M.S.: The strength of weak ties. Am. J. Sociol. **78**, 1360–1380 (1973)
15. Henrich, J., Boyd, R., Bowles, S., Camerer, C., Fehr, E., Gintis, H., McElreath, R.: In search of homo economicus: behavioral experiments in 15 small-scale societies. Am. Econ. Rev. **91**, 73–78 (2001)
16. John, O.P., Srivastava, S.: The big five trait taxonomy: history, measurement, and theoretical perspectives. In: Handbook of Personality: Theory and Research, 1999, vol. 2, pp. 102–138 (1999)
17. Kohonen, T.: The self-organizing map. Neurocomputing **21**(1), 1–6 (1998)
18. Krueger, A., Mas, A., Niu, X.: The evolution of rotation group bias: will the real unemployment rate please stand up? Technical report, National Bureau of Economic Research (2014)
19. Lazer, D., Pentland, A.S., Adamic, L., Aral, S., Barabasi, A.L., Brewer, D., Christakis, N., Contractor, N., Fowler, J., Gutmann, M., et al.: Life in the network: the coming age of computational social science. Science **323**(5915), 721 (2009). (New York, NY)
20. Llorente, A., Garcia-Herranz, M., Cebrian, M., Moro, E.: Social media fingerprints of unemployment. Plos One **10**(5), e0128692 (2015). http://dx.doi.org/10.1371%2Fjournal.pone.0128692
21. Montjoye, Y.-A., Quoidbach, J., Robic, F., Pentland, A.S.: Predicting personality using novel mobile phone-based metrics. In: Greenberg, A.M., Kennedy, W.G., Bos, N.D. (eds.) SBP 2013. LNCS, vol. 7812, pp. 48–55. Springer, Heidelberg (2013). doi:10.1007/978-3-642-37210-0_6
22. Nielsen, H.B., Lophaven, S.N., Sondergaard, J.: Dace, a matlab kriging toolbox. Technical University of Denmark, DTU, Informatics and mathematical modelling. Lyngby-Denmark (2002)
23. Pan, W., Ghoshal, G., Krumme, C., Cebrian, M., Pentland, A.: Urban characteristics attributable to density-driven tie formation. Nat. Commun. **4**, 1–35 (2013)
24. Pappalardo, L., Vanhoof, M., Gabrielli, L., Smoreda, Z., Pedreschi, D., Giannotti, F.: An analytical framework to nowcast well-being using mobile phone data. Int. J. Data Sci. Anal., 1–18 (2016). Springer
25. Pentland, A.: Social Physics: How Good Ideas Spread-The Lessons from a New Science. Penguin, New York (2014)
26. Phithakkitnukoon, S., Smoreda, Z., Olivier, P.: Socio-geography of human mobility: a study using longitudinal mobile phone data. Plos One **7**(6), e39253 (2012). http://dx.doi.org/10.1371%2Fjournal.pone.0039253
27. Schneider, C.M., Belik, V., Couronné, T., Smoreda, Z., González, M.C.: Unravelling daily human mobility motifs. J. R. Soc. Interface **10**(84), 20130246 (2013)
28. Song, C., Koren, T., Wang, P., Barabási, A.L.: Modelling the scaling properties of human mobility. Nat. Phys. **6**(10), 818–823 (2010)
29. Toole, J.L., Lin, Y.R., Muehlegger, E., Shoag, D., González, M.C., Lazer, D.: Tracking employment shocks using mobile phone data. J. R. Soc. Interface **12**(107), 20150185 (2015)
30. Wehrens, R., Buydens, L.M., et al.: Self-and super-organizing maps in R: the Kohonen package. J. Stat. Softw. **21**(5), 1–19 (2007)
31. Welch, W.J., Buck, R.J., Sacks, J., Wynn, H.P., Mitchell, T.J., Morris, M.D.: Screening, predicting, and computer experiments. Technometrics **34**(1), 15–25 (1992)

Identifying Stereotypes in the Online Perception of Physical Attractiveness

Camila Souza Araújo[✉], Wagner Meira Jr., and Virgilio Almeida

Universidade Federal de Minas Gerais, Belo Horizonte, MG, Brazil
{camilaaraujo,meira,virgilio}@dcc.ufmg.br

Abstract. Stereotyping can be viewed as oversimplified ideas about social groups. They can be positive, neutral or negative. The main goal of this paper is to identify stereotypes for female physical attractiveness in images available in the Web. We look at the search engines as possible sources of stereotypes. We conducted experiments on Google and Bing by querying the search engines for beautiful and ugly women. We then collect images and extract information of faces. We propose a methodology and apply it to analyze photos gathered from search engines to understand how race and age manifest in the observed stereotypes and how they vary according to countries and regions. Our findings demonstrate the existence of stereotypes for female physical attractiveness, in particular negative stereotypes about black women and positive stereotypes about white women in terms of beauty. We also found negative stereotypes associated with older women in terms of physical attractiveness. Finally, we have identified patterns of stereotypes that are common to groups of countries.

Keywords: Discrimination · Algorithm bias · Beauty stereotypes

1 Introduction

Prejudice, discrimination and stereotyping often go hand-in-hand in the real world. While stereotyping can be viewed as oversimplified ideas about social groups, discrimination refers to actions that threat groups of people unfairly or put them at a disadvantage with other groups. Stereotypes can be positive, neutral or negatives. For example, tiger moms are considered a positive stereotype that refers to Asian-American mothers that keep focus on achievement and performance in the education of their children. However, negative stereotypes based on gender, religion, ethnicity, sexual orientation and age can be harmful, for they may foster bias and discrimination. As a consequence, they can lead to actions against groups of people [1,2].

Some appearance stereotypes associated with women in the physical world follow them in the online world. A recent study by Kay et al. [2] shows a systematic under representation of women in image search results for occupations. This kind of stereotype affects people's ideas about professional gender ratios in the real world and may create conditions for bias and discrimination.

© Springer International Publishing AG 2016
E. Spiro and Y.-Y. Ahn (Eds.): SocInfo 2016, Part I, LNCS 10046, pp. 419–437, 2016.
DOI: 10.1007/978-3-319-47880-7_26

In the past, television, movies, and magazines have played a significant role in the creation and dissemination of stereotypes related to the physical appearance or physical attractiveness of women [3]. The concepts of beauty, ugly, young and old have been used to create categories of cultural and social stereotypes. The idealized images of beautiful women have contributed to created negative consequences such as eating disorders, low self esteem and job discrimination. These stereotypes have been a serious problem among teenage girls.

With the ongoing growth of Internet and social media, people are constantly exposed to steady flows of news, information and subjective opinions of others about cultural trends, political facts, economic ideas, social issues, etc. In addition to information that come from different sources, people use Google to obtain answers and information in order to form their own opinion on various social issues. Every day, Google processes over 3.5 billion search queries. Google decides which of the billions of web pages are included in the search results, and it also decides how to rank the results. Google provides images as the result of queries. Thus, in order to understand the existence of global stereotypes, we need to start by looking at the search engines, as possible sources of stereotypes. In this paper we focus our analysis on the following research questions:

- Can we identify stereotypes for female physical attractiveness in the images available in the Web?
- How do race and age manifest in the observed stereotypes?
- How do stereotypes vary according to countries and regions?

In our analyses, we look for patterns of women's physical features that are considered aesthetically pleasing or beautiful in different cultures. We also look at the reverse, i.e., patterns are considered aesthetically ugly [4]. In order to answer the research questions, we conduct a series of experiments on the two most popular search engines, Google and Bing. We start the experimentation by querying the search engines for beautiful and ugly women. We then collect the top 50 image search results for up to 42 different countries. Once we have verified the images, we use Face++, which is an online API that detects faces in a given photo. Face++ infers information about each face in the photo such as age, race and gender. Its accuracy is known to be over 90 % [5] for face detection. The images collected from Google and Bing, classified by Face++, form the datasets used to conduct the stereotype analyses. Based on the data we collected, we have the following observations, which are explained throughout the paper.

- we have observed the existence of both negative stereotypes for black women and positive stereotypes for white women in terms of beauty;
- we have noticed that there are negative stereotypes about older women in terms of physical attractiveness;
- we have identified patterns of stereotypes that are common to groups of countries. For example, US and several Hispanic countries share negative stereotypes about black women, positive stereotypes for white women and almost neutral about Asian women.

The first step in solving a problem is to recognize that it does exist. Our findings demonstrate the existence of stereotypes for female physical attractiveness. An important way to fight gender and age discrimination is to discourage stereotypes.

2 Related Work

In this section we present some related work on characterization studies of search engines, bias and discrimination in the media, as well as physical attractiveness.

Characterization of Search Engines. Because of its scope and impact power, Google has become an object of study in the field of digital media and key to understand how the results of queries affect people who use search engines. Previous studies investigated the existence of bias in specific scenarios. [6] shows how racial and gender identities may be misrepresented, when, in this context, there is commercial interest. The result of a query to Google typically prioritizes some kind of advertisement, which should - ideally - be related to the query. But search engines are often biased, so it is important to assess how the result ranking is built and how it affects the access to information [7]. Some more recent results argue that discriminating a certain group is inappropriate, since search engines are 'information environments' that may affect the perception and behavior of people [2]. One example of such discrimination is, when searching the names of people with black last names, the higher likelihood of getting ads suggesting that these people were arrested, or face a problem with justice, even when it did not happen [8]. In this case, the search algorithm supposedly discriminates a certain group of people while looking for profit from advertising. [9] has questioned the commercial search engines because the way they represent women, especially black women, and other marginalized groups, regardless of cultural issues. This behavior masks and perpetuate unequal access to social, political and economic life of some groups.

Bias and Discrimination in the Media. Media influences people's perceptions about ethnic issues [10]. In the USA, media tends to propagate stereotypes that benefit dominant groups. Black men, for example, are often stereotyped as violent. Even though much of the black population does not agree with the way they are represented and believe that this construction is harmful, unpleasant or distasteful. Uber drivers who have African American last names tend to get more negative reviews. Just as black tenants have less chances of getting a vacancy at rented apartments on Airbnb site [11]. In the medical scenario, because of false judgments, black patients may receive inferior treatment compared to the treatment given to white people [12]. Many health-care professionals believe in biological differences with respect to black and white people, for example, black skin to be more resistant.

Beauty as a Concept. The reasons why beauty standards exist and how they are built are topics that are broadly discussed from the biological and evolutionary point of view. In the book "The Analysis of Beauty" [4] published in 1753, the author describes theories of visual beauty and grace. For the authors in [13] the aesthetic preference of the human beings is a case of *gene-culture co-evolution.* In other words, our standards of beauty are shaped, simultaneously, by a genetic and cultural evolution. Other studies [14,15] argue that the beauty standards are part of human evolution and therefore reinforce characteristics related to health, among other features that may reflect the search for more 'qualified' partners for reproduction. Some works are concerned to understand how, despite cultural differences, the concept of beauty seems to be built in the same way worldwide. Diverse ethnic groups agree consistently over the beauty of faces [16], although they disagree regarding the attractiveness of female bodies. It is even possible to indicate which features are most desirable: childish face features for women - big eyes, small nose, etc. In [17], the authors conclude that: people tend to agree more with respect to faces that are more familiar and in some cultures the skin tone is more important in the classification of beautiful people, but, in other cases, it is the face shape. In Computer Science, using methods of machine learning, it is possible to predict, 0.6 of correlation, a face attractiveness score, showing that it is possible for a machine to learn what is beautiful from the point of view of a human [18].

3   Data Gathering and Analysis

In this section we describe the methodology used for characterizing stereotypes. The first step of the methodology involves the data collection process: what and how to collect the data. Then we extract information about the collected photos, using computer vision algorithms to identify race, age and gender of the people in each picture. The second part of the research refers to the use of the collected information to identify stereotypes.

3.1   Data Gathering

Data gathering was carried through two search engines APIs for images: Google and Bing. Once gathered, we extract features from the photos using Face++[1].
    The data gathering process is depicted in Fig. 1 and is summarized next:

1. **Define search queries**
    For each context, in our case beauty, define the relevant search queries and translate[2] the query to the target languages.
2. **Gathering**
    Using the search engines APIs, perform the searches with the defined queries. Then, filter photos that contain the face of just one person.

---

[1] http://www.faceplusplus.com/.
.[2] Using Google Translator: http://translate.google.com.br/.

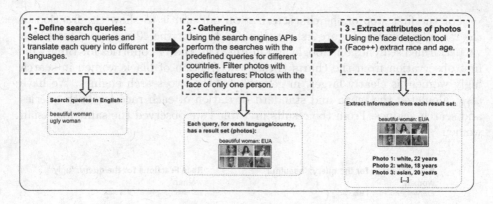

Fig. 1. Data gathering framework.

3. Extract attributes of photos
Using face detection tools estimate race and age.

Beauty is a property, or set of properties, that makes someone capable of producing a certain sort of pleasurable experience in any suitable perceiver [19]. For the beauty context we collected the top 50 photos of the results of the following queries (in different languages): beautiful woman and ugly woman. It is known that what is defined as beautiful or ugly might change from person to person, then we chose these two antonyms adjectives that are commonly used to describe the quality of beauty of people.

Bing's API offers the option of 22 countries to perform the searches, we collected data for all these countries. For Google we collected data for the same 22 countries and added more countries with different characteristics, providing better coverage in terms of regions and internet usage. The searches were performed for the following countries and their official languages:

Google: Afghanistan, South Africa, Algeria, Angola, Saudi Arabia, Argentina, Australia, Austria, Brazil, Canada, Chile, Denmark, Egypt, Finland, France, Germany, Greece, Guatemala, India, Iraq, Ireland, Italy, Japan, Kenya, United Kingdom, South Korea, Malaysia, Mexico, Morocco, Nigeria, Paraguay, Peru, Portugal, Russia, Spain, United States, Sweden, Turkey, Ukraine, Uzbekistan, Venezuela and Zambia.

Bing: Saudi Arabia, Denmark, Austria, Germany, Greece, Australia, Canada, United Kingdom, United States, South Africa, Argentina, Spain, Mexico, Finland, Italy, Japan, South Korea, Brazil, Portugal, Russia, Turkey and Ukraine.

Now we present a brief characterization of the datasets collected for this work. As mentioned, we picked the top 50 photos for each query but we consider as valid only images for which Face++ was able to detect a single face (see Appendix A). The characterization and analysis will be performed for all query responses that contain at least 20 valid photos.

For the first step of the characterization our aim is to show the fraction of the races by country. Figure 2 shows this fraction for the 42 countries for which we performed searches on Google and in Fig. 3 for the 17 countries for Bing. Our first observation from the charts is that the fraction of black women in search 'ugly women' is clearly larger, in general, for the two search engines. We have also calculated the mean and standard deviation of each race for both queries and search engines. From the results in Table 1 we observed the same for Asian women.

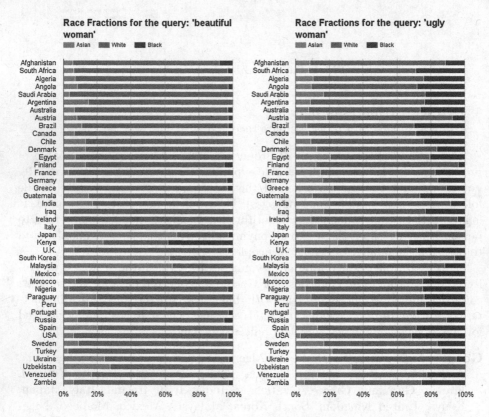

Fig. 2. Race fractions for Google.

The second step of the characterization shows the difference between the age distribution of women in photos by query and search engine through boxplots (Figs. 4 and 5). In the x-axis we have the analyzed countries and the y-axis represents ages. Analyzing the median and upper quartile, we noticed that beautiful women tend to be younger than the ugly women.

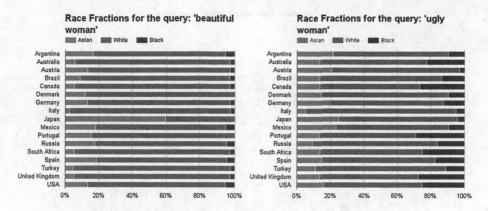

Fig. 3. Race fractions for Bing.

Table 1. Mean and standard deviation of fractions

Google											
Beautiful woman						Ugly woman					
Asian		Black		White		Asian		Black		White	
Mean	Stdv	Mean	Stdv	Mean	Stdv	Mean	Stdv	Mean	Stdv	Mean	Stdv
13.77	15.65	2.37	5.99	83.86	16.96	15.36	11.48	19.20	9.23	65.44	9.48
Bing											
Beautiful woman						Ugly woman					
Asian		Black		White		Asian		Black		White	
Mean	Stdv	Mean	Stdv	Mean	Stdv	Mean	Stdv	Mean	Stdv	Mean	Stdv
12.96	11.82	03.09	2.59	83.94	11.78	15.35	5.19	15.63	8.54	69.02	5.19

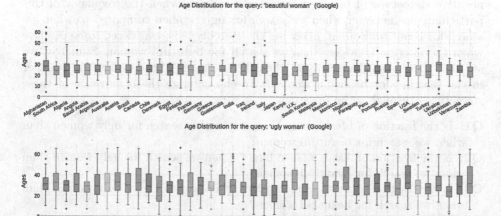

Fig. 4. Age distribution for Google.

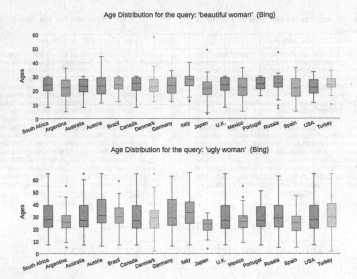

Fig. 5. Age distribution for Bing.

3.2 Data Analysis

Our main purpose is to identify whether there is a stereotype in the perception of physical attractiveness. For sake of our analysis, we distinguish two characteristics extracted from the pictures: race and age. As discussed, stereotype is a subjective concept and quantifying it through objective criteria is a challenge. In our case, we employed a *contrast*-based strategy. Considering race as a criteria, we check the difference between the fractions of each race for opposite queries, that is, beautiful woman and ugly woman. We consider that there is a negative stereotype of beauty in relation to a race, when the frequency of this particular race is larger when we search for ugly women compared to when we search for beautiful woman. Likewise, the stereotype is considered to be positive when the fraction is larger when we search for beautiful woman. Similarly, we say that there is a age stereotype when the age range of the women are younger in the searches for beautiful women. We characterize the occurrence of these stereotypes through seven questions:

Q1: Is the fraction of black women larger when we search for ugly women than when we search for beautiful women?

Q2: Is the fraction of Asian women larger when we search for ugly women than when we search for beautiful women?

Q3: Is the fraction of white women larger when we search for ugly women than when we search for beautiful women?

Q4: Is the fraction of black women smaller when we search for ugly women than when we search for beautiful women?

Q5: Is the fraction of Asian women smaller when we search for ugly women than when we search for beautiful women?

Q6: Is the fraction of white women smaller when we search for ugly women than when we search for beautiful women?

Q7: Are the women's ages when we search for beautiful women younger than the ages of the women when we search for ugly women?

Each of these questions is associated with a test hypothesis. For the questions **Q1**, **Q2** and **Q3**, the test hypothesis is:

H_0(**null hypothesis**): The fraction of women of the specific race (i.e., black, white, Asian) is smaller when we search for ugly women, than when we search for beautiful women.

H_a(**alternative hypothesis**): The fraction of women of the specific race (i.e., black, white, Asian) is larger when we search for ugly women than when we search for beautiful women.

For the questions **Q4**, **Q5** and **Q6**:

H_0: The fraction of women of the specific race (black, white, Asian) is larger when we search for ugly women than when we search for beautiful women.

H_a: The fraction of women of the specific race (black, white, Asian) is smaller when we search for ugly women than when we search for beautiful women.

For the question **Q7**:

H_0: The age range of the beautiful women is older than the age range of the ugly women.

H_a: The age range of the beautiful women is younger than the age range of the ugly women.

We assume that there is a negative stereotype when the fraction of a given race is significantly larger when we search for ugly woman than when we search for beautiful woman and there is a positive stereotype when the fraction associated with a search for ugly woman is significantly smaller. We then calculate the difference between these two fractions for each race and each country and verify the significance of each difference through the **two-proportion z-test**, with a significance level of 0.05. This test determines whether the difference between the fractions is significant, as follows.

Racial Stereotype. For the first three questions, (**Q1**, **Q2** and **Q3**), with confidence of 95 % we reject the null hypothesis when the z-score is smaller than -0.8289 and we accept the alternative hypothesis, which is the hypothesis in study. For example, considering Afghanistan, the z-score calculated for the hypothesis associated with question **Q1** was -0.48, -0.53 for **Q2** and 0.74 for **Q3**. Since none of these values is smaller than -0.8289 we can not reject the null hypothesis and we can not answer positively to any of the 3 questions. On the other hand, for Italy, the z-score associated with question **Q1** was -2.51 and

−1.05 for **Q2**, then we can answer positively to both questions and consider that there is a negative stereotype associated with blacks and Asians.

For questions (**Q4**, **Q5** and **Q6**), under the same conditions, we reject the null hypothesis when the z-score is greater than 0.8289. Detailed results of the tests and z-scores for each country and each search engine are in the Appendix B.

Age Stereotype. For characterizing the age stereotype, we verify our hypothesis through the unpaired Wilcoxon test [20]. The null hypothesis is rejected when p-value is less than 0.05, and with 95 % of confidence we can answer positively to question **Q7** (see Appendix C for detailed results). Once again, considering Afghanistan, the p-value found was 0.1819 then we can not reject the null hypothesis. For South Africa the p-value was 0.0001 and we accept the alternative hypothesis that demonstrates the existence of a stereotype that gives priority to younger women.

Table 2. Summary of results for questions **Q1**, **Q2**, **Q3**, **Q3**, **Q4**, **Q5**, **Q6** and **Q7**

	Results	
	Google	Bing
Q1 (Black)	**85.71 %**	**76.47 %**
Q2 (Asian)	**26.19 %**	**29.41 %**
Q3 (White)	4.76 %	5.88 %
Q4 (Black)	2.38 %	0.00 %
Q5 (Asian)	4.76 %	11.76 %
Q6 (White)	**78.57 %**	**82.35 %**
Q7 (Age)	**69.05 %**	**82.35 %**

Table 2 summarizes the test results with the fraction of countries that we answer positively to each of the 7 questions (reject the null hypothesis). For instance, column 'Google' and line 'Q1' indicates that for 85.71 % of countries we rejected the null hypothesis and we answered positively to the question **Q1**. That is, for almost 86 % of the countries the fraction of black women is larger when we search for ugly women than when we search for beautiful women. We can see that the results of the two search engines agree. There is a beauty stereotype in the perception of physical attractiveness, that is, we can say that significantly the fraction of black and Asian women is greater when we search for ugly women compared to the fraction of those races when we search for beautiful women (negative stereotype). The opposite occurs for white women (positive stereotype).

3.3   Clustering Stereotypes

After identifying the existence of stereotypes in the perception of physical attractiveness, we want to discover whether there is a cohesion among these beauty stereotypes across countries. For this we will use a clustering algorithm to identify the countries that have the same racial stereotype of beauty. The results for each country and search engine is represented by a 3D point where the dimensions are Asian, black and white z-scores.

There are several strategies for clustering, however a hierarchical clustering strategy was used in this paper because it outputs a hierarchy that can be very useful for our analysis. We used the Ward's minimum variance method[3] which is briefly described next. Using a set of dissimilarities for the objects being clustered, initially, each object is assigned to its own cluster and then the algorithm proceeds interactively. At each stage it joins the two most similar clusters, continuing until there is just a single cluster. The method aims at finding compact and spherical clusters [21]. Another advantage of employing a hierarchical clustering strategy is that it is not necessary to set in advance parameters such as the number of clusters of minimal similarity thresholds, allowing us to investigate various clusters configurations easily.

The clusters we are looking for should be cohesive and also semantically meaningful. Cohesion is achieved by the Ward's minimum variance method, but the semantic of the clusters should take into account cultural, political and historical aspects. In our case, the variance is taken in its classical definition, that is, it measures how far the entities, each one represented by a numeric triple (Q1, Q2 and Q6), that compose a cluster are spread out from their mean. For the results presented here we traversed the dendrogram starting from the smallest variance to the maximum variance, which is the root of the dendrogram. For each group of entities, we verify what they do have in common so that we may understand why they behaved similarly or not. As we show next, we are able to identify relevant and significant stereotypes across several entities (e.g., countries).

Figure 6 presents the dendrograms for both search engines, we use a cutoff of 6 clusters to illustrate the process of clustering from the dendrogram structure. The centroids of the clusters are shown in Table 3. It is important to emphasize that when analyzing the centroids of each cluster the dimensions represent the per race average z-score. In our previous analysis we have shown that for black and Asian women, a more negative score represents a stronger negative stereotype regarding the two races. For white women, a more positive score represents a stronger positive stereotype.

4   Discussion

In this section, we discuss the stereotypes identified in the previous section. A positive stereotype exists when the fraction of beautiful women for a given race is larger than the fraction of ugly women for same race. The opposite defines a negative stereotype.

Our results point out that, for the majority of countries analyzed, there is a positive stereotype for white women and a negative one for black and Asian

---

[3] R library: https://stat.ethz.ch/R-manual/R-devel/library/stats/html/hclust.html

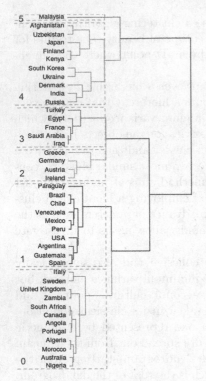

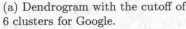

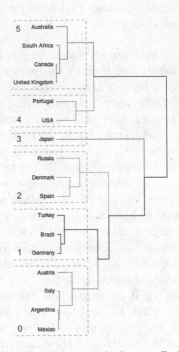

(a) Dendrogram with the cutoff of 6 clusters for Google.

(b) Dendrogram with the cutoff of 6 clusters for Bing.

Fig. 6. Clusters

women. The number of countries for which there is a negative stereotype for black women dominates our statistics, i.e., 85.71 % of the countries collected in Google and 76.47 % in Bing display this type of stereotype. In the same way we show that there is a negative stereotype about older women. In 69.05 % of the countries in Google and 82.35 % in Bing, the concept of beauty is associated with young women and ugly women are associated with older women. Countries have different configurations of stereotypes, and they can be grouped accordingly. For example, some countries have a very negative stereotype against black women, but can be 'neutral' with respect to the other races.

In the Google dendrogram (Fig. 6(a)), we can highlight cluster 1 - Spain, Guatemala, Argentina, USA, Peru, Mexico, Venezuela, Chile, Brazil and Paraguay - which has a geographical semantic meaning. They are countries from the Americas and Spain. Their population (or a large fraction of it as in the US) speak Latin languages, Spanish and Portuguese. Thus, these are countries with a strong presence of the Hispanic and Latino cultures. The centroid of this cluster (black: −3.28, Asian: 0.60, white: 2.02) indicates that for this group of countries there is a very negative stereotype regarding black women and a

Table 3. Clusters centroids

	Google					
	Black		Asian		White	
	Mean	STDEV	Mean	STDEV	Mean	STDEV
Cluster 0	−2.85	0.16	−0.41	0.38	2.65	0.12
Cluster 1	−3.28	0.21	0.60	0.31	2.02	0.34
Cluster 2	−1.23	0.22	−1.72	0.92	2.23	0.71
Cluster 3	−2.68	0.24	−1.87	0.38	3.40	0.19
Cluster 4	−0.60	0.28	0.05	0.44	0.28	0.73
Cluster 5	−2.24	0.00	2.86	0.00	−1.84	0.00
	Bing					
	Black		Asian		White	
	Mean	STDEV	Mean	STDEV	Mean	STDEV
Cluster 0	−0.53	0.31	−0.66	0.10	0.90	0.09
Cluster 1	−1.60	0.03	−0.83	0.21	1.75	0.15
Cluster 2	−1.82	0.02	0.47	0.74	0.78	0.44
Cluster 3	−1.33	0.00	2.70	0.00	−2.37	0.00
Cluster 4	−2.85	0.65	0.00	0.42	2.27	0.19
Cluster 5	−2.85	0.22	−1.23	0.05	3.20	0.23

positive stereotype for white women. In Cluster 2 - Ireland, Austria, Germany and Greece - we have European countries and a different ·stereotype (black: −1.23, Asian: −1.72, white: 2.23) since Asians have a more negative stereotype than blacks. For Cluster 4 - Russia, India, Denmark, Ukraine, South Korea, Kenya, Finland, Japan, Uzbekistan and Afghanistan - we could not identify a clear semantic meaning for the group. However, the cluster has an interesting stereotype of beauty (black: −0.60, Asian: 0.05, white: 0.28) in which the stereotype, positive or negative, regarding the races do not exist or are small. There is a coherence between the proportions of the races for the two queries, that is, for most of these countries there is no significant difference between the fractions of the races when we search for beautiful women or ugly women.

In the clustering process of data collected from Bing, Cluster 3 (black: −1.33, Asian: 2.70, white: −2.37), composed only by Japan, has the same stereotype of beauty than cluster 5 of Google (black: −2.24, Asian: 2.86, white: −1.84), composed only by Malaysia. Both are composed of just an Asian country and therefore have a very positive stereotype regarding Asian and negative stereotype regarding black women and white women.

In order to deepen the understanding of the stereotypes, we looked at the race composition of some countries to verify if they may explain some of the identified patterns. In Japan, Asians represent 99.4 % of population[4], in Argentina 97 %

---

[4] http://www.indexmundi.com/japan/demographics_profile.html.

of population are white[5], in South Africa 79.2 % are blacks and 8.9 % white[6], at last, in EUA racial composition is 12 % of blacks and 62 % of whites[7]. Although the racial composition of these countries indicate different fractions of black people, the search engine results show for all of them the presence of the negative stereotype of beauty about black women, with the exception of Japan in Google and Argentina in Bing. We did not find any specific relation between the racial composition of a country and the patterns of stereotypes identified for the country.

5   Conclusions and Future Work

To the best of our knowledge, this is the first study to systematically analyze differences in the perception of physical attractiveness of women in the online world. Using a combination of face images obtained by search engine queries plus face's characteristics inferred by a facial recognition system, the study shows the existence of appearance stereotypes for women in the online world. These findings result from applying a methodology we propose for analyzing stereotypes in online photos that portray people. As future work we plan to expand the analysis to the male gender as well.

Overall, we found negative stereotypes for black and older women. We have demonstrated that this pattern of stereotype is present in almost all the continents, Africa, Asia, Australia/Oceania, Europe, North America, and South America. Our experiments allowed us to pinpoint groups of countries that share similar patterns of stereotypes. The existence of stereotypes in the online world may foster discrimination both in the online and real world. This is an important contribution of this paper towards actions to reduce bias and discrimination in the online world.

It is important to emphasize that we do not know exactly the reasons for the existence of the identified stereotypes. They may stem from a combination of the stocks of available photos and characteristics of the indexing and ranking algorithms of the search engines. The stock of photos online may reflect prejudices and bias of the real world that transferred from the physical world to the online world by the search engines. Given the importance of search engines as source of information, we suggest that they analyze the problems caused by the prominent presence of negative stereotypes and find algorithmic ways to minimize the problem.

We know that using Face++, even though it is a widely used tool, implies some limitations. The set of photos used for the algorithm training can introduce itself a racial bias since the concept of racial identity is not the same around the world. Therefore, follow-up studies will employ a crowdsourcing annotation - for example, Amazon Mechanical Turk - for racial analysis and extraction of characteristics of face images to generate a more detailed description of classes

---

[5] http://www.indexmundi.com/argentina/ethnic_groups.html.

[6] http://www.southafrica.info/about/people/population.htm#.V4koMR9yvCI.

[7] http://kff.org/other/state-indicator/distribution-by-raceethnicity/.

of stereotypes and compare them with the results of different facial recognition systems. Using the same service we will validate the translation of search queries used in this work.

Acknowledgments. This work was partially funded by Fapemig, CNPq, CAPES, and by projects InWeb, MASWeb, and EUBra-BIGSEA.

A   Data Gathering Statistics

Tables 4 and 5 present the number of photos that Face++ was able to detect a single face per country and for Google and Bing, respectively.

Table 4. Useful photos from Google.

Google								
Country	Beautiful	Ugly	Country	Beautiful	Ugly	Country	Beautiful	Ugly
Afghanistan	37	35	France	37	34	Morocco	30	28
South Africa	34	38	Germany	29	25	Nigeria	34	36
Algeria	30	29	Greece	31	27	Paraguay	41	33
Angola	37	32	Guatemala	41	30	Peru	41	30
Saudi Arabia	30	30	India	29	35	Portugal	38	31
Argentina	41	34	Iraq	30	30	Russia	37	36
Australia	33	38	Ireland	30	22	Spain	40	31
Austria	39	27	Italy	36	31	USA	35	38
Brazil	37	30	Japan	39	27	Sweden	46	36
Canada	33	38	Kenya	34	24	Turkey	40	28
Chile	39	33	United Kingdom	34	39	Ukraine	41	37
Denmark	31	24	South Korea	32	33	Uzbekistan	40	46
Egypt	30	29	Malaysia	39	33	Venezuela	36	31
Finland	39	25	Mexico	41	32	Zambia	36	39

B   Results of Z-score Tests

In the Tables 6 and 7 the results highlighted are those which we reject the null hypothesis and accept the alternative hypothesis. In other words, we can answer YES to the questions **Q1**, **Q2** and/or **Q3**.

In the Tables 8 and 9 the results highlighted are those which we keep the alternative hypothesis and we can answer YES to the questions **Q4**, **Q5** and/or **Q6**.

C   Results of Wilcoxon tests

Results highlighted in the Tables 10 and 11 show those countries for which we keep the alternative hypothesis.

Table 5. Useful photos from Bing.

Bing Country	Beautiful	Ugly	Country	Beautiful	Ugly
South Africa	38	44	Italy	43	40
Saudi Arabia	28	<20	Japan	42	24
Argentina	37	42	United Kingdom	37	44
Australia	36	45	South Korea	<20	<20
Austria	37	34	Mexico	39	43
Brazil	34	37	Portugal	32	38
Canada	38	45	Russia	45	44
Denmark	34	21	Spain	43	41
Finland	37	<20	USA	37	44
Germany	38	40	Turkey	42	36
Greece	37	<20	Ukraine	25	<20

Table 6. Z-score table associated with the questions Q1, Q2 and Q3 (Google)

Z-score table (Google) Country	Q1 (Black)	Q2 (Asian)	Q3 (White)	Country	Q1 (Black)	Q2 (Asian)	Q3 (White)
Afghanistan	−0.48	−0.53	0.74	Italy	−2.51	−1.05	2.59
South Africa	−2.96	−0.34	2.79	Japan	0.84	0.62	−0.84
Algeria	−2.87	−0.51	2.65	Kenya	0.38	−0.13	−0.26
Angola	−2.99	−0.19	2.62	United Kingdom	−2.91	0.14	2.53
Saudi Arabia	−2.81	−1.72	3.45	South Korea	−1.41	0.65	−0.16
Argentina	−3.29	0.77	1.82	Malaysia	−2.24	2.86	−1.84
Australia	−2.70	−0.30	2.53	Mexico	−3.15	0.26	1.98
Austria	−0.93	−1.33	1.68	Morocco	−2.92	−0.55	2.73
Brazil	−3.12	0.59	2.21	Nigeria	−2.64	−0.54	2.65
Canada	−2.91	−0.30	2.73	Paraguay	−3.34	1.25	1.35
Chile	−3.26	0.50	2.09	Peru	−3.26	0.16	2.15
Denmark	−2.36	0.04	1.50	Portugal	−3.10	−0.26	2.76
Egypt	−2.63	−1.57	3.13	Russia	−0.89	−0.03	0.68
Finland	0.21	0.10	-0.20	Spain	−3.65	0.79	2.01
France	−2.67	−1.82	3.33	USA	−3.20	0.66	2.65
Germany	−1.59	−1.06	1.97	Sweden	−2.88	−1.09	2.79
Greece	−1.18	−2.77	3.03	Turkey	−2.46	−2.53	3.66
Guatemala	−3.51	0.58	2.15	Ukraine	−0.68	0.58	−0.25
India	−1.61	−0.28	1.07	Uzbekistan	0.00	−0.51	0.51
Iraq	−2.81	−1.72	3.45	Venezuela	−3.01	0.43	1.76
Ireland	−1.18	−1.68	2.08	Zambia	−2.80	0.08	2.43

Table 7. Z-score table associated with the questions Q1, Q2 and Q3 (Bing)

Z-score table (Google)

Country	Q1 (Black)	Q2 (Asian)	Q3 (White)	Country	Q1 (Black)	Q2 (Asian)	Q3 (White)
South Africa	−2.86	−1.28	3.23	Japan	−1.33	2.70	−2.37
Argentina	−0.69	−0.59	0.94	United Kingdom	−3.00	−1.24	3.36
Australia	−2.54	−1.16	2.87	Mexico	−0.72	−0.59	0.95
Austria	−0.06	−0.80	0.77	Portugal	−3.32	0.29	2.41
Brazil	−1.60	−0.62	1.62	Russia	−1.79	1.20	0.31
Canada	−3.00	−1.24	3.35	Spain	−1.84	0.49	0.87
Denmark	−1.83	−0.27	1.17	USA	−2.39	−0.30	2.13
Germany	−1.64	−0.81	1.72	Turkey	−1.57	−1.05	1.91
Italy	−0.65	−0.65	0.94				

Table 8. Z-score table associated with the questions Q4, Q5 and Q6 (Google)

Z-score table (Google)

Country	Q4 (Black)	Q5 (Asian)	Q6 (White)	Country	Q4 (Black)	Q5 (Asian)	Q6 (White)
Afghanistan	−0.48	−0.53	0.74	Italy	−2.51	−1.05	2.59
South Africa	−2.96	−0.34	2.79	Japan	0.84	0.62	−0.84
Algeria	−2.87	−0.51	2.65	Kenya	0.38	−0.13	−0.26
Angola	−2.99	−0.19	2.62	United Kingdom	−2.91	0.14	2.53
Saudi Arabia	−2.81	−1.72	3.45	South Korea	−1.41	0.65	−0.16
Argentina	−3.29	0.77	1.82	Malaysia	−2.24	2.86	−1.84
Australia	−2.70	−0.30	2.53	Mexico	−3.15	0.26	1.98
Austria	−0.93	−1.33	1.68	Morocco	−2.92	−0.55	2.73
Brazil	−3.12	0.59	2.21	Nigeria	−2.64	−0.54	2.65
Canada	−2.91	−0.30	2.73	Paraguay	−3.34	1.25	1.35
Chile	−3.26	0.50	2.09	Peru	−3.26	0.16	2.15
Denmark	−2.36	0.04	1.50	Portugal	−3.10	−0.26	2.76
Egypt	−2.63	−1.57	3.13	Russia	−0.89	−0.03	0.68
Finland	0.21	0.10	−0.20	Spain	−3.65	0.79	2.01
France	−2.67	−1.82	3.33	USA	−3.20	0.66	2.65
Germany	−1.59	−1.06	1.97	Sweden	−2.88	−1.09	2.79
Greece	−1.18	−2.77	3.03	Turkey	−2.46	−2.53	3.66
Guatemala	−3.51	0.58	2.15	Ukraine	−0.68	0.58	−0.25
India	−1.61	−0.28	1.07	Uzbekistan	0.00	−0.51	0.51
Iraq	−2.81	−1.72	3.45	Venezuela	−3.01	0.43	1.76
Ireland	−1.18	−1.68	2.08	Zambia	−2.80	0.08	2.43

Table 9. Z-score table associated with the questions Q4, Q5 and Q6 (Bing)

Z-score table (Bing)

Country	Q1 (Black)	Q2 (Asian)	Q3 (White)	Country	Q1 (Black)	Q2 (Asian)	Q3 (White)
South Africa	−2.86	−1.28	3.23	Japan	−1.33	2.70	−2.37
Argentina	−0.69	−0.59	0.94	United Kingdom	−3.00	−1.24	3.36
Australia	−2.54	−1.16	2.87	Mexico	−0.72	−0.59	0.95
Austria	−0.06	−0.80	0.77	Portugal	−3.32	0.29	2.41
Brazil	−1.60	−0.62	1.62	Russia	−1.79	1.20	0.31
Canada	−3.00	−1.24	3.35	Spain	−1.84	0.49	0.87
Denmark	−1.83	−0.27	1.17	USA	−2.39	−0.30	2.13
Germany	−1.64	−0.81	1.72	Turkey	−1.57	−1.05	1.91
Italy	−0.65	−0.65	0.94				

Table 10. P-value table associated with the questions **Q7** (Google)

Google					
Wilcoxon test (Q7)					
Country	P-value	Country	P-value	Country	P-value
Afghanistan	0.1819	France	0.0572	Morocco	**0.0036**
South Africa	**0.0001**	Germany	**0.0107**	Nigeria	**0.0000**
Algeria	**0.0023**	Greece	**0.0040**	Paraguay	**0.0471**
Angola	**0.0072**	Guatemala	0.0512	Peru	**0.0499**
Saudi Arabia	**0.0131**	India	0.1221	Portugal	**0.0014**
Argentina	**0.0271**	Iraq	**0.0196**	Russia	**0.0146**
Australia	**0.0003**	Ireland	0.0703	Spain	**0.1869**
Austria	**0.0017**	Italy	0.2288	USA	**0.0000**
Brazil	**0.0298**	Japan	0.0520	Sweden	**0.0071**
Canada	**0.0001**	Kenya	**0.0041**	Turkey	**0.0093**
Chile	**0.0134**	United Kingdom	**0.0000**	Ukraine	0.1699
Denmark	0.3731	South Korea	0.1363	Uzbekistan	0.8407
Egypt	**0.0122**	Malaysia	**0.0005**	Venezuela	**0.0218**
Finland	0.1759	Mexico	**0.0174**	Zambia	**0.0002**

Table 11. P-value table associated with the questions **Q7** (Bing)

Bing			
Wilcoxon test (Q7)			
Country	P-value	Country	P-value
South Africa	**0.0179**	Japan	0.1058
Argentina	0.0612	United Kingdom	**0.0226**
Australia	**0.0077**	Mexico	**0.0257**
Austria	**0.0001**	Portugal	**0.0314**
Brazil	**0.0002**	Russia	**0.0302**
Canada	**0.0211**	Spain	0.0553
Denmark	**0.0168**	USA	**0.0021**
Germany	**0.0012**	Turkey	**0.0040**
Italy	**0.0025**		

References

1. Cash, T.F., Brown, T.A.: Gender and body images: stereotypes and realities. Sex Roles **21**(5), 361–373 (1989)
2. Kay, M., Matuszek, C., Munson, S.A.: Unequal representation and gender stereotypes in image search results for occupations. In: Proceedings of the 33rd Annual ACM Conference on Human Factors in Computing Systems, CHI 2015, pp. 3819–3828. ACM, New York (2015)

3. Downs, A.C., Harrison, S.K.: Embarrassing age spots or just plain ugly? Physical attractiveness stereotyping as an instrument of sexism on american television commercials. Sex Roles **13**(1), 9–19 (1985)

4. William, H.: The Analysis of Beauty: Written with a View of Fixing the Fluctuating Ideas of Taste. Samuel Bagster & Sons, London (1753)

5. Bakhshi, S., Shamma, D.A., Gilbert, E.: Faces engage us: photos with faces attract more likes and comments on instagram. In: Proceedings of the 32nd Annual ACM Conference on Human Factors in Computing Systems, pp. 965–974. ACM (2014)

6. Umoja Noble, S.: Google search: hyper-visibility as a means of rendering black women and girls invisible. InVis. Cult. J. Vis. Cult. Univ. Rochester. (2013). http://ivc.lib.rochester.edu/google-search-hyper-visibility-as-a-means-of-rendering-black-women-and-girls-invisible/

7. Introna, L.D., Nissenbaum, H.: Shaping the web: why the politics of search engines matters. Inf. Soc. **16**(3), 169–185 (2000)

8. Sweeney, L.: Discrimination in online ad delivery. Queue **11**(3), 10 (2013)

9. Umoja Noble, S.: Missed connections: what search engines say about women (2012)

10. Mazza, F., Da Silva, M.P., Le Callet, P.: Racial identity and media orientation: exploring the nature of constraint. J. Black Stud. **29**, 367–397 (1999)

11. Allibhai, A.: On racial bias and the sharing economy (2016). https://goo.gl/mhpr6C. Accessed 13 May 2016

12. Hoffman, K.M., Trawalter, S., Axt, J.R., Oliver, M.N.: Racial bias in pain assessment and treatment recommendations, and false beliefs about biological differences between blacks and whites. Proc. Nat. Acad. Sci. **113**(16), 4296–4301 (2016). doi:10.1073/pnas.1516047113

13. van den Berghe, P.L., Frost, P.: Skin color preference, sexual dimorphism and sexual selection: a case of gene culture co-evolution? Ethn. Racial Stud. **9**(1), 87–113 (1986)

14. Grammer, K., Fink, B., Moller, A.P., Thornhill, R.: Darwinian aesthetics: sexual selection and the biology of beauty. Biol. Rev. **78**, 385–407 (2003)

15. Fink, B., Grammer, K., Matts, P.J.: Visible skin color distribution plays a role in the perception of age, attractiveness, and health in female faces. Evol. Hum. Behav. **27**(6), 433–442 (2006)

16. Cunningham, M.R., Roberts, A.R., Barbee, A.P., Druen, P.B., Wu, C.H.: Their ideas of beauty are, on the whole, the same as ours. J. Pers. Soc. Psychol. **68**, 261–279 (1995)

17. Coetzee, V., Greeff, J.M., Stephen, I.D., Perrett, D.I.: Cross-cultural agreement in facial attractiveness preferences: the role of ethnicity and gender. PLoS ONE **9**(7), 1–8 (2014)

18. Eisenthal, Y., Dror, G., Ruppin, E.: Facial attractiveness: beauty and the machine. Neural Comput. **18**(1), 119–142 (2006)

19. Rationality: The Cambridge Dictionary of Philosophy, 2nd edn. Cambridge University Press (1999)

20. Wilcoxon, F.: Individual comparisons by ranking methods. Biom. Bull. **1**(6), 80–83 (1945)

21. Murtagh, F., Legendre, P.: Ward's hierarchical agglomerative clustering method: Which algorithms implement ward's criterion? J. Classif. **31**(3), 274–295 (2014)

Analysing RateMyProfessors Evaluations Across Institutions, Disciplines, and Cultures: The Tell-Tale Signs of a Good Professor

Mahmoud Azab, Rada Mihalcea$^{(\boxtimes)}$, and Jacob Abernethy

University of Michigan, 2260 Hayward Street, Ann Arbor, MI 48109, USA
{mazab,mihalcea,jabernet}@umich.edu

Abstract. Can we tell a good professor from their students' comments? And are there differences between what is considered to be a good professor by different student groups? We use a large corpus of student evaluations collected from the RateMyProfessors website, covering different institutions, disciplines, and cultures, and perform several comparative experiments and analyses aimed to answer these two questions. Our results indicate that (1) we can reliably classify good professors from poor professors with an accuracy of over 90 %, and (2) we can separate the evaluations made for good professors by different groups with accuracies in the range of 71–89 %. Furthermore, a qualitative analysis performed using topic modeling highlights the aspects of interest for different student groups.

1 Introduction

Assessing teaching quality is a difficult and subjective task. Most if not all schools evaluate their professors by asking students to provide course feedback, which often consists of ratings as well as open-ended comments in response to several prompts. With few exceptions, this feedback is kept confidential and is shared with neither current nor prospective students. It is therefore not surprising that the Web 2.0 wave has brought several sites that encourage students to share their in-class experiences and the opinions they hold on the professors teaching their courses. Among these sites, the one that is by far the most popular is RateMyProfessors[1] (RMP), where students can anonymously rate different aspects of their professors (i.e., clarity, helpfulness, easiness), and also provide open-ended comments. The site currently has approximately 15 million evaluations for 1.4 million professors from 7,000 schools in the United States, Canada, and United Kingdom. Students appear to have confidence in the RMP ratings and there is evidence that they use the site to make academic decisions [5].

In this paper, we analyze the language used by students when discussing their professors. Using a large collection of 908,903 RMP comments collected for 71,404 professors from 33 different institutions, our study aims to answer

---

[1] http://www.ratemyprofessors.com/.

© Springer International Publishing AG 2016
E. Spiro and Y.-Y. Ahn (Eds.): SocInfo 2016, Part I, LNCS 10046, pp. 438–453, 2016.
DOI: 10.1007/978-3-319-47880-7_27

the following two questions. First, can we use automatic text classification to distinguish between professors regarded as good vs. professors regarded as poor? After several feature selection experiments, we show that we can reliably separate good professors from poor professors with an accuracy of over 90 %.

Second, and perhaps more interestingly, we ask whether there are differences between what characterizes a good professor across different groups. To answer this question, we focus exclusively on the good professors in our dataset, and specifically look for differences across disciplines (e.g., Sociology vs. Computer Science), across institutions (top-ranked vs. low-ranked schools), and across cultures (U.S. vs. Canada). We perform a quantitative analysis of these differences by performing automatic classification of good professor comments contributed by different groups using domain-independent features, and show that we can achieve classification accuracies in the range of 71–89 %, suggesting that different students value different aspects of a good professor. To understand these differences, we use topic modeling to perform a qualitative analysis through comparisons between the distributions of several topics in the students comments. This analysis leads us to several interesting findings, e.g., computer science students appear to exhibit greater appreciation for a professor's clarity, while philosophy students are more concerned with readings and discussions, and so on.

2   Related Work

While there is no previous work that we are aware of in the field of natural language processing focusing on the analysis of RMP student evaluations, there are several studies in fields such as education and sociology. These studies confirmed the validity of RMP evaluations and found significant correlations between RMP rating scores and their corresponding scores in official student evaluations of teaching for professors from different schools [5,6,14,19]. There are also studies on the intercorrelations among RMP rating scores. For instance, RMP overall quality score is highly correlated to the easiness and the physical attractiveness of the professor [7,8]. Freng and Webber [9] also showed that the attractiveness is responsible of 8 % of the variance in the data.

The study that is closest to our, although not computational, is the one by Helterbran [10], who manually analyzed RMP comments for 283 instructors from three universities in Pennsylvania, and identified certain personal attitudes and instructional behaviors that are most beneficial to students, such as being knowledgeable and approachable. This study was limited in terms of the numbers and institutions studied, and did not have discipline and cultural diversity.

Also related to our work is research on opinion mining and sentiment analysis, which is a well-established area in natural language processing. It has been approached at different levels of granularity from document- to sentence- to phrase-level sentiment classification [1,11,15,20,21]. The nature of the examined data varied from online products and movie reviews to opinions posted on microblogs like Twitter [2]. These studies used different machine learning techniques for classification such as Naive Bayes and Support Vector Machines with different sets of features such as unigrams and bigrams. To the best of our knowledge, no previous work has tackled students' evaluations.

High-level classification of students opinions is not enough to understand what are the instructional behaviors that students care about the most. We found inspiration in recent work on topic modeling, which has been successfully used to extract personal values and behaviors from open-ended text [4], or to integrate expert reviews with opinions scattered over the Web in a semi-supervised approach [12].

3   Dataset

The study reported in this paper is based on a corpus compiled from the RMP site. Our goal was to build a dataset of professors and their evaluations from a diverse pool, covering institutions with different academic rankings, covering different countries, and also covering different disciplines.

The crawl, made during the summer of 2015, was started by specifying a list of 33 schools. When constructing this list, we considered the academic ranking of the schools according to the U.S. News ranking. We included 10 U.S. top-ranked public schools, such as the University of California Berkeley and the University of Michigan, 10 low-ranked public schools, as well as 4 additional U.S. public schools.

We also considered the country of each institution, and in addition to the 24 U.S. schools, we included 9 schools from Canada, such as the University of Toronto and University of Montreal.

We collected the records of every professor affiliated with each school, covering all 33 schools, which in aggregate provided a very diverse set of faculty disciplines. For each professor, we then collected the entire set of their students' ratings. Finally, we removed ratings that had the comment field left blank and also the professors who received no comments. The resulting dataset consists of 908,903 evaluations with textual comments for 71,404 professors from 33 schools. Table 1 shows the distribution of professors and comments in our dataset.

Table 1. Statistics on the RMP dataset.

	Professors	Evaluations
U.S. top-ranked	21,119	245,553
U.S. low-ranked	15,631	195,728
Canada	19,672	313,868

In addition to specifying the professor and the class, each evaluation includes an optional comment, as well as several attributes, such as helpfulness, clarity, and easiness scores. These attributes can have a value between $[1, 5]$, where 1 is the worst score and 5 is the best score. Each evaluation also receives an

Table 2. Sample RMP evaluations.

Overall	Helpfulness	Clarity	Easiness	Department	Comment
Good	4	5	4	Economics	Uses real world examples to make lectures more interesting. Clear and concise. Recommended.
Poor	1	2	1	Computer Science	Bad at explaining material, doesn't seem to care about individuals.
Good	5	3	2	Statistics	Statistics requires that you work for it, so be prepared to work for this

overall classification of good, average or poor, determined by RMP based on the helpfulness and clarity scores. For each professor, overall helpfulness and clarity scores are also calculated, as the average of all the helpfulness and clarity scores given to this professor by the students. Finally, RMP calculates the overall quality score of a professor as the average of her overall helpfulness and clarity scores. Table 2 shows examples of RMP evaluations.

In all our experiments, we use a random split of the dataset into training and test, consisting of 57,150 and 14,254 professors respectively. The comments are also split based on the professors they belong to. Therefore, a professor and her corresponding comments exist in either the training or the test set, but not in both. We do not balance the data because in our analyses we want to capture as many aspects and concerns in students' comments as possible. Balancing the data might result in a loss of important information.

4   Can We Tell a Good Professor?

Our first set of experiments is concerned with determining whether the textual comments from RMP can be used to automatically predict the overall classification of an individual comment or of a professor as either "good" or "poor" (see below for an explanation of these labels). This task is akin to that of sentiment analysis, in that we use the text of a comment to predict whether that comment is reflective of a "good" or a "poor" student evaluation (comment-level classification); or, we use the text of all the comments submitted for a professor to predict if that professor is rated as "good" or not (professor-level classification). These experiments, along with the feature selection discussed in Sect. 4.1, allow us to determine the words that have high predictive power in students' textual comments, which are necessary for our analyses to understand the characteristics of good professors.

To represent the text, we extract features consisting of unigrams, bigrams, and a mix of unigrams and bigrams. Each instance in our dataset (whether an individual comment or a professor) is thus represented as a feature vector encoding the counts of the n-grams in the representation.

In addition to raw n-gram features, we also experiment with the use of sentiment/emotion lexical resources. Specifically, we use the following lexicons: OpinionFinder [21], which includes 2,570 words labeled as positive and 4,581 words as negative; a subset of WordNet Affect [18], with 1,128 words grouped into six basic emotions: anger, disgust, fear, joy, sadness and surprise; and General Inquirer [17], with 29,090 words mapped to 96 categories. We first filter the input text based on these lexicons by removing words that do not exist in the lexical resources and then generate unigram and bigram features from this filtered text.

To identify the most distinctive lexical features in the students' comments, we use feature selection, as described below. The features are then used in a multi-nomial Naive Bayes classifier; we also ran experiments using a Support Vector Machine classifier, but its performance was significantly below that of the multi-nomial Naive Bayes.

Note that all our experiments exclusively rely on the text in the comments, and are not making use of the other attributes available on the RMP site (helpfulness, clarity, easiness) in any ways.

4.1 Feature Selection

We experiment with two feature selection methods to identify the most useful features for our task. The methods are compared by using five-fold cross-validation on training data, and the best method is selected and applied on the test set.[2]

The first feature selection method is linear regression which, for each feature, uses uni-variate linear regression tests to compute the correlation between a target class and the data.

The second one is chi-square, which measures the degree of dependence between two stochastic variables: in our case, for each feature, we determine if there is a significant difference between the observed and expected frequencies in one or more target classes. For each feature selection method, we use their scores to rank the features, and keep the top K-percent features for the classification.

4.2 Comment-Level Classification

In this initial experiment, we classify the individual comments as either "good" or "poor". We use the RMP overall *quality* rating, which is associated with each comment and can have one of the following values: good, average or poor. We only consider comments that are labeled as good or poor, and ignore those labeled as average.

---

[2] The feature selection methods and the machine learning algorithms used in this study have been implemented in Python using the Sci-kit Learn machine learning library [16]. We use a maximum document frequency of 0.5 and lowercased text. We also experimented with stemming but it was found to degrade performance.

Table 3. The distribution of the training and test data in the comment- and professor-level classification experiment.

	Training	Test
Comment-level		
Good	471,566	117,816
Poor	165,593	40,631
Professor-level		
Good	36,958	9,265
Poor	8,615	2,152

To determine the training and test datasets, we use the random split mentioned in Sect. 3, ensuring that all the comments belonging to a professor are either in training or in test. Table 3 shows the distribution of the good and poor comments in the data. As seen in this table, the distribution is similar in both training and test, with 74 % of the comments being labeled as good.

In order to tune the classifier and select the best set of features, we use five-fold cross-validation on the training data, and compare the accuracies obtained with the two feature selection methods (linear regression and chi-square) and different features (unigrams, bigrams, unigrams+bigrams, unigrams+bigrams pre-filtered based on the lexicons). Figure 1 shows the average accuracy obtained in this cross-validation experiments on the training data for the top-K percentile of the features with an incremental step of size 2. The best accuracy is achieved using the top 18 % of the mixed raw unigrams+bigrams features, ranked according to the chi-square test. Interestingly, the features based on the sentiment/emotion lexicons do not perform as well as the raw features, which may suggest that student comments are different from the opinions/reviews previously used in sentiment analysis research. We use these top 18 % features to train and test our final classifier. Tables 4 and 5 show that our classifier achieves significantly higher accuracy, precision, recall, and f-score than a majority class baseline.

4.3 Professor-Level Classification

In a second experiment, instead of classifying individual comments, we now classify professors as either "good" or "poor". To represent a professor, we use all the comments submitted for that professor. To label a professor as good or poor, we use the overall score field that is calculated by RMP for each professor. We consider a professor with an overall rating score of ≥ 3.5 as good, and a professor with an overall rating score of ≤ 2.5 as poor.

As before, we use the training/test split described in Sect. 3. Table 3 shows the distribution of professors labeled as good/poor in the data. Once again, the numbers indicate that the class distribution is similar in training and test, with 81 % of the professors being labeled as good. We use the same approach as in the

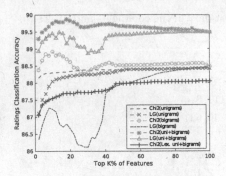

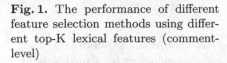

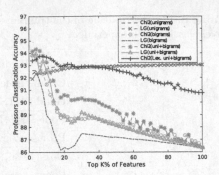

Fig. 1. The performance of different feature selection methods using different top-K lexical features (comment-level)

Fig. 2. The performance of different feature selection methods using different top-K lexical features (professor-level)

Table 4. Comment- and Professor-level classification accuracy on test data.

	Majority class	Multinomial Naive Bayes
Comment-level	74.35 %	90.09 %
Professor-level	81.15 %	94.14 %

Table 5. Comment- and Professor-level precision, recall and f-score of each class on test data.

	Majority class			Multinomial Naive Bayes		
	Precision	Recall	F-score	Precision	Recall	F-score
Comment-level						
Good	74.36	100	85.29	91.44	95.62	93.48
Poor	-	-	-	85.36	74.04	79.30
Professor-level						
Good	81.15	100	89.59	95.84	96.99	96.41
Poor	-	-	-	86.33	81.88	84.04

comment-level experiment to tune the parameters of this classifier, and run five-fold cross validation experiments on the training data. Figure 2 shows the average accuracy for different methods using the top-K percentile of the features with an incremental step of size 2. The best accuracy is achieved using the top 4 % of the unigrams+bigrams features with a chi-square test. This suggests that there are words that are not included in the lexical resources that can distinguish good from poor professors. We use this setting to train our final classifier, and evaluate it on the test data. The final result, shown in Tables 4 and 5, indicates that we can reliably distinguish between good and poor professors, with an accuracy, precison, recall, and f-score significantly higher than the majority class baseline.

Table 6. Top ten features associated with professors with a good/poor rating.

Rating	Top features
Good	interesting, best, awesome, fun, funny, helpful, amazing, great teacher, great professor, highly
Poor	worst, avoid, horrible, terrible, teach, worst professor, worst teacher, useless, does, costs

Not surprisingly, the accuracy obtained in the professor-level classification is higher than the one obtained by the comment-level classifier. Although the number of training instances is larger in the comment-level classifier, the professor-level classifier benefits from more data available for each instance, and also from a higher baseline.

To provide some insight into the features that play a significant role in the classification, Table 6 lists the top ten features for each class obtained from the professor-level classifier, ranked in reverse order of their chi-square weight. The Naive Bayes probability (i.e., P(feature|good), P(feature|poor)) was used to determine the class that each feature "belongs" to.

5 Can We Tell the Group Behind the Comments of a Good Professor?

The results presented in the previous section have shown that we can accurately classify a comment or a professor as either good or poor based on student language. While this is an interesting result in itself, we are also interested in finding whether there are differences between what is regarded as a good professor by different groups.

If we condition on professor quality, all else being equal, how well can we determine other particular factors of the faculty member in question, such as the rank of their institution, their discipline, or the country in which they teach? Our answers to these questions provide some insight into the complex attribute-specific components that determine the perception of professor quality. For instance, are there differences between good professors in Canada vs. U.S.? Or good professors in Computer Science vs. Sociology?

In these experiments, we specifically focus on the "good" professors in our dataset, with an overall rating of 3.5 or higher similar to RMP criteria. We perform three different analyses: (1) cross-culture, where we separate good professors from U.S. schools vs. good professors at schools in Canada; (2) cross-institution, where we classify good professors from top-ranked vs. low-ranked public U.S. schools, according to the U.S. News ranking; and (3) cross-discipline, where we try to see if there are differences between good professors in different disciplines. For this third analysis, we work with three pairs of disciplines that are unrelated (Sociology vs. Computer Science), (Philosophy vs. Physics), and (Fine Arts vs. Biological Sciences); and one pair that is somewhat related (Management vs. Business Administration).

To create the experimental datasets for these analyses, we use the original training and test sets described in Sect. 3, and filter for the group of interest.

Table 7. Number of good professors in different groups.

	Training	Test
Cultures		
Canada	9, 463	2, 395
U.S	27, 495	6, 870
Institutions		
Low-ranked	8, 139	2, 059
Top-ranked	11, 261	2, 884
Disciplines		
Biological Sciences	203	49
Business Administration	122	29
Computer Science	674	182
Fine Arts	372	79
Management	236	45
Philosophy	793	195
Physics	539	141
Sociology	872	229

For instance, to obtain the training dataset for Canada, we extract all the good professors from the large training dataset that are affiliated with a Canadian institution, and so forth. For the discipline datasets, we determine the discipline of the professor using the department name that the professor is affiliated with. Table 7 shows the number of good professors in our dataset, broken down for each of the groups mentioned above.

One difficulty with the classification of such groups is the presence of confounding factors: while our main goal is to identify differences between these groups in terms of what they appreciate in a good professor, the groups are also distinct because of culture-, institution-, or discipline- specific words. For example, the word "programming" is more likely to appear in comments made about Computer Science professors than in comments on Biology professors. Similarly, French words are more likely to be used in comments on professors at schools in Canada than in comments on professors at schools in the U.S. In order to disallow the classifier to use such words in the classification process, we impose on all these group classifiers the same set of features, consisting of the top 500 unigram features reversely sorted according to their chi-square weight obtained from the good vs. poor professor experiments, described in Sect. 4. Moreover, we manually revised these features, removing by hand all culture-, institution-, or discipline-specific words, to ensure that the feature set includes only general attribute words, e.g. "good," "humorous," or "knowledgeable. "We also normalized the words that are spelled differently in both Canada and the US, e.g. "favorite" and "favourite".

Table 8. Classification accuracy for different groups.

Group Pair	Majority class	Multinomial Naive Bayes
Canada vs. U.S	74.15 %	89.49 %
Top- vs. low-ranked	58.35 %	74.71 %
Philosophy vs. Physics	58.03 %	82.14 %
Biological Sciences vs. Fine Arts	61.72 %	89.06 %
Sociology vs. Computer Science	55.72 %	84.43 %
Business Administration vs. Management	60.81 %	71.62 %

Table 9. Precision, recall and f-score for each group.

Group Pair	Majority class			Multinomial Naive Bayes		
	Precision	Recall	F-score	Precision	Recall	F-score
U.S	74.15	100	85.16	91.91	94.1	92.99
Canada	-	-	-	81.85	76.24	78.95
Top-ranked	58.35	100	73.69	77.88	79.13	78.50
Low-ranked	-	-	-	70.09	68.53	69.30
Sociology	55.72	100	71.56	84.23	88.65	86.38
Computer Science	-	-	-	84.71	79.12	81.82
Philosophy	58.04	100	73.45	83.92	85.64	84.77
Physics	-	-	-	79.56	77.3	78.42
Fine Arts	61.72	100	76.33	90.12	92.41	91.25
Biological Studies	-	-	-	87.23	83.67	85.42
Business Administration	60.81	100	75.63	75.63	37.93	51.16
Management	-	-	-	70.00	93.33	80.00

Table 8 shows the classification accuracy that our classification models achieve for each experiment. Table 9 shows the precision, recall, and f-score for each group in each classification experiment. The classification accuracies between these groups are statistically significant except for Business Administration vs. Management. Thus, it seems that the differences between the comments of different groups changes according to the (dis)similarity of the two disciplines they represent. These results indicate that the groups writing comments about good professors can be separated with an accuracy significantly higher than the baseline, which, given that the features used in the classification do not include any group-specific words, suggest that there are indeed differences between what is considered to be a good professor by different groups. For additional insight into these differences, Table 10 shows the top ten features for each group, according to their chi-square weight.

Table 10. Top ten features associated with good professors rated by different groups

Group	Top ten features
Canada	prof, marker, profs, notes, textbook, fair, excellent, clear, approachable, best
U.S	homework, credit, grader, book, papers, interesting, extra, guides, material, reading
Top-ranked	lecturer, office, ta, readings, clear, reading, interesting, engaging, fair, slides
Low-ranked	attendance, credit, help, extra, gives, work, study, notes, willing, book
Sociology	readings, reading, papers, paper, study, discussion, essay, attendance, loved, passionate
Computer Science	homework, comments, teach, guy, excellent, office, time, help, explains, mistakes
Philosophy	papers, readings, reading, essays, paper, essay, marker, discussion, discussions, boring
Physics	homework, problems, exams, curve, help, accent, office, book, solutions, extra
Biological Sciences	notes, exams, material, questions, prof, clear, study, understand, fair, textbook
Fine Arts	work, nice, inspiring, comments, does, help, awesome, teaching, best, little
Management	paper, boring, book, papers, excellent, essay, kept, teachers, dr, instructor
Business Admin	prof, arrogant, curve, fair, extremely, lecturer, clear, engaging, approachable, definitely

6   What Are the Tell-Tale Signs of a Good Professor?

The results of the experiments described in the previous section show clear differences between what is considered to be a good professor by different groups. However, the numbers by themselves do not say much about what the actual differences are. In order to gain a better understanding of what each group looks for in a good professor, we use topic modeling to determine the main topics of interest in the students comments, and consequently compare the distribution of these topics in different groups.

To perform topic modeling, we use the Latent Dirichlet Allocation (LDA) implementation provided in Mallet (a machine learning for language toolkit) [13], applied on the professor-level representation of the data. LDA is a generative model that in our case considers each professor as a mixture of a small number of topics, and assumes that each word in this professor's data are associated with one of the topics [3]. Consistent with the analyses in the previous section, aiming at identifying differences among good professors as regarded by different groups,

Table 11. Ten main topics addressed in students comments, along with sample words.

Topic	Sample words
Approachability	prof, fair, clear, helpful, teaching, approachable, nice, organized, extremely, friendly, super, amazing
Clarity	understand, hard, homework, office, material, clear, helpful, problems, explains, accent, questions, extremely
Course Logistics	book, study, boring, extra, nice, credit, lot, hard, attendance, make, fine, attention, pay, mandatory
Enthusiasm	teaching, passionate, awesome, enthusiastic, professors, loves, cares, wonderful, fantastic, passion
Expectations	hard, work, time, lot, comments, tough, expects, worst, stuff, avoid, horrible, classes
Helpfulness	helpful, nice, recommend, cares, super, understanding, kind, extremely, effort, sweet, friendly, approachable
Humor	guy, funny, fun, awesome, cool, entertaining, humor, hilarious, jokes, stories, love, hot, enjoyable
Interestingness	interesting, material, recommend, lecturer, engaging, classes, knowledgeable, enjoyed, loved, topics
Readings/ Discussions	readings, papers, writing, ta, interesting, discussions, grader, essays, boring, books, participation
Study Material	exams, notes, questions, material, textbook, hard, slides, study, answer, clear, tricky, attend, long, understand

we extract ten topics using the data corresponding to the "good" professors. Table 11 shows these topics, along with several sample words for each topic.

Starting with these ten topics, we determine their distribution in each of the groups considered in the previous section. Figures 3, 4, 5, 6, 7 and 8 show these distributions, leading to interesting findings.[3] For instance, students in Canada seem to be more concerned with Approachability and Study Materials, whereas students from U.S. schools appear to talk more about Readings/Discussions and Clarity (Fig. 3). Students at top- and low-ranked U.S. public schools appear to be concerned with similar aspects of their good professors, with a somehow higher interest for Readings/Discussions and Clarity among students in top-ranked institutions, and more interest in Course Logistics among students in low-ranked schools (Fig. 4).

There are also differences among the aspects of interest for different disciplines. Sociology students talk more about Readings/Discussions, whereas Computer Science students focus more on Clarity (Fig. 5). A similar difference is observed between Philosophy and Physics (Fig. 6). Fine Arts students are more concerned with the Enthusiasm of their professors and tend to talk more about

---

[3] In each of these figures, the topic distributions for a group add up to 100 % (e.g., the blue/dark and yellow/light columns in Fig. 3 each add up to 100 %).

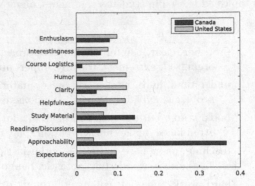

Fig. 3. Top topic distribution among good professors from U.S. schools vs. good professors from Canadian schools

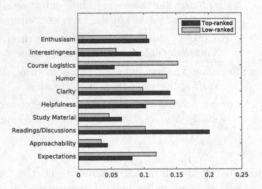

Fig. 4. Top topic distribution among good professors from top-ranked vs. low-ranked U.S. public schools

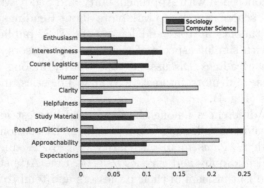

Fig. 5. Top topic distribution among good professors from Sociology vs. Computer Science

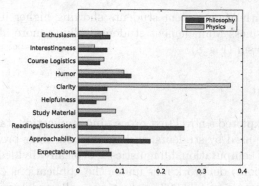

Fig. 6. Top topic distribution among good professors from Philosophy vs. Physics

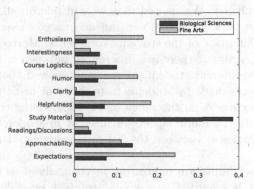

Fig. 7. Top topic distribution among good professors from Biological Sciences vs. Fine Arts

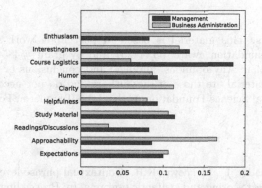

Fig. 8. Top topic distribution among good professors from Management vs. Business Adiminstration

the Expectations of their classes; on the other hand, Biological Sciences students primarily talk about Course Logistics and Study Materials (Fig. 7). Finally, although Management and Business Administration are related disciplines, we note differences with Management students showing higher interest in Course Logistics, and Business Management students talking more about Approachability and Enthusiasm (Fig. 8).

7 Conclusion

In this paper, we explored a novel text processing application, targeting an analysis of the language used by students when evaluating their professors. Research work in the field of computational linguistics is typically divided into algorithms, data, and applications; our work falls under the applications category. We constructed a new dataset of 908,903 evaluations collected for 71,404 professors from 33 different institutions, covering different disciplines, different institutions, and two different cultures. We showed that we can reliably distinguish between good professors and poor professors with an accuracy of over 90 %, by relying exclusively on the language of the students comments. Moreover, we performed experiments to determine if there are differences between what is regarded as a good professor by different student groups, and showed that we can separate between the comments made by students from different institutions, disciplines, or cultures, with accuracies in the range of 71–89 %. Using topic modeling, we were able to identify the main aspects of interest in student evaluations, and highlighted the differences between the aspects appreciated more by different student groups.

We believe these results are interesting in themselves, as they clearly show differences in what is regarded as a good professor by different groups. Our findings can also be useful to professors, by enabling them to identify the aspects that matter to their students, so that they can improve the overall teaching quality.

Acknowledgments. This material is based in part upon work supported by the National Science Foundation award #1344257 and by grant #48503 from the John Templeton Foundation. Any opinions, findings, and conclusions or recommendations expressed in this material are those of the authors and do not necessarily reflect the views of the National Science Foundation or the John Templeton Foundation.

References

1. Agarwal, A., Biadsy, F., Mckeown, K.R.: Contextual phrase-level polarity analysis using lexical affect scoring and syntactic n-grams. In: Proceedings of the 12th Conference of the European Chapter of the Association for Computational Linguistics, pp. 24–32. Association for Computational Linguistics (2009)
2. Bermingham, A., Smeaton, A.F.: Classifying sentiment in microblogs: is brevity an advantage? In: Proceedings of the 19th ACM International Conference on Information and Knowledge Management, pp. 1833–1836. ACM (2010)
3. Blei, D.M., Ng, A.Y., Jordan, M.I.: Latent Dirichlet allocation. J. Mach. Learn. Res. **3**, 993–1022 (2003)

4. Boyd, R.L., Wilson, S.R., Pennebaker, J.W., Kosinski, M., Stillwell, D.J., Mihalcea, R.: Values in words: using language to evaluate and understand personal values. In: Ninth International AAAI Conference on Web and Social Media (2015)
5. Brown, M.J., Baillie, M., Fraser, S.: Rating ratemyprofessors.com: a comparison of online and official student evaluations of teaching. Coll. Teach. **57**(2), 89–92 (2009)
6. Coladarci, T., Kornfield, I.: Ratemyprofessors.com versus formal in-class student evaluations of teaching. Pract. Assess. Res. Eval. **12**(6), 1–15 (2007)
7. Felton, J., Koper, P.T., Mitchell, J., Stinson, M.: Attractiveness, easiness and other issues: student evaluations of professors on ratemyprofessors.com. Assess. Eval. High. Educ. **33**(1), 45–61 (2008)
8. Felton, J., Mitchell, J., Stinson, M.: Web-based student evaluations of professors: the relations between perceived quality, easiness and sexiness. Assess. Eval. High. Educ. **29**(1), 91–108 (2004)
9. Freng, S., Webber, D.: Turning up the heat on online teaching evaluations: does hotness matter? Teach. Psychol. **36**(3), 189–193 (2009)
10. Helterbran, V.R.: The ideal professor: student perceptions of effective instructor practices, attitudes, and skills. Education **129**(1), 125 (2008)
11. Kim, S.M., Hovy, E.: Determining the sentiment of opinions. In: Proceedings of the 20th International Conference on Computational Linguistics, p. 1367. Association for Computational Linguistics (2004)
12. Lu, Y., Zhai, C.: Opinion integration through semi-supervised topic modeling. In: Proceedings of the 17th International Conference on World Wide Web, pp. 121–130. ACM (2008)
13. McCallum, A.K.: Mallet: a machine learning for language toolkit (2002). http://mallet.cs.umass.edu
14. Otto, J., Sanford, D.A., Ross, D.N.: Does ratemyprofessor.com really rate my professor? Assess. Eval. High. Educ. **33**(4), 355–368 (2008)
15. Pang, B., Lee, L.: A sentimental education: sentiment analysis using subjectivity summarization based on minimum cuts. In: Proceedings of the 42nd Annual Meeting on Association for Computational Linguistics, p. 271. Association for Computational Linguistics (2004)
16. Pedregosa, F., Varoquaux, G., Gramfort, A., Michel, V., Thirion, B., Grisel, O., Blondel, M., Prettenhofer, P., Weiss, R., Dubourg, V., et al.: Scikit-learn: machine learning in Python. J. Mach. Learn. Res. **12**, 2825–2830 (2011)
17. Stone, P., Earl, B.: A computer approach to content analysis: studies using the general inquirer system. In: Proceedings of the Spring Joint Computer Conference, ACM (1963)
18. Strapparava, C., Valitutti, A.: Wordnet affect: an affective extension of wordnet. In: Proceedings of the 4th International Conference on Language Resources and Evaluation, pp. 1083–1086 (2004)
19. Timmerman, T.: On the validity of ratemyprofessors.com. J. Educ. Bus. **84**(1), 55–61 (2008)
20. Turney, P.D.: Thumbs up or thumbs down? Semantic orientation applied to unsupervised classification of reviews. In: Proceedings of the 40th Annual Meeting on Association for Computational Linguistics, pp. 417–424. Association for Computational Linguistics (2002)
21. Wilson, T., Wiebe, J., Hoffmann, P.: Recognizing contextual polarity in phrase-level sentiment analysis. In: Proceedings of the Conference on Human Language Technology and Empirical Methods in Natural Language Processing, pp. 347–354. Association for Computational Linguistics (2005)

Detecting Coping Style from Twitter

Jennifer Golbeck[✉]

College of Information Studies, University of Maryland,
College Park, MD 20742, USA
jgolbeck@umd.edu

Abstract. *Coping styles* are psychological and behavioral strategies people use to deal with stressful situations. They may be adaptive (helping to reduce stressors), or maladaptive (which tend to reduce symptoms without addressing the underlying problem). Some coping styles—particularly maladaptive ones—are tied to specific conditions.

This study explores whether coping style can be predicted by analyzing user behavior on Twitter. Our results show that a combination of text analysis and behavioral information can be used to build a classifier that can accurately determine whether individuals use primarily adaptive or maladaptive coping styles. Furthermore, we show this can be predicted using a small feature set of psycholinguistic measures, which directly map to core elements of coping as identified in the psychological literature.

In addition to the results contributing to the literature on individual attribute prediction, information about coping strategies is useful for understanding more complex psychological phenomena (like addiction and PTSD). Understanding such attributes is of growing interest to the research community, and our results add a tool to support further work in that area. Our results may also be useful in contributing to personalization, especially in health-related topics, and to a personal analysis tool to guide people toward building healthier coping styles if their current actions are maladaptive.

1   Introduction

Coping styles are psychological and behavioral strategies people use to deal with stressful situations. They may be *adaptive*, helping to reduce stressors; or they may be *maladaptive*, which tend to alleviate symptoms without addressing the underlying problem. Adaptive coping encompasses an analytic approach to problem solving and use of healthy relationships for support [33].

Some coping styles are tied to specific conditions. For example, alcoholism is often tied to and even predicted by an avoidance-based coping style [21,34]. Similarly, PTSD is linked to a dissociative coping style [23].

This study addresses whether someone's coping style can be predicted by their behavior on Twitter. We primarily investigate the linguistic attributes of tweets, including grammatical, cognitive, emotional, and personality-based traits. Using a Naïve Bayes classifier, our results show that we can accurately

E. Spiro and Y.-Y. Ahn (Eds.): SocInfo 2016, Part I, LNCS 10046, pp. 454–467, 2016.
DOI: 10.1007/978-3-319-47880-7_28

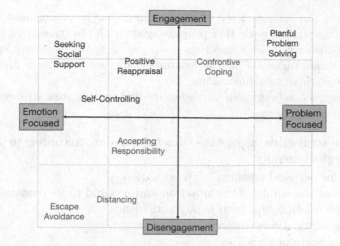

Fig. 1. Various coping styles plotted against dimensions of engagement and focus. Adaptive styles are shown in green and maladaptive in red. (Color figure online)

label people as having primarily *adaptive* or *maladaptive* coping styles. Furthermore, we investigated using only three psycholinguistic features—"thinking style", "analytic thinking", and "adjustment"—which directly map to the core analytical and relationship elements of adaptive coping styles.

Our results show we can achieve good performance with only these three features, *as well as* without them. This indicates that we can achieve the methodologically desirable outcome of building accurate classifiers from features which directly relate to known psychological elements of the attribute being predicted. Furthermore, the fact that achieving good results doesn't *require* these elements also shows that the fundamental psycholinguistic insights are themselves sufficient to gain more complex insights.

In addition to the results contributing to the growing body of work on individual attribute prediction, coping strategies are useful elements of understanding more complex psychological phenomena, like addiction or depression. Understanding these latter attributes is of growing interest to the research community, and our results support further work in that area.

2 Background and Related Work

2.1 Coping Styles

Coping styles describe how people deal with stressful situations [33]. The concept originates with Freud, and the ego's processes for defense and reducing emotional tension [19,20]. Coping styles are conscious choices people make to reduce, overcome, or endure difficult events.

Coping styles can be evaluated on many dimensions. One of the most common is *emotion-focused vs. problem-focused* coping. With emotion-focused styles, people address their emotional reaction to a stressor, where problem-focused copers try to address the stressor itself. Another dimension is *engagement vs. disengagement*. These indicate whether people engage with the stressor or if they try to ignore or push away the stressor.

Figure 1 shows various coping styles, further discussed below in *Methodology*, plotted against these two dimensions.

Coping styles can be broken into *adaptive* and *maladaptive* strategies. There are understood as follows:

– **Adaptive strategies** manage the stressful situation. According to [33], adaptive strategies generally:
  - look for personal meaning in the situation
  - confront the reality of the situation and respond to its demands
  - rely on relationships with family and friends
  - maintain emotional balance
  - and preserve one's self image

These strategies all center around recognizing the problem and dealing with it to resolve the issue.

– **Maladaptive strategies**, on the other hand, push off the stressful situation. People with these strategies may:
  - shift all blame to others
  - assume responsibility through severe self-criticism
  - ignore the problem, or simply hope it will go away

These approaches may temporarily reduce the stress of the situation, but do nothing to handle it long-term.

The importance of coping styles in maintaining psychological health through difficult situations has been widely studied. Since health crises can be major stressors, the medical community has looked at coping styles and their impact across a range of conditions. Studies of coping styles' effect on quality of life and mental health in patients with cancer [37], rheumatoid arthritis [28], HIV/AIDS [2], tinnitus [7], fibromialgya [16], and health threats more broadly [42] are just a few examples of this.

However, coping styles are not limited to health-related crises. Coping styles research has examined how adults deal with serious negative events at work [6,35]. It also is an important factor in how young people deal with bullying [30,45] and childhood sexual abuse [15]. More broadly, how children deal with stress affects their health and happiness as they grow [5].

Maladaptive coping styles are tied to substance abuse [3,4,9,18,29]. Drugs and alcohol are common means of avoiding or distancing oneself from stress. They are also common in people with PTSD, whose coping styles tend toward distancing themselves from trauma [41,44].

This list is far from exhaustive, but it shows the range of issues for which coping styles are important for quality of life.

2.2   Psychological Attribute Prediction from Social Media

There is a growing body of work on predicting personal psychological attributes from social media. These techniques generally rely on machine learning models for prediction, but they work with a variety of different inputs.

Personality traits are the most common in this space. Most research uses the *Big Five* personality traits:

- Openness to new experiences
- Agreeableness
- Conscientiousness (which deals with planning vs. procrastination)
- Neuroticism
- Extroversion

The first work in this space predicted Big Five traits from Facebook profiles, and achieved accuracy rates within about 10 % [25]. Followup work has achieved similar results [32].

Personality was among many traits that [31] showed could be predicted by using just Facebook likes as input. Their followup studies showed that the machine-learning based predictions could actually be more accurate than human predictions [46].

Facebook is not the only data source that supports these predictions. Network analysis has also been effective for predicting personality. In [43], the predictions were made from cellular phone-based social networks. Twitter behavior [1] and interaction patterns [24] have also been effective inputs. Even Instagram pictures have been useful for understanding users' personality traits [14].

Across social media platforms, text analysis is a powerful tool. Many of the references mentioned above used some type of text analysis, and it was the core data input for personality prediction studies on [13,36,40,43].

These tools also support analysis of other psychological traits and states. In [11], authors showed that depression could be predicted from social media. Postpartum depression can also be predicted with surprisingly high accuracy, even before mothers give birth, by analyzing their Twitter behavior patterns [10]. Language analysis of Twitter also allows researchers to measure post-traumatic stress disorder (PTSD) [12].

Taken together, this work shows that social media is a potent source of information for understanding people's psychological conditions and backgrounds. Furthermore, it suggests many more traits may be predictable from data on these platforms—which, in turn, can be used to support tools or more advanced diagnostic or inference techniques.

3   Methodology

3.1   Subjects

We recruited subjects on social media to participate in our study. We posted primarily on Twitter, but also posted some messages on Facebook. Our messages read "Take a quick survey to help find your coping style, and help science along the way! URL". When users completed the study, they were also presented with the option to create a tweet announcing their own coping style and inviting others to find theirs with the message, e.g. "I just found out my coping style is COPING STYLE. Find yours here! URLo".

When new subjects clicked on the study URL, they were presented with a short description of the study, including notification that we would collect their Twitter user ID "so we can see if your coping style comes through in your tweets.". The page contained a link to the IRB consent form, a notification that all information would be kept confidential, and asked subjects to read and agree to the consent form before continuing. They were also informed that the requirement to continue included that they were over 18 and fluent in English.

After agreeing, the subjects took the Ways of Coping survey.

We collected data from 260 people. We eliminated any subjects who did not have valid or publicly accessible Twitter accounts, and whose Twitter accounts were not primarily in English. We further excluded any subjects with fewer than 50 tweets, in order to ensure there was sufficient text available for analysis.

This left us with a remaining total of 105 subjects.

3.2   Classification of Subjects

We used the Ways of Coping survey [17] to determine subjects' coping styles. This is a standard and widely accepted psychological test to measure coping styles.

Subjects are asked to think about a stressful situation they dealt with in the last week. They are then asked to rate how strongly each of 66 statements described the strategy they used, which together indicate how they handled the situation. Ratings were on a 4-point Likert scale with options:

0 does not apply or not used
1 used somewhat
2 used quite a bit
3 used a great deal

Example statements include:

- "I tried to analyze the problem in order to understand it better"
- "I tried to forget the whole thing"
- "I thought about how a person I admire would handle this situation and used that as a model"
- "I refused to believe it had happened"

Each of the 66 statements is associated with one of eight coping styles:

Confrontive Coping
    taking aggressive efforts to change the situation to the point of being risky and aggressive

Distancing

detaching from the situation and minimizing its significance

Escape-Avoidance

avoiding dealing with the situation

Accepting Responsibility

blaming oneself for the situation

Planful Problem Solving

coping through analysis and planning to resolve the situation

Self-Controlling

attempting to control one's own feelings about the stressor

Seeking Social Support

looking to friends for emotional or other types of support

Positive Reappraisal

trying to grow from the experience of dealing with the stress

Scores for each type are the sum of subject's ratings for the associated questions. The raw scores for each coping style indicate how much the subject relied on that strategy. Relative scores, averaged for the number of statements in each category, can indicate which coping style is most prominent for the subject.

Confrontive Coping, Distancing, Accepting Responsibility, and Escape-Avoidance are considered *negative* or *maladaptive* coping styles. The rest are considered *positive* or *adaptive styles* [47].

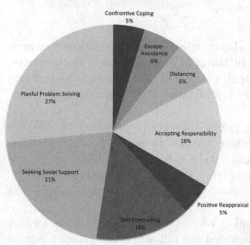

Coping Styles Among Subjects

Confrontive Coping 5%

Escape-Avoidance 6%

Distancing 6%

Planful Problem Solving 27%

Accepting Responsibility 16%

Seeking Social Support 21%

Positive Reappraisal 5%

Self-Controlling 14%

Relative scores for each style are computed by averaging the rating a subject gave to the questions associated with a given coping style. A higher relative score for a given coping style means a person used that style more than others. When assigning a coping style to our subjects based on relative scores, we found adaptive coping was more common than maladaptive styles. Figure 2 shows that distribution.

However, for this study, were interested in classifying on a coarser level—into either adaptive or maladaptive styles, rather than into more specific styles. To classify a subject as having a predominantly adaptive or maladaptive coping style, we summed their ratings for all adaptive coping style

Fig. 2. Frequency of various coping styles among subjects. Green slices are adaptive coping styles and red are maladaptive styles.

questions and all maladaptive coping style questions. We then assigned subjects to the "adaptive" or "maladaptive" coping styles category based on whichever score was highest.

Our features were developed from subjects' tweets. The Twitter API limits the number of tweets available for any one person to 3,200, which set an upper bound on the data we could collect. Based on data from the Twitter API, our subjects averaged 1,904.0 tweets (SD 12,701 / median 2,790), with an average of 30,178.1 words (SD 21,057 / median 36,256).

Each subject's tweets were compiled into a single document. We used the 2015 version Linguistic Inquiry and Word Count (LIWC) [38] tool to analyze subjects' tweets. LIWC is a psycholinguistic text analysis tool that measures how frequently a document uses words in different categories, like "cognitive processes", "emotional words", or different types of grammar.

We chose to consider the following LIWC categories in our analysis:

- language metrics
- function words
- basic grammar
- cognitive processes
- perception words
- punctuation
- time orientation

We also included LIWC's "Receptiviti" traits for Big Five personality traits, and the following features:

- thinking style
- adjustment
- independence
- insecurity

Many previous studies, discussed above, have found correlations between personal attributes and language use in these LIWC categories. Furthermore, research has shown that LIWC-based text analysis of non-social media text can be tied to coping styles. For example, [39] showed the "cognitive processes" category was tied to adaptive coping.

Our analysis included features on personality traits provided by LIWC's Receptiviti API, since research has shown personality is tied to coping styles. Neuroticism is tied to maladaptive coping styles. Meanwhile, Openness, Agreeableness, Conscientiousness, and Extroversion are tied to adaptive coping [8]. We used a total of 58 categories and sub-categories from this automated text analysis.

In addition, we considered the number of friends a person had on Twitter (i.e. a user's "following" count). We excluded the follower count, since a person cannot control who follows them.

3.3   Machine Learning Algorithms

We began by selecting a machine learning algorithm. After testing SVM, regression algorithms, rule-based, and Bayesian algorithms, we found the best performance from the Naïve Bayes implemented as NaïveBayesSimple in Weka [26]. We used a five-fold validation over 100 randomized trials.

4   Results

4.1   Correlations

While our primary goal did not include searching for correlations in the data, several correlations stood out even while statistically correcting for the many attributes in our feature set. Values for the tests discussed in this section are shown in Table 1

For $p < 0.01$ with a Bonferroni correction for significance over 58 attributes, we found several significant differences between subjects with adaptive and maladaptive coping styles. There were differences in values for "Analytical Thinking" and "Thinking Style", which both measure the degree to which someone is an analytic thinker, making decisions based on facts and data, as well as "Adjustment", which measures the degree to which someone is grounded with healthy relationships and life goals. Adaptive copers scored significantly higher for these attributes. This echoes foundational research in this area, which shows that adaptive coping styles are tied to an analytical approach to problem solving and maintaining healthy relationships [33].

4.2   Classification

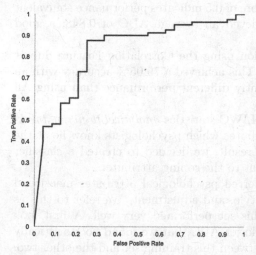

Fig. 3. ROC for prediction of coping styles.

In the grammatical measures of verb, adverb, auxiliary verb, and negation use, maladaptive copers scored significantly higher in each category. These results echo those in the postpartum depression study mentioned above [10]. In that work, researchers found important differences for all of these grammatical measures between mothers in their two groups. While we do not have an explanation for the connection between these language attributes and these two psychological studies, their repeated appearance as important class distinguishers suggests a meaningful connection. This is an interesting area for future psycholinguistic research.

To determine if and how well coping style could be predicted from Twitter, we trained a classifier using the feature set described above. We calculated five performance metrics: accuracy, precision, recall, F1, and ROC Area Under Curve (AUC).

Results are shown in Table 2, and are averaged over all trials.

Table 1. Values for statistically significant differences in language use between subjects with adaptive and maladaptive coping styles.

	Adaptive	Maladaptive
	Avg Percentile	
Analytical thinking	76.8	55.5
Thinking style	84.5	55.0
Adjustment	80.7	49.8
	Avg LIWC score	
Verb	0.11	0.15
Auxiliary verb	0.06	0.08
Adverb	0.04	0.05
Negation	0.01	0.02

Table 2. Performance metrics for Naïve Bayes classifier predicting coping style of subjects.

	Accuracy	Precision	Recall	F1	AUC
All features	77.90 %	0.918	0.787	0.843	0.848
CFS subset	79.05 %	0.856	0.790	0.805	0.861
Thinking/Adjustment	80.19 %	0.811	0.802	0.806	0.779

Using all features returned by the LIWC Receptiviti API, We achieved an accuracy of 77.9 %. The Receiver Operating Characteristic (ROC) curve also illustrates the good performance.

The curve shows the relationship between the true positive rate and false positive rate of a binary classifier. A score of 0.5 indicates performance equivalent to random guessing. Overall, we achieved an average AUC of 0.848: a good performance (see Fig. 3).

We then performed feature selection using the Correlation Feature Subset Evaluation with a Best First method. This achieved a 79.05 % accuracy with an AUC of 0.848. This was not significantly different performance than using All Features over 100 runs.

We found strong correlations with LIWC statistics *analytical thinking, thinking style*, and *adjustment*—three attributes which psychologists know lie at the heart of adaptive coping styles. As a result, we decided to created a classifier that used features known to be relevant to the coping attribute.

We used only the three LIWC-inferred psychological attributes mentioned above: analytical thinking, thinking style, and adjustment. We refer to this as "thinking/adjustment" in Table 2. This set performed very well. A first look suggests it had higher accuracy with lower AUC, but we found no statistically significant differences over 100 runs between this training set and the other two.

Table 3. Confusion matrix for algorithmically selected attributes

Prediction		
NR	R	True Class
62	19	Adaptive
3	21	Maladaptive

5   Discussion

5.1   Prediction from Psychological Foundations

In the space of attribute prediction, it would be methodologically most satisfying to build classifiers from features that directly represent known features related to the attribute. In this case, we are building upon decades of psychological research into coping styles. Foundational psychological research on coping strategies states that analytical thinking and maintaining healthy relationships are at the core of adaptive coping. Thus, we found it particularly encouraging that we could build a classifier with good performance from only the three LIWC features that directly related to the core psychological bases of adaptive coping.

Furthermore, these attributes are themselves developed from additional fundamental psycholinguistic research that ties them to existing word categories in LIWC. This indicates that high-performing classifiers can be combined and used as features for new classifiers that predict increasingly sophisticated traits, which holds promise for researchers continuing work in this area.

The fact that we can achieve results of the same performance *without* those three attributes is also encouraging. As discussed, the three features are built on top of LIWC dictionaries that reflect their respective psycholinguistic components. Thus, it would seem that a learning algorithm could construct a proxy for these composite attributes when the base elements are part of the feature set. Nevertheless, we confirmed equal performance with the core elements alone. This supports the popular approach of research on inferring psychological attributes from social media. It is common to analyze the available text and use the results in a feature set, even in the absence of a mapping from those features to known psychological foundations. Our results support this technique.

5.2   Applications

Computationally, inferred coping style would be a useful input value for any research into predicting psychological attributes and conditions from social media. For example, one of our current projects is examining the recovery rates of alcoholics participating in AA, and hoping to predict success from data present in their social media profiles before they enter recovery. Since we know alcoholism is tied to maladaptive coping styles, inferred coping style could improve prediction. Our early results show coping style, predicted using this classifier, is a highly predictive attribute for inferring alcoholism recovery.

Also, coping affects the way people accept and deal with information (e.g. [42]). Further research shows that presenting stressful information, like risk for developing breast cancer, along with information that guides people away maladaptive coping styles and toward adaptive ones improves their behavior [22]. This suggests that coping style could be used for personalization, particularly on apps and websites related to health, relationships, and other stressful topics. As an example, those with adaptive styles could be given information to encourage their natural tendencies, while those with maladaptive styles could receive direction on how to handle the related stress.

Because coping style is a conscious choice, maladaptive coping styles can be adjusted with therapy. Cognitive-Behavioral Therapy has been used to work with patients on improving their coping styles in the process of treating addiction [27]. While a social media-based predictor should never replace a trained human therapist, insights from social media may be able to augment or bootstrap the treatment process and used to personalize related online content. Furthermore, we described the many medical conditions where an adaptive coping style has been linked to higher quality of life. Offering a one-click analysis of a patient's social media profile may help clinicians recognize when their patients may benefit from referral to therapists to help improve their coping.

5.3   Limitations

The main limitation of this work is a sample size. While we were able to achieve good results in our classifier, we would ultimately like to test this approach with tens of thousands of users. We are encouraged—especially because the classification works well with the psycholinguistic attributes that directly map to known psychological underpinnings of adaptive coping—but ultimately, we want a stronger validation. We hope to build partnerships to collect and analyze larger data sets.

As with all Twitter research, there is the potential for difficulty when people maintain accounts that are professional, themed, or otherwise do not represent their natural communication style. Though many of the linguistic features of LIWC are unconscious and therefore difficult to control, even when the topic is carefully managed, differences in the subject matter on non-personal accounts, where stressful issues potentially being hidden, may create noise for our classifiers. More research is necessary to determine whether these types of accounts cause difficulties for classifiers.

6   Conclusions

We have shown that people's coping styles can be inferred by analyzing their Twitter accounts. Our subjects took the Ways of Coping survey. Based on their results, we labeled their coping style as adaptive or maladaptive. We then collected their most recent 3,200 accessible tweets and analyzed the language using LIWC. We built a feature set from a combination of grammatical features, cognitive and perceptive traits, and inferred personality attributes. Using a Naïve Bayes algorithm, we created a high-performing classifier with 78 % accuracy and an ROC AUC of 0.85.

Further analysis showed that there was no significant loss in performance when the feature set was limited to only three features—thinking style, analytic thinking, and adjustment—which directly mapped to core personality traits tied to coping style. This supports the methodologically attractive approach to building feature sets directly from known psychological or sociological elements related to the attribute being predicted.

Finally, the fact that these three features were not necessary to achieve results also shows that combinations of the underlying features that they were built upon can be used effectively for classification as well. We hope this will be an active area of future work in the attribute prediction research community.

References

1. Adalı, S., Golbeck, J.: Predicting personality with social behavior: a comparative study. Soc. Netw. Anal. Min. **4**(1), 1–20 (2014)
2. Ashton, E., Vosvick, M., Chesney, M., Gore-Felton, C., Koopman, C., Oshea, K., Maldonado, J., Bachmann, M.H., Israelski, D., Flamm, J., et al.: Social support and maladaptive coping as predictors of the change inphysical health symptoms among persons living with hiv/aids. AIDS Patient Care STDs **19**(9), 587–598 (2005)
3. Amy, S., Badura, R.C.R., Altmaier, E.M., Rhomberg, A., Elas, D.: Dissociation, Somatization, Substance Abuse, and Coping in Women with Chronic Pelvic Pain. Obstet. Gynecol. **90**(3), 405–410 (1997)
4. Mark, A.B., Iguchi, M.Y., Lamb, R.J., Lakin, M., Terry, R.: Coping strategies and continued drug use among methadone maintenance patients. Addict. Behav. **21**(3), 389–401 (1996)
5. Boekaerts, M.: Coping with Stress in Childhood and Adolescence. Wiley, New York (1996)
6. Steven, P.B., Westbrook, R.A., Challagalla, G.: Good cope, bad cope: adaptive and maladaptive coping strategies following a critical negative work event. J. Appl. Psychol. **90**(4), 792 (2005)
7. Richard, J.B., Pugh, R.: Tinnitus coping style and its relationship to tinnitus severity and emotional distress. J. Psychosom. Res. **41**(4), 327–335 (1996)
8. Charles, S.C., Connor-Smith, J.: Personality and coping. Ann. Rev. Psychol. **61**, 679–704 (2010)
9. Lynne, M.C., Frone, M.R., Russell, M., Mudar, P.: Drinking to regulate positive and negative emotions: a motivational model of alcohol use. J. Personal. Soc. Psychol. **69**(5), 990 (1995)
10. De Choudhury, M., Counts, S., Horvitz, E.: Predicting postpartum changes in emotion and behavior via social media. In: Proceedings of the SIGCHI Conference on Human Factors in Computing Systems, pp. 3267–3276. ACM (2013)
11. De Choudhury, M., Gamon, M., Counts, S., Horvitz, E.: Predicting depression via social media. In: Proceedings of the 7th International AAAI Conference on Weblogs and Social Media (ICWSM 2013), Boston, MA, USA (2013)
12. Coppersmith, G., Harman, C., Dredze, M.: Measuring post traumatic stress disorder in twitter (2014)
13. Farnadi, G., Zoghbi, S., Moens, M.-F., De Cock, M.: Recognising personality traits using facebook status updates. In: Proceedings of WCPR, pp. 14–18 (2013)
14. Ferwerda, B., Schedl, M., Tkalcic, M.: Predicting personality traits with instagram pictures. In: Proceedings of the 3rd Workshop on Emotions and Personality in Personalized Systems 2015, EMPIRE 2015, pp. 7–10. ACM, New York (2015)
15. Henrietta, H.F., Ullman, S.E.: Child sexual abuse, coping responses, self-blame, posttraumatic stress disorder, and adult sexual revictimization. J. Interpers. Violence **21**(5), 652–672 (2006)
16. Patrick, H., Finan, A.J.Z., Davis, M.C., Lemery-Chalfant, K., Covault, J., Tennen, H.: COMT moderates the relation of daily maladaptive coping and pain in fibromyalgia. PAIN **152**(2), 300–307 (2011)

17. Folkman, S., Lazarus, R.S.: Ways of Coping Questionnaire. Consulting Psychologists Press, Palo Alto (1988)
18. Franken, I.H.A., Hendriks, V.M., Haffmans, P.M., van der Meer, C.W.: Coping style of substance-abuse patients: effects of anxiety and mood disorders on coping change. J. Clin. Psychol. **57**(3), 299–306 (2001)
19. Freud, A.: The Ego and the Mechanisms of Defence. Karnac Books, London (1946)
20. Freud, S.: Splitting of the ego in the defensive process. Int. J. Psychoanal. **22**, 65–68 (1938)
21. Fromme, K., Rivet, K.: Young adults' coping style as a predictor of their alcohol use and response to daily events. J. Youth Adolesc. **23**(1), 85–97 (1994)
22. Rachel, B.F., Prentice-Dunn, S.: Effects of coping information and value affirmation on responses to a perceived health threat. Health Commun. **17**(2), 133–147 (2005)
23. Gil, S., Caspi, Y.: Personality traits, coping style, and perceived threat as predictors of posttraumatic stress disorder after exposure to a terrorist attack: a prospective study. Psychosom. Med. **68**(6), 904–909 (2006)
24. Golbeck, J., Robles, C., Edmondson, M., Turner, K.: Predicting personality from Twitter. In: 2011 IEEE Third Inernational Conference on Social Computing (SocialCom), 2011 IEEE Third International Conference on Privacy, Security, Risk and Trust (PASSAT), pp. 149–156. IEEE (2011)
25. Golbeck, J., Robles, C., Turner, K.: Predicting personality with social media. In: CHI 2011 Extended Abstracts on Human Factors in Computing Systems, pp. 253–262. ACM (2011)
26. Hall, M., Frank, E., Holmes, G., Pfahringer, B., Reutemann, P., Witten, I.H.: The weka data mining software: an update. ACM SIGKDD Explor. Newslett. **11**(1), 10–18 (2009)
27. Kadden, R.: Cognitive-Behavioral Coping Skills Therapy Manual: A Clinical Research Guide for Therapists Treating Individuals with Alcohol Abuse and Dependence. DIANE Publishing, Collingdale (1995)
28. Keefe, F.J., Brown, G.K., Wallston, K.A., Caldwell, D.S.: Coping with rheumatoid arthritis pain: catastrophizing as a maladaptive strategy. Pain **37**(1), 51–56 (1989)
29. Kelley, J.: Stress and coping behaviors of substance-abusing mothers. J. Special. Pediatr. Nurs. **3**(3), 103 (1998)
30. Kochenderfer-Ladd, B.: Peer victimization: the role of emotions in adaptive and maladaptive coping. Soc. Dev. **13**(3), 329–349 (2004)
31. Kosinski, M., Stillwell, D., Graepel, T.: Private traits and attributes are predictable from digital records of human behavior. Proc. Natl. Acad. Sci. **110**(15), 5802–5805 (2013)
32. Markovikj, D., Gievska, S., Kosinski, M., Stillwell, D.: Mining facebook data for predictive personality modeling. In: Proceedings of the 7th International AAAI Conference on Weblogs and Social Media (ICWSM 2013), Boston, MA, USA (2013)
33. Moos, R.: Coping with Life Crises: An Integrated Approach. Springer, Heidelberg (1976)
34. Rudolf, H.M., Moos, B.S.: Rates and predictors of relapse after natural and treated remission from alcohol use disorders. Addiction **101**(2), 212–222 (2006)
35. Parasuraman, S., Hansen, D.: Coping with work stressors in nursing effects of adaptive versus maladaptive strategies. Work Occup. **14**(1), 88–105 (1987)
36. Park, G., Schwartz, H.A., Eichstaedt, J.C., Kern, M.L., Kosinski, M., Stillwell, D.J., Ungar, L.H., Seligman, M.E.P.: Automatic personality assessment through social media language. J. Personal. Soc. Psychol. **108**(6), 934 (2015)
37. Parle, M., Jones, B., Maguire, P.: Maladaptive coping and affective disorders among cancer patients. Psychol. Med. **26**(04), 735–744 (1996)

38. Pennebaker, J.W., Francis, M.E., Booth, R.J.: Linguistic Inquiry and Word Count: LIWC 2001, vol. 71. Lawrence Erlbaum Associates, Mahwah (2001)
39. James, W.P., Mayne, T.J., Francis, M.E.: Linguistic predictors of adaptive bereavement. J. Personal. Soc. Psychol. **72**(4), 863 (1997)
40. Quercia, D., Kosinski, M., Stillwell, D., Crowcroft, J., Our twitter profiles, our selves: Predicting personality with twitter. In: 2011 IEEE Third Inernational Conference on Social Computing (SocialCom), 2011 IEEE Third International Conference on Privacy, Security, Risk and Trust (PASSAT), pp. 180–185. IEEE (2011)
41. Reynolds, M., Brewin, C.R.: Intrusive cognitions, coping strategies and emotional responses in depression, post-traumatic stress disorder and a non-clinical population. Behav. Res. Ther. **36**(2), 135–147 (1998)
42. Patricia, A.R., Rogers, R.W.: Effects of components of protection-motivation theory on adaptive and maladaptive coping with a health threat. J. Personal. Soc. Psychol. **52**(3), 596–604 (1987)
43. Staiano, J., Lepri, B., Aharony, N., Pianesi, F., Sebe, N., Pentland, A.: Friends don't lie: Inferring personality traits from social network structure. In: Proceedings of the 2012 ACM Conference on Ubiquitous Computing, UbiComp 2012, pp. 321–330. ACM, New York (2012)
44. Sarah, E., Ullman, M.R., Peter-Hagene, L., Vasquez, A.L.: Trauma histories, substance use coping, PTSD, and problem substance use among sexual assault victims. Addict. Behav. **38**(6), 2219–2223 (2013)
45. Vosvick, M., Gore-Felton, C., Koopman, C., Thoresen, C., Krumboltz, J., Spiegel, D.: Maladaptive coping strategies in relation to quality of life among HIV+ adults. AIDS Behav. **6**(1), 97–106 (2002)
46. Youyou, W., Kosinski, M., Stillwell, D.: Computer-based personality judgments are more accurate than those made by humans. Proc. Natl. Acad. Sci. **112**(4), 1036–1040 (2015)
47. Zeidner, M., Saklofske, D.: Adaptive and Maladaptive Coping. Wiley, NewYork (1996)

User Privacy Concerns with Common Data Used in Recommender Systems

Jennifer Golbeck(⊠)

College of Information Studies, University of Maryland,
College Park, MD 20742, USA
jgolbeck@umd.edu

Abstract. Recommender systems, and personalization algorithms more broadly, have become an integral part of modern e-commerce, streaming, and social media services. Collaborative filtering in particular leverages users' ratings to compute new items of interest. The algorithms that drive them use a variety of data, from user ratings to measures of social relationships. As a field, we have built more effective, accurate algorithms with the available data. However, recommender systems are often opaque to users, and users' privacy concerns about the data these algorithms use is unknown.

In this project, we administered a survey to nearly 1,000 subjects to gauge their opinions about privacy issues tied to a variety of common personal data points used in making recommendations and the ways that data is used. We found that data collected within in an application is generally of low concern, while the use of social data and data obtained from third parties is often considered a privacy violation. Furthermore, users expressed discomfort with their data being used anonymously to help personalize content for others - a common practice in collaborative filtering. We discuss the survey results and implications for creating privacy-respecting recommender systems.

1 Introduction

Recommender systems have become a critical component of the modern web, supporting users interacting in a vast space of online information. Whether it's media, shopping, or other online content, there are often too many items for users to sort through themselves, and personalization algorithms provide a valuable tool for finding items of interest. They also help businesses attract customers to new content, increasing the value of their services.

These algorithms are driven by information about users. Whether it is user ratings, purchase histories, or profiles that support measures of similarity with others, user data is critical for personalization. With that comes privacy concerns. Users may find it creepy, intrusive, or unsettling that their data is used in this way, often without their explicit consent, awareness, or understanding. At its core, privacy is the ability to control what information we share, when, and with whom. In online systems, users may have varying levels of awareness about

© Springer International Publishing AG 2016
E. Spiro and Y.-Y. Ahn (Eds.): SocInfo 2016, Part I, LNCS 10046, pp. 468–480, 2016.
DOI: 10.1007/978-3-319-47880-7_29

what information is collected and how it is used. As a result, they may find certain aspects of recommendation algorithms to be privacy violating because they use data the user may never have chosen to share.

In this study, we investigate users' privacy concerns regarding common personal data used by recommender systems. This includes item ratings and reviews, purchase and browsing history, social media data, and social networks. We administered a survey through Amazon Mechanical Turk and obtained valid results from 983 subjects. For 19 different types of personal information, they rated their level of comfort and feelings of privacy violation when the data was used for personalization. Subjects also rated their feelings about the ways their data might be used within an application.

Our results show that users have a range of feelings about the data recommender systems use, from generally low levels of concern about ratings and reviews to very serious concerns about common social network information used in some algorithms. We present the results of this survey, highlight important points of concern for users, and discuss the implications for creating privacy-respecting recommender systems going forward.

2   Related Work

Questions of privacy issues in recommender systems are not new; back in 2001 John Riedl edited a special issue of IEEE Internet Computing [12] that included a focus on the privacy risks associated with recommender systems and personalization. Articles, including [11], looked at risks to user privacy from these algorithms.

Work on users' experiences with recommender systems has highlighted the tension users feel when balancing the benefits of personalization and the desire to keep their personal information private [8]. Some privacy concerns are based in worries about *exposure* of personal information [13], though users may also have fundamental concerns about their personal data being collected and used even without the exposure risk.

User privacy preferences were analyzed in depth a decade ago in [1], which looked at how people's perception of information transparency related to their preferences regarding personalization. They found that those who wanted more transparency were, in turn, less willing to be profiled. The authors recommended companies focus on personalization for those who were more willing to be profiled, yet we have not seen such a discerning approach put into practice. No major systems limit personalization for more privacy-sensitive customers.

3   Methodology

3.1   Survey Instrument

We began by presenting subjects with the following scenario:

Imagine you are using a website or have installed an app on your smartphone or tablet. With it, you can watch movies, listen to music, and read e-books. Apps (and websites) like this often collect data about you to personalize your experience. The personalization might be suggesting things for you to buy, showing genres you might like, or changing the order of search results to show things you might like first. We want to know how you feel about apps like this collecting and using data about you. When answering the following questions, please consider only how the data would be used within the app (not how it could be shared with anyone outside the app). Please rate how you would feel about the following types of data being used by the app in the following ways:

We then presented subjects with a list of data types. These are shown in Table 1 with the same phrasing used in the survey. For each of these data types, they were shown the following statements and asked to rate them on a 7-point Likert scale.

- How sensitive is this information to you? (not at all sensitive to very sensitive)
- I am ok with the app collecting and using this data to personalize my experience. (strongly disagree - strongly agree)
- It violates my privacy for the app to collect this data. (strongly disagree - strongly agree)
- I benefit from the app using this data to personalize my experience. (strongly disagree - strongly agree)
- Please select any of the following words that describe how it feels to know the app is collecting this data and using it for personalization: (check all that apply): *creepy helpful, intrusive, valuable, insightful, reasonable*

For the word-choice item, the order of the words was randomized.

We also wanted to know how users felt about uses of their personal data created in the app/website. We presented the following scenario:

Now imagine this app is only collecting data from how you use the app and not from any other sources. The collected data would include your search history in the app, what products you viewed and for how long, what products you purchased, and everything you've rated and reviewed.

We asked subjects "How you would feel if the app used that data for the following purposes", and we showed them the following applications:

- Recommending products I might be interested in
- Customizing my search results
- For internal analysis or research
- Anonymously, as background information to personalize other people's experiences.

For each application type, rated the following statements on a 7-point Likert scale.

Table 1. The list of data types presented to subjects, and their ratings. All values are on a 1–7 scale, with 4 representing neutral. Values range from "Not at all sensitive" to "Very Sensitive" for the sensitivity statement and from "Strongly Disagree" to "Strongly Agree" for the three statements.

	How sensitive	I'm Ok with this	Violates privacy	I benefit
App data				
The products I have purchased through the app	3.55	4.60	3.22	4.72
My search history on the app	4.19	3.96	3.79	4.07
The products I have viewed (but not necessarily purchased) on the app	3.38	4.31	3.42	4.03
My product ratings in the app	2.21	5.50	2.36	4.64
My product reviews in the app	2.34	5.38	2.41	4.56
The app using data from other sites				
The history of products I have purchased on other sites	5.02	2.59	5.19	2.98
My web search history	5.78	1.97	5.81	2.38
The history of products I have viewed on other sites (but not necessarily purchased)	4.83	2.59	5.16	2.75
My reviews of products on other sites	3.33	3.71	3.98	3.32
The app using data from social media				
Data from my social media profiles and posts (e.g. Facebook, Twitter, Instagram, etc.)	5.68	1.84	5.89	2.08
Observations recorded by a representative from the company who comes and sits in my house, watching me use the app 24 h a day	6.67	1.11	6.71	1.42
(Note: this question was included as a check on Mechanical Turk workers. We will discuss it more below.)				
The app using data about my friends				
Lists of my friends on social media	5.59	1.81	5.86	1.73
Information about my relationships with each of my friends, like who I trust most	6.47	1.33	6.47	1.42
My friends purchasing, browsing, and/or search history in the app	5.48	1.62	5.51	1.63
My friends ratings and reviews on this app	3.58	3.49	3.74	2.96
My friends ratings and reviews on other websites or apps	3.35	3.76	3.99	3.34
Other data				
My contact list	6.27	1.53	6.25	1.62
My location	5.28	2.79	5.08	3.06
Data purchased from data brokers (i.e. companies who sell information about you)	6.13	1.53	6.24	1.80

- I am ok with the app using my data in this way. (strongly disagree - strongly agree)
- It violates my privacy for the app to use my data in this way. (strongly disagree - strongly agree)
- I benefit from the app using my data in this way. (strongly disagree - strongly agree)

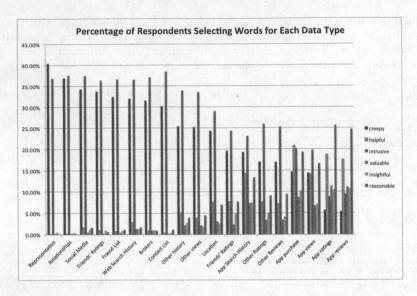

Fig. 1. Percentage of subjects who selected each word - *creepy, helpful, intrusive, valuable, insightful, reasonable* - for each data type. Data types are sorted from highest to lowest value of "creepy".

– Please select any of the following words that describe how it feels to know the app is collecting this data and using it for personalization: (check all that apply): *creepy, helpful, intrusive, valuable, insightful, reasonable*

3.2   Subjects

We recruited subjects from Amazon Mechanical Turk. They were compensated $1.75 for their participation, which worked out to an average hourly rate of $7/h.

We included a question to test attention (asking users to enter all "3" values in place of showing them a data point to rate). After rejecting users who failed that test, we had 1,086 responses. However, as mentioned above, we included a data point that described sending a representative to sit in their house and observe them 24-h a day. This was to discover users who had passed our attention test but were otherwise answering randomly or incorrectly. 103 subjects gave a rating of neutral to positive for the statement "I am ok with the app collecting and using this data to personalize my experience."

We believe basically no one would truly be ok with (or neutral about) a stranger sitting in their house and watching them all day. While there may be exceptions, we believe it was sound to exclude these subjects from consideration either because they were not answering honestly or they may have been confused about how to answer. We also note that while we believe it is methodologically sound to exclude these users who failed the test, it had little effect on the data. The correlation between average ratings of the included and excluded subjects was >.98 for all data types and statements (excluding the differentiating question), and the average difference in their ratings for a statement-data type combination was less than 7 %.

This left us with a total of 983 subjects. The average age was 35.3 (stdev = 11.3) and 52.2 % were male.

We asked subjects to rate how tech savvy they felt themselves to be on a scale from 1 to 5. The average was 3.7 (stdev = 0.9) and skewed heavily to the high end, with only 2.3 % of users rating themselves below average.

We also asked the following three questions to classify users as Privacy Fundamentalists, Privacy Pragmatists, or Privacy Unconcerned using a slightly modified version of the Westin-Harris Privacy Inventory. Our modifications added "online" before mentions of businesses and changed "consumer privacy" to "online user privacy". Subjects rated each on a 5-point Likert scale from strongly disagree (1) to strongly agree (5):

- People have lost all control over how personal information is collected and used by online companies.
- Most online businesses handle the personal information they collect about users in a proper and confidential way.
- Existing laws and organizational practices provide a reasonable level of protection for online user privacy today.

Overall 23.3 % of subjects were Privacy Fundamentalists, 70.1 % were Privacy Pragmatists, and only 5.6 % were Privacy Unconcerned.

We found no statistically significant relationships among the Westin-Harris Privacy Type, age, and gender. Self-rated tech savviness did not have any significant relationship to subjects' ratings.

4   Results

4.1   Data Collection

The average values for the four statements we asked subjects to rate were highly correlated. The sensitivity of data and the perception that it violates subjects' privacy were almost perfectly correlated ($\rho = 0.98$), as were the average ratings for subjects being ok with data points being collected and their perception that it benefits them ($\rho = 0.97$). The negative correlations were also nearly perfect, with all remaining pairs of Pearson correlation coefficients <-0.95, except for the relationship between perceived sensitivity and benefit ($\rho = -0.91$). Because of this, we will generally talk about perceived privacy violations in the following sections, and this will track with all the other scores.

Overall, ratings and reviews as well as in-app activity was perceived to be on the less sensitive and less privacy violating end of the scale, with users generally stating they were ok with the data being collected and seeing a benefit in it. The one exception was in-app search history, which had average ratings right at 4 (a neutral score on our scale) for all four statements.

Table 2. Box plots for subject ratings of four data collection questions. The rating scale is 1–7, with 1 representing "strongly disagree" on the three statements and "not at all sensitive" on the sensitivity question. A rating of 4 would represent a neutral opinion. Each data type is shown on the x-axes. Each graph has the data types sorted from lowest average score to highest average score.

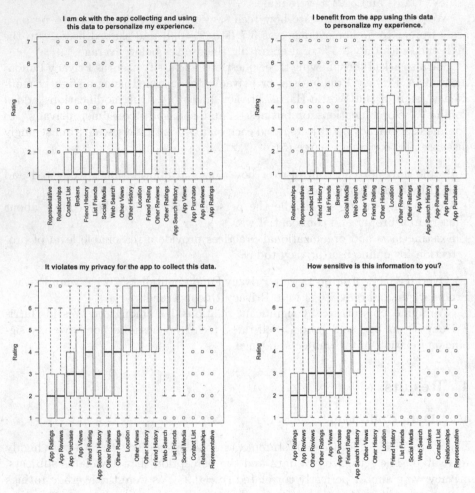

Not surprisingly, our test question about a company representative visiting subjects at home to gather data was viewed as a strong privacy violation. We have kept that question in the data presented here because it provides an interesting point of comparison. Users rated data collection about their relationships with others, access to their contact lists, and data purchased from data brokers among the most privacy violating data that an app could collect. Data from social media, including profiles and friend lists, along with subjects' web search history were also given high ratings for their privacy violating nature. Scores for all data points and statements are shown in Table 1. Boxplots of the rating distributions for each data type and statement are shown in Table 2.

Table 3. Box plots for subject ratings of four usage questions. X-axis labels correspond to the data usage questions described above. The rating scale is 1–7, with 1 representing "strongly disagree". A rating of 4 would represent a neutral opinion. Each data type is shown on the x-axis. Each graph has the data types sorted from lowest average score to highest average score.

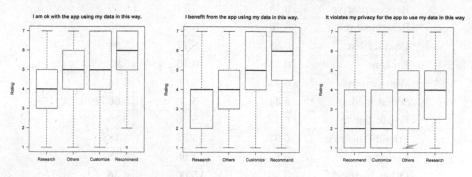

The words people chose to describe collection of each data type was strongly correlated with their ratings for that item. The absolute value of the correlation between the average privacy violation ratings and the percentage of users who used a word was >0.92 for all words. The data types perceived as more privacy violating received more frequent descriptions as "creepy" (our most popular term, by far) and intrusive, and lower ratings of "helpful", "valuable", "insightful", and "reasonable". Figure 1 shows the percentage of users who selected each word for each data type, as well as the relationship among the word choices.

4.2 Data Use

The survey also asked subjects to share their feelings about how uses of data collected within the app itself. Overall, people were not deeply concerned about use of this data. Average ratings were above 4 (neutral) for "I am ok with the app collecting and using this data to personalize my experience." except for using the data to personalize others' experiences. In that case, the average rating was 3.60, slightly below neutral. Similarly, people tended to disagree that any uses violated their privacy, with average scores under 4, except on the use for personalizing others' experiences, where it was 4.07. Subjects saw the benefits of customized search and recommending products, agreeing that they benefitted with averages of 4.93 and 5.25 respectively, but did not see a benefit in using their data to help recommend products to others (2.84) or for research (3.51). Boxplots of these results are shown in Table 3.

Table 4. Average ratings from three types of privacy concern levels on the question of whether collecting these data points is a privacy violation. With the exception of a visiting representative where there was no difference, Privacy Fundamentalists have higher ratings than Privacy Pragmatists on every data point. They also have higher ratings than Privacy Unconcerned subjects on all points except for the Friend List. A * indicates data types where Privacy Pragmatists have significantly higher ratings than the Privacy Unconcerned. These occur on the lower-concern points.

Data type	Fundamentalist	Pragmatist	Unconcerned
Representative	6.78	6.71	6.50
Relationships	6.70	6.40	6.45
Contact list	6.58	6.17	5.84
Brokers	6.53	6.16	5.95
Web search	6.42	5.66	5.14
Social media	6.31	5.76	5.66
Friend list	6.18	5.76	5.86
Friends' history	5.97	5.38	5.18
Location*	5.66	4.95	4.38
Other history*	5.96	5.00	4.36
Other views*	5.85	4.99	4.38
App search history *	4.45	3.63	3.02
Friends' ratings*	4.19	3.65	3.04
App purchases	3.77	3.10	2.55
App views*	3.97	3.30	2.56
Other ratings*	4.67	3.85	3.05
Other reviews *	4.61	3.86	2.91
App reviews*	2.74	2.36	1.75
App ratings*	2.69	2.29	1.77

4.3 Breakdown by Privacy Class

As described above, we had subjects take a modified version of the Westin-Harris Privacy Index to rate them as "Privacy Fundamentalists", who are most concerned with protecting and controlling their data; "Privacy Pragmatists", who tend to believe there are compromises to be made between keeping data private and receiving benefits from its use; and "Privacy Unconcerned", people who generally are not worried about their personal data privacy. In our results, 23.6 % of people were Privacy Fundamentalists, 70.9 % were Privacy Pragmatists, and 5.6 % were Privacy Unconcerned.

We broke down the survey responses based on which group subjects fell into. Not surprisingly, Privacy Fundamentalists expressed more concern about data collection than Privacy Pragmatists, who in turn had greater concern than the

Privacy Unconcerned. Table 4 shows the average answers to the statement "It violates my privacy for the app to collect this data." for each data point.

Overall, Privacy Fundamentalists were more likely to believe that data collection was a privacy violation for all points except for our test question about the representative visiting (where all users felt equally violated by the idea). An ANOVA showed significant differences among the populations for each other datapoint. We performed pairwise t-tests between each group. Privacy Fundamentalists gave significantly higher ratings of privacy violation vs Privacy Pragmatists for every data type ($p < 0.05$). Fundamentalists' ratings were also higher than Privacy Unconcerned on every data type except for Friend Lists, where there was no significant difference. On the items of higher privacy concern, there was no significant difference between Privacy Pragmatists and Privacy Unconcerned subjects, but there were significant differences on all the lower-concern items. These are indicated with a * in Table 4.

On average, Privacy Fundamentalists tended to view every data point as a privacy violation except for ratings and reviews and in-app views and purchases. The Privacy Pragmatists and Privacy Unconcerned track the Privacy Fundamentalists' ratings, but with lower levels of concern. Both of these latter groups' average ratings indicate that friends' ratings of items and in-app search history are not considered privacy violations.

5 Discussion

Overall, in-app activity was considered less privacy violating than other data, and ratings and reviews were generally not considered privacy violating. This is good news for recommender systems researchers, since this data lies at the core of many algorithms. Similarly, using someone's friends' ratings and reviews was not considered privacy violating, which is good news for many socially-oriented recommender systems. That said, when asked about their own ratings and reviews being used anonymously to personalize content for *other* users, subjects were less enthusiastic. On average, they disagreed that they would be ok with this kind of data being used, they were neutral with a slight lean toward believing this use violated their privacy, and they disagreed that such usage benefitted them. 19.9 % of users said this application would be "intrusive" and 14.3 % said it was "creepy".

This is a point worth exploring further in future research. Are people resistant to their own data being used in applications like collaborative filtering because they did not explicitly consent? Is it based on a misunderstanding of the technology? Do they want their personal data used only for personal applications? Do they have concerns that their personal data will be revealed to others? Further work will be necessary to understand the underlying issues.

We were most surprised by the consistent result that subjects believed that using information about their relationships with their friends was a serious privacy violation. When asked about whether this was a privacy violation, this received an average rating of 6.47 on a 7 point scale, scoring just 0.24 lower than the privacy violation of having a company representative sitting in your house 24 h a day. When asked if they were ok with this data being collected, it received a score of only 1.33, with a minimum of 1 - again, only a scant 0.22 higher than the rating given for the visiting representative. Subjects believed this data was intensely sensitive, rating it a 6.47 on a 1–7 scale.

This data is, of course, the basis for trust-based recommender systems (e.g. [2,6,9,10,14] to list just a few). This type of information is also commonly used in social recommender systems that suggest friends and filter content in social networking systems (e.g. [4,7]). This level of user anxiety about relationship data is concerning, as trust-based recommender systems grow increasingly popular as a way to leverage the wealth of data available in social networks to improve the user experience.

Subjects also believe that their phones' contact lists were highly sensitive and that their use by apps was a privacy violation. Yet this is one of the most commonly accessed "sensitive" data points, according to PrivacyGrade[1], accessed by thousands of apps. These are not necessarily using contacts for personalization, though some are.

Subjects also found the use of information from data brokers highly sensitive and privacy violating. Again, companies that use this data are not always using it for making recommendations or doing personalization. Many do, however, especially in personalizing advertising content. This includes very large players, like Facebook [3].

Ultimately, we believe the results of this large survey raise questions about how the personalization community should respond to people's privacy concerns. Many of the data types we commonly collect and use in our algorithms are seen by users as privacy violating. It is not just the data points with high concern levels that we need to worry about. Even when the average ratings for a data type were favorable, a significant portion of subjects disagreed. For example, 12 % of subjects felt that using in-app ratings was a privacy violation, 13 % felt using in-app reviews was a violation, and 25 % believed using an in-app purchase history was a privacy violation. These are among the most common types of data we use for generating recommendations, and a large minority sees their use as invasive.

Do we ignore these preferences, believing the benefit our systems offer outweighs users' privacy concerns? Do we violate their privacy in pursuit of system performance, subscriber retention, or sales? Do we try to be more transparent about how these algorithms work in order to convince users that we are not intruding on a private space? These are open questions, but the results we found in this survey make a strong point that average users have many privacy concerns about how recommender systems work - concerns that many of us may have been unaware of.

We note that a sizable minority of users objected to nearly all data uses; 4.6 % felt that collection of *every* data type was a privacy violation, and 8.6 % rated every data type neutral to privacy violating. These users could still benefit

---

[1] http://www.privacygrade.org/.

from item-based recommender systems while having their privacy preferences respected. Privacy-personalized personalization would mark a new approach for companies, who would be choosing to ignore data that could improve recommender performance in order to respect users' preferences.

A privacy-aware approach to building recommender systems could require new evaluation metrics that include privacy-respecting behavior as a factor in determining performance. Ultimately, though, this is a philosophical question before it is a technical one. Are user's preferences important enough that recommender systems should respect them at the cost of traditional performance? If yes, then there is work to be done in developing new algorithms and performance measures. If not, we must acknowledge and accept that our algorithms are not so much designed to improve the whole user experience (of which privacy is a part) as they are designed to optimize a mathematical measure.

6   Conclusions

Recommendations and personalization are important in the ever-growing space of online content. At the same time, because these algorithms use personal data in their computations, there are privacy implications for users. In this study, we report on the results of a survey of 983 people who stated their privacy concerns regarding a variety of common data types used for personalization.

Subjects had a range of concerns about the data used in many recommender systems. Data collected within an application along with ratings and reviews were generally considered less sensitive and less privacy violating, while users felt data obtained from social media and third parties was particularly intrusive. Subjects also expressed hesitation at the idea of their in-app data being used anonymously in algorithms that personalize content for other users - the heart of collaborative filtering algorithms.

We believe these results should prompt a discussion about the role of privacy in the recommender systems community. We were surprised at the level of concern that subjects had about data commonly used in recommendation algorithms. Should we continue to use this data with no regard for user preferences? Should systems develop mechanisms for users to permit or block algorithm's access to certain data points? Should we start considering privacy-respecting behavior as a component of recommender system evaluation? These are questions the community should strive to answer, especially since recommender systems are considered a feature that enhances the user experience.

Acknowledgements. Thanks to Michael Ekstrand and Ingo Burghardt for their comments on early drafts of this survey, and to Jessica Vitak and Katie Shilton for advice on how to handle mturk workers who want company representatives in their houses.

References

1. Awad, N.F., Krishnan, M.S.: The personalization privacy paradox: an empirical evaluation of information transparency and the willingness to be profiled online for personalization. MIS Q. **30**, 13–28 (2006)

2. Bedi, P., Kaur, H., Marwaha, S.: Trust based recommender system for semantic web. IJCAI **7**, 2677–2682 (2007)
3. Bodle, R.: Predictive algorithms and personalization services on social network sites. Ubiquit. Internet: User Ind. Perspect. **25**, 130 (2014)
4. Chen, J., Geyer, W., Dugan, C., Muller, M., Guy, I., Make new friends, but keep the old: recommending people on social networking sites. In: Proceedings of the SIGCHI Conference on Human Factors in Computing Systems, pp. 201–210. ACM (2009)
5. Elmisery, A.M., Botvich, D.: An agent based middleware for privacy aware recommender systems in IPTV networks. In: Watada, J., Phillips-Wren, G., Jain, L.C., Howlett, R.J. (eds.) Intelligent Decision Technologies. SIST, vol. 10, pp. 821–832. Springer, Heidelberg (2011)
6. Golbeck, J., Hendler, J., et al.: Filmtrust: movie recommendations using trust in web-based social networks. In: Proceedings of the IEEE Consumer Communications and Networking Conference, vol. 96, pp. 282–286. Citeseer (2006)
7. Guy, I.: Social recommender systems. In: Ricci, F., Rokach, L., Shapira, B. (eds.) Recommender Systems Handbook, pp. 511–543. Springer, New York (2015)
8. Knijnenburg, B.P., Willemsen, M.C., Gantner, Z., Soncu, H., Newell, C.: Explaining the user experience of recommender systems. User Model. User-Adap. Inter. **22**(4–5), 441–504 (2012)
9. Massa, P., Avesani, P.: Trust-aware collaborative filtering for recommender systems. In: Meersman, R. (ed.) OTM 2004. LNCS, vol. 3290, pp. 492–508. Springer, Heidelberg (2004)
10. O'Donovan, J., Smyth, B.: Trust in recommender systems. In: Proceedings of the 10th International Conference on Intelligent User Interfaces, pp. 167–174. ACM (2005)
11. Ramakrishnan, N., Keller, B.J., Mirza, B.J., Grama, A.Y., Karypis, G.: Privacy risks in recommender systems. IEEE Internet Comput. **5**(6), 54 (2001)
12. Riedl, J.: Personalization and privacy. IEEE Internet Comput. **5**(6), 29–31 (2001)
13. Lam, S.K.T., Frankowski, D., Riedl, J.: Do you trust your recommendations? An exploration of security and privacy issues in recommender systems. In: Müller, G. (ed.) ETRICS 2006. LNCS, vol. 3995, pp. 14–29. Springer, Heidelberg (2006)
14. Walter, F.E., Battiston, S., Schweitzer, F.: A model of a trust-based recommendation system on a social network. Auton. Agents Multi-Agent Syst. **16**(1), 57–74 (2008)

How a User's Personality Influences Content Engagement in Social Media

Nathan O. Hodas[1(✉)], Ryan Butner[2], and Court Corley[1]

[1] Pacific Northwest National Laboratory, Richland, WA, USA
{nhodas, court}@pnnl.gov
[2] Monsanto, St. Louis, MO, USA
ryan.scott.butner@monsanto.com

Abstract. Social media presents an opportunity for people to share content that they find to be significant, funny, or notable. No single piece of content will appeal to all users, but are there systematic variations between users that can help us better understand information propagation? We conducted an experiment exploring social media usage during disaster scenarios, combining electroencephalogram (EEG), personality surveys, and prompts to share social media, we show how personality not only drives willingness to engage with social media, but also helps to determine what type of content users find compelling. As expected, extroverts are more likely to share content. In contrast, one of our central results is that individuals with depressive personalities are the most likely cohort to share informative content, like news or alerts. Because personality and mood will generally be highly correlated between friends via homophily, our results may be an import factor in understanding social contagion.

1 Introduction

Whether for disaster response, advertising campaigns, or general entertainment, people leverage social media to spread information to wide and varied audiences. When crafting a message on social media, authors may attempt to consider humor (Evers et al. 2013), trustworthiness (Kietzmann et al. 2011), or timeliness (Lee and Ma 2012), among other factors, to increase the reach of their message. Authors may not consider the personality or mood of target users when anticipating the impact and propagation of their messages. Systematic biases in target populations will confound attempts to understand social contagion (Hodas and Lerman 2014). Because of homophily, personality types will not be randomly distributed in the social network, and users will be exposed to content biased by the personality of their friends (Hodas et al. 2013). It is important to better understand the link between personality, mood and social contagion.

In this paper, we reveal a systematic link between personality type and mood, brain response, and the type of content people choose to share online. Although it comes as no surprise that there is a relationship between how someone uses social media and their personality (Ryan and Xenos 2011; Correa et al. 2010; Hughes et al. 2012), this is the first experiment that measured both the user's present mood and personality, quantitative measures of engagement and interest, as well as their final reactions to the content. We originally conducted this research in the context of understanding user's responses

© Springer International Publishing AG 2016
E. Spiro and Y.-Y. Ahn (Eds.): SocInfo 2016, Part I, LNCS 10046, pp. 481–493, 2016.
DOI: 10.1007/978-3-319-47880-7_30

to natural disasters via social media. In the methods section below we explore this experiment in detail, including how users were assessed for personality and mood, were shown videos describing the disasters, then asked to share (or not) tweets and emergency alerts, all while being continuously monitored via electroencephalogram (EEG). In this way, we have quantitative measures of personality, attention, and action.

The main finding of this paper is that users systematically prefer different types of content, and that this content depends on their personality and mood in significant ways. The different types of content, such as "informative", "social", or "sympathetic," which we describe below, each resonate differently depending on personality and mood. For example, as one would expect, extroverts are more likely to share any content, consistent with previous findings. However, we also find that users that score highest on measures of depression were more likely to share informative messages, compared to the least depressed users. Because of correlation between content, type of information, and personality, we show that different types of personalities will be more responsive to different kinds of information campaigns.

The paper is presented as follows. First, we discuss the unique experiment we conducted and describe the methods we used to understand user behavior. Next, we describe the results of our experiments and discuss their importance to understanding how personality impacts information transmission. Lastly, we compare our work to the existing literature. Our unique contribution is to separate personality from engagement using brain monitoring, revealing that the personality and mood of targeted users plays a significant role in determining the type of information that gets selected by users to share.

2   Methods

The purpose of this project, using an electroencephalogram (EEG) data-driven approach, was to evaluate the physiological response of individuals to social media content within the context of emergency situations. This approach allowed an analysis of how subjects perceive disaster alerts, observation of the level of attention elicited in subjects, and observation of subjects' response to the question of whether to share such alerts with their peers over a social media platform. The authors chose Twitter as the target platform because its 140-character limit is most representative of the current 90-character allowance for cell phone-based alerts about weather-related emergencies and because it is a ubiquitous platform.

The experiment evaluated test subjects' willingness to share messages to their own personal social network. These messages included Wireless Emergency Alerts (WEA) and tweets associated with five different types of disasters. Among the tweets were messages conveying sympathy for the victims of disasters, and other forms of sociable communication over the social network. We asked subjects, within the context of a natural disaster, to evaluate how important they perceived various forms of communication about several disasters (specifically, a blizzard, flood, gas leak, hurricane, and tornado). By evaluating their responses alongside their physiological response to the messages, the experiment measured their willingness to disseminate information about disasters and analyze the underlying cognitive models that drive their perceptions and reactions about different types of disasters.

The Twitter messages used in this study were a combination of real messages posted on Twitter during that disaster and disaster alerts sent by news stations and other emergency alert services within a defined geographic region surrounding the site of a declared emergency. For disasters that were declared at a definite point (such as tornados), tweets were collected from within the surrounding 25-mile radius. For disasters that affected broader swathes of land (such as hurricanes or blizzards), tweets were selected from within the entire region being alerted for a weather emergency. We collected tweets associated from the following disasters:

- blizzard, a winter storm that struck South Dakota in October 2013.
- flash flood, an episode of flooding that occurred in southern California in the summer of 2013.
- gas leak, an incident that occurred in Alamo, California on July 24, 2013.
- hurricane, across the northeastern United States, where Hurricane Sandy made landfall in late October 2012.
- tornado, a tornado that struck Moore, Oklahoma on May 20, 2013.

Experimental Data Collection.

Scientists at the Advanced Brain Monitoring (ABM) laboratory in Carlsbad, CA acquired electroencephalography (EEG) data from 51 participants during an experiment to evaluate the ways in which people perceive different kinds of disasters. The ABM wireless B-Alert® EEG sensor headset, a lightweight, easy-to-apply system was used to acquire 20 channels of data from sites: Fz, Fp1, Fp2, F3, F4, F7, F8, Cz, C3, C4, Pz, P3, P4, POz, T3, T4, T5, T6, O1, and O2, all referenced to linked mastoids.

The experiment presented each subject with five disasters in randomized order. A random benchmark assessment of six neutral tweets (i.e., tweets that were not related to any disaster) was presented either immediately before or after the set of disaster blocks (i.e., first or last). Immediately before each disaster block, subjects were first shown a 5-min newsreel video depicting news coverage for the disaster type they were about to evaluate.

After the newsreel ended, subjects were presented with a series of 50 WEA and Twitter messages for each of the five types of disasters. The testbed presented each message for a minimum of six seconds before the user was permitted to answer, to give the user time to read the message and reduce impulsive responses. The testbed then asked participants if they would share the message on social media. The subjects were required to use the keyboard to respond "yes" or "no" before moving on to the next message. Each subject received the disasters and associated messages in random order.

2.1 Description of the EEG Data

EEG data was time-locked to both the onset of each stimulus (messages or videos) and to each response (yes/no). All EEG data were acquired at a 256 Hz sampling rate (i.e., there are 256 measurements of brain activity taken every second) to provide a high level of fidelity in the analysis. The analysis of this data focused on measuring the brain's electrical response resulting from exposure to a particular cognitive or sensory event. ABM filtered the signal to remove blinks and other known signal confounders, and the remaining raw signal was used for analysis.

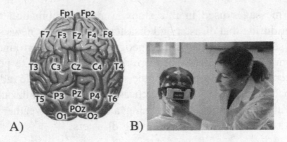

Fig. 1. B-Alert X24 EEG system. (A) standard 10–20 montage (B) 24 channel wireless headset

ABM measured each participant's head to ensure proper sizing and positioning of the 20-channel sensor cap, shown in Fig. 1. ABM designed the headset to position sensors over all cortical regions in accordance with the International 10–20 system, also shown in Fig. 1. While the association between a specific location on the human scalp and the precise activity occurring in the brain beneath it is not an exact correlation in all cases, in practice, EEG activity recorded over specific regions has been associated with particular functions or responses in subjects.

2.2 Personality Assessment

Each user completed multiple surveys designed to assess their personality and mood. Here, we report on two of those surveys. This includes the NEO personality inventory to assess the "Big 5" personality traits: openness, conscientiousness, extraversion, agreeableness and neuroticism. To assess current, transient mood, the subjects completed the Profile of Mood States (POMS). The mood states include anxiety, depression, anger, fatigue, vigor, and confusion.

2.3 Content Annotation

Each message, either an emergency alert or a tweet, was hand annotated by subject matter experts to be in one of three categories: informative, social, or sympathetic. Although these categories were chosen to be particularly apropos for disaster scenarios, they may be generalized to other domains as follows:

Informative – These messages contain objective information related to an event intended for a user to factor into their decision-making. Examples include:

- `12" of snow so far just NW of Rapid City, SD. Sustained winds over 40 mph. Blizzard warning until tomorrow morning.`
- `Flash flood warning #palmsprings #coachellavalley @ [username]`
- `Superstorm Sandy will hit east coast USA - 140 km/h winds Monday _ Connecticut, New York, New Jersey #amsterdam #haarlem #rotterdam`

- Tornado emergency for Moore #OK from @[username] Take shelter now
- Still a tornado warning for Paul's Valley area. Continue to be taking cover.

Social – These messages convey information about an event but in a way that emphasizes the social aspects of the event or is used as a means to communicate informal information about the event. Examples include:

- @[username] we had the worst blizzard in the history of souf dadoka
- And we have power! 26 h w/o electricity and heat is #funtimes #BlackHills #blizzard
- Blizzard is going on and also lighting and thunder
- Blizzard still raging on. No power for almost two hours. Hello October.
- Another flash flood warning? Uh ohh I hope there isn't any more thunder storms

Sympathetic – These messages convey explicit emotions or sympathy specifically related to others involved in the event. Examples include:

- @[username]: South Dakota's state veterinarian believes up to 20,000 cattle died in a blizzard. sadly, we made cnn
- @[username]: West river South Dakota cattle losses may total 25 % of herd. 25 % of a herd of +2,000,000 head. #blizzard2013â €□ sad sad deal here
- a mother was trying to drive her 2 young sons to Brooklyn because she was scared about the storm & a huge wave hit them & two baby boys gone
- Death toll now up to 96 from #Sandy. #RIP to the beautiful souls.
- Please pray for all those in Moore Oklahoma #tornados have devastated the area.

3 Results and Discussion

The results of this study reveal that the personality types of social media users impact their preferences or willingness to share certain forms of content. During the experiment, users exhibited distinct preferences for specific types of content that corresponded to their scores on various personality dimensions. These preferences became most apparent when cohorts of subjects with the highest scores for specific personality dimensions compared against cohorts of subjects with the lowest scores. These preferences were observed both in the levels of brain activity when viewing the disaster context videos, the frequency at which the subjects shared content, and in their corresponding EEG signatures when choosing to share content.

The frontal regions of the brain (designated by sites named with an "F") were of particular interest in this study, as this region is most strongly associated with executive function and decision-making. Among the 20 channels examined in the study, the subject response to the disaster context videos and the brain activity during decision-making for the individual tweets were observed to be most prominent in the F7, F8, Fp1, and Fp2 channels (all odd number regions are left hemisphere and even numbers are right hemisphere). Channels associated with other brain regions did not exhibit any noteworthy response to either the context videos or messages.

Specifically, subjects exhibited the higher levels of brain activity over the Fp1 and Fp2 regions channels during presentation of the context videos. Subjects similarly exhibited greater levels of activity over the F7 and F8 regions when determining if they would share specific messages with their social network. The left and right frontal regions appear to drive the decision making process to share specific messages, a finding consistent with prior evidence linking these regions to motivation and mood regulation (Davidson 2004). Conversely, EEG activity over the Fp1 and Fp2 regions is commonly associated with logical or emotional attention, judgment, and decision making (Chen et al. 2015).

As shown in Fig. 2, during presentation of the context videos, subjects with the highest scores for depressive, fatigued, or confused personalities from the POMS personality test battery exhibited the lower engagement EEG scores. Similarly, subjects with the most extroverted, open, and agreeable personality scores according to the NEO personality test battery exhibited stronger signs of engagement and attention than their counterparts with the lowest scores on these metrics. All differences were significant to at least $p < 0.05$.

Because the activity during the video is indicative of attention and engagement in the task—users whose mood is characterized as depressed, fatigued or confused—will generally be less engaged with the content on social media, even during controlled conditions (Fig. 2a). Conversely, highly agreeable, extroverted or open users appear to more readily engage in the videos than their lower scoring counterparts (Fig. 2b). As an aside, empirical studies of social contagion have difficulty distinguishing if users don't spread a message because they didn't like it or if they didn't see it, i.e. low visibility (Hodas and Lerman 2014). The present results show that underlying personality and mood may play a significant role in moderating the engagement levels of users.

We calculated the power spectral density estimation on the subject EEG data to quantify subject attention with the disaster context videos, analyzing the gamma band (30–100 Hz). Gamma waves are strongly associated with intentional attention and cognition, thus by estimating the power density of subjects during each period of video presentation it was possible to quantify different levels of engagement between cohorts of subjects based upon the strength of NEO and POMS personality traits. Subjects in the top quartile for each respective personality trait were compared against their peers in the bottom quartile to determine which group had a greater power density in the gamma wave range, and thus, which group was the most engaged during the videos.

The gamma band had the greatest differences between the cohorts at opposing extremes of the Fatigue and Vigor traits identified by the POMS test. The subjects with the lowest fatigue scores had greater power densities in the gamma wave band across the Fp2, Fp1, F7, and F8 channels, all of which were statistically significant from their

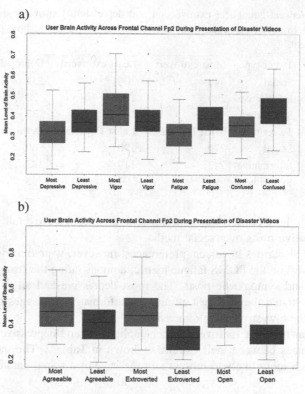

Fig. 2. Mean level of RMS brain activity while subjects watch the videos, for users in the lowest (blue) quartile, i.e., least, and highest (red) quartile, i.e., most. All differences are significant to $p < 0.05$. a) for POMS mood traits and b) NEO personality traits (Color figure online)

peers in the bottom quartile. Likewise, the most vigorous subjects had a statistically significant and greater power density relative to their least vigorous peers. The least depressive subjects only had statistically significant differences in gamma power densities for the F8 and Fp2 channels relative to their most depressive peers (whom had lower power densities). No traits identified by the NEO personality test revealed a statistically significant difference between cohorts, but all traits had overall trends consistent with Fig. 2. For this reason, we do not plot gamma bands, but based on this analysis, we can conclude that certain personality traits indeed confer users with a greater or lesser predisposition to pay attention to content, as shown in Fig. 2.

After the users watch the videos, the testbed presented each user with the relevant messages and alerts in randomized order. Table 1 shows a summary of the relative preference for each type of content. We may safely assume that a user has a "preference" for a specific type of content if they are more likely to retweet that content than other types. Subjects with the most extroverted personalities demonstrated the strongest preference for dismissive messages in the study, as well as the strongest preference for social messages from among the NEO personality types. The most conscientious individuals similarly demonstrated the second strongest preference for social content

Table 1. Total retweet counts for extroversion and depression. Extroverts show significantly more willingness to tweet, particularly social tweets. Depressive users show preferences from informative content.

Tweet type	Most extrovert	Least extrovert	Delta
Social	440	206	234
Informative	209	189	20
Sympathetic	143	106	37
Tweet type	**Most Depressed**	**Least Depressed**	**Delta**
Social	270	190	80
Informative	242	119	123
Sympathetic	94	91	3

relative to their least-conscientious peers, and the strongest overall preference for sharing informative posts over social media.

The largest disparities in content preferences, however, were observed for subjects scoring the lowest on the POMS fatigue metric, demonstrating the strongest preference against social and sympathetic posts. The most depressive and angry subjects, conversely, demonstrated the strongest affinity for informative messages relative to their peers scoring the lowest on these dimensions.

Of particular note, the extroversion personality trait and depressive mood showed notable differences between their extremes, shown in Table 2. The most extroverted users were much more likely to retweet any message of a social nature.

Table 2. Personality and Mood (Trait) states along with their preferred content (Preference). All noted preferences are statistically significant to at least $p < 0.05$. The "retweet difference between cohorts" is the difference between the total number retweets made by users in the 1st quartile (i.e., the most) and 4th quartile (i.e., the least) score from each trait. We analyzed content categorized as "social", "informative", or "sympathetic."

Trait	Preference	Retweet difference between cohorts		
		Social	Informative	Sympathetic
Agreeable	More social	78	−18	20
Conscientious	Less content in general, fewer social and informative	−164	−208	−97
Extroversion	All, particularly social	234	20	37
Anger	More informative	27	114	12
Confusion	Less content in general, and social content in particular	−104	−31	−64
Depression	More social and much more informative	80	123	3
Fatigue	Fewer social and sympathetic content, and moderately less informative	−324	−85	−139
Vigor	All	96	40	53

Analysis of the EEG data collected during the window of time when subjects were asked if they would retweet a message to their social network reveal that the preferences observed above were often accompanied by significantly different EEG responses as compared to their peers, with the notable exception of depressive personalities, shown Figs. 3, 4, 5, 6, 7 and 8. These figures show the instantaneous power in the F8 channel in users with the lowest and the highest scoring quartiles for each trait on messages they chose to share. The x-axis shows a scaled time such that t = 0 % is the time of exposure to the message, and t = 100 % is the moment the user replied. Each user's axis is scaled individually and averaged together with the other users, allowing us to understand engagement over the decision making process.

The levels of relative EEG activity shown in Figs. 3, 4, 5, 6, 7 and 8 demonstrate that the subject's preferences for certain forms of content exhibited by their responses correspond to particularly high levels of activity over the F8 region of the brain relative to their less responsive peers. As noted earlier, this lone exception to this observation was that the most depressive subjects, which were generally more responsive to informative and social content than their peers, who did not exhibit similarly elevated levels of EEG activity prior to endorsing messages for sharing over their social network. Thus, we see that the notion that users show preferences for sharing messages – and that this preference may be highly sensitive to personality and mood – is

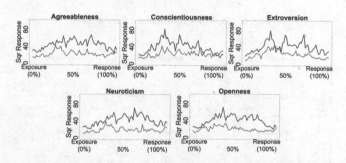

Fig. 3. NEO - informative messages. All time is scaled such that 0 % is time of exposure; 100 % is time of response.

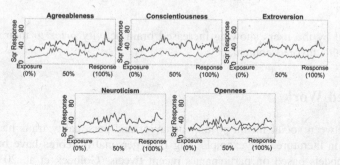

Fig. 4. NEO - social messages.

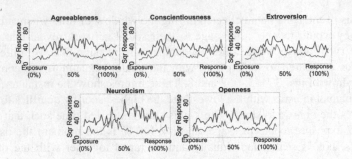

Fig. 5. NEO – sympathetic messages

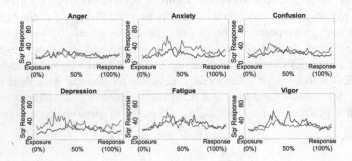

Fig. 6. POMS - informative messages

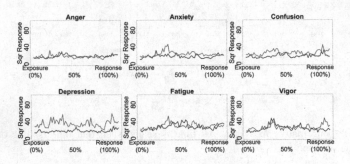

Fig. 7. POMS - social messages

corroborated by the users showing increased brain activity *prior to* their decision to retweet for this favored content.

4 Related Work

The link between social media posting behavior and personality traits has been well established in literature. For example, Big Five personality scores have been used in predicted models based on participant's recent tweets (Golbeck et al. 2011). Similar

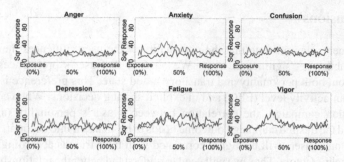

Fig. 8. POMS - sympathetic

calculations were run with social graph and interactions between users taken into consideration (Adali and Golbeck 2012). Big Five personality traits were also modeled on abstract groups of users (such as 'listeners', popular', 'highly-read' and 'influential') based on user behavior (Golbeck et al. 2011). Anti-social traits such as narcissism, psychopathy and Machiavellianism (the "Dark Triad") were predicted and compared with the Big Five personality traits, using language features of tweets (Sumner et al. 2012).

Examination of emotion, personality and brain modeling techniques such as EEG and fMRI has been similarly well established, from predicting patterns of regional brain activity related to extraversion and neuroticism (Schmidtke and Heller 2004), to EEG based emotion recognition when listening to music (Lin et al. 2010) or stories designed to evoke specific emotions (Correa et al. 2015; Stikic et al. 2014). Broader emotional recognition with EEG has also been examined with high accuracy (Petrantonakis and Hadjileontiadis 2010; Correa, et al. 2015; Stikic et al. 2014), as well as a functional MRI study of the neuroanatomy of grief (Gündel et al. 2003).

A previous effort at fusing EEG, emotion, and social media focused on producing tweets reflecting a user's emotions at certain physical locations. These tweets included both an emotion component and geotagged location component ("I am Frustrated at this location (Bus Station)") (Almehmadi et al. 2013). Work has been done to tag content based on neurophysiological signals, a technique described in (Yazdani et al. 2009) to produce implicit tagging of emotional states represented in multimedia via EEG and a brain computer interface.

Our present work demonstrates that personality and mood significantly effect that type of content users choose to share under controlled conditions. This shows there is need for models to better characterize user's mood and personality to understand them in live social media feeds. In addition, a broader model of personality and social media would allow us to understand better the friendship paradox (Hodas et al. 2013) and how user-user correlation in personality traits and mood drives social contagion (Kramer et al. 2014).

5 Conclusions

Our experiments demonstrate that the personality of the user influences their behavior online in subtle, yet significant, ways. We observe that user's preferences might be predicted from both personality and transitory mood state. This preference is evident in both the brain-activity level (EEG) and in explicit sharing decisions. When constructing an information campaign, the correlation between a user's personality (and mood), interests, and the desired campaign outcome needs to be all taken into account. Because of homophily, most users will be highly correlated with their friends according these very same personality factors. Thus, we will need to understand better the relationship between personality and content preference to better understand and model social behavior online.

It is not surprising that some personalities and moods are more attracted to certain kinds of content. However, this is one of the first results to systematically compare personality measures with content produced during an event – natural disasters, in this case. We also controlled for some of the common confounders that take place during empirical experiments; we were able to account for the correlation between user engagement and preference.

Future work will allow us to further investigate not only the statistical preferences of different users, but also which types of events different personality or moods may be drawn toward when the actively engage with social media. Future modeling may reveal that systematic correlation between the personality of friends may significantly bias local information propagation and information awareness.

Appendix

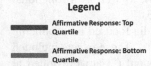

Legend

Affirmative Response: Top Quartile

Affirmative Response: Bottom Quartile

The following are plots of average squared EEG on the F8 channel for response for messages the users decided to share. We believe F8 the most discriminative channel during retweeting. Each trait is broken down according to the top quartile (users have the "most" of that trait), and bottom quartile (users with the "least" of that trait). *Higher signals indicate more engagement and attention to the message.*

References

Adali, S., Golbeck, J.: Predicting personality with social behavior, pp. 302–309. IEEE (2012)

Almehmadi, A., Bourque, M., El-Khatib, K.: A tweet of the mind: automated emotion detection for social media using brain wave pattern analysis, pp. 987–991. IEEE (2013)

Chen, J., Luo, X., Ren, B., Song, X., Chen, J., Luo, X., Ren, B., Song, X.: Revealing the 'invisible gorilla' in construction: assessing mental workload through time-frequency analysis. In: ISARC Proceedings 2015 Proceedings of the 32nd ISARC, Oulu, Finland, pp. 1–8 (2015)

Correa, K.A., Stone, B.T., Stikic, M., Johnson, R.R., Berka, C.: Characterizing donation behavior from psychophysiological indices of narrative experience. Decis. Neurosci. **301** (2015)

Correa, T., Hinsley, A.W., de Zúñiga, H.G.: Who interacts on the web?: the intersection of users' personality and social media use. Comput. Hum. Behav. **26**, 247–253 (2010)

Davidson, R.J.: What does the prefrontal cortex 'do' in affect: perspectives on frontal EEG asymmetry research. Biolo. Psychol. **67**, 219–234 (2004)

Evers, C.W., Albury, K., Byron, P., Crawford, K.: Young people, social media, social network sites and sexual health communication in Australia: 'this is funny, you should watch it'. Int. J. Commun. **7**, 18 (2013)

Golbeck, J., Robles, C., Edmondson, M., Turner, K.: Predicting personality from twitter. In: 2011 IEEE Third International Conference on Privacy, Security, Risk and Trust (PASSAT) and 2011 IEEE Third Inernational Conference on Social Computing (SocialCom), pp. 149–156. IEEE (2011)

Gündel, H., O'connor, M.-F., Littrell, L., Fort, C., Lane, R.D.: Functional neuroanatomy of grief: an FMRI study. Am. J. Psychiatry (2003)

Hodas, N.O., Kooti, F., Lerman, K.: Friendship paradox redux: your friends are more interesting than you. In: ICWSM 2013 (2013)

Hodas, N.O., Lerman, K.: The simple rules of social contagion. Sci. Rep. **4**, 4343 (2014)

Hughes, D.J., Rowe, M., Batey, M., Lee, A.: A tale of two sites: twitter vs. facebook and the personality predictors of social media usage. Comput. Hum. Behav. **28**, 561–569 (2012)

Kietzmann, J.H., Hermkens, K., McCarthy, I.P., Silvestre, B.S.: Social media? Get serious! understanding the functional building blocks of social media. Bus. Horiz. **54**, 241–251 (2011)

Kramer, A.D.I., Guillory, J.E., Hancock, J.T.: Experimental evidence of massive-scale emotional contagion through social networks. Proc. Natl. Acad. Sci. **111**, 8788–8790 (2014)

Lee, C.S., Ma, L.: News sharing in social media: the effect of gratifications and prior experience. Comput. Hum. Behav. **28**, 331–339 (2012)

Petrantonakis, P.C., Hadjileontiadis, L.J.: Emotion recognition from eeg using higher order crossings. IEEE Trans. Inf. Technol. Biomed. **14**, 186–197 (2010)

Quercia, D., Kosinski, M., Stillwell, D., Crowcroft, J.: Our twitter profiles, our selves: predicting personality with twitter. In: 2011 IEEE Third International Conference on Privacy, Security, Risk and Trust (PASSAT) and 2011 IEEE Third Inernational Conference on Social Computing (SocialCom), pp. 180–185. IEEE (2011)

Ryan, T., Xenos, S.: Who uses facebook? An investigation into the relationship between the big five, shyness, narcissism, loneliness, and facebook usage. Comput. Hum. Behav. **27**, 1658–1664 (2011)

Schmidtke, J.I., Heller, W.: Personality, affect and EEG: predicting patterns of regional brain activity related to extraversion and neuroticism. Pers. Individ. Differ. **36**, 717–732 (2004)

Stikic, M., Johnson, R.R., Tan, V., Berka, C.: EEG-based classification of positive and negative affective states. Brain-Comput. Interfaces **1**, 99–112 (2014)

Sumner, C., Byers, A., Boochever, R., Park, G.J.: Predicting dark triad personality traits from twitter usage and a linguistic analysis of tweets, pp. 386–393. IEEE (2012)

Yazdani, A., Lee, J.-S., Ebrahimi, T.: Implicit emotional tagging of multimedia using EEG signals and brain computer interface. In: Proceedings of the first SIGMM workshop on Social media, pp. 81–88. ACM (2009)

Lin, Y.-P., Wang, C.-H., Jung, T.-P., Wu, T.-L., Jeng, S.-K., Duann, J.-R., Chen, J.-H.: EEG-based emotion recognition in music listening. IEEE Trans. Biomed. Eng. **57**, 1798–1806 (2010)

Semi-supervised Knowledge Extraction for Detection of Drugs and Their Effects

Fabio Del Vigna[1,2], Marinella Petrocchi[2(✉)], Alessandro Tommasi[3],
Cesare Zavattari[3], and Maurizio Tesconi[2]

[1] Department of Information Engineering, University of Pisa, Pisa, Italy
[2] Institute of Informatics and Telematics (IIT-CNR), Pisa, Italy
{f.delvigna,m.petrocchi,m.tesconi}@iit.cnr.it
[3] LUCENSE SCaRL, Lucca, Italy
{alessandro.tommasi,cesare.zavattari}@lucense.it

Abstract. New Psychoactive Substances (NPS) are drugs that lay in a grey area of legislation, since they are not internationally and officially banned, possibly leading to their not prosecutable trade. The exacerbation of the phenomenon is that NPS can be easily sold and bought online. Here, we consider large corpora of textual posts, published on online forums specialized on drug discussions, plus a small set of known substances and associated effects, which we call seeds. We propose a semi-supervised approach to knowledge extraction, applied to the detection of drugs (comprising NPS) and effects from the corpora under investigation. Based on the very small set of initial seeds, the work highlights how a contrastive approach and context deduction are effective in detecting substances and effects from the corpora. Our promising results, which feature a F1 score close to 0.9, pave the way for shortening the detection time of new psychoactive substances, once these are discussed and advertised on the Internet.

Keywords: Text mining · NPS detection · NPS data mining · Drugs forums · Social media analysis · Machine learning · Automatic classification

1 Introduction

US and European countries are facing a raising emergency: the trade of substances that lay in a grey area of legislation, known as New Psychoactive Substances (NPS). The risks connected to this phenomenon are high: every year, hundreds of consumers get overdoses of these chemical substances and hospitals have difficulties to provide effective countermeasures, given the unknown nature of NPS. Government and health departments are struggling to monitor the market to tackle NPS diffusion, forbid NPS trade and sensitise people to the harmful effects of these drugs[1]. Unfortunately, legislation is typically some

---

[1] http://www.emcdda.europa.eu/start/2016/drug-markets#pane2/4; All URLs in the paper have been accessed on July 10, 2016.

© Springer International Publishing AG 2016
E. Spiro and Y.-Y. Ahn (Eds.): SocInfo 2016, Part I, LNCS 10046, pp. 494–509, 2016.
DOI: 10.1007/978-3-319-47880-7_31

steps back and newer NPS quickly replace old generation of substances. Also, the abuse of certain prescription drugs, like opioids, central nervous system depressants, and stimulants, is a widespread as an alarming trend, which can lead to a variety of adverse health effects, including addiction[2].

The described phenomena are being exacerbating by the fact that online shops and marketplaces convey NPS through the Internet [21]. Moreover, specialised forums offer a fertile stage for questionable organisations to promote NPS, as a replacement of well known drugs. Forums are contact points for people willing to experiment with new substances or looking for alternatives to some chemicals.

In this work, we consider the myriads of posts published on two big drugs forums, namely Bluelight[3] and Drugsforum[4]. Posts consist of natural language, unstructured text, which, generally speaking, can be analysed with text mining techniques to discover meaningful information, useful for some particular purposes [25]. We propose DAGON (DAta Generated jargON), a novel, semi-supervised knowledge extraction methodology, and we apply it to the posts of the drugs forums, with the main goals of: (i) detecting substances and their effects; (ii) put the basis for linking each substance to its effects. A successful application of our technique is paramount: first, we envisage the possibility to shorten the detection time of NPS; then, it will be possible to group together different names that refer to the same substance, as well as to distinguish between different substances, commonly referred to with the same name (such as "Spice" [20]) and timely detect changes in drug composition over time [8]. Finally, knowing the effects tied to novel substances, first-aid facilities may overcome the current difficulties to provide effective countermeasures.

While traditional supervised techniques usually require large amount of hand-labeled data, our proposal features a semi-supervised learning approach in order to minimize the work required to build an effective detection system. Semi-supervised learning exploits unlabeled data to mitigate the effect of insufficient labeled data on the classifier accuracy. This specific approach attempts to automatically generate high-quality training data from an unlabeled corpus. With very little information, our solution is able to achieve excellent detection results on drugs and their effects, with an FMeasure close to 0.9.

The paper is structured as follows. The next section describes our data sources. In Sect. 3, we introduce our semi-supervised methodology. Section 4 presents a set of experiments and results. Section 5 provides related work on mining drugs over the Internet and it discusses text analysis approaches, highlighting differences and similarities with our proposal. Finally, Sect. 6 concludes the paper.

---

[2] https://www.drugabuse.gov/publications/research-reports/prescription-drugs/director.

[3] http://www.bluelight.org.

[4] https://drugs-forum.com.

Table 1. Drug forums: Posts and Users

Forum	First post	Last post	Tot posts	Users
Bluelight	22-10-1999	09-02-2016	3,535,378	347,457
Drugsforum	14-01-2003	26-12-2015	1,174,759	220,071

2   Datasets

The approach in this work is tested over two different large data sources, in order to consider a variety of contents and information, and to push the automatic detection of drugs. We collected more than a decade of posts from Bluelight and Drugsforum. As shown in Table 1, the available data comprises more than half million users and more than 4.6 million posts. Data was collected through web scraping and stored in a relational database for further querying. These forums were early and partially analysed in [23] and then explored in detail [9]. Here, we present the very same datasets to show how it is possible to extract knowledge from text using few seeds as the starting point for the algorithm introduced in Sect. 3.

2.1   Seeds

We have downloaded a list of 416 drug names of popular psychoactive substances, including the slang which is adopted among consumers to commonly name them, from the website of the project *Talk to Frank*[5] and a dataset containing 8206 pharmaceutical drugs retrieved from Drugbank[6]. This list constitutes a ground truth for known drugs.

Also, we collected a list of 129 symptoms that are typically associated to substance assumption.

3   The DAGON Methodology (DAta Generated jargON)

In this section, we introduce DAGON, a methodology that will be applied in Sect. 4 for the task of identifying new "street names" for drugs and their effects. A street name is the name a substance is usually referred to amongst users and pushers.

The task of name identification can be split into two subtasks:

(a) Identifying text chunks in the forums, which represent candidate drug names (and candidate drug effects);
(b) Classifying those chunks as drugs, effects, or none of the above.

---

[5] http://www.talktofrank.com.
[6] http://www.drugbank.ca.

The first subtask - identification of candidates - could be tackled with different approaches, including a noun-phrase identifier[7], usually based on a simple part-of-speech-based grammar, or on a technique akin to the identification of named entities, as in [14].

In this work, the identification of candidates is based on domain terminology extraction techniques based on a contrastive approach similar to [16]. Essentially, we identify chunks of texts that appear to be especially significant in the context of drug forums. Based on the frequency in which terms appear both in the posts of drugs forums and in contrastive datasets dealing with different topics, we extract the most relevant terms for the forums. We have extracted unigrams, 2-grams, and 3-grams. This approach does not require English specific annotated resources and, thus, it can scale easily to different languages.

The second subtask is a classification problem. Following a supervised approach would have required to have annotated posts and use them as the training set for our classifier. Instead, we have chosen to work on unlabeled data (i.e., the posts on the drugs forums, see Sect. 2) and to exploit the external list of seeds introduced in Sect. 2.1.

We represent a candidate by means of the words found along with it when it was used in a post, selecting windows of N characters surrounding the candidate whenever it was used in the dataset. Hereafter, we call *context* (of a candidate) the text surrounding the term of interest.

Thus, we have shifted the problem: from classifying candidate street names to the classification of their contexts, which are automatically extracted from the unlabeled forum datasets.

It is worth noting that, in the drugs scenario, there would be at least 3 classes, i.e., Substance, Effect, and "none of the above" - the latter to account for the cases where the candidate does not represent substances and effects. However, the seed list at our disposal consists of flat lists of substances/effects names, provided with no additional information (Sect. 2.1). Therefore, in the following, we will first automatically identify positive examples for the two classes (Substance and Effect), training a classifier on them, and then we will tune the classifier settings to determine when a candidate does not fall in either.

Summarising, we have split the task of classifying a candidate into the following sub-tasks:

(a) Fetch a set of occurrences of the term along with the surrounding text (forming in such a way the so called contexts).
(b) Classify each context along the 2 known classes (Sect. 3.3).
(c) Determine a classification for the term given the classification result for the context related to that term (obtained at step (b)).

The single context classification task [1] falls within the realm of standard text categorization, for which there is a rich literature.

---

[7] A noun-phrase is a phrase that plays the role of a noun such as "the kid that Santa Claus forgot".

Hereafter, we detail the training phase for our classifier (Sect. 3.1), we give detail on the choice of seeds (Sect. 3.2), we specify the procedure for classifying a new candidate (Sect. 3.3), and we illustrate a simple approach to link substances to their effects (Sect. 3.4).

3.1 Training Phase

We are equipped with a list of examples for both the drugs and the effects, as described in Sect. 2.1. This list of entry terms is the training set for the classification task and we call it *list of seeds*.

Each post in the target drug forums was indexed by a full-text indexer (Apache Lucene[8]) as a single document.

The training phase is as follows:

(i) Let T_S and T_E be the set of example contexts, for the Substance and Effects classes respectively, initialized empty.

(ii) From the lists of seeds, we pick a new seed (a drug name) for the Substance class and one (an effect name) for the Effects class. A seed is therefore an example of the corresponding class taken from the seed list (Sect. 2.1). See Sect. 3.2 for the heuristic to select a seed out of the list.

(iii) We use the full-text index to retrieve M posts containing the seed s; we only use the bit of text surrounding the seed. In Sect. 4, we will show how results change by varying M. We pick a window of 50 characters surrounding the searched seed.

(iv) We strip s from the text, replacing it always with the same unlikely string (such as "CTHULHUFHTAGN"), in order to avoid the bias carried by the term itself, but maintaining the position of the term in the phrase for classification purposes. We call the texts thus obtained ctx_s (context of seed s).

(v) We add the texts thus generated to the set of training examples for the category C the seed belongs to (either T_S or T_E)

(vi) We use the training examples to train a multiclass classification model M^{ctx}, which can be any multiclass model, as long as it features a measure (e.g., a probability) interpretable as a confidence score of the classification. In Sect. 4 we will show results when using SVM with linear kernel [5].

At the end of these steps, we have obtained a classifier of contexts (M^{ctx}), but as seeds (not contexts) are labeled, we are unable to assess its performance directly. We therefore define a classifier of candidate terms (M^{trm}) using the method described later in Sect. 3.3, the performance of which we can assess against the seed list. This allows us to optionally iterate back to step (ii), in order to provide additional seeds to extend the training sets, and improve performances.

The rationale behind this process is that drug (and effects) mentions will likely share at least part of their immediate contexts. Clearly, when a very small

---

[8] http://lucene.apache.org/.

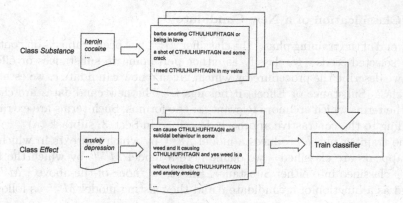

Fig. 1. Training phase

number of seeds is provided (e.g., 1 per class) there will be a strong bias in the examples ultimately used for training, which means that the resulting model will be overly specific to the type of drug used in the training. By providing more seeds, and with enough variety, the model will eventually become more generic to encompass the various drug types, and the relative differences in the contexts in which they are mentioned in the dataset (Fig. 1).

3.2 Choosing a Seed

Obtaining a large seed list is often costly, since it may require to manually annotate texts, or to provide to the algorithm a initial set of words. Thus it is important to design a system with high performances that uses the minimum amount possible of seeds for the train phase. Choosing an effective seed is paramount, and, in doing so, there are various aspects to consider:

(a) Is the seed mentioned verbatim enough times in the data collection? Failing this, the seed will only serve to collect a small number of additional training elements, and it will not impact the model enough;

(b) Is the seed adding new information? The most effective seeds are those whose contexts are misclassified by the current iteration of the classification model. In order to pick the most useful one, we could select, from the list of available unused seeds, those whose contexts are frequently misclassified. Using these seeds, the model is modified to address a larger number of potential errors.

In information retrieval, Inverse Document Frequency [19] (idf) is often used along with term frequency (tf) as a measure of relevance of a term, capturing the fact that a term is frequent, but not so frequent to be essentially meaningless (non-meaning words, such as articles and conjunctions, are normally the most frequent ones). A common way to address point (a) would therefore be using a standard tf·idf metric. However, because our seeds list is guaranteed to only contain meaningful entries, we can safely select the terms occurring in more documents first (i.e., with an increasing idf). We leave point (b) for future work.

3.3   Classification of a New Candidate

At the end of the training phase, the classifier M^{ctx} has been trained - on contexts of the selected seeds - to classify as either pertaining to substances or effects. Here, we describe the procedure by which, given a new candidate c, we establish what class (Substance or Effect) it belongs to. The new candidates are chosen from the terms which are more relevant for the forums. Such terms are extracted according to the contrastive approach described in Sect. 3, subtask (a) .

The training phase produces a model M^{ctx} by which contexts in which the term appears are classified – we define here a model M^{trm} by which the term itself is classified into either Substance, Effect, or "none-of-the-above". M^{trm} is defined as a function of a candidate c and the existing model M^{ctx} as follows:

1. We apply steps (iii) and (iv) of the algorithm described in Sect. 3.1 to obtain the contexts for c (ctx_c).
2. We classify the elements of ctx_c using M^{ctx}. We discard all categorizations whose confidence, according to the model, falls below a threshold θ_p, which we have experimentally set to 0.8 as a reference value.
3. We consider the remaining categorizations thus obtained. If a sizeable portion of them (θ_c, initially set to 0.6, we will show how results vary along with its value) belongs to the same class C, then c belongs to C; otherwise it is left unassigned.

In Fig. 2 we give a high level graphical description of this process.

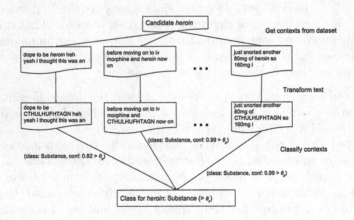

Fig. 2. Classification of a new term

3.4   Linking Substances to Effects

We outline here a simple procedure by which we can associate the substances mentioned in the drugs forums to the effects they produce.

When indexing a post, the significant terminology elements found in the post are linked to it as metadata. As introduced, the terminology elements have been extracted following a contrastive approach, as in [16].

We assume to have already tagged the terminology elements found in each post as referring to substances or effects, using the method described in Sect. 3.3. Thus, when searching for mentions of a particular substance, we can correspondingly fetch, for each post the substance mention is found in, the relative metadata. Then, from the matadata, we can sort the list of effects by frequency – it is very likely that those effects are related to the searched substance.

As a simple example, let's suppose to have a single post, with Text: *heroin gave me a terrible headache*; Substances: [heroin]; Effects: [headache].

Intuitively, we can assume that [headache] is an effect of [heroin]. If we consider all the posts in our datasets where the substance [heroin] is among the metadata, and we count the most frequent metadata effects associated to [heroin], we can have an indication of the links between substances and effects. However many substances may appear in the same text. Thus, it is necessary to filter out the rarest links substance-effect since they are often due by chance. Section 4 will report on some findings we were able to achieve for our datasets about drugs and their effects.

4   Experiments

We show a set of experiments on the data described in Sect. 2. First, from all the posts, we need to identify a list of candidates (unless we want to try and classify every term – a possible, but undesirable strategy, to pinpont substances or effects out of which. Candidates are selected using a contrastive terminology extraction [16], to identify terms and phrases common within the community and yet specific to it; this is the first subtask outlined in Sect. 3. Then, we apply the M^{trm} classifier, described in Sect. 3.3, to assign to candidates either the class Substance or Effect or none of the above, and evaluate the performance of the classification. The intermediate M^{ctx} classifier was trained using SVM with linear kernel [5].

We report experiments and results for the Bluelight forum. The lists used to select seeds and to validate results have been described in Sect. 2.1. These lists represent 2 classes: Substance and Effect.

It is worth noting that, for our experiments, we consider the intersection between the lists of seeds and the extracted terminology. This is necessary because: (i) items that are present in the lists may not be present in the downloaded dataset; (ii) many terminological entries might be neither drug names nor drug effects. The intersection contains 226 substances and 89 effects. Some of these will be used as seeds, the rest of the entries to validate the results.

The results are given in terms of three standard metrics in text categorization, based on true positives (TP - items classified in category C, actually belonging to C), false positives (FP - items classified in C, actually not belonging to C) and false negatives (FN - items not classified in C, actually belonging to C),

Table 2. Classification results for substances and effects, varying the number of seeds

# of seeds	Recall	Precision	F1
1	0.502	0.649	0.566
2	0.576	0.734	0.645
3	0.65	0.827	0.728
4	0.769	0.891	0.826
5	0.823	0.909	0.864
6	0.832	0.926	0.876

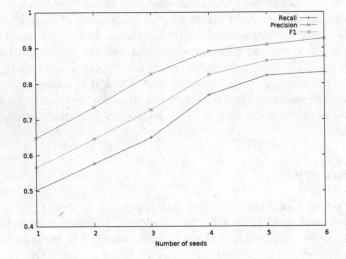

Fig. 3. Recall, precision and F1 varying the number of seeds

computed over the decisions taken by the classifier: precision[9], recall[10] and F1-micro averaged[11].

The first results are in Table 2 and Fig. 3. Even though the training set is limited to a small number of entries, the results are interesting: with only 6 seeds, the proposed methodology achieves a F1 score close to 0.88 (on the 2 classes - Substance and Effect). With the aim of monitoring the diffusion of new substances, the result is quite promising, since it is able to detect unknown substances without human supervision.

Dealing with "the rest". Finding mentions of new substances or effects means classifying candidates terms in either one class. Playing with thresholds, we can discard some candidates, as belonging to none of the two classes (see Sect. 3.3).

---

[9] $precision = \frac{TP}{TP+FP}$.

[10] $recall = \frac{TP}{TP+FN}$.

[11] harmonic mean of *precision* and *recall*: $F1 = 2 \cdot \frac{precision \cdot recall}{precision+recall}$.

Table 3. Classification results for substances and effects, including the "rest" category

# of seeds	Recall	Precision	F1
1	0.502	0.502	0.502
2	0.576	0.563	0.569
3	0.650	0.628	0.639
4	0.769	0.694	0.730
5	0.823	0.723	0.770
6	0.832	0.733	0.779

Thus, within the extracted terminology, we have manually labeled about 100 entries as neither drugs nor effects, and we have used them as candidates. This has been done to evaluate the effectiveness of using the parameter θ_c to avoid classifying these terms as either substances or effects. Performance-wise, this resulted in few more false positives given by terms erroneously assigned to the substance and effect classes, when instead these 100 candidates should ideally all be discarded. The results are in Table 3 and Fig. 4. We can observe that, when we include in the evaluation also those data that are neither substances nor effects, with no training data other than the original seeds, and operating only on the thresholds, the precision drops significantly.

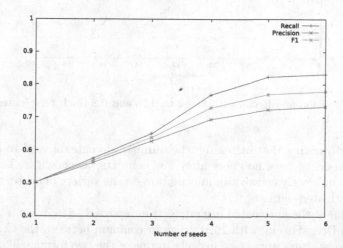

Fig. 4. Recall, precision and F1 including the "rest" category

To achieve comparable performances, we have conducted experiments changing the number of seeds and θ_c used to keep relevant terms. The results are shown in Table 4 and Fig. 5. The higher the threshold, the higher the precision, while increasing the number of seeds improves the recall, which is to be expected:

Table 4. Precision, Recall and F1 with θ_c set to 0.75 and 0.8 (incl. "rest" category)

# of seeds	Recall 0.75	Precision 0.75	F1 0.75	Recall 0.8	Precision 0.8	F1 0.8
5	0.607	0.755	0.673	0.508	0.787	0.618
10	0.759	0.852	0.803	0.654	0.889	0.754
15	0.811	0.837	0.824	0.705	0.874	0.781
20	0.833	0.854	0.843	0.753	0.866	0.805

adding seeds "teaches" the system more about the variety of the data. More-over, recall augments when we increase the number of contexts per seed used to train the system (Table 5 and Fig. 6).

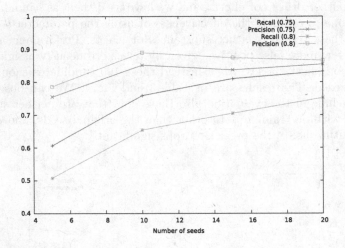

Fig. 5. Precision and Recall with θ_c set to 0.75 and 0.8 (incl. "rest" category)

It is worth noting that increasing the number of contexts used to classify a new term seems to have no effect after few contexts, as shown in Table 6 and Fig. 7). This indirectly conveys an information on the variety of contexts present on the investigated datasets.

Interestingly, the automated drug detection reported 1846 drugs in Bluelight and 1857 in DrugsForum, with 1520 drugs in common between the two forums. Moreover, some drugs appear exclusively in one of the two forums, like the *triptorelin, candesartan* and *thiorphan* in Bluelight and the *lymecycline, boceprevir* and *imipenem* in Drugsforum, although the majority is shared.

Finally, upon training the system with the seeds, for every post it is possible to link the drugs to their effects. An example of links is in Table 7.

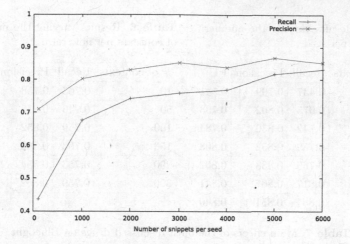

Fig. 6. Recall and precision varying the number of contexts (snippets) per seed, 10 seeds used

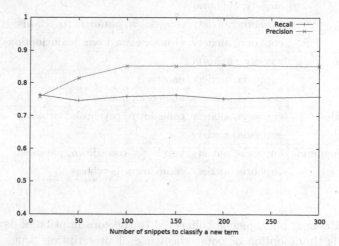

Fig. 7. Recall and precision varying the number of contexts (snippets) per new term, 10 seeds used

5   Related Work

Recently, Academia has started mining online communities, to seek for comments on drugs and drugs reactions [27]. Indeed, forums and social networks offer spontaneous information, with abundance of data about experiences, doses, assumption methods [7,9]. Authors in [15] realized ADRMine, a tool for adverse drugs reaction detection. The tool relies on advanced machine learning algorithms and semantic features based on word clusters - generated from pre-trained word representation vectors using deep learning techniques. Also, intelligence analysis has been applied to social media to detect new outbreaking trends in drug markets,

Table 5. Results varying the number of contexts per seed

# of contexts	Recall	Precision	F1
100	0.437	0.709	0.541
1000	0.675	0.802	0.733
2000	0.742	0.830	0.784
3000	0.759	0.852	0.803
4000	0.769	0.838	0.802
5000	0.817	0.867	0.841
6000	0.831	0.851	0.840

Table 6. Results varying the number of contexts per new term

# of contexts	Recall	Precision	F1
10	0.763	0.758	0.760
50	0.746	0.815	0.779
100	0.759	0.852	0.803
150	0.763	0.852	0.805
200	0.753	0.854	0.800
300	0.759	0.852	0.803

Table 7. Main effects of the most discussed drugs on Bluelight

Drug	Effects
heroin	anxiety, euphoria
cocaine	euphoria, anxiety, comedown, paranoia, psychosis
ketamine	euphoria, anxiety, visuals, comedown, hallucinations, nausea
methadone	anxiety, euphoria
codeine	euphoria, anxiety, nausea
morphine	euphoria, anxiety, analgesic, nausea
amphetamine	euphoria, anxiety, comedown, psychosis, visuals
oxycodone	euphoria, anxiety
methamphetamine	euphoria, anxiety, psychosis, comedown, paranoia
dopamine	euphoria, anxiety, comedown, psychosis

as in [24]. A raising phenomenon connected to the consumption of psychoactive substances is the adoption of nonmedical use of prescription drugs [13], such as sedatives, opioids, and stimulants. Even these drugs are often traded and advertised online by fake pharmacies [11,12].

The amount of data available nowadays has made automated text analysis veer towards more machine learning-based approaches. Because complex tasks might require many training examples, however, there is a vivid study on unsupervised and semi-supervised approaches. Our task encompasses identifying names in text, something often associated with named-entity extraction. Unsupervised methods such as [22] use unlabeled data contrasted with other data assumed irrelevant - to use as negative examples - in order to build a classification model. Instead, we use seeds, a small set of examples, because the writers on forums often attempt not to mention drugs explicitly, resorting to paraphrases or nicknames, making a purely contrastive approach difficult to apply. Also, multi-level bootstrapping proved to be a valid improvement in information extraction [17]; this techniques feature an iterative process to gradually enlarge and refine a dictionary of common terms. Our approach, instead, splits the problem of finding candidate terms and classifying them in two separate subproblems, the second of which is fed with a small number of annotated examples, i.e., the seeds. Co-training is a common technique [3] to evaluate whether to use an unlabeled piece of data as a training example: the idea is building different classifiers, and use the label assigned by one as a training example for another. In our case, we instead leverage the redundancy among the data, to ensure candidate examples are selected with a high degree of confidence. Relation extraction is an even more complex task which seeks for the relationships among the entities. This is relevant here, because substances can only be identified basing on their role in the sentence (since common names are often used to refer to them). Work in [18] proposes a method based on corpus statistics that requires no human supervision and no additional corpus resources beyond the corpus used for relation extraction. Our approach does not explicitly address relation extraction, but it exploits the redundancy of a substance (or effect) being often associated with other entities to identify them. KnowItAll [10] is a tool for unsupervised named entity extraction with improved recall, thanks to the pattern learning, the subclass extraction and the list extraction features that still includes bootstrapping to learn domain independent extraction patterns. For us, common mention patterns are also strong indicators of the substance or effect class; however, we do not use patterns to extract, but only, implicitly, for classification purposes. Furthermore, [4] pursues the thesis that much greater accuracy can be achieved by further constraining the learning task, by coupling the semi-supervised training of many extractors for different categories and relations; we use a single multiclass classifier to achieve the same goal. Under the assumption that the number of labeled data points is extremely small and the two classes are highly unbalanced, the authors of [26] realized a stochastic semi-supervised learning approach that was used in the 2009–2010 Active Learning Challenge. While the task is similar, our approach is different, because we do not need to use unlabeled data as negative examples. The framework proposed in [6] suggests to use domain knowledge, such as dictionaries and ontologies, as a way to guide semi-supervised learning, so as to inject knowledge into the learning process. We have not relied on rare expert knowledge for our task, arguing that a few labeled seeds are easier to produce than dictionaries or other forms of expert knowledge representations. A mixed case of learning extraction patterns, relation extraction and injecting expert knowledge is in [2], which also shows the challenge of evaluating a technique when few labeled examples are available. As shown above, the problem of building a model with a limited set of information, but with a large enough amount of data, has been tackled by various angles. Our main staples were: (a) the availability of a large set of unlabeled data, and (b) the availability of a small set of labeled substance and effect names.

6   Conclusions

We have automatically identified and classified substances and effects from posts of drugs forums, making use of a semi-supervised text mining approach. Human intervention is required for the creation of a small training set, but the algorithm is able to automatically discover substances and effects with such a very few initial information. We believe our proposal will help sensitizing drug consumers about the risks of their choices and will contrast the diffusion of NPS, which spread on the online market at an impressive high rate.

Acknowledgements. This publication arises from the project CASSANDRA, (Computer Assisted Solutions for Studying the Availability aNd Distribution of novel psychoActive substances)" which has received funding from the European Union under the ISEC programme.

Prevention of and fight against crime [JUST2013/ISEC/DRUGS/AG/6414].

References

1. Attardi, G., Gull, A., Sebastiani, F.: Theseus: categorization by context. Univ. Comput. Sci. (1998)
2. Bellandi, A., Nasoni, S., Tommasi, A., Zavattari, C.: Ontology-driven relation extraction by pattern discovery. In: Information, Process, and Knowledge Management, pp. 1–6. IEEE Computer Society (2010)
3. Blum, A., Mitchell, T.: Combining labeled and unlabeled data with co-training. In: Computational Learning Theory. pp. 92–100. ACM (1998)
4. Carlson, A., Betteridge, J., Wang, R.C., Hruschka Jr., E.R., Mitchell, T.M.: Coupled semi-supervised learning for information extraction. In: Web Search and Data Mining, pp. 101–110. ACM (2010)
5. Chang, C.C., Lin, C.J.: LIBSVM: a library for support vector machines. ACM Trans. Intell. Syst. Technol. **2**(3), 27:1–27:27 (2011)
6. Chang, M.W., Ratinov, L., Roth, D.: Guiding semi-supervision with constraint-driven learning. In: Annual Meeting - Association for Computational Linguistics, pp. 280–287 (2007)
7. Davey, Z., Schifano, F., Corazza, O., Deluca, P.: e-Psychonauts: conducting research in online drug forum communities. J. Ment. Health **21**(4), 386–394 (2012)
8. Davies, S., et al.: Purchasing legal highs on the Internet - is there consistency in what you get? QJM **103**(7), 489–493 (2010)
9. Del Vigna, F., Avvenuti, M., Bacciu, C., Deluca, P., Marchetti, A., Petrocchi, M., Tesconi, M.: Spotting the diffusion of new psychoactive substances over the internet. arXiv preprint arXiv:1605.03817 (2016)
10. Etzioni, O., Cafarella, M., Downey, D., Popescu, A.M., Shaked, T., Soderland, S., Weld, D.S., Yates, A.: Unsupervised named-entity extraction from the web: an experimental study. Artif. Intell. **165**(1), 91–134 (2005)
11. Freifeld, C.C., Brownstein, J.S., Menone, C.M., Bao, W., Filice, R., Kass-Hout, T., Dasgupta, N.: Digital drug safety surveillance: monitoring pharmaceutical products in Twitter. Drug Saf. **37**(5), 343–350 (2014)
12. Katsuki, T., Mackey, T.K., Cuomo, R.: Establishing a link between prescription drug abuse and illicit online pharmacies: analysis of Twitter data. J. Med. Internet Res. **17**(12) (2015)

13. Mackey, T.K., Liang, B.A., Strathdee, S.A.: Digital social media, youth, and non-medical use of prescription drugs: the need for reform. J. Med. Internet Res. **15**(7), e143 (2013)
14. Marsh, E., Perzanowski, D.: MUC-7 evaluation of IE technology: overview of results. In: Seventh Message Understanding Conference (MUC-7) (1998)
15. Nikfarjam, A., Sarker, A., OConnor, K., Ginn, R., Gonzalez, G.: Pharmacovigilance from social media: mining adverse drug reaction mentions using sequence labeling with word embedding cluster features. J. Am. Med. Inform. Assoc. **22**(3), 671–681 (2015)
16. Penas, A., Verdejo, F., Gonzalo, J.: Corpus-based terminology extraction applied to information access. In: Corpus Linguistics, pp. 458–465 (2001)
17. Riloff, E., Jones, R., et al.: Learning dictionaries for information extraction by multi-level bootstrapping. In: AAAI/IAAI, pp. 474–479 (1999)
18. Rosenfeld, B., Feldman, R.: Using corpus statistics on entities to improve semi-supervised relation extraction from the web. In: Annual Meeting - Association for Computational Linguistics, pp. 600–607 (2007)
19. Salton, G., Buckley, C.: Term-weighting approaches in automatic text retrieval. Inf. Process. Manag. **24**(5), 513–523 (1988)
20. Schifano, F., Corazza, O., Deluca, P., Davey, Z., Furia, L.D., Farre', M., Flesland, L., Mannonen, M., Pagani, S., Peltoniemi, T., Pezzolesi, C., Scherbaum, N., Siemann, H., Skutle, A., Torrens, M., Kreeft, P.V.D.: Psychoactive drug or mystical incense? Overview of the online available information on Spice products. Int. J. Cult. Ment. Health **2**(2), 137–144 (2009)
21. Schmidt, M.M., Sharma, A., Schifano, F., Feinmann, C.: Legal highs on the net-Evaluation of UK-based websites, products and product information. Forensic Sci. Int. **206**(1), 92–97 (2011)
22. Smith, N.A., Eisner, J.: Contrastive estimation: training log-linear models on unlabeled data. In: Annual Meeting - Association for Computational Linguistics, pp. 354–362 (2005)
23. Soussan, C., Kjellgren, A.: Harm reduction and knowledge exchange–a qualitative analysis of drug-related Internet discussion forums. Harm Reduct. J. **11**(1), 1–9 (2014)
24. Watters, P.A., Phair, N.: Detecting illicit drugs on social media using automated social media intelligence analysis (ASMIA). In: Xiang, Y., Lopez, J., Kuo, C.-C.J., Zhou, W. (eds.) CSS 2012. LNCS, vol. 7672, pp. 66–76. Springer, Heidelberg (2012)
25. Witten, H.I., Don, J.K., Dewsnip, M., Tablan, V.: Text mining in a digital library. Int. J. Digit. Libr. **4**(1), 56–59 (2004)
26. Xie, J., Xiong, T.: Stochastic semi-supervised learning on partially labeled imbalanced data. In: Active Learning Challenge Challenges in Machine Learning (2011)
27. Yang, C.C., Yang, H., Jiang, L.: Postmarketing drug safety surveillance using publicly available health-consumer-contributed content in social media. ACM Trans. Manage. Inf. Syst. **5**(1), 2:1–2:21 (2014)

Using Social Media to Measure Student Wellbeing: A Large-Scale Study of Emotional Response in Academic Discourse

Svitlana Volkova[1]([✉]), Kyungsik Han[2], and Courtney Corley[1]

[1] Data Sciences and Analytics Group, Pacific Northwest National Laboratory,
902 Battelle Blvd, Richland, WA 99354, USA
{svitlana.volkova,court}@pnnl.gov
[2] Visual Analytics Group, Pacific Northwest National Laboratory,
902 Battelle Blvd, Richland, WA 99354, USA
kyungsik.han@pnnl.gov

Abstract. Student resilience and emotional wellbeing are essential for both academic and social development. Earlier studies on tracking students' happiness in academia showed that many of them struggle with mental health issues. For example, a 2015 study at the University of California Berkeley found that 47 % of graduate students suffer from depression, following a 2005 study that showed 10 % had considered suicide. This is the first large-scale study that uses signals from social media to evaluate students' emotional wellbeing in academia. This work presents fine-grained emotion and opinion analysis of 79,329 tweets produced by students from 44 universities. The goal of this study is to qualitatively evaluate and compare emotions and sentiments emanating from students' communications across different academic discourse types and across universities in the U.S. We first build novel predictive models to categorize academic discourse types generated by students into personal, social, and general categories. We then apply emotion and sentiment classification models to annotate each tweet with six Ekman's emotions – joy, fear, sadness, disgust, anger, and surprise and three opinion types – positive, negative, and neutral. We found that emotions and opinions expressed by students vary across discourse types and universities, and correlate with survey-based data on student satisfaction, happiness and stress. Moreover, our results provide novel insights on how students use social media to share academic information, emotions, and opinions that would pertain to students academic performance and emotional well-being.

Keywords: Social media analytics · Opinion and emotion prediction · Student wellbeing · Academic discourse

1 Introduction

Social media has been widely used by people as a way of sharing what they do, how they live, where they visit, whom they interact with, what they are interested in, etc. The use of social media goes beyond simply sharing one's personal

© Springer International Publishing AG 2016
E. Spiro and Y.-Y. Ahn (Eds.): SocInfo 2016, Part I, LNCS 10046, pp. 510–526, 2016.
DOI: 10.1007/978-3-319-47880-7_32

life or interests and has been extensively used in various contexts [23]. Especially in the context of education and academia, a great body of research has demonstrated its positive influences. For example, class instructors use Twitter to notify students of any class updates or additional class-related information, and students use Facebook to discuss course materials or issues and have peer-to-peer, social interactions outside the class [6,8]. Research has indicated that instructors' and students' online discussions or activities in a classroom environment show a positive relationship with course engagement and grades [9]. In addition, leveraging social media in Massive Open Online Courses (MOOCs) has been found to increase students retention in class [34]. Scholars (i.e., professors, researchers) use Twitter for professional purposes to access research-related information, share their thoughts or updates related to their research interests, and build professional networks [30]. With social media platforms, people are not only information consumers but also active co-producers of academic content.

Given that heavy use of social media by young generations, including students [23], it is important to understand the emerging practices of social media use and engagement for academic purposes, because those insights can be related to students overall academic engagement, satisfaction, goals, well-being, and career expectation within or after their degree. Although prior research has extensively presented social media influence on education and academia, we realized there are missing components in the effort to better understand student communications in social media as follows.

First, there is a lack of studies that have paid close attention to academic discourse by students by looking into the actual content shared by them. Much prior research has primarily relied on survey responses [1,6,8–10], and it appears that very few studies have looked into and qualitatively analyzed emotions and opinions emanating from the actual content that students share online.

Second, there is little understanding of academic discourse by students by means of a large-scale data analysis. Prior research has mostly relied on small sample sizes (e.g., 100–200 students) and small-scale contexts (e.g., single classroom) which would fail to deliver a comprehensive picture of students' academic engagement through social media platforms.

With these motivations, we study how students use social media broadly, and Twitter specifically, for academic purposes through a mixture of qualitative and quantitative approaches. To do that we collected 26,710 academic-related tweets posted by 133 students and annotated their content as having three main categories (general, personal, and social) and six sub-categories of academic discourse. We then used these data to build classification models to predict academic-discourse types, and applied emotion and sentiment prediction models to predict affects. We applied these models to label 79,329 tweets produced by students from 44 universities with academic-discourse types, sentiments, and emotions. This gives us an opportunity to measure not only a level of academic engagement, but also affects in academic discourse that would pertain to students' academic performance and wellbeing. Our study is original in the following ways:

- Building models to classify academic-discourse types as social, personal, and general in social media.
- Analyzing emotions and sentiments (affects) emanating from social, personal, and general academic discourse.
- Measuring variations in emotions and sentiments expressed by students in different academic discourse across universities.
- Correlating affects expressed in social media with public survey data on student satisfaction rates, the level of happiness, or stress across universities.

Our novel findings demonstrate how emotions and opinions vary across discourse types e.g., sadness is expressed more in personal discourse (students share achievements, activities, thoughts), disgust in general discourse (students report academic information), positive and neutral opinions in social communications (students are involved in academic dialogs), and across universities e.g., students from Ohio University (OU) produce the most joy and positive opinions, and the least sadness, anger, and negative opinions compared to other colleges.

Moreover, by correlating our affect signals in social media with public survey data on student satisfaction rates and university tuition, we found that lower tuition correlates with positive affects, and higher tuition with negative emotions and sentiments. The more students report to be satisfied with their schools the more positive emotions and sentiments are being observed in social media. Thus, similar to recent studies on large-scale public opinion polling [17,28], the results of this work imply that signals from public social media could be a faster and less expensive way to understand public opinions [2].

2   Related Work

2.1   Social Media Use for Academic Purposes

A great body of research has presented how scholars and students use social media for accessing and sharing academic-related information. We have identified two primary research efforts – describing positive effects of social media in this context and articulating how people use social media for academic purposes.

First, understanding the effects of leveraging social media usually refers to its positive outcomes for students and scholars. Researchers studied the role of Twitter used for educational purposes how it would impact students' engagement and grades [9]. From a total of 125 students, they conducted a comparative analysis between control and experimental groups. They found a greater level of engagement and higher grades from the students in the experimental group. Similarly, another work presented the use of Facebook by comparing two groups – higher education faculty ($n = 62$) and students ($n = 120$) through the survey [27]. The results indicate that students are much more likely than faculty to use Facebook and are significantly more open to the possibility of using Facebook and similar technologies to support classroom work. A recent study found that social media use is positively associated with their academic engagement and satisfaction [7]. Other works found that using Facebook for collecting and sharing information was positively predictive of overall GPA, while using Facebook for socializing was negatively predictive [9]. Another study found scholars' positive attitudes and practices toward finding and citing articles through Twitter because of faster speed of citation and showing scholarly impact [24]. Overall, using social media for academic purposes appears to positively influence students' and scholars' academic engagement and achievement.

Second, the practice of using social media for academic purposes incorporates individual and collaborative standpoints. Researchers detailed scholars' practices on Twitter including information, and media sharing, expanding learning opportunities, requesting assistance, connecting and networking [30]. Results in [7] indicated that students mainly use social media for broadcasting and keeping up with up-to-date academic information. However, making connections and developing networks is also one of the primary reasons for social media use.

Although such prior research has presented many aspects of utilizing social media for academic purposes from various populations (e.g., students, instructors) with respect to its effects, practices, and user motivations, there exists a lack of detailing academic activities and discourse through the analysis of content posted by students and presenting large-scale data-driven and comprehensive results. Thus, in this paper, we aim to address these limitations by conducting content analysis and applying the results of the analysis to a large set of tweets. Especially for the large-scale analysis, we developed classification models and obtained the outcomes of academic discourse, emotions, and opinions emanating from 79,329 tweets posted by students in social media across 44 universities.

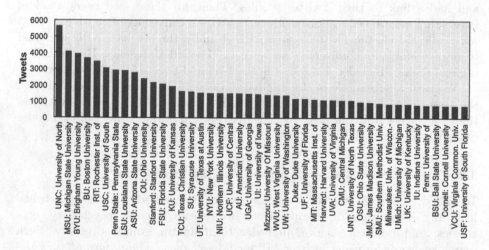

Fig. 1. The list of 44 U.S. universities and the number of tweets per university.

2.2   Sentiment and Emotion Analysis

Emotion analysis[1] has been successfully applied to many kinds of informal and short texts including emails, blogs [11], and news headlines [29]. Although *sentiment* classification in social media has been extensively studied [16,18–20,35], *emotions* in social media, including Twitter and Facebook, have only been investigated recently [32].

Researchers have used supervised models trained on word ngram features, synsets, emoticons, topics, and lexicons to determine which emotions are being expressed on Twitter [15,25,26,33]. Most of this line of work focused on capturing six high-level emotions proposed by Ekman – joy, anger, sadness, fear, disgust, and surprise [5]. Other papers studied moods, including tension, depression, fatigue, and issues such as politeness, rudeness, embarrassment, and formality. To the best of our knowledge, this is the first large-scale study of emotions and opinions expressed in different types of academic discourse produced by students from 44 universities in social media.

3   Data and Methods

3.1   Tweets Annotated with Universities

We base our analysis on a large corpus of tweets tagged with education attributes e.g., universities collected by [13].[2] Education labels were obtained by crawling user profiles on Google Plus[3] to find seed users who reported their education and had a link to their Twitter profiles. Then, for these seed users a set of tweets that mention education entities e.g., #USC (University of South Carolina) were collected via the Twitter search API.[4] Finally, the Freebase API[5] was used to resolve ambiguous university names e.g., Harvard University, Harvard. The original dataset included 124,801 education-related tweets from 7,208 users.

We used a subsample of the original data of 79,329 tweets associated with the most frequently mentioned 44 universities. We excluded universities with less than 750 tweets from our analysis. The distribution of the number of tweets per university is shown in Fig. 1.

3.2   Tweets Anotated with Academic-Discourse Type

To collect academic-relevant discourse, we considered several research fields, including bioinformatics, computer science, social science, psychology, economics, and political science, in order to maintain sample diversity. Such search keywords for using Twitter API include *#biology, #bioinformatics, #bioengineering #computerscience, #hci, #socialscience, #sociology, #psychology,*

---

[1] EmoTag Project: http://nil.fdi.ucm.es/index.php?q=node/186.
[2] http://web.stanford.edu/~jiweil/ACL_profile_data.zip.
[3] Google+ API –https://developers.google.com/+/web/api/rest/.
[4] Twitter API – https://dev.twitter.com/rest/public.
[5] Freebase – https://www.freebase.com/.

#economics, and *#politicalscience*. Through the official Twitter API, we then collected the profiles of the users who posted the tweets with the corresponding hashtags. We chose users who indicated that they are currently a student in their profile. We also only considered current, active Twitter users who posted more than 500 tweets. As a result, we had 133 unique samples. We next collected their tweets posted in 2015, yielding 46,648 tweets in total. We excluded retweets, because we were mostly interested in the content posted by the users, although retweets imply ones similar opinion. We finally used 26,710 tweets for the analysis.

Table 1. The example tweets annotated with three types of academic discourse – social, personal, and general. The total number of annotated tweets is 1,569.

Main category	Sub-category (Tweet Example)
GENERAL (850)	(1) GENERAL ACADEMIC INFORMATION [46]
	New 2 year postgraduate positions [name] in social studies of algorithms and data! 1st deadline Sept 26th.
	(2) OTHER RESEARCH STUDIES, REPORTS, OR ACTIVITIES (804)
	Insightful comparison of the two statistical cultures (data & algorithmic) by [name]. [URL]
PERSONAL (495)	(3) PERSONAL ACADEMIC ACHIEVEMENTS (32)
	I'm excited to intern at @Microsoft this summer!
	(4) PERSONAL ACADEMIC ACTIVITY UPDATES (210)
	Submission complete... First conference submission as a first author. Woo!
	(5) PERSONAL ACADEMIC THOUGHTS OR QUESTIONS (253)
	Is it weird that while I'm so tired w/work, completing tasks somehow makes me feel good? #school
SOCIAL (224)	(6) ACADEMIC INTEREST DIALOGUES (224)
	@ [name1] @ [name2] Seems there is a theoretical limit on the number of citations you can fit into 10 pages...

We used a qualitative method to analyze the data. We manually read over tweets and checked if each tweet contained any academic-related content. First, from a small number of users' Twitter activities, we took an initial data analysis session to identify core themes. We then employed axial coding to further generate categories. Next, categories were refined by an iterative coding process that involved two coders. The preliminary results offered a coding guideline for the next round of coding for new datasets. We continued this coding process until the following round of analysis was not able to discover any more new themes or categories. As a result, we identified a total of 2,074 tweets related to users' academic interests or activities. Our analysis revealed three main and six secondary categories relating to academic activities in social media as shown in Table 1.

3.3 Discourse Type Classification Models

We trained discourse type classification models from the manually annotated data described above to automatically label tweets with three types of academic discourse – general, personal, and social. We used logistic regression implemented in scikit-learn [21] to train models that can predict personal, social, and general academic discourse. We relied on a variety of features including, words ngrams (unigrams, bigrams, and trigrams), tf-idf (term frequency-inverse document frequency), and text embeddings (described below) to learn the mapping between each tweet t and the most likely academic-discourse type value assignment $A(t) = a$ as shown below:

$$\Phi_A(t) = \text{argmax}_a P(A(t) = a \mid t) \tag{1}$$

We relied on pretrained text embeddings such as GLoVe[6] [22], Normalized Pointwise Mutual Information (NPMI) [12] and Word2Vec[7] [14]. We varied the number of embedded clusters $c = [30, 50, 100, \ldots, 2000]$ to estimate the best classification accuracy.

3.4 Sentiment and Emotion Classification Models

Emotions and sentiments directly or indirectly imply the way we feel and think, and what we say or do in online social networks. Though both are affective states, there are important differences between them. Emotions are the states of consciousness in which various internal sensations are experienced. They can be triggered by events in the external environment. Sentiments are our likes and dislikes, and they involve a person-object relationship e.g., people express sentiments towards people, products, or services. Emotions are relatively short in duration, while sentiments display themselves over longer periods of time [4].

We used publicly available emotion and sentiment classification models developed by [31][8] that rely on lexical (word ngrams), syntactic, and stylistic (e.g., elongations, positive and negative emoticons, hashtags, punctuation, and negation) features. The sentiment classifier was trained on 19,555 tweets annotated with three sentiment classes – positive, negative, and neutral. The emotion model was trained on 52,925 tweets annotated with six Ekman's emotions – joy, fear, sadness, surprise, anger, and disgust. Sentiment prediction quality was estimated on 3,223 tweets released as an official SemEval-2013 test set [16]. Emotion prediction quality was evaluated using 10-fold cross validation on their emotion dataset of 52,925 tweets. Prediction performance was reported in terms of weighted F-score – F1 = 0.6 for sentiment (3 categories) and F1 = 0.78 for emotion (6 categories).

---

[6] http://nlp.stanford.edu/projects/glove/.
[7] https://radimrehurek.com/gensim/models/word2vec.html.
[8] Pretrained models were released during NAACL Tutorial on Social Media Predictive analytics: http://naacl.org/naacl-hlt-2015/tutorial-social-media.html.

3.5   Analysis

We report our findings using two types of analyses – university-based and discourse-based as described below. Each tweet $t \in T$, $|T| = 79,329$ is annotated with an emotion $e \in E$, sentiment $s \in S$ and academic discourse $a \in A$:

$$E \to \{\text{joy, sad, fear, disgust, anger, surprise}\},$$
$$S \to \{\text{positive, negative, neutral}\},$$
$$A \to \{\text{social, personal, general}\}.$$

For the *discourse-based analysis* we calculate proportions of emotions and sentiments aggregated over three discourse types as shown for the example emotion below:

$$p_{a=\text{personal}}^{e=\text{joy}} = \frac{\sum_t t_a^e}{\sum_t t_a}. \tag{2}$$

For the *university-based analysis* we aggregate tweets by university $u \in U$, $|U| = 44$ and measure proportions of affects as shown for the example opinion below:

$$p_{u=\text{Harvard}}^{s=\text{positive}} = \frac{\sum_t t_u^s}{\sum_t t_u}. \tag{3}$$

For the *correlation analysis* of student emotions and sentiments expressed in social media and public survey data we relied on several public university rankings: (1) student satisfaction rate reported at myPlan.org,[9] (2) Forbes college ranking as of 2014,[10] (3) the list of top 10 most happy[11] and stressed[12] colleges.

4   Results

This section presents the results of academic-discourse type classification and reports novel findings on emotions and sentiments expressed in different academic discourse across universities in social media.

4.1   Discourse Type Classification

We applied binary vs. frequency-based word ngrams (unigrams and bigrams), tf-idf, and word embedding features to learn and evaluate models for academic-discourse type prediction (Table 2). We found that (a) binary ngrams outperform frequency-based ngrams, and (b) bigrams yield higher performance compared to unigrams – F1 = 0.60. Table 2 reports classification results obtained using word

---

[9] http://www.myplan.com/education/colleges/college_rankings_1.php.

[10] http://www.forbes.com/top-colleges/list/#tab:rank.

[11] http://www.huffingtonpost.com/2013/12/31/happiest-colleges-daily-beast-2013_n_4521921.html.

[12] http://www.universityprimetime.com/top-50-colleges-with-the-most-stressed-out-student-bodies/.

Table 2. Discourse type classification results obtained using 10-fold cross validation. We used tf-idf features to train models.

Features	Precision	Recall	F1
Tweet ngrams	0.65	0.66	0.60
Tweet TDIDF	**0.71**	**0.69**	**0.64**
GLoVe	0.54	0.59	0.56
Word2Vec	0.56	0.61	0.58
NPMI	0.55	0.60	0.57

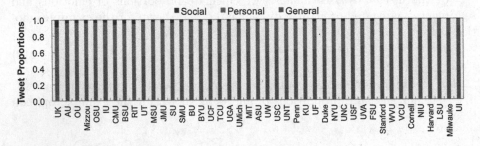

Fig. 2. Proportion of tweets per university classified as general, personal, and social. Proportions are sorted by general type in an ascending order.

embeddings – GLoVe, Word2Vec and NPMI. We found that all embedding types yield comparable performance (F1 is between 0.56 and 0.58), which is significantly lower than tf-idf features. We observed that tf-idf features boost classification performance to F1 = 0.64. We realize that the best model performance in not ideal (F1 = 0.64), but it can be potentially improved by annotating more tweets with academic discourse types e.g., via crowdsourcing.

We thus used the best models learned using tf-idf features to label 79,329 academic-related tweets with three classes – general, social and personal and report the discourse type distribution per university in Fig. 2. We observed that the most representative classes were general (62 %) and personal (34 %), and only 4 % of all tweets were labeled as social.

We observed that students from several universities produced significantly more general than personal tweets, for example, University of Iowa (UI), University of Wisconsin-Milwaukee (Milwaukee), Louisiana State University (LSU), and Harvard. On the other hand, students from some other universities generated an equal amount of general and personal communications, for example, University of Kentucky (UK), American University (AU), and Ohio University (OU). Overall, the result indicates that students from different university utilize social media as a way of describing their academic discourse in different ways.

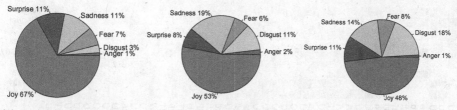

(a) General: 53,190 tweets (b) Personal: 28,461 tweets   (c) Social: 2,498 tweets

Fig. 3. Discourse-based analysis results: emotion proportions extracted from 79,329 tweets aggregated over 44 universities.

Table 3. Discourse-based analysis results: sentiment proportions extracted from 79,329 tweets over 44 universities.

Discourse type	General	Personal	Social
Neutral	29 %	**38 %**	**44 %**
Negative	**51 %**	36 %	24 %
Positive	20 %	25 %	32 %
Subjective	**71 %**	62 %	56 %

4.2 Discourse-Based Analysis

Figure 3 reports the proportion of six basic emotions emanating from three communication types. We found that: *joy* was the most prevalent emotion across all discourse types; *sadness* was expressed more frequently and *surprise* was expressed less frequently in personal compared to other discourse types; *disgust* was expressed more frequently in social discourse than other tweets; *anger* and *fear* were expressed equally across all three communication types.

We report our results on sentiment proportions across three discourse types in Table 3 and outline our key findings below. *Positive* and *neutral* sentiments were expressed more frequently in social tweets. *Negative* opinions were generated more in general tweets. *Subjective opinions (positive and negative)* were expressed more in general discourse compared to social and personal tweets.

Joy was the prevalent emotion across all three discourse types, but when it comes to sentiment, positive was generally lower than others types of opinions. This finding further confirms that emotions and sentiment are both affective states but there are important differences between them.

4.3 University-Based Analysis

For the university-based analysis we aggregated emotions and sentiments expressed in all discourse types by university. Figure 4 demonstrates that affect proportions not only vary across discourse types e.g., personal (achievements, activities, thoughts) vs. general (academic informations, studies) as discussed in

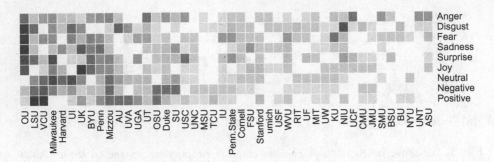

Fig. 4. University-based analysis results: sentiment and emotion proportions expressed across universities extracted from all discourse types. Red↑ represents high, blue↓ represents low sentiment and emotion proportion values. (Color figure online)

a previous section, but also differ by university. We outline our key findings on affect differences across 44 universities and report schools with the most↑ and the least↓ expressed affect proportions below.

- *Anger:* U. of Central Florida (UCF)↑, Ohio University (OU)↓.
- *Disgust:* American University (AU)↑, Northern Illinois University (NIU)↓.
- *Fear:* University of Kansas (KU)↑, Ohio University (OU)↓.
- *Sadness:* University of Kentucky (UK)↑, Ohio University (OU)↓.
- *Surprise:* Louisiana State University (LSU)↑, Milwaukee↓.
- *Joy:* Ohio University (OU)↑, University of Kentucky (UK)↓.
- *Neutral:* University Of Missouri (Mizzou)↑, University of Iowa (UI)↓.
- *Negative:* Virginia Commonwealth Univ. (VCU)↑, Ohio State University (OSU)↓, Duke Univ. (Duke)↓, Syracuse University (SU)↓.
- *Positive:* Ohio State University (OSU)↑, Louisiana State University (LSU)↓, and Virginia Commonwealth Univ. (VCU)↓.

In addition, we found that there are some differences in the degree of expressing emotions and sentiments across universities. For example, BU, NYU, UNT, and ASU only showed few emotions or sentiments compared to others. With these results, we further investigated how the emotions are related to other university-based, publicly available results.

4.4   Correlation Analysis with Public Survey Data

Figure 5 presents Pearson correlation of affects expressed by students in social media and (a) students' satisfaction and (b) university tuition.[13] Our results demonstrate that: (a) the more satisfied students are with their school, the significantly higher positive sentiment and emotions they showed in their academic-related tweets; the less satisfied they are with their school, the higher negative sentiments and emotions they expressed in their academic discourse (Fig. 5, left),

---

[13] We found that correlations with Forbes university ranking are not significant.

(b) the higher tuition schools have, more negative tweets posted by students; the lower tuition schools have, more positive tweets posted by students (Fig. 5, right). Our findings show that students' tweets about their academic life could be potentially used to understand their attitudes toward their school e.g., student satisfaction.

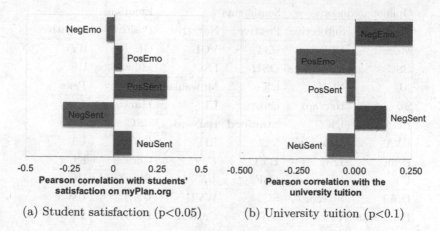

(a) Student satisfaction (p<0.05)   (b) University tuition (p<0.1)

Fig. 5. Comparing university-specific emotion and opinion distributions with public survey data: student satisfaction rate reported on myPlan.org (left) and university tuition (right).

4.5   Top Positive and Negative University Ranking

Table 4 presents top 10 colleges that are most opinionated (express positive and negative sentiments) vs neutral, and most positive vs negative in terms of sentiments and emotions. We found that several schools (highlighted in bold in Table 4) – OSU, Stanford, BYU, FSU and Harvard are among top 10 universities with the most positive sentiments and emotions expressed according to our analysis and are among the top happiest schools in the U.S. according to public survey data.[11] Similarly, we found that several universities (highlighted in italic in Table 4) – UNC, UF, Penn, AU and JMU among top 10 universities with the most negative sentiments and emotions expressed according to our analysis and are among the top stressed schools in the U.S. according to public survey data.[12] This result indicates that affects depicted in students' tweets could be used to understand their happiness (or stress) in their school life.

Lastly, we looked into affects at a state level. Figure 6 aggregates affects across universities for 25 U.S. states (the rest of the states with no affect proportion observations are shown in grey). We observe that students from the universities located in Ohio, Arizona, and Utah express the most positive emotions, and students from the universities located in Virginia, New York, and North Carolina

Table 4. Top 10 colleges that are the most subjective vs. neutral, produce most positive vs. negative opinions, and produce most positive vs. negative emotions. Colleges in bold show the overlap with the top happiest schools[11], and schools in italic show the overlap with the top stressed schools[12] obtained using survey data, demonstrating that affects depicted in students tweets could be used to understand their happiness (or stress) in their school life.

Opinionatedness		Sentiments		Emotions	
Neutral	Subjective	Positive	Negative	Positive	Negative
Mizzou	UI	UST	VCU	OU	BYU
USC	Milwaukee	**OSU**	LSU	Milwaukee	UK
AU	VCU	UK	Milwaukee	**FSU**	*Penn*
SU	Harvard	Duke	UI	**Harvard**	*UNC*
UGA	UK	**Stanford**	Harvard	USC	*AU*
UVA	Carolina	TCU	RIT	LSU	UT
Penn	LSU	**BYU**	*UNC*	Cornell	KU
UT	UW	USF	*UF*	**Stanford**	UGA
Duke	Stanford	SU	WVU	UNC	UW
Penn State	UF	JMU	Cornell	UMich	*JMU*

express the most negative emotions. Similarly, students from the universities located in Kentucky, Ohio, and DC are the most opinionated, and students from the universities located in Indiana and Utah are the most neutral.

5   Discussion and Limitations

Our findings on variations of social, personal, and general academic discourse across universities call for a follow-up study of understanding the relationship between social media use for academic purposes and some other academic perspectives. Prior research has shown how personality affects social media use and engagement [3] and the positive association between social media engagement and final grades for undergraduates [9]. With the categories identified, we can specifically investigate how students academic activities in social media are associated with their personality, motivations of pursuing degrees, academic achievements, or satisfaction, etc. In addition, we can compare academic activities in social media between scholars (e.g., faculty, instructors) and students, and see how each group utilizes social media for academic purposes in different fashions. This could be further used to build models that reflect additional yet important academic aspects.

According to the American College Health Association, 32 % of students say they have felt highly depressed "that it was difficult to function." Even so, the rate of suicide among college students is lower than that of the general population e.g., between 6 and 8 % of students report having suicidal thoughts, but

(a) Positive emotion proportions   (b) Negative emotion proportions

(c) Subjective opinion proportions   (d) Neutral opinion proportions

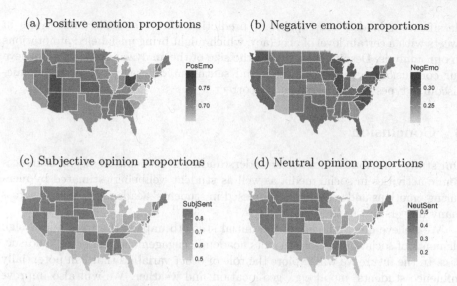

Fig. 6. Geo-located positive and negative emotion proportions, neutral and subjective sentiment proportions aggregated across universities and U.S. states (grey color shows states with no emotion and sentiment data available).

only between 1 and 2 % will actually attempt suicide each year.[12] A recent 2015 study at the UC Berkeley found that 47 % of graduate students suffer from depression, following a 2005 study that showed 10 % had considered suicide.[14] Models designed in this study not only allow understanding students' wellbeing at scale. We found that emotion and opinion signals expressed on Twitter correlate with university tuition, and student satisfaction.

Moreover, our findings on emotions and sentiments expressed in academic discourse across universities not only correlate with public survey data on the most happy and stressed colleges in the U.S., but go beyond that. We report novel findings on how universities are different in terms of their students expressing anger, disgust, fear, and surprise, in addition to joy and sadness emotions. Moreover, in addition to measuring students' emotional responses, we estimate subjective (e.g., positive and negative) vs. neutral opinions expressed in academic discourse. These results further reflect on how students from difference colleges use social media to express their opinions vs. share information.

We acknowledge some limitations of our study. First, for the data collection in the content analysis, our sample may not represent a larger group of students on Twitter. Second, although we had a rigorous data manipulation process, there may still exist human-biases from a manual categorization. We could mitigate these concerns by collecting a large number of students in different domains and inviting more annotators to label the tweets and only use samples which achieved agreement from all coders for the analysis. Next, we note that our

---

[14] UC Berkeley report: http://ga.berkeley.edu/wellbeingreport/.

classification models are capable of predicting affects and discourse types in tweets with a certain level of accuracy, which might bring mislabeled annotations to our analysis. Despite that, due to the size of the analyzed dataset, we believe our conclusions regarding emotion and sentiment differences across academic-discourse types and universities are correct.

6   Conclusion

Our study contributes to a better understanding of students engagement in academic activities in social media as well as student wellbeing estimated by measuring emotions and sentiments expressed in students' academic discourse across many universities in the U.S.

We believe our findings are the initial steps to unpack the roles and design elements of social media for students academic engagement, wellbeing, and success. In the future we will explore the role of other variables that can potentially influence students' mood e.g., geo-location and weather. We will also improve our models to measure emotions and opinions over time, and capture differences in affects expressed by users of different demographics e.g., male vs. female and departments e.g., computer science vs. mathematics across multiple social media platforms e.g., Facebook and Twitter.

References

1. Benson, V., Morgan, S., Filippaios, F.: Social career management: social media and employability skills gap. Comput. Hum. Behav. **30**, 519–525 (2014)
2. Chang, L., Krosnick, J.A.: National surveys via RDD telephone interviewing versus the internet comparing sample representativeness and response quality. Public Opin. Q. **73**(4), 641–678 (2009)
3. Correa, T., Hinsley, A.W., De Zuniga, H.G.: Who interacts on the web? The intersection of users personality and social media use. Comput. Hum. Behav. **26**(2), 247–253 (2010)
4. Desmet, P.M.: Designing emotion (2002)
5. Ekman, P.: An argument for basic emotions. Cogn. Emot. **6**(3–4), 169–200 (1992)
6. Grosseck, G., Bran, R., Tiru, L.: Dear teacher, what should I write on my wall? A case study on academic uses of Facebook. Procedia Soc. Behav. Sci. **15**, 1425–1430 (2011)
7. Han, K., Volkova, S., Corley, C.D.: Understanding roles of social media in academic engagement and satisfaction for graduate students. In: Proceedings of the ACM CHI Conference on Human Factors in Computing Systems, pp. 1215–1221. ACM (2016)
8. Jacobsen, W.C., Forste, R.: The wired generation: academic and social outcomes of electronic media use among university students. Cyberpsychol. Behav. Soc. Netw. **14**(5), 275–280 (2011)
9. Junco, R., Heiberger, G., Loken, E.: The effect of Twitter on college student engagement and grades. J. Comput. Assist. Learn. **27**(2), 119–132 (2011)

10. Knight, C.G., Kaye, L.K.: To tweet or not to tweet? A comparison of academics and students usage of Twitter in academic contexts. Innovations Educ. Teach. Int. pp. 1–11 (2014)
11. Kosinski, M., Stillwell, D., Graepel, T.: Private traits and attributes are predictable from digital records of human behavior. Proc. Natl. Acad. Sci. **110**, 5802–5805 (2013)
12. Lampos, V., Aletras, N., Preoţiuc-Pietro, D., Cohn, T.: Predicting and characterizing user impact on Twitter. In: Proceedings of the 14th Conference of the European Chapter of the Association for Computational Linguistics (EACL), pp. 405–413 (2014)
13. Li, J., Ritter, A., Hovy, E.: Weakly supervised user profile extraction from Twitter. In: Proceedings of ACL (2014)
14. Mikolov, T., Chen, K., Corrado, G., Dean, J.: Efficient estimation of word representations in vector space. arXiv preprint arXiv:1301.3781 (2013)
15. Mohammad, S.M., Kiritchenko, S.: Using hashtags to capture fine emotion categories from tweets. Comput. Intell. **31**, 301–326 (2014)
16. Nakov, P., Rosenthal, S., Kozareva, Z., Stoyanov, V., Ritter, A., Wilson, T.: Semeval-2013 task 2: sentiment analysis in Twitter. In: Proceedings of SemEval, pp. 312–320 (2013)
17. O'Connor, B., Balasubramanyan, R., Routledge, B.R., Smith, N.A.: From tweets to polls: linking text sentiment to public opinion time series. In: Proceedings of ICWSM, pp. 122–129 (2010)
18. Pak, A., Paroubek, P.: Twitter as a corpus for sentiment analysis and opinion mining. In: Proceedings of LREC (2010)
19. Pang, B., Lee, L.: Opinion mining and sentiment analysis. Found. Trends IR **2**(1–2), 1–135 (2008)
20. Pang, B., Lee, L., Vaithyanathan, S.: Thumbs up? Sentiment classification using machine learning techniques. In: Proceedings of EMNLP, pp. 79–86 (2002)
21. Pedregosa, F., Varoquaux, G., Gramfort, A., Michel, V., Thirion, B., Grisel, O., Blondel, M., Prettenhofer, P., Weiss, R., Dubourg, V., Vanderplas, J., Passos, A., Cournapeau, D., Brucher, M., Perrot, M., Duchesnay, E.: Scikit-learn: machine learning in Python. J. Mach. Learn. Res. **12**, 2825–2830 (2011)
22. Pennington, J., Socher, R., Manning, C.D.: Glove: global vectors for word representation. In: Proceedings of EMNLP, vol. 14, pp. 1532–1543 (2014)
23. Perrin, A.: Social media usage: 2005–2015 (2015)
24. Priem, J., Costello, K.L.: How and why scholars cite on Twitter. Proc. Am. Soc. Inf. Sci. Technol. **47**(1), 1–4 (2010)
25. Qadir, A., Riloff, E.: Bootstrapped learning of emotion hashtags# hashtags4you. In: Proceedings of WASSA (2013)
26. Roberts, K., Roach, M.A., Johnson, J., Guthrie, J., Harabagiu, S.M.: Empatweet: annotating and detecting emotions on Twitter. In: Proceedings of LREC (2012)
27. Roblyer, M.D., McDaniel, M., Webb, M., Herman, J., Witty, J.V.: Findings on Facebook in higher education: a comparison of college faculty and student uses and perceptions of social networking sites. Internet High. Educ. **13**(3), 134–140 (2010)
28. Salathé, M., Khandelwal, S.: Assessing vaccination sentiments with online social media: implications for infectious disease dynamics and control. PLoS Comput. Biol. **7**(10), e1002199 (2011)
29. Strapparava, C., Mihalcea, R.: Semeval-2007 task 14: affective text. In: Proceedings of SemEval, pp. 70–74 (2007)

30. Veletsianos, G.: Higher education scholars' participation and practices on Twitter. J. Comput. Assist. Learn. **28**(4), 336–349 (2012)
31. Volkova, S., Yoram, B.: On predicting sociodemographic traits and emotions from communications in social networks and their implications to online self-disclosure. Cyberpsychol. Behav. Soc. Netw. **18**(12), 726–736 (2015)
32. Volkova, S., Bachrach, Y., Armstrong, M., Sharma, V.: Inferring latent user properties from texts published in social media (demo). In: Proceedings of AAAI (2015)
33. Wang, W., Chen, L., Thirunarayan, L., Sheth, A.P.: Harnessing Twitter big data for automatic emotion identification. In: Proceedings of SocialCom, pp. 587–592 (2012)
34. Zheng, S., Han, K., Rosson, M.B., Carroll, J.: The role of social media in MOOCs: how to use social media to enhance student retention. In: Proceedings of the ACM Conference on Learning@ Scale, pp. 419–428. ACM (2016)
35. Zhu, X., Kiritchenko, S., Mohammad, S.M.: NRC-Canada-2014: recent improvements in the sentiment analysis of tweets. In: Proceedings of SemEval (2014)

EmojiNet: Building a Machine Readable Sense Inventory for Emoji

Sanjaya Wijeratne[✉], Lakshika Balasuriya, Amit Sheth, and Derek Doran

Kno.e.sis Center, Wright State University, Dayton, OH, USA
{sanjaya,lakshika,amit,derek}@knoesis.org
http://www.knoesis.org

Abstract. Emoji are a contemporary and extremely popular way to enhance electronic communication. Without rigid semantics attached to them, emoji symbols take on different meanings based on the context of a message. Thus, like the word sense disambiguation task in natural language processing, machines also need to disambiguate the meaning or 'sense' of an emoji. In a first step toward achieving this goal, this paper presents EmojiNet, the first machine readable sense inventory for emoji. EmojiNet is a resource enabling systems to link emoji with their context-specific meaning. It is automatically constructed by integrating multiple emoji resources with BabelNet, which is the most comprehensive multilingual sense inventory available to date. The paper discusses its construction, evaluates the automatic resource creation process, and presents a use case where EmojiNet disambiguates emoji usage in tweets. EmojiNet is available online for use at http://emojinet.knoesis.org.

Keywords: EmojiNet · Emoji analysis · Emoji sense disambiguation

1 Introduction

Pictographs commonly referred to as 'emoji' have grown from their introduction in the late 1990's by Japanese cell phone manufacturers to an incredibly popular form of computer mediated communication (CMC). Instagram reported that as of April 2015, 40 % of all messages posted on Instagram consist of emoji [6]. From a 1 % random sample of all tweets published from July 2013 to July 2016, the service Emojitracker reported its processing of over 15.6 billion tweets with emoji[1]. Creators of the SwiftKey Keyboard for mobile devices also report that 6 billion messages per day contain emoji [15]. Even authorities on language use have acknowledged emoji; the American Dialect Society selected 'eggplant' 🍆 as the most notable emoji of the year[2], and The Oxford Dictionary recently

---

The rights of this work are transferred to the extent transferable according to title 17 U.S.C. 105.

[1] http://www.emojitracker.com/api/stats.

[2] http://www.americandialect.org/2015-word-of-the-year-is-singular-they.

© Springer International Publishing AG 2016
E. Spiro and Y.-Y. Ahn (Eds.): SocInfo 2016, Part I, LNCS 10046, pp. 527–541, 2016.
DOI: 10.1007/978-3-319-47880-7_33

	😂			🔫			💰	
Sense	**Example**		**Sense**	**Example**		**Sense**	**Example**	
Laugh (noun)	I can't stop laughing 😂		Kill (verb)	He tried to kill one of my brothers last year. 🔫		Costly (Adjective)	Can't buy class la 💰	
Happy (noun)	Got all A's but 1😂 😂		Shot (noun)	Oooooooh shots fired! 🔫 🔫 🔫		Work hard (noun)	Up early on the grind 💰	
Funny (Adjective)	Central Intelligence was damn hilarious!😂		Anger (noun)	Why this the only emotion I know to show anger? 🔫		Money (noun)	Earn money when one register /w ur link 💰	

Fig. 1. Emoji usage in social media with multiple senses.

awarded 'face with tears of joy' 😂 as the word of the year in 2015[3]. All these reports suggest that emoji are now an undeniable part of the world's electronic communication vernacular.

People use emoji to add color and whimsiness to their messages [7] and to articulate hard to describe emotions [1]. Perhaps by design, emoji were defined with no rigid semantics attached to them[4], allowing people to develop their own use and interpretation. Thus, similar to words, emoji can take on different meanings depending on context and part-of-speech [8]. For example, consider the three emoji 😂, 🔫, and 💰 and their use in multiple tweets in Fig. 1. Depending on context, we see that each of these emoji can take on wildly different meanings. People use the 😂 emoji to mean laughter, happiness, and humor; the 🔫 emoji to discuss killings, shootings or anger; and the 💰 emoji to express that something is expensive, working hard to earn money or simply to refer to money.

Knowing the meaning of an emoji can significantly enhance applications that study, analyze, and summarize electronic communications. For example, rather than stripping away emoji in a preprocessing step, sentiment analysis application reported in [11] uses emoji to improve its sentiment score. However, knowing the meaning of an emoji could further improve the sentiment score. A good example for this scenario would be the 😂 emoji, where people use it to describe both happiness (using senses such as laugh, joy) and sadness (using senses such as cry, tear). Knowing under which sense the emoji is being used could help to understand its sentiment better. But to enable this, a system needs to understand the particular meaning or *sense* of the emoji in a particular instance. However, no resources have been made available for this task [8]. This calls for the need of a machine readable *sense inventory for emoji* that can provide information such as: (i) the plausible part-of-speech tags (PoS tags) for a particular use of emoji; (ii) the definition of an emoji and the senses it is used in; (iii) example uses of emoji for each sense; and (iv) links of emoji senses to other inventories or knowledge bases such as BabelNet or Wikipedia. Current research on emoji analysis has been limited to emoji-based sentiment analysis [11], emoji-based emotion

---

[3] http://blog.oxforddictionaries.com/2015/11/word-of-the-year-2015-emoji/.

[4] http://www.unicode.org/faq/emoji_dingbats.html#4.0.1.

analysis [17], and Twitter profile classification [2,18] etc. However, we believe introduction of an emoji sense inventory can open up new research directions on emoji sense disambiguation, emoji similarity analysis, and emoji understanding.

This paper introduces **EmojiNet**, the first machine readable sense inventory for emoji. EmojiNet links emoji represented as Unicode with their English language meanings extracted from the Web. To achieve this linkage, EmojiNet integrates multiple emoji lexicographic resources found on the Web along with BabelNet [10], a comprehensive machine readable sense inventory for words, to infer sense definitions. Our contributions in this work are threefold:

1. We integrate four openly available emoji resources into a single, query-able dictionary of emoji definitions and interpretations;
2. We use word sense disambiguation techniques to assign senses to emoji;
3. We integrate the disambiguated senses in an open Web resource, EmojiNet, which is presently available for systems to query.

The paper also discusses the architecture and construction of EmojiNet and presents an evaluation of the process to populate its sense inventory.

This paper is organized as follows. Section 2 discusses the related literature and frames how this work differs from and furthers existing research. Section 3 discusses our approach and explains the techniques we use to integrate different resources to build EmojiNet. Section 4 reports on the evaluation of the proposed approach and the evaluation results in detail. Section 5 offers concluding remarks and plans for future work.

2   Related Work

Emoji was first introduced in the late 1990s but did not become a Unicode standard until 2009 [5]. Following standardization, emoji usage experienced major growth in 2011 when the Apple iPhone added an emoji keyboard to iOS, and again in 2013 when the Android mobile platform began emoji support [6]. In an experiment conducted using 1.6 million tweets, Novak *et al.* report that 4 % of them contained at least one emoji [11]. Their recent popularity explains why research about their use is not as extensive as the research conducted on emoticons [8], which are the predecessor to emoji [11] that used to represent facial expression, emotion or to mimic nonverbal cues in verbal speech [13] in CMC.

Early research on emoji focuses on understanding its role in computer-aided textual communications. From interviews of 20 participants in close personal relationships, Kelly *et al.* reported that people use emoji to maintain conversational connections in a playful manner [7]. Pavalanathan *et al.* studied how emoji compete with emoticons to communicate paralinguistic content on social media [12]. They report that emoji were gaining popularity while emoticons were declining in Twitter communications. Miller *et al.* studied whether different emoji renderings would give rise to diverse interpretations [8], finding disagreements based on the rendering. This finding underscores the need for tools to help machines disambiguate the meaning and interpretation of emoji.

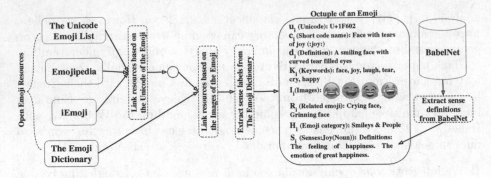

Fig. 2. Building EmojiNet by integrating multiple open resources.

The Emoji Dictionary[5] is a promising Web resource for emoji sense disambiguation. It is a crowdsourced emoji dictionary that provides emoji definitions with user defined sense labels, which are `word(PoS tag)` pairs such as `laugh(noun)`. However, it cannot be utilized by a machine for several reasons. First, it does not list the Unicode or short code names for emoji, which are common ways to programmatically identify emoji characters in text. Secondly, it does not list sense definitions and example sentences along with different sense labels for emoji. Typically, when using machine readable dictionaries, machines use such sense definitions and example sentences to generate contextually relevant words for each sense in the dictionary. Thirdly, the reliability of the sense labels is unclear as no validation of the sense labels submitted by the crowd is performed. With EmojiNet, we address these limitations by linking The Emoji Dictionary with other rich emoji resources found on the Web. This allows sense labels to be linked with their Unicode and short code name representations and discards human-entered sense labels for emoji that are not agreed upon by the resources. EmojiNet also links sense labels with BabelNet to provide definitions and example usages for different senses of an emoji.

3   Building EmojiNet

We formally define EmojiNet as a collection of octuples representing the senses of an emoji. Let E be the set of all emoji in EmojiNet. For each $e_i \in E$, EmojiNet records the octuple $e_i = (u_i, c_i, d_i, K_i, I_i, R_i, H_i, S_i)$, where u_i is the Unicode representation of e_i, c_i is the short code name of e_i, d_i is a description of e_i, K_i is the set of keywords that describe basic meanings attached to e_i, I_i is the set of images that are used in different rendering platforms such as the iPhone and Android, R_i is the set of related emoji for e_i, H_i is the set of categories that e_i belongs to, and S_i is the set of different senses in which e_i can be used within a sentence. An example for an octuple notation is shown as part of Fig. 2. Each element in the octuple provides essential information for sense disambiguation.

---

[5] http://emojidictionary.emojifoundation.com/home.php?learn

EmojiNet uses unicode u_i and short code name c_i of an emoji $e_i \in E$ to uniquely identify e_i, and hence, to search EmojiNet. d_i is needed to understand what is represented by the emoji. It can also help to understand how an emoji should be used. K_i is essential to understand different human-verified senses that an emoji could be used for. I_i is needed to understand the rendering differences in each emoji based on different platforms. Images in I_i can also help to conduct similar studies as [8], where the focus is to disambiguate the different representations of the same emoji on different rendering platforms. R_i and H_i could be helpful in tasks such as calculating emoji similarity and emoji sense disambiguation. Finally, S_i is the key enabler of EmojiNet as a tool to support emoji sense disambiguation as S_i holds all sense labels and their definitions for e_i based on crowd and lexicographic knowledge. Next, we describe the open information EmojiNet extracts and integrates from the Web to construct the octuples.

3.1 Open Resources

Several emoji-related open resources are available on the Web, each carrying their own strengths and weaknesses. Some have overlapping information, but none has all of the elements required for a machine readable sense inventory. Thus, EmojiNet collects information across multiple open resources, linking them together to build the sense inventory. We describe the resources EmojiNet utilizes below.

Unicode Consortium. Unicode is a text encoding standard enforcing a uniform interpretation of text byte code by computers[6]. The consortium maintains a complete list of the standardized Unicodes for each emoji[7] along with manually curated keywords and images of emoji. Let the set of all emoji available in the Unicode emoji list be E_U. For each emoji $e_u \in E_U$, we extract the Unicode character u_i of e_u, the set of all images I_{e_u} associated with e_u that are used to display e_u on different platforms, and the set of keywords $K_{U_{e_u}} \subset K_{e_u}$ associated with e_u, where K_{e_u} is the set of all manually-assigned keywords available for the emoji e_u.

Emojipedia. Emojipedia is a human-created emoji reference site[8]. It provides Unicode representations for emoji, images for each emoji based on different rendering platforms, short code names, and other emoji manually-asserted to be related. Emojipedia organizes emoji into a pre-defined set of categories based on how similar the concepts are represented by each emoji, i.e., Smileys & People, Animals & Nature, or Food & Drink. Let the set of all emoji available in Emojipedia be E_P. For each emoji $e_p \in E_P$, we extract the Unicode representation u_p, the short code name c_p, and the emoji definition d_p of e_p, the set of related emoji R_{e_p}, and its category set H_{e_p}.

---

[6] http://www.unicode.org/.
[7] http://www.unicode.org/emoji/charts/full-emoji-list.html.
[8] https://en.wikipedia.org/wiki/Emojipedia.

iEmoji. iEmoji[9] is a service tailored toward understanding how emoji are being used in social media posts. For each emoji, it provides a human-generated description, its Unicode character representation, short code name, images across platforms, keywords describing the emoji, its category within a manually-built hierarchy, and examples of its use in social media (Twitter) posts. Let the set of all emoji available in iEmoji be E_{IE}. For each emoji $e_{ie} \in E_{IE}$, we collect the Unicode representation u_{ie} of e_{ie} and the set of keywords $K_{IE_{e_{ie}}} \subset K_{e_{ie}}$ associated with e_{ie}, where $K_{e_{ie}}$ is the set of all keywords available for e_{ie}.

The Emoji Dictionary. The Emoji Dictionary[10] is a crowdsourced site providing emoji definitions with sense labels based on how they could be used in sentences. It organizes meanings for emoji under three part-of-speech tags, namely, nouns, verbs, and adjectives. It also lists an image of the emoji and its definition with example uses spanning multiples sense labels. Let the set of all emoji available in The Emoji Dictionary be E_{ED}. For each emoji $e_{ed} \in E_{ED}$, we extract its image $i_{e_{ed}} \in I_{ED}$, where I_{ED} is the set of all images of all emoji in E_{ED} and the set of crowd-generated sense labels $S_{e_{ed}}$.

BabelNet. BabelNet is the most comprehensive multilingual machine readable semantic network available to date [10]. It is a dictionary with lexicographic and encyclopedic coverage of words tied to a semantic network that connects concepts in Wikipedia to the words in the dictionary. It is built automatically by merging lexicographic data in WordNet with the corresponding encyclopedic knowledge extracted from Wikipedia[11]. BabelNet has been shown effective in many research areas including word sense disambiguation [10], semantic similarity, and sense clustering [4]. For the set of all sense labels $S_{e_{ed}}$ in each $e_{ed} \in E_{ED}$, we extract the sense definitions and examples (if available) for each sense label $s_{e_{ed}} \in S_{e_{ed}}$ from BabelNet.

Table 1. Emoji data available in open resources

Emoji resource	u	c	d	K	I	R	H	S
Unicode Consortium	✓	✓	✗	✓	✓	✗	✗	✗
Emojipedia	✓	✓	✓	✗	✓	✓	✓	✗
iEmoji	✓	✓	✓	✓	✓	✗	✓	✗
The Emoji Dictionary	✗	✗	✗	✗	✓	✗	✗	✓

Table 1 summarizes the data about an emoji available across the four open resources. A '✓' denotes the availability of the information in the resource where

---

[9] http://www.iemoji.com/.
[10] http://emojidictionary.emojifoundation.com/home.php?learn.
[11] http://babelnet.org/about.

'✕' denotes the non-availability. It is important to note that unique crowds of people deposit information about emoji into each resource, making it important to integrate the same type of data across many resources. For example, the set of keywords K_i, the set of related emoji R_i, and the set of categories H_i for an emoji e_i are defined by the crowds qualitatively, making it necessary to compare and scrutinize them to determine the elements that should be considered by EmojiNet. Data types that are 'fixed', e.g. the Unicode u_i of an emoji e_i, will also be useful to link data about the same emoji across the resources. We also note that The Emoji Dictionary uniquely holds the sense labels of an emoji, yet does not store its Unicode u_i. This requires EmojiNet to link to this resource via emoji images, as we discuss further in the next section.

3.2 Integrating Emoji Resources

We now describe how EmojiNet integrates the open resources described above. The integration, illustrated in Fig. 2, starts with the Unicode's emoji characters list as it is the official list of 1,791 emoji accepted by the Unicode Consortium for support in standardized software products. Using Unicode character representation in the emoji list, we link these emoji along with the information extracted from Emojipedia and the iEmoji websites. Specifically, for each emoji $e_u \in E_U$, we compare u_u, with all Unicode representations of the emoji in E_P and E_{IE}. If there is an emoji $e_u \in E_U$ such that $u_u = u_p = u_{ie}$, we merge the three corresponding emoji $e_u \in E_U$, $e_p \in E_P$, and $e_{ie} \in E_{IE}$ under a single emoji representation $e_i \in E$. In other words, we merge all emoji where they share the same Unicode representation. We store all the information extracted from the merged resources under each emoji e_i as the octuple described in Sect. 3.

Linking to the Emoji Dictionary. Unfortunately, The Emoji Dictionary does not store the Unicode of an emoji. Thus, we merge this resource into EmojiNet by considering emoji *images*. We created an index of multiple images of the 1,791 Unicode defined emoji in the Unicode Consortium website. We have downloaded a total of 13,387 images for the 1,791 emoji. These images are referred to as our example image dataset I_x. We additionally downloaded images of all emoji listed on The Emoji Dictionary website, which resulted in a total of 1,074 images. We refer to this set of images as the test image dataset I_t.

To align the two datasets, we implement a nearest neighborhood-based image matching algorithm [14] that matches each image in I_t with the images in I_x. Because images are of different resolutions, we normalize them into a 300×300 px space and then divide them along a lattice of 25 non-overlapping regions of size 25×25 px. We then find an average color intensity of each region by averaging its R, G and B pixel color values. To calculate the dissimilarity between two images, we sum the L_2 distance of the average color intensities of the corresponding regions. The final accumulated value that we receive for a pair of images will be a measure of the dissimilarity of the two images. For each image in I_t, the least dissimilar image from I_x is chosen and the corresponding emoji octuple information is merged.

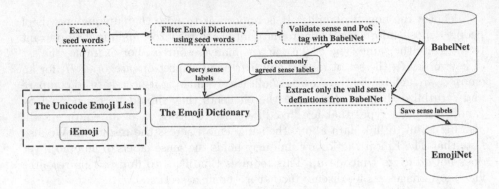

Fig. 3. Emoji sense and part-of-speech filtering.

Emoji Sense and Part-of-Speech Filtering. With The Emoji Dictionary linked to the rest of the open resources via images, EmojiNet can now integrate its sense and part-of-speech information (sense labels). However, as mentioned in Sect. 2, The Emoji Dictionary does not validate the sense labels collected from the crowd. Thus, EmojiNet must pre-process the sense labels from The Emoji Dictionary to verify its reliability. This is done in a three step process and it is elaborated in Fig. 3. First, we use the set of keywords K_i of emoji e_i collected from the Unicode Consortium and iEmoji as seed words to identify reliable sense labels. These keywords are human-generated and represent the meanings in which an emoji can be used. For example, the 😂 emoji has been tagged with the keywords *face, joy, laugh, and tear* in the Unicode emoji list and *tear, cry, joy, and happy* in the iEmoji website. Taking the union of these lists as a set of seed words, we filter the crowdsourced sense labels of an emoji from The Emoji Dictionary. For each keyword $k_i \in K_i$, we extract crowdsourced sense labels. For example, for the emoji 😂, The Emoji Dictionary lists three sense labels for the sense *laugh* as `laugh(noun)`, `laugh(verb)` and `laugh(adjective)`. However, the word *laugh* cannot be used as an adjective in the English language. Therefore, in the second step, we cross-check if the sense labels extracted from The Emoji Dictionary are valid using the information available in BabelNet. In this step, BabelNet reveals that *laugh* cannot be used as an adjective, so we discard `laugh(adjective)` and use `laugh(noun)`, `laugh(verb)` in EmojiNet. We do this for all seed keywords we obtain from the Unicode emoji list and iEmoji websites. In the final step, for any sense label in The Emoji Dictionary that is not a seed word but was submitted by more than one human (commonly agreed senses in Fig. 3), EmojiNet validates these sense labels using BableNet. For example, the sense label `funny(adjective)` has been added by more than one user to The Emoji Dictionary as a possible sense for 😂 emoji. This was not in our seed set; however, since there is common agreement on the sense label `funny(adjective)` and the word *funny* can be used as an adjective in the English language, EmojiNet extracts `funny(adjective)` from The Emoji Dictionary and

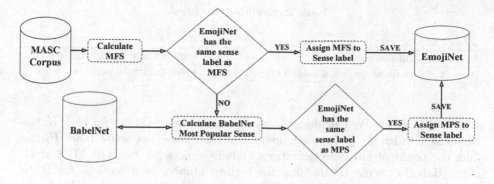

Fig. 4. Using BabelNet to assign sense definitions.

adds it to its sense inventory under 😊 emoji. Note that EmojiNet does not assign BabelNet sense IDs (or sense definitions) to the extracted sense labels yet. That process will require a word sense disambiguation step, which we will discuss in the next section.

3.3 Linking Emoji Resources with BabelNet

Having access to sense labels extracted from The Emoji Dictionary for each emoji, EmojiNet can now link these sense labels with BabelNet. This linking allows EmojiNet to interpret each sense label on how it can be used in a sentence. For example, the current version of BabelNet lists 6 different sense definitions for the sense label laugh(noun). Thus, EmojiNet must select the most appropriate sense definition out of the six. As we described in Sect. 2, The Emoji Dictionary does not link its sense labels with example sentences. Therefore, we cannot directly perform WSD on the sense labels or example sentences available in The Emoji Dictionary. Thus, to align the two resources, we use the MASC[12] corpus with a most frequent sense (MFS) baseline for WSD. MASC corpus is a balanced dataset that represents different text categories such as tweets, blogs, emails, letters, essays, and speech; words in the MASC corpus are already disambiguated using BabelNet [9]. We use these disambiguated words to calculate MFS for each word in the MASC corpus. Once the MFS of each word is calculated, for every sense label in EmojiNet, we assign its definition as the MFS of that same sense label retrieved from MASC corpus. We use an MFS-based WSD baseline due to the fact that MFS is a very strong, hard-to-beat baseline model for WSD tasks [3]. Figure 4 depicts the steps followed in our WSD approach.

EmojiNet has a total of 3,206 sense labels that need to be sense disambiguated using BabelNet. However, not all sense labels in EmojiNet were assigned BabelNet senses in the above WSD task. There were sense labels in EmojiNet which were not present in MASC corpus, hence they were not disambiguated. To disambiguate such sense labels which were left out, we define a

---

[12] https://en.wikipedia.org/wiki/Manually_Annotated_Sub-Corpus_(MASC).

Table 2. EmojiNet statistics

Emoji statistic	u	c	d	K	I	R	H	S
Number of emoji with each feature	1,074	845	1,074	1,074	1,074	1,002	705	875
Amount of data stored for each feature	1,074	845	1,074	8,069	28,370	9,743	8	3,206

second WSD task. We calculate the most popular sense (MPS) for each Babel-Net sense, which we define as follows. For each BabelNet sense label B_s, we take the count of all sense definitions BabelNet lists for B_s. The MPS of B_s is the BabelNet sense ID that has the highest number of definitions for B_s. If there are more than one MPS available, a sense ID will be picked randomly out of the set of MPS sense IDs as the MPS. Once the MPS is calculated, those will be assigned to their corresponding sense labels in EmojiNet which were left out in the first WSD task. Note that BabelNet holds multiple definitions that come from multiple resources such as WordNet, VerbNet, Wikipedia, etc. which are integrated into it. Hence, MPS of B_s gives an indication of the popularity of B_s. With this step, we complete the integration of open resources to create EmojiNet.

3.4   EmojiNet Web Application

We expose EmojiNet as a web application at http://emojinet.knoesis.org/. The current version of EmojiNet supports searching for emoji based on Unicode character and short code name. It also lets the user search emoji by specifying a part-of-speech tagged sense and returns a list of emoji that are tagged with the searched sense. Table 2 lists statistics for EmojiNet. It currently holds a total of 1,074 emoji. It has a total of 3,206 valid sense definitions that are shared among 875 emoji, with an average of 4 senses per emoji. The resource is freely available to the public for research use[13].

4   Evaluation

Note that the construction of EmojiNet is based on linking multiple open resources together in an automated fashion. We thus evaluate the automatic creation of EmojiNet. In particular, we evaluate the nearest neighborhood-based image processing algorithm that we used to integrate emoji resources and the most frequent sense-based and most popular sense-based word sense disambiguation algorithms that we used to assign meanings to emoji sense labels. Note that we do not evaluate the usability of EmojiNet based on its performance on a selected task or a benchmark dataset. While sense inventories such as BabelNet have been evaluated on benchmark datasets for WSD or word similarity calculation performance, emoji sense disambiguation and finding emoji similarity are two research problems on their own that have not been explored yet [8]. The

---

[13] http://emojinet.knoesis.org/.

focus of this paper is not to study or solve those problems. Evaluating the usefulness of EmojiNet should, and will, be addressed once emoji similarity tasks and emoji sense disambiguation tasks are defined with baseline datasets. In lieu of task evaluation, we demonstrate the usefulness of EmojiNet with a use case of how it can be used to address the emoji sense disambiguation problem.

4.1 Evaluating Image Processing Algorithm

We next evaluate how well the nearest neighborhood-based image processing algorithm could match each image in I_t with the images in I_x. I_x could contain multiple images for a given emoji (7 images per emoji on average), based on different rendering platforms on which the emoji could appear. The set of all different images $I_i \in I_x$ that belongs to e_i are tagged with the Unicode representation u_i, which is the Unicode representation of e_i. For us to find a match between I_t and I_x, we only require one of the multiple images that represents an emoji from I_x match with any image from I_t. Once the matching process is done, we pick the top ranked match based on the dissimilarity of the two matched images and manually evaluate them for equality. While the image processing algorithm we used is naive, it works well in our use case due to several reasons. First, the images of emoji are not complex as they represent a single object (e.g. eggplant) or face (e.g. smiling face). Second, the emoji images do not contain very complex color combinations as in textures and they are small in size. The image processing algorithm combines color (spectral) information with spatial (position/distribution) information and tends to represent those features well when the images are simple. Third, Euclidean distance ($L2$ distance) prefers many medium disagreements to one large disagreement as in $L1$ distance. Therefore, this nearest neighborhood-based image processing algorithm fits well for our problem.

Manual evaluation of the algorithm revealed that it achieves 98.42 % accuracy in aligning images in I_t with I_x. Out of the 1,074 image instances we checked, our algorithm could correctly find matching images for 1,057 images in I_t and it could not find correct matches for 17 images. We checked the 17 incorrect alignments manually and found that eight were clock emoji that express different times of the day. Those images were very similar in color despite the fact that the two arms in the clocks were at different positions. There were three incorrect alignments involving people characters present in the emoji pictures. Those images had minimal differences, which the image processing algorithm could not identify correctly. There were two instances where flags of countries were aligned incorrectly. Again, those flags were very similar in color (e.g. Flag of Russia and Flag of Solvenia). In our error analysis, we identified that the image processing algorithm does not perform correctly when the images are very similar in color but have slight variations in the object(s) it renders. Since the color of the image plays a huge role in this algorithm, the same picture taken in different lighting conditions (i.e. changes in the background color, while the image color stays the same) could decrease the accuracy of the program. However, that does not apply in our case as all the images we considered have a transparent background.

4.2 Evaluating Word Sense Disambiguation Algorithm

Here we discuss how we evaluate the most frequent sense-based and most popular sense-based word sense disambiguation algorithms that we used to link Emoji senses with BabelNet sense IDs. We use a manual evaluation approach based on human judges to validate whether a BabelNet sense ID assigned to an emoji sense is valid. We sought the help of two human judges in this task and our judges were graduate students who had either worked on or taken a class on natural language processing. We provided them with all the emoji included in EmojiNet, listing all the valid sense labels extracted from The Emoji Dictionary and their corresponding BabelNet senses (BabelNet sense IDs with definitions) extracted from each WSD approach. The human judges were asked to mark whether they thought that the suggested BabelNet sense ID was the correct sense ID for the emoji sense label listed. If so, they would mark the sense ID prediction as correct, otherwise they would mark it as incorrect. We calculated the agreement between the two judges for this task using Cohen's kappa coefficient[14] and obtained an agreement value of 0.7355, which is considered to be a good agreement.

Out of the 3,206 sense labels to disambiguate, the MFS-based method could disambiguate a total of 2,293 sense labels. Our judges analysed these sense labels manually and marked 2,031 of them as correct, with an accuracy of 88.57 % for the MFS-based WSD task. There were 262 cases where the emoji sense was not correctly captured. The correctly dissambiguated sense labels belong to 835 emoji. The 913 sense labels which were not disambiguated in the MFS-based WSD task were considered in a second WSD task, based on the MPS. Our evaluation revealed that the MPS-based WSD task could correctly disambiguate 700 sense labels, with an accuracy of 76.67 %. There were 213 cases where our MPS-based approach failed to correctly disambiguate the sense label. The correctly dissambiguated sense labels belong to 446 emoji.

Table 3 integrates the results obtained by both word sense disambiguation algorithms for different part-of-speech tags. The results shows the two WSD approaches we used have performed reasonably well in disambiguating the sense labels in EmojiNet. They have sense-disambiguated with an combined accuracy of 85.18 %. These two methods combined have assigned BabelNet sense IDs to a total of 875 emoji out of the 1,074 emoji we extracted from The Emoji Dictionary website. It shows that our WSD approaches combined have disambiguated senses for 81.47 % of the total number of emoji present in The Emoji Dictionary. However, we do not report on the total number of valid sense labels that we did not extract in our data extraction process since The Emoji Dictionary had 16,392 unique sense labels, which were too big for one to manually evaluate.

4.3 EmojiNet at Work

We also provide an illustration of EmojiNet in action with a disambiguation of the sense of the 🙏 emoji as it is used in two example tweets. We choose this

---
[14] https://en.wikipedia.org/wiki/Cohen's_kappa

Table 3. Word sense disambiguation statistics

	Correct	Incorrect	Total
Noun	1,271 (83.28 %)	255 (16.71 %)	1,526
Verb	735 (84.00 %)	140 (16.00 %)	875
Adjective	725 (90.06 %)	80 (9.93 %)	805
Total	2,731 (85.18 %)	475 (14.81 %)	3,206

emoji since it is reported as one of the most misused emoji on social media[15]. The tweets we consider are:

T_1: Pray for my family 🙏 God gained an angel today.

T_2: Hard to win, but we did it man 🙏 Lets celebrate!

EmojiNet lists `high five(noun)` and `pray(verb)` as valid senses for the above emoji. For `high five(noun)`, EmojiNet lists three definitions and for `pray(verb)`, it lists two definitions. We take all the words that appear in their corresponding definitions as possible context words that can appear when the corresponding sense is being used in a sentence (tweet in this case). For each sense, EmojiNet extracts the following sets of words:

`pray(verb)` : {*worship, thanksgiving, saint, pray, higher, god, confession*}
`highfive(noun)` : {*palm, high, hand, slide, celebrate, raise, person, head, five*}

To calculate the sense of the 🙏 emoji in each tweet, we calculate the overlap between the words which appear in the tweet with words appearing with each emoji sense listed above. This method is called the Simplified Lesk Algorithm [16]. The sense with the highest word overlap is assigned to the emoji at the end of a successful run of the algorithm. We can see that 🙏 emoji in T_1 will be assigned `pray(verb)` based on the overlap of words {god, pray} with words retrieved from the sense definition of `pray(verb)` and the same emoji in T_2 will be assigned `high five(noun)` based on the overlap of word {celebrate} with words retrieved from the sense definition of `high five(noun)`. In the above example, we have only shown the minimal set of words that one could extract from EmojiNet. Since we link EmojiNet senses with their corresponding BabelNet senses using BabelNet sense IDs, one could easily utilize other resources available in BabelNet such as related WordNet senses, VerbNet senses, Wikipedia, etc. to collect an improved set of context words for emoji sense disambiguation tasks. It should be emphasized that this example was taken only to show the usefulness of the resource for research directions.

---

[15] http://www.goodhousekeeping.com/life/g3601/surprising-emoji-meanings/.

5   Conclusion and Future Work

This paper presented the construction of EmojiNet, the first ever machine read-able sense inventory to understand the meanings of emoji. It integrates four different emoji resources from the Web to extract emoji senses and align those senses with BabelNet. We evaluated the automatic creation of EmojiNet by eval-uating (i) the nearest neighborhood-based image processing algorithm used to align different emoji resources and (ii) the most frequent sense-based and the most popular sense-based word sense disambiguation algorithms used to align different emoji senses extracted from the Web with BabelNet. We plan to extend our work in the future by expanding the sense definitions extracted from Babel-Net with words extracted from tweets, using a word embedding model trained on tweets that contain emoji. We also plan to evaluate the usability of EmojiNet by first defining the emoji sense disambiguation and emoji similarity finding problems, and then applying EmojiNet to disambiguate emoji senses based on different contexts in which they appear. We are working on applying EmojiNet to improve sentiment analysis and exposing EmojiNet as a web service.

Acknowledgments. We are grateful to Sujan Perera for thought-provoking discus-sions on the topic. We acknowledge partial support from the National Institute on Drug Abuse (NIDA) Grant No. 5R01DA039454-02: "Trending: Social Media Analysis to Monitor Cannabis and Synthetic Cannabinoid Use", National Institutes of Health (NIH) award: MH105384-01A1: "Modeling Social Behavior for Healthcare Utilization in Depression", and Grant No. 2014-PS-PSN-00006 awarded by the Bureau of Justice Assistance. The Bureau of Justice Assistance is a component of the U.S. Department of Justice's Office of Justice Programs, which also includes the Bureau of Justice Sta-tistics, the National Institute of Justice, the Office of Juvenile Justice and Delinquency Prevention, the Office for Victims of Crime, and the SMART Office. Points of view or opinions in this document are those of the authors and do not necessarily represent the official position or policies of the U.S. Department of Justice, NIH or NIDA.

References

1. Emogi research team - 2015 emoji report (2015)
2. Balasuriya, L., Wijeratne, S., Doran, D., Sheth, A.: Finding street gang members on Twitter. In: The 2016 IEEE/ACM International Conference on Advances in Social Networks Analysis and Mining (ASONAM 2016), vol. 8, San Francisco, CA, USA, pp. 685–692 (08 2016)
3. Basile, P., Caputo, A., Semeraro, G.: An enhanced lesk word sense disambiguation algorithm through a distributional semantic model. In: COLING, pp. 1591–1600 (2014)
4. Camacho-Collados, J., Pilehvar, M.T., Navigli, R.: Nasari: a novel approach to a semantically-aware representation of items. In: Proceedings of NAACL, pp. 567–577 (2015)
5. Davis, M., Edberg, P.: Unicode emoji - unicode technical report #51. Technical report 51(3) (2016)
6. Dimson, T.: Emojineering part 1: machine learning for emoji trends. Instagram Engineering Blog (2015)

 7. Kelly, R., Watts, L.: Characterising the inventive appropriation of emoji as relationally meaningful in mediated close personal relationships. Experiences of Technology Appropriation: Unanticipated Users, Usage, Circumstances, and Design (2015)
 8. Miller, H., Thebault-Spieker, J., Chang, S., Johnson, I., Terveen, L., Hecht, B.: Blissfully happy or ready to fight: varying interpretations of emoji. In: ICWSM 2016 (2016)
 9. Moro, A., Navigli, R., Tucci, F.M., Passonneau, R.J.: Annotating the MASC corpus with BabelNet. In: LREC, pp. 4214–4219 (2014)
10. Navigli, R., Ponzetto, S.P.: BabelNet: the automatic construction, evaluation and application of a wide-coverage multilingual semantic network. Artif. Intell. **193**, 217–250 (2012)
11. Novak, P.K., Smailović, J., Sluban, B., Mozetič, I.: Sentiment of emojis. PLOS One **10**(12), e0144296 (2015)
12. Pavalanathan, U., Eisenstein, J.: Emoticons vs. emojis on Twitter: a causal inference approach. arXiv preprint arXiv:1510.08480 (2015)
13. Rezabek, L., Cochenour, J.: Visual cues in computer-mediated communication: supplementing text with emoticons. J. Vis. Lit. **18**(2), 201–215 (1998)
14. Santos, R.: Java image processing cookbook (2010). http://www.lac.inpe.br/JIPCookbook
15. SwiftKey, P.: Most-used emoji revealed: Americans love skulls, Brazilians love cats, the French love hearts [blog] (2015). http://bit.ly/2c5biPU
16. Vasilescu, F., Langlais, P., Lapalme, G.: Evaluating variants of the lesk approach for disambiguating words. In: LREC (2004)
17. Wang, W., Chen, L., Thirunarayan, K., Sheth, A.P.: Harnessing Twitter big data for automatic emotion identification. In: 2012 International Conference on Privacy, Security, Risk and Trust (PASSAT), 2012 International Conference on Social Computing (SocialCom), pp. 587–592. IEEE (2012)
18. Wijeratne, S., Balasuriya, L., Doran, D., Sheth, A.: Word embeddings to enhance Twitter gang member profile identification. In: IJCAI Workshop on Semantic Machine Learning (SML 2016), vol. 07, pp. 18–24. CEUR-WS, New York City (2016)

Author Index

Printed in the United States
By Bookmasters